Plasticity of Muscle

Plasticity of Muscle

Proceedings of a Symposium
held at the University of Konstanz, Germany
September 23-28, 1979

Editor Dirk Pette

Walter de Gruyter · Berlin · New York 1980

Editor
Dirk Pette, Dr. med.
Professor of Biochemistry
Department of Biology
University of Konstanz
D-7750 Konstanz 1
West Germany

CIP-Kurztitelaufnahme der Deutschen Bibliothek

Plasticity of muscle: proceedings of a symposium, held at the Univ. of Konstanz, Germany, September 23-28, 1979/ed. Dirk Pette. – Berlin, New York: de Gruyter, 1980.
ISBN 3-11-007961-5
NE: Pette, Dirk (Hrsg.); Universität Konstanz

Library of Congress Cataloging in Publication Data

Main entry under title: Plasticity of muscle.
Bibliography: p. Includes indexes. 1. Muscle-Congresses.
I. Pette, Dirk, 1933-
QP321.P55 599.01'852 80-15502
ISBN 3-11-007961-5

 Printing: Karl Gerike, Berlin. – Binding: Dieter Mikolai, Berlin. – Printed in Germany.

To the memory of Ernest Gutmann

PREFACE

Ever since 1844, when Justus von Liebig[1] described his studies on compounds extracted from "meat", muscle tissue has attracted the interest of chemists and biochemists. Generations of biochemists have used skeletal muscle as a rich source from which to purify numerous biomolecules and as a tissue most suitable for studying energy metabolism and energy transformation. In most of these investigations, muscle has been treated as a uniform and static tissue. The complexity of muscle tissue, its variability and fibre architecture, has been generally realized only lately. With this realization research on muscle has evolved in a completely new direction and has challenged many disciplines in science and medicine to produce a compelling, fruitful collaboration. The understanding of the relationships between structure and function has been greatly expanded by the recognition of the dynamic properties of muscle. The classical experiments of A.J. Buller, J.C. Eccles and R.M. Eccles in which the nerve was shown to modulate the properties of muscle, opened an exciting new era in muscle research and neurobiology. The concept of plasticity of muscle has stimulated many lines of investigation since then.

The idea of assembling scientists active in this field originated from a Conference on Muscle held at Szeged, Hungary, in July 1977 and was enthusiastically maintained by Wilfried Mommaerts. The symposium "Plasticity of Muscle" was held under the honorary Chairmanship of Arthur J. Buller at the University of Konstanz, September 23-28, 1979. The broad concept of this symposium was to bring together investigators from various disciplines in science and medicine who deal with the phenomenon of plasticity and its underlying mechanisms. Specifically the topics to be treated were heterogeneity of muscle as a tissue, the programme of its ontogeny and differentiation, and the regulation of its properties by innervation, usage, hormones and artificial modulation of neural input.

The enthusiastic responses to my invitations for participation indicated that the time was mature for this symposium. Its organization was a pleasure for me and in this regard I wish to express my gratitude to Dr. Patti M. Nemeth for her continuous and valuable assistance. The material realization of this symposium was made possible by the generous financial contributions of the Deutsche Forschungsgemeinschaft, the Ministerium für Kunst und Wissenschaft des Landes Baden-Württemberg and the European Molecular Biology Organization. Further support was given by the Gesellschaft der Freunde und Förderer der Universität Konstanz, by the Paul-Martini-Stiftung, Frankfurt, by Byk Gulden Lomberg Chemische Fabrik GmbH, Konstanz, by the Ernst Leitz Wetzlar GmbH, and by Chemiewerk Homburg.

The occasion of the symposium would not have proceeded so efficiently and smoothly without the competent assistance of all my coworkers. I wish to thank all of them and especially my secretary, Ms. Trude Streicher. The preparation of the proceedings of the symposium for this book was accomplished by the commitment and untiring efforts of Ms. Jane Vincent and the excellent technical assistance of Mrs. Norella Harris.

As I am writing this preface, I recall the lively and warm atmosphere at the symposium. In my closing remarks at the symposium I expressed the wish that all participants could stay in Konstanz and work together in harmony. I feel that this book in some way serves this idea. At this time I wish to extend the idea of unincumbered collaboration to those who could not be present. In this connection it is my personal desire to keep alive the memory of my unforgettable friend Ernest Gutmann whose scientific life has been dedicated to this ideal and who also has contributed so much to our understanding of muscle plasticity.

Konstanz, January 1980 Dirk Pette

[1]Justus von Liebig: Bestandteile des Fleischextractes. In "Chemische Briefe", 32. Brief, Leipzig, Heidelberg 1844.

CONTENTS

SECTION II. DEVELOPMENT AND GROWTH

SECTION III. NERVE MUSCLE INTERACTION

SECTION IV. INFLUENCE OF USAGE

SECTION V. EFFECTS OF CHRONIC STIMULATION

SECTION VI. MECHANISMS OF HYPERTROPHY AND ATROPHY

LIST OF CONTRIBUTORS

T. Aitman, Department of Physiology, The Medical School, University of Birmingham, Birmingham B15 2TJ, Great Britain

G. Bennett, Department of Anatomy, Medical School, University of Pennsylvania, Philadelphia, PA 19104, USA

R. Betto, Centro di Studio per la Biologia e la Fisiopatologia Muscolare del C.N.R., Istituto di Patologia generale, Università di Padova, I-35100 Padova, Italy

R. Billeter, Institut für Pharmakologie und Biochemie, Veterinärmedizinische Fakultät der Universität Zürich, CH-8057 Zürich, Switzerland

D. Biral, Centro di Studio per la Biologia e la Fisiopatologia Muscolare del C.N.R., Istituto di Patologia generale, Università di Padova, I-35100 Padova, Italy

R. Bischoff, Department of Anatomy and Neurobiology, Washington University School of Medicine, St. Louis, MO 63110, USA

F.W. Booth, Department of Physiology, University of Texas Medical School, Houston, TX 77030, USA

M. Brooke, Department of Neurology, Washington University School of Medicine, St. Louis, MO 63110, USA

A.J. Buller, Department of Health & Social Security, Alexander Fleming House, London SE1 6BY, Great Britain

V. Chen, Departments of Biology and Physical Education, York University, Toronto, Ontario, Canada M3J 1P3

M.M.-Y. Chi, Department of Pharmacology, Washington University School of Medicine, St. Louis, MO 63110, USA

T.C. Cope, Department of Physiology, Duke University, Durham, NC 27710, USA

J. Croop, Department of Anatomy, Medical School, University of Pennsylvania, Philadelphia, PA 19104, USA

A. D'Albis, Laboratoire de Biologie Physico-chimique, Université de Paris-Sud, F-91405 Orsay-Cedex, France

R. Dabrowska, Department of Biochemistry of Nervous System and Muscle, Nencki Institute of Experimental Biology, PL-02-093 Warsaw 22, Poland

L. Dalla Libera, Centro di Studio per la Biologia e la Fisiopatologia Muscolare del C.N.R., Istituto di Patologia generale, Università di Padova, I-35100 Padova, Italy

G.K. Dhoot, Department of Immunology, University of Birmingham, Birmingham B15 2TJ, Great Britain

W. Drabikowski, Department of Biochemistry of Nervous System and Muscle, Nencki Institute of Experimental Biology, PL-02-093 Warsaw 22, Poland

C. Dunlop, Department of Physiology, University of Sydney, Sydney, NSW 2006, Australia

V.R. Edgerton, Department of Kinesiology, UCLA, Los Angeles, CA 90024, USA

E. Eldred, Department of Anatomy, School of Medicine, UCLA, Los Angeles, CA 90024, USA

L. Eldridge, Department of Physiology, School of Medicine, UCLA, Los Angeles, CA 90024, USA

L. Engebretsen, Institute of Neurophysiology, University of Oslo, Oslo 1, Norway

J.D. Etlinger, Department of Anatomy and Cell Biology, State University of New York, Downstate Medical Center, Brooklyn, NY 11203, USA

S. Felder, Department of Pharmacology, Washington University School of Medicine, St. Louis, MO 63110, USA

S. Fellini, Department of Anatomy, Medical School, University of Pennsylvania, Philadelphia, PA 19104, USA

R. Fitts, Department of Biology, Marquette University, Milwaukee, WI, 53233, USA

I. Furka, Department of Experimental Surgery, Medical School, University of Debrecen, H-4012 Debrecen, Hungary

G.F. Gauthier, Department of Anatomy, University of Massachusetts Medical School, Worcester, MA 01605, USA

I. Gesztelyi, Central Research Laboratory, University of Debrecen, H-4012 Debrecen, Hungary

A. L. Goldberg, Department of Physiology, Harvard Medical School, Boston, MA 02115, USA

D.F. Goldspink, Department of Physiology, Medical Biology Centre, The Queen's University, Belfast, Northern Ireland

T. Gordon, Department of Physiology, University of Alberta, Edmonton, Canada T6G 2H7

L. Gorza, Centro di Studio per la Biologia e la Fisiopatologia Muscolare del C.N.R., Istituto di Patologia generale, Università di Padova, I-35100 Padova, Italy

F. Guba, Institute of Biochemistry, University of Medical Sciences, H-6701 Szeged, Hungary

A. Heilig, Fakultät für Biologie, Universität Konstanz, D-7750 Konstanz, Germany

C. Heilmann, Medizinische Universitätsklinik, D-7800 Freiburg, Germany

C.S. Hintz, Department of Pharmacology, Washington University School of Medicine, St. Louis, MO 63110, USA

J.F.Y. Hoh, Department of Physiology, University of Sydney, Sydney, NSW 2006, Australia

J. Holloszy, Department of Preventive Medicine, Washington University School of Medicine, St. Louis, MO 63110, USA

H. Holtzer, Department of Anatomy, Medical School, University of Pennsylvania, Philadelphia, PA 19104, USA

O. Hudlicka, Department of Physiology, The Medical School, University of Birmingham, Birmingham B15 2TJ, Great Britain

G.R. Hugman, Baylor School of Medicine, Houston, TX 77030, USA

C.D. Ianuzzo, Departments of Biology and Physical Education, York University, Toronto, Ontario, Canada M3J 1P3

E. Jenny, Institut für Pharmakologie und Biochemie, Veterinärmedizinische Fakultät der Universität Zürich, CH-8057 Zürich, Switzerland

M.A. Johnson, Muscular Dystrophy Group Research Laboratories, Newcastle General Hospital, Newcastle upon Tyne, NE4 6BE, Great Britain

K. Kaiser, Department of Neurology, Washington School of Medicine, St. Louis, MO 63110, USA

T. Kameyama, Department of Anatomy and Cell Biology, State University of New York, Downstate Medical Center, Brooklyn, NY 11203, USA

S. Kaufman, Laboratory of Neurochemistry, National Institute of Mental Health, Bethesda, MD 20205, USA

A.M. Kelly, Department of Pathology, School of Veterinary Medicine, University of Pennsylvania, Philadelphia, PA 19174, USA

B.H. Kim, Department of Physiology, University of Sydney, Sydney, NSW 2006, Australia

Z. Kiss, Institute of Biochemistry, University of Medical Sciences, H-6701 Szeged, Hungary

A. Köver, Central Research Laboratory, University of Debrecen, H-4012, Debrecen, Hungary

E. Kugelberg, Department of Neurology, Karolinska Sjukhuset, S-104 01 Stockholm 60, Sweden

B.T.S. Kwan, Department of Physiology, University of Sydney, Sydney, NSW 2006, Australia

G. Lacombe, U 127 INSERM, Hôpital Lariboisière, F-75010 Paris, France

T. Lømo, Institute of Neurophysiology, University of Oslo, Oslo 1, Norway

A.M. Lompré, U 127 INSERM, Hôpital Lariboisière, F-75010 Paris, France

S. Lowey, Department of Biochemistry, The Rosenstiel Basic Medical Sciences Research Center, Brandeis University, Waltham, MA 02254, USA

C.V. Lowry, Department of Pharmacology, Washington University School of Medicine, St. Louis, MO 63110, USA

O.H. Lowry, Department of Pharmacology, Washington University School of Medicine, St. Louis, MO 63110, USA

H. Lutz, Institut für Pharmakologie und Biochemie, Veterinär-medizinische Fakultät der Universität Zürich, CH-8057 Zürich, Switzerland

K. Mabuchi, Department of Muscle Research, Boston Biomedical Research Institute, Boston, MA 02114, USA

A. Margreth, Centro di Studio per la Biologia e la Fisiopatologia Muscolare del C.N.R., Istituto di Patologia generale, Università di Padova, I-35100 Padova, Italy

F.L. Mastaglia, Muscular Dystrophy Group Research Laboratories, Newcastle General Hospital, Newcastle upon Tyne, NE4 6BE, Great Britain

K. Matsumoto, Department of Anatomy and Cell Biology, State University of New York, Downstate Medical Center, Brooklyn, NY 11203, USA

L.M. Mendell, Brain Research Institute, UCLA, Los Angeles, CA 90024, USA

W.F.H.M. Mommaerts, Department of Physiology, School of Medicine, UCLA, Los Angeles, CA 90024, USA

A. Montgomery, Muscular Dystrophy Group Research Laboratories, Newcastle General Hospital, Newcastle upon Tyne, NE4 6BE, Great Britain

Z. Nagy, Central Research Labortory, University of Debrecen, H-4012, Debrecen, Hungary

P.M. Nemeth, Fakultät für Biologie, Universität Konstanz, D-7750 Konstanz, Germany

P. O'Brien, Departments of Biology and Physical Education, York University, Toronto, Ontario, Canada M3J 1P3

R.A.D. O'Brien, Department of Anatomy and Embryology, University College London, London WC1E 6BT, Great Britain

P. Patel, Departments of Biology and Physical Education, York University, Toronto, Ontario, Canada M3J 1P3

S.V. Perry, Department of Biochemistry, University of Birmingham, Birmingham B15 2TT, Great Britain

D. Pette, Fakultät für Biologie, Universität Konstanz, D-7750 Konstanz, Germany

S. Pierobon Bormioli, Centro di Studio per la Biologia e la Fisiopatologia Muscolare del C.N.R., Istituto di Patologia generale, Università di Padova, I-35100 Padova, Italy

M.G. Pluskal, Department of Muscle Research, Boston Biomedical Research Institute, Boston, MA 02114, USA

B. Pope, MRC Laboratory of Molecular Biology, Cambridge CB2 2QH, Great Britain

M.A. Reis, Department of Life Sciences, Bar-Ilan University, Ramat-Gan, Israel

R.K. Roy, Department of Muscle Research, Boston Biomedical Research Institute, Boston, MA 02114, USA

N.A. Rubinstein, Department of Anatomy, School of Medicine, University of Pennsylvania, Philadelphia, PA 19104, USA

S. Salmons, Department of Anatomy, The Medical School, University of Birmingham, Birmingham, B15 2TJ, Great Britain

S. Salvatori, Centro di Studio per la Biologia e la Fisiopatologia Muscolare del C.N.R., Istituto di Patologia generale, Università di Padova, I-35100 Padova, Italy

G. Salviati, Centro di Studio per la Biologia e la Fisiopatologia Muscolare del C.N.R., Istituto di Patologia generale, Università di Padova, I-35100 Padova, Italy

S. Sarkar, Department of Muscle Research, Boston Biomedical Research Institute, Boston, MA 02114, USA

S. Sartore, Centro di Studio per la Biologia e la Fisiopatologia Muscolare del C.N.R., Istituto di Patologia generale, Università di Padova, I-35100 Padova, Italy

M.G. Sarzala, Department of Biochemistry of Nervous System and Muscle, Nencki Institute of Experimental Biology, PL-02-093 Warsaw 22, Poland

S. Schiaffino, Centro di Studio per la Biologia e la Fisiopatologia Muscolare del C.N.R., Istituto di Patologia generale, Università di Padova, I-35100 Padova, Italy

H. Schmidt, Phyiologisches Institut der Universität des Saarlandes, D-6650 Homburg/Saar, Germany

K. Schwartz, U 127 INSERM, Hôpital Lariboisière, F-75010 Paris, France

M.J. Seider, Department of Anesthesiology, University of Texas Medical School, Houston, TX 77030, USA

A. Shainberg, Department of Life Sciences, Bar-Ilan University, Ramat-Gan, Israel

L.A. Smith, Department of Kinesiology, UCLA, Los Angeles, CA 90024, USA

J. Sosiński, Department of Biochemistry of Nervous System and Muscle, Nencki Institute of Experimental Biology, PL-02-093 Warsaw 22, Poland

C. Spamer, Medizinische Universitätsklinik, D-7800 Freiburg, Germany

N.C. Spurway, Department of Physiology, University of Glasgow, Glasgow G12 8QQ, Great Britain

F.A. Sreter, Department of Muscle Research, Boston Biomedical Research Institute, Boston, MA 02114, USA

R.B. Stein, Department of Physiology, University of Alberta, Edmonton, Canada T6G 2H7

B. Swynghedauw, U 127 INSERM, Hôpital Lariboisière, F-75010 Paris, France

T. Szilágyi, Institute of Physiology and Pathophysiology, University of Medical Sciences, H-4012 Debrecen, Hungary

Á. Szöör, Institute of Physiology and Pathophysiology, University of Medical Sciences, H-4012 Debrecen, Hungary

Ö. Takács, Institute of Biochemistry, University of Medical Sciences, H-6701 Szeged, Hungary

N.V. Thiem, U 127 INSERM, Hopital Lariboisière, F-75010 Paris, France

K. Toner, Department of Anatomy and Cell Biology, State University of New York, Downstate Medical Center, Brooklyn, NY 11203, USA

Y. Toyama, Department of Anatomy, Medical School, University of Pennsylvania, Philadelphia, PA 19104, USA

K.R. Tyler, Department of Physiology, The Medical School, University of Birmingham, Birmingham B15 2TJ, Great Britain

H.H. Vandenburgh, Laboratory of Neurochemistry, National Institute of Mental Health, Bethesda, MD 20205, USA

G. Vrbová, Department of Anatomy and Embryology, University College London, London WC1E 6BT, Great Britain

H. Weber, Institut für Pharmakologie und Biochemie, Veterinärmedizinische Fakultät der Universität Zürich, CH-8057 Zürich, Switzerland

A. Weeds, MRC Laboratory of Molecular Biology, Cambridge CB2 2QH, Great Britain

C. West, Department of Anatomy, Medical School, University of Pennsylvania, Philadelphia, PA 19104, USA

R.H. Westgaard, Institute of Neurophysiology, University of Oslo, Oslo 1, Norway

D. van der Westhuyzen, Department of Anatomy and Cell Biology, State University of New York, Downstate Medical Center, Brooklyn, NY 11203, USA

R. G. Whalen, Département de Biologie Moléculaire, Institut Pasteur, F-75724 Paris, France

W. Winder, Division of Biochemistry, Physiology and Pharmacology, University of South Dakota School of Medicine, Vermillion, SD 57069, USA

E. Zubrzycka-Gaarn, Department of Biochemistry of Nervous System and Muscle, Nencki Institute of Experimental Biology, PL-02-093 Warsaw 22, Poland

INTRODUCTION

A. J. Buller

It gives me as much pleasure to write a brief introduction to these Proceedings of the International Symposium on Plasticity of Muscle as it gave me to act as the gathering's Honorary Chairman. The meeting, held at the University of Konstanz between September 23 and 26, 1979, was sponsored by Deutsche Forschungsgemeinschaft and Land Baden Württemberg and wonderfully organized by Professor Dr. Dirk Pette. Approximately eighty participants gathered together from some sixteen countries to discuss work of mutual interest not only during fourteen formal sessions but also outside the lecture hall and late into the night. In his address of welcome Dirk Pette spoke of what was in the minds of many of us, namely the absence from our meeting of two men who, each in his own way, had pioneered much of the research which was pivotal to our subsequent discussion. First, Ernest Gutmann, the most modest of men, yet possessed of enormous insight into the metabolic complexities of skeletal muscle and, second, Jack Eccles, that ebullient and gifted experimenter whose sorties into the field of nerve-muscle physiology produced such remarkable observations. Sadly the former is no longer with us, the latter was unable to attend our meeting. However the participants dispatched to him a message of good will.

It cannot be pretended that the meeting at Konstanz solved all our problems or differences, it would have been unthinkable that it could. What it did was to provide an opportunity for free, unbridled discussion (with more than a trace of humor) of those remaining enigmas which beset this fascinating field. Such opportunities are the life blood of modern scientific exchange. The written paper has its place - including within a volume such as this - but the face to face exchanges with scientific friends and colleagues from other parts of the world is the essence of the small Symposium. The meeting at Konstanz was a model of its kind. It sent us back to our respective corners of the globe refreshed, excited and greatly appreciative of our sponsor's and host's generosity.

I hope you enjoy this account of our formal proceedings. My only regret is that a written account cannot fully convey the enjoyment of those few days the participants spent together.

Section I. Heterogeneity of Metabolic and Molecular Properties of Muscle Fibres

ENZYMOLOGICAL HETEROGENEITY OF HUMAN MUSCLE FIBERS

Oliver H. Lowry, Charles V. Lowry, Maggie M.-Y. Chi, Carol S. Hintz and Stephen Felder

Department of Pharmacology, Washington University School of Medicine, St. Louis, Missouri 63110, USA

Introduction

Most muscles appear to consist of mixtures of at least three different fiber types. The nature of the differences among these types has been studied in many laboratories by histochemical staining and by analysis of different whole muscles in which one or another type predominates. It has been usually assumed that each type is relatively homogeneous in composition. However, measurements of a number of enzymes in individual rabbit muscle fibers by Spamer and Pette (1,2) and in human fibers by Essén et al. and our own group (3,4) do not support this. Major groupings are observed, but within each group there is a broad and almost continuous spectrum of enzyme activities.

Fig. 1 illustrates this for lactic dehydrogenase and adenylokinase each measured in duplicate on the same 89 fibers from a small portion of a human biceps muscle. Except for a distinct gap in the middle, there is an almost continuous spectrum of activity for both enzymes over a 10-fold range.

Fig. 2 illustrates the interrelationships among 10 enzymes within a set of 20 fibers from another individual. In spite of the diversity in enzyme levels, a consistency in interrelationship among the enzymes is evident. For example, both E and D groups are probably slow twitch fibers, but the E group is consistently lower in enzymes of glycogenolysis and higher in mitochondrial enzymes.

Plasticity of Muscle

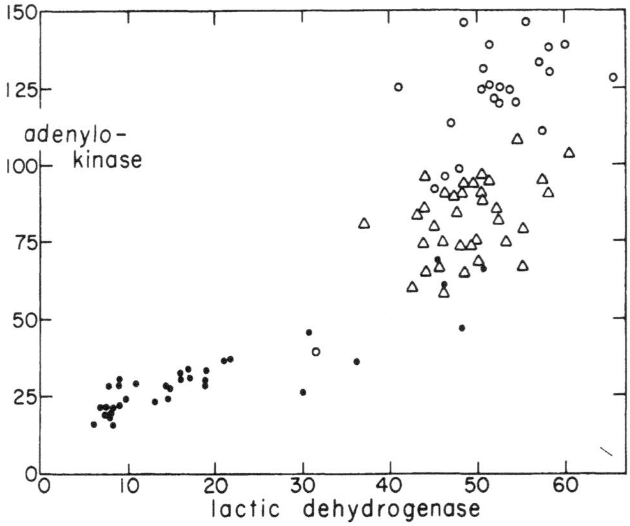

Figure 1. Adenylokinase and lactic dehydrogenase on the same 89 muscle fibers. The numbers indicate activity as mol kg^{-1} (dry weight) h^{-1}. The symbols indicate the range of β-hydroxyacyl-CoA dehydrogenase activity for the same fibers in the same units: O, <1.60; Δ, 1.61 to 2.75; ●, >2.75 (from reference 4).

It is an attractive but unproven hypothesis (5) that the fibers of a single motor unit are similar in composition but that there is a considerable diversity in composition among different motor units of the same major fiber group.

Results such as these make it desirable to examine other facets of muscle metabolism on an individual fiber basis. For example, how do metabolites vary among individual fibers at rest or during activity? Does intensive training shift fiber composition into a narrower band at one end of the range?

For studies of abnormal muscle, single fiber analysis is even more important. Some diseases appear to affect only one fiber type, or one type more than another. In other diseases, particularly muscular dystrophy, the fibers not only visibly vary more than normal, but are so admixed with connective tissue as to greatly distort analytical results. This can be avoided by analysis of individual fibers.

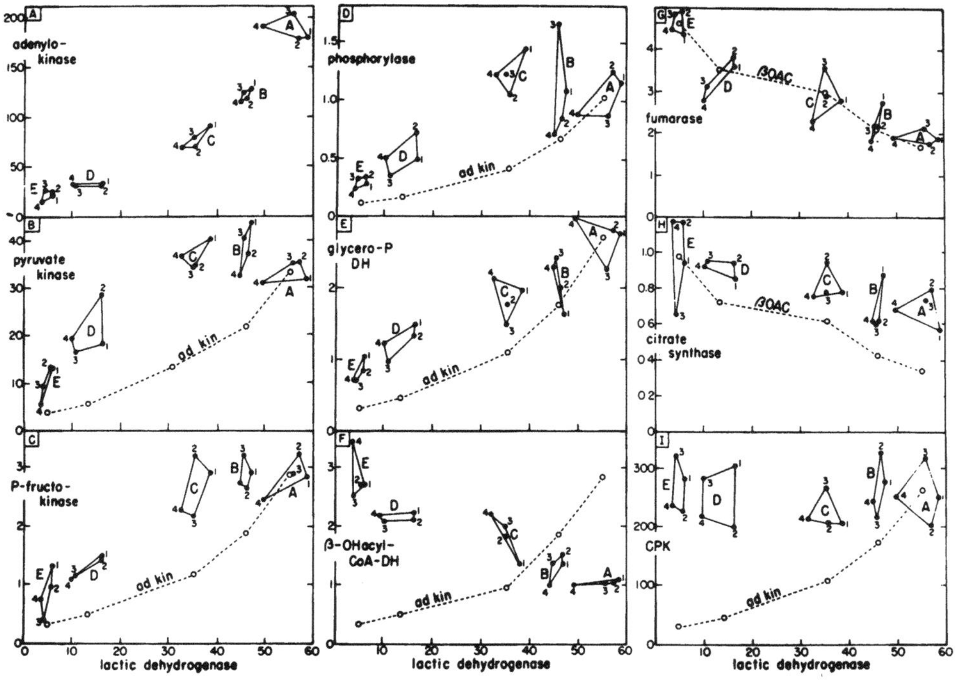

Figure 2. Correlation of nine enzymes with lactic dehydrogenase among individual fibers. Twenty muscle fibers from one person were analyzed in duplicate for the 10 enzymes, and the activities of each of the other nine were plotted in turn against lactic dehydrogenase activity. Each fiber is identified by the group letter and the number (1 to 4) within that group. All activities are recorded as mol kg^{-1} (dry weight) h^{-1}. In six of the panels, the average adenylokinase values from panel A have been repeated for comparison (ad kin). The scale for adenylokinase has been adjusted in each case so that the mean value for group A coincides with that of the enzyme in question. Similarly, in two panels β-hydroxyacyl-CoA dehydrogenase values from panel F have been repeated (βOAC). Creatine kinase is abbreviated CPK (from reference 4).

Single Fiber Analysis

Sampling technique

The data given above were obtained with samples obtained by a technique originating with Essén *et al.* (3). A biopsy specimen is quick frozen and a small portion freeze dried and then splintered apart into segments of single fibers several mm long. We remove duplicate samples, 10 or 20 ng in weight, for a particular analysis and store the remainder under vacuum at -70° for other assays. A 2 mm segment of a normal skeletal muscle fiber can provide at least 50 duplicate 20 ng samples for as many different assays. Moreover, under the prescribed storage conditions, enzymes and metabolites in general appear to be indefinitely stable. Therefore, one can return months or years later to the same fibers to measure some component that hindsight showed to be important.

Analytical methods

Our work so far has utilized a somewhat standardized methodology, capable of nearly unlimited sensitivity. It is applicable to any substance which can be made to oxidize or reduce NAD or NADP, directly or indirectly. For example, P-creatine or creatine kinase is measured by the NADPH formed in the following sequence (enzymes not shown):

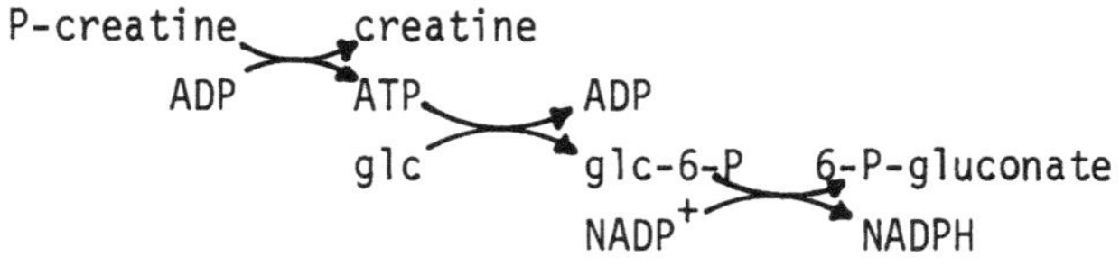

In the case of creatine kinase, enough NADPH is produced to measure by its fluorescence. However, for P-creatine, considerable amplification is necessary, since the P-creatine present in 20 ng of dry muscle fiber is only about 2 pmoles, whereas the creatine kinase in the same amount of muscle can breakdown 2000 times this amount of P-creatine in an hour.

Fortunately, amplification is easy: the excess $NADP^+$ is destroyed with NaOH and heat, and the sample is added to reagent with components to carry out the following rapidly repeating enzymatic cycle:

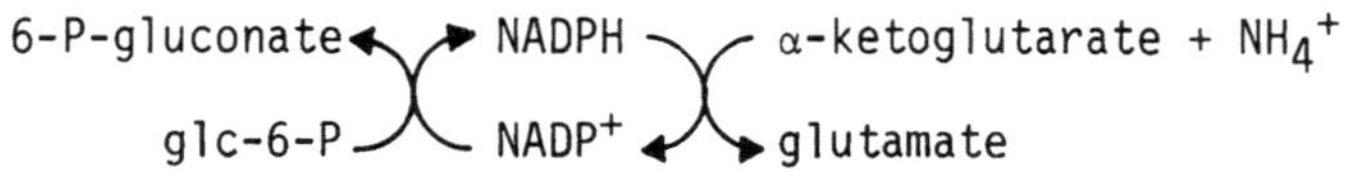

After a sufficient number of cycles, the enzymes (not shown) are killed with heat or alkali and the 6-P-gluconate is measured by the fluorescence of the NADPH formed from extra $NADP^+$ in a final enzymatic step:

The amplification that can be achieved in this way is rather impressive. Cycling rates up to 30,000 per hour can be obtained by using high levels of the two enzymes (in this case glutamic and glucose-6-P dehydrogenases). By incubating for long periods, overall amplication of 400,000-fold has been achieved (6).

Amplification by enzymatic cycling can also be applied when the first step results in formation of $NADP^+$ rather than NADPH, or when the reactants are NAD^+ or NADH. If $NADP^+$ or NAD^+ is the product the excess NADPH or NADH is easily destroyed with acid before cycling, and if the primary reaction concerns NAD rather than NADP, an equally rapid cycling system for it is available (7).

The amplification process might appear complicated and likely to introduce additional errors because so many factors could influence the cycling rate. In fact, an enzymatic cycling step adds only a minute or two per sample to the overall assay time, and the results are highly reproducible and reliable. Standards are carried through the entire procedure in the same racks as the samples. Therefore, during cycling, except for a minute amount of tissue (usually less than 1 part in a million) every sample and standard is identical in composition, temperature and time of incubation. Therefore, substantial differences in any of these, including enzyme activity, will not distort the calculated results.

Results

A. Normal adult muscle fibers

(All of the enzyme data are recorded in the figures as mol kg^{-1} (dry wt) hr^{-1} at 20°C.)

Fig. 3 compares the relationships between lactic dehydrogenase and adenylokinase among female and male fibers. Individual fiber values are shown for one woman. The lines are curves drawn through the center of single fiber data for each of three women and three men. Curves for two other normal men fell within the limits of these three (4). The relationships between the two enzymes are similar in the two sexes except that lactic dehydrogenase is substantially lower in the female fibers, and adenylokinase is if anything higher.

Glycogen phosphorylase, in contrast to lactic dehydrogenase, was much higher in female than in male fibers (Fig. 4). The ratio was bout 3 to 1

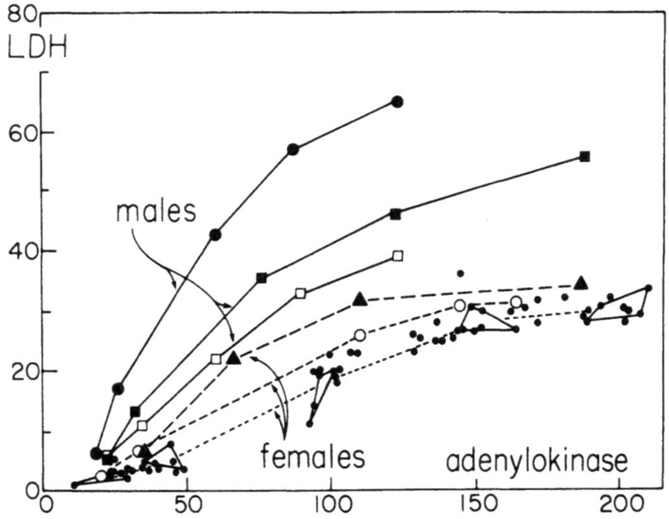

Figure 3. Correlation of lactic dehydrogenase and adenylokinase for muscle fibers from three normal men and women. The points represent the midpoints of fiber groups such as seen in Fig. 2. The individual values are shown for one woman.

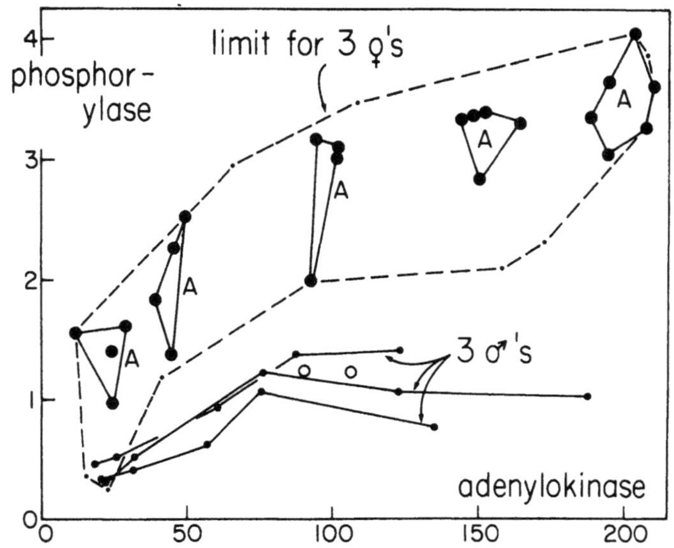

Figure 4. Correlation of glycogen phosphorylase and adenylokinase for fibers from three normal men and women. Lines through midpoints of fiber groups are shown for the men. The large area enclosed by a dashed line encompasses all the data for the three women, except for the two fibers shown below by open circles. Small circumscribed areas marked "A" represent individual fibers from one of the three women. These are the same fibers indicated by small circumscribed areas in Fig. 3.

across the whole range. Activities in "slow twitch" female fibers with lowest adenylokinase were at least as high as in "fast twitch" male fibers with highest adenylokinase levels.

Three other enzymes have been assayed in fibers from both sexes: fructose bisphosphatase was similar in male and female fibers, P-fructokinase was on the average a little higher among female fibers, and β-hydroxyacyl-CoA dehydrogenase covered a somewhat greater range, averaging about 50% higher in the female among "slow twitch" fibers (none of these data shown).

Whether the observed differences between the fibers from the men and the women are true sex differences is hard to prove, considering the limited number of individuals and their random backgrounds. The differences probably can not be explained by difference in muscle activity. Baldwin *et al.* observed only modest changes (decreases) in both phosphorylase and lactic dehydrogenase in rat quandriceps muscle with adaptation to exercise (8).

B. Atrophy of disuse

Three women with Type II fiber atrophy of disuse, without obvious cause, have been studied. Fig. 5 shows the relationship between fiber size and adenylokinase for one of these patients. (Fiber size was determined by simply weighing a measured length of the dry fiber. A circular fiber with a dry weight of 1 μg/mm would have a diameter of about 70 μm). All of the Type II fibers (adenylokinase less than 50 mol kg^{-1} (dry) hr^{-1}) were smaller than normal, and the size was inversely related to adenylokinase activity. Type I fibers fell in the normal range, but were somewhat more variable than normal. Similar results were obtained for the other two patients, except that in one patient the small fibers were almost limited to the higher adenylokinase range (not shown).

Lactic dehydrogenase activites were inconsistent among the three women with disuse atrophy. In one case, activities for nearly all Type II fibers, whether atrophic or not, were higher than for any of the normal women (Fig. 6). In another case the atrophic Type II fibers had much

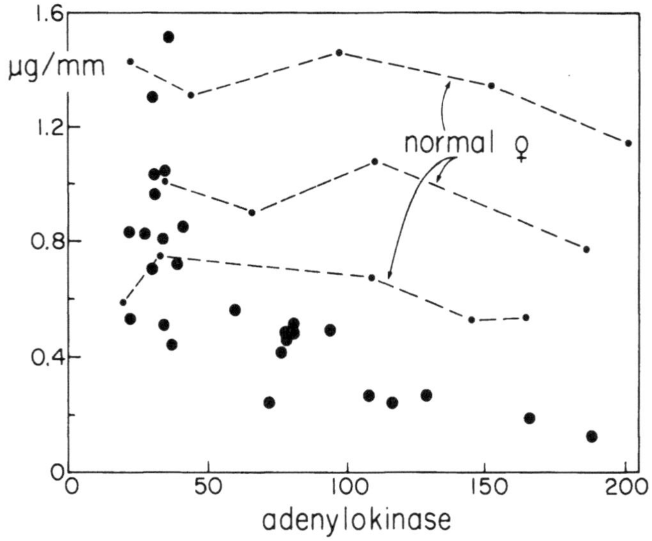

Figure 5. Relationship between fiber size (μg dry wt per mm) and adenylokinase for fibers from three normal women and one woman with Type II fiber atrophy of disuse. The lines join midpoints of groups of fibers in each enzyme range.

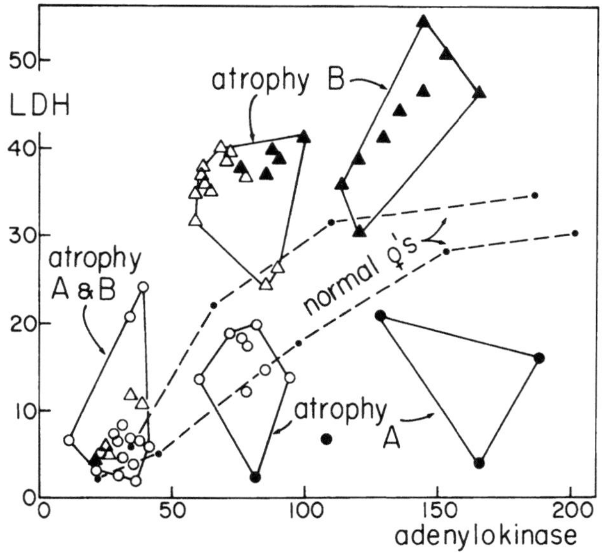

Figure 6. Relationship between lactic dehydrogenase and adenylokinase for fibers from two women with Type II atrophy of disuse and two normal women. The values for individual fibers are shown for the atrophy patients. Solid symbols indicate fibers weighing less than 0.4 μg (dry) per mmm (about 43 μm diameter if circular). The normal values are shown by dashed lines drawn through midpoints of arbitrary fiber groups of each woman. Values for a third normal woman fell between the two lines shown.

lower lactic dehydrogenase activities. In the third case all activities fell in the normal female range (not shown).

Phosphorylase activities were low on the average in one case in particular; none of the Type II fibers and only half of the Type I fibers fell in the normal range (Fig. 7). In the case of the other two patients, half the values were below the normal range (not shown).

P-fructokinase, unlike phosphorylase, was slightly higher than normal (not shown). Consequently the ratio of P-fructokinase to phosphorylase was higher than normal in almost all fibers from the patients with disuse atrophy. The ratio between the two enzymes averaged 1.21, 1.42 and 1.69 in the fibers from the three normal women and 2.27, 2.80 and 4.46 in the fibers from the three patients.

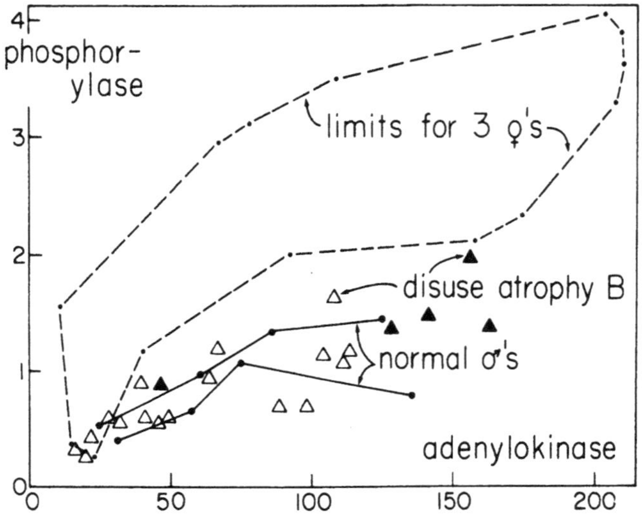

Figure 7. Relationship between glycogen phosphorylase and adenylokinase for a patient with Type II fiber atrophy of disuse. The values for individual fibers from this patient are superimposed on what is essentially the same figure as Fig. 4, which see for details. Solid symbols indicate fibers weighing less than 0.4 μg (dry) per mm.

β-hydroxybutyryl-CoA dehydrogenase was normal in two of the three cases of disuse atrophy, but in the third case activites were lower by a factor of two to three across the whole fiber range (Fig. 8).

It is clear that there are a number of minor and major differences to be found in Type II fiber atrophy of disuse. But it is also evident that there are few clear consistencies. Fiber atrophy has many causes. The low phosphorylase activity, seen in one case in particular, is consistent with the decrease found in experimental denervation and tenotomy (9). It will require a great deal more study to determine the differential effects of different causative agents or conditions.

C. Fibers in a normal male child

As a control for assays of fibers from Duchenne dystrophy patients, fibers from the quadriceps of a normal 8 year old boy were analyzed.

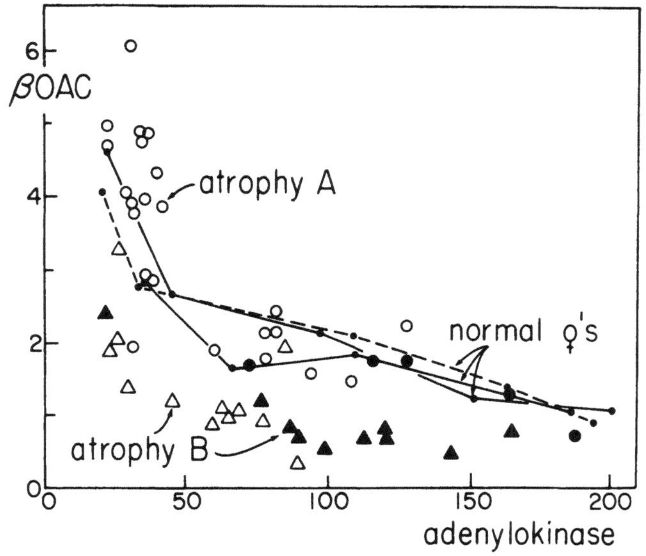

Figure 8. Relationship between β-hydroxyacyl-CoA dehydrogenase and adenylokinase for three normal women (represented by lines through midpoints of fiber groups) and two women with Type II fiber atrophy of disuse. Solid symbols indicate fibers weighing less than 0.4 μg (dry) per mm.

Lactic dehydrogenase and adenylokinase values were much more tightly gathered into two clusters than in normal adults (Fig. 9). Lactic dehydrogenase activities in Type II fibers were higher on the average than either adult male or female fibers. The same was true for adenylokinase in Type I fibers.

β-Hydroxyacyl-CoA dehydrogenase was also higher in Type II fibers than in adult males (Fig. 10), and phosphorylase was much higher in all fiber types (compare Fig. 12 and Fig. 4). The phosphorylase activities are very similar to those in adult female fibers.

Creatine kinase was decidedly higher in the boy's fibers than in any adult male (Fig. 11). As in the adult, there was at most a minor difference in activity (in this case about 20%) between Type I and Type II fibers.

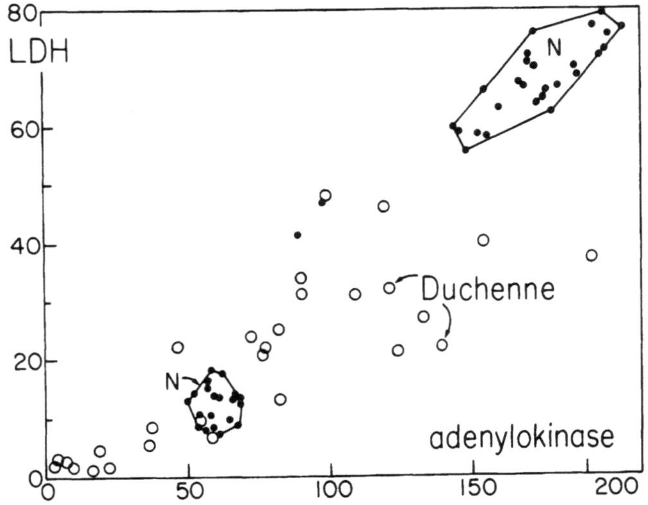

Figure 9. Relationship between lactic dehydrogenase and adenylokinase for fibers from a normal 8 year old boy (points) and a 9 year old boy with Duchenne dystrophy (open circles).

D. Duchenne dystrophy

Studies have been made of five patients with muscular dystrophy. Four were diagnosed as Duchenne dystrophy, the fifth may be Becker's dystrophy. The five patients were all boys, age 5 to 9. The patient represented in Fig. 9 is typical of lactic dehydrogenase and adenylokinase results for four of the five patients. In contrast to the results for the normal child: a) activities were not bunched into two groups; b) There were fewer fibers with high levels of either enzyme and upper limits were lower in all but one case; and c) there were a number of fibers with enzyme activities close to zero. The fifth child differed in that fibers with the extremely low activities were missing.

The β-hydroxyacyl-CoA data in Fig. 10 indicate the significance of the very low activity fibers of Fig. 9. Instead of showing high activity of

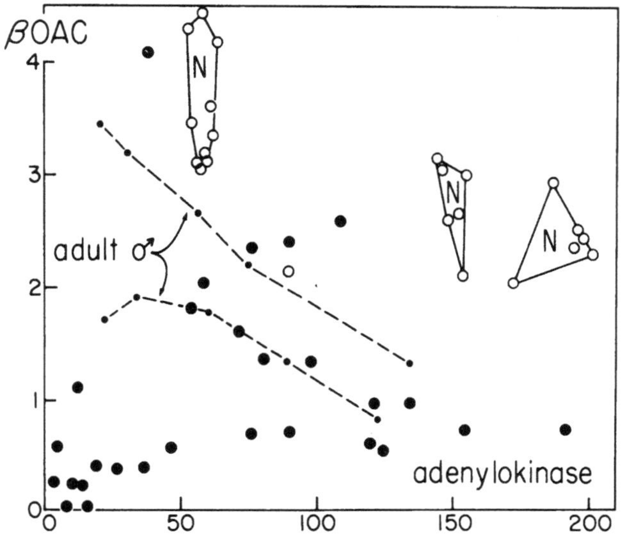

Figure 10. Relationship between β-hydroxyacyl-CoA dehydrogenase and adenylokinase in fibers from an 8 year old normal boy "N" (open circles) and a 9 year old boy with Duchenne dystrophy (solid circles); also shown are the two extremes among five normal adult males, represented by lines drawn through midpoints for groups of fibers covering the range of enzyme activities.

this mitochondrial enzyme, as expected for Type I fibers, these fibers had very low activity of this enzyme as well. It seems clear that these fibers were dead or nearly so. Only in the fifth boy mentioned above were these "dead" fibers not seen. Among the fibers with higher adenylokinase there was in most cases more than normal scatter. In three of the other four dystrophy cases, highest activities of β-hydroxyacyl-CoA dehydrogenase exceeded those for the case shown in Fig. 10. In the case mentioned with no "dead" fibers, Type II fibers had on the average 50% higher activity than the normal fibers in Fig. 10, and Type I fiber activities were nearly twice normal.

<u>Creatine kinase</u> activities in Type II fibers were near normal, but in Type I fibers they were lower and more scattered (Fig. 11). In fibers that were low in both adenylokinase and β-hydroxyacyl-CoA dehydrogenase, creatine kinase activity was nearly zero, another sign these fibers were dead.

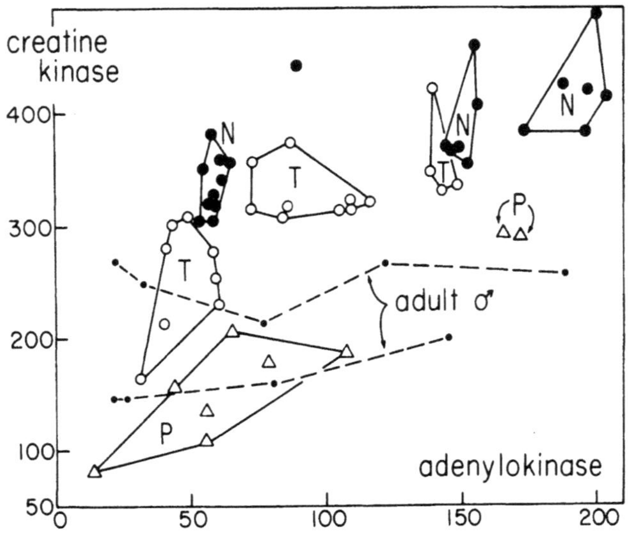

Figure 11. Relationship between creatine kinase and adenylokinase for fibers from a normal 8 year old boy (N), two boys with Duchenne dystrophy (T, age 9 and P, age 5) and two normal adult males. The last were the extreme among five adults and are represented as in Fig. 10.

Glycogen phosphorylase is the enzyme found to be most consistently and severely affected. In all cases of Duchenne dystrophy, activities, especially in the Type II enzyme range, were far below normal. Two of these cases are shown in Fig. 12. Two of the others had activities comparable to those of case A in Fig. 12; the fifth had somewhat higher levels, but still well below those of the normal child.

It will be necessary to analyze fibers from other normal boys before being sure of the significance of these results.

Conclusion

The data presented represent only a step toward the study of individual muscle fibers of normal and abnormal muscles. However, they suffice to

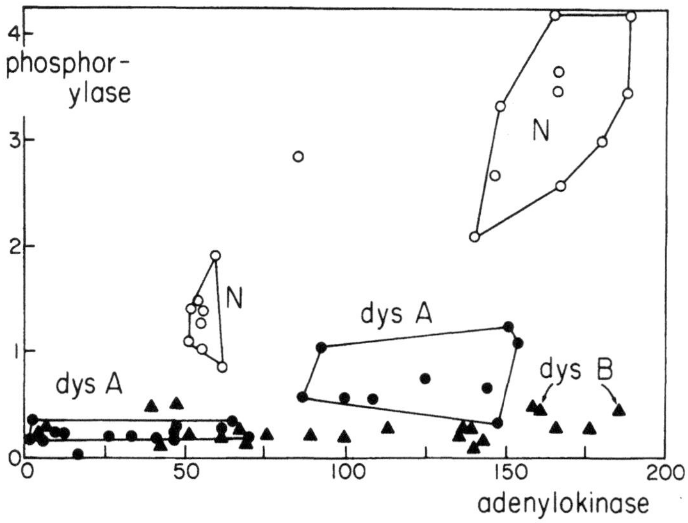

Figure 12. Relationship between glycogen phosphorylase and adenylokinase in fibers from a normal boy age 9 (N) and two boys with muscular dystrophy, "dys A" age 5, and "dys B" age 7. The first had Duchenne dystrophy, the second had either Duchenne or Becker's dystrophy.

show that there may be considerable value in examining individual fibers rather than the muscle as a whole. This is dramatized by the great variability encountered among Duchenne dystrophy fibers, which would be completely obscured in an assay of the whole muscle.

Acknowledgement. This work was supported in part by a Research Center Grant from the Muscular Dystrophy Association and grants from the American Cancer Society (BC-4T) and the National Institutes of Health (NS-08862).

References

1. Spamer, C., Pette, D.: Histochemistry 52, 201-216 (1977).

2. Spamer, C., Pette, D.: Histochemistry 60, 9-19 (1979).

3. Essén, B., Jansson, E., Henriksson, J., Taylor, A.W., Saltin, B.: Acta Physiol. Scand. 95, 153-165 (1975).

4. Lowry, C.V., Kimmey, J.S., Felder, S., Chi, M.M.-Y., Kaiser, K.K., Passonneau, P.N., Kirk, K.A., Lowry, O.H.: J. Biol. Chem. 253, 8269-8277.

5. Edström, L., Kugelberg, E.: J. Neurol. Neurosurg. Psychiatry 31, 424-433 (1968).

6. Chi, M.M.-Y., Lowry, C.V., Lowry, O.H.: Anal. Biochem. 89, 119-129 (1978).

7. Kato, T., Berger, S.J., Carter, J.A., Lowry, O.H.: Anal. Biochem. 53, 86-97 (1973).

8. Baldwin, K.M., Winder, W.W., Terjung, R.L., Holloszy, J.O.: Am. J. Physiol. 225, 962-966 (1973).

9. Engel, .WK., Brooke, M.H., Nelson, P.G.: Annals N. Y. Acad. Sci. 138, 160-185 (1966).

METABOLIC SUBPOPULATIONS OF RABBIT SKELETAL MUSCLE FIBRES

C. Spamer and D. Pette
Fakultät für Biologie, Universität Konstanz, D-7750 Konstanz, Germany

Introduction

Various muscles of the same animal and even those acting upon the same limb, reveal metabolic differences which exceed by far those existing between parenchymal cells of other organs. It is assumed that the composition of vertebrate muscle by different fibre types at various percentages reflects a response to the functional needs of the specific locomotional patterns.

Although qualitative histochemical studies have provided important insights, the full extent of metabolic specialization at the level of single muscle fibres has become evident only recently by the application of microdissection and quantitative enzymological microanalyses (1-5).

Materials and Methods

The present study of enzyme activity patterns in single fibres is based on the microdissection of single fibres from freeze-dried rabbit psoas and soleus muscle. According to Essén et al.(1), fibres were typed by staining small pieces for myofibrillar myosin ATPase activity after preincubations at pH 4.3, 4.6 and 10.3. This histochemical typing leads to the distinction of type I, type IIA, type IIB and type IIC fibres (6). All fibres of psoas muscle studied proved to be IIB's. Similar to the pro-

Plasticity of Muscle

cedure recently described by Lowry et al. (5), several enzymes were measured in samples of the same fibre. Individual fibres were cut into several pieces which were stored separately in small capillaries under vacuum at -40°C. Activity determinations were performed by using the oil well and cycling techniques of Lowry and Passonneau (7). Details have been described elsewhere (2,3). The following enzymes were investigated: Phosphofructokinase (PFK), hexokinase (HK), fructose-1,6-diphosphatase (FDPase), malate dehydrogenase (MDH) and 3-OH-acyl-CoA dehydrogenase (HAD).

Results

Enzyme Activity Patterns of Single Fibres.

Fig.1 depicts enzyme activity patterns which were measured in three selected fibres of rabbit psoas and soleus muscle. The three patterns illustrate differences in enzyme activity which might be related to the fibre classification derived by Peter et al. (8) from qualitative histochemical findings. The two specimen of type II fibres in Fig.1 would thus represent fast-twitch-glycolytic (FG) and fast-twitch-oxidative-glycolytic (FOG) fibres with high glycolytic capacity and varying levels of mitochondrial enzymes of aerobic substrate oxidation. It is seen, that the "FOG" type fibre has a lower activity of PFK. On the other hand, this fibre has a higher activity level of HK. The type I fibre shows an enzyme activity pattern similar to that characteristic of the "slow-twitch-oxidative" type of Peter et al. (8). It has the lowest activity level of PFK and also of FDPase Its HK activity is highest. The activities of MDH and HAD are similar to those in the "FOG" type fibre.

However, measurements on a great number of fibres have shown, that a great variability in metabolic properties exists which cannot be restricted to the three fibre types shown by Peter et al. (8).

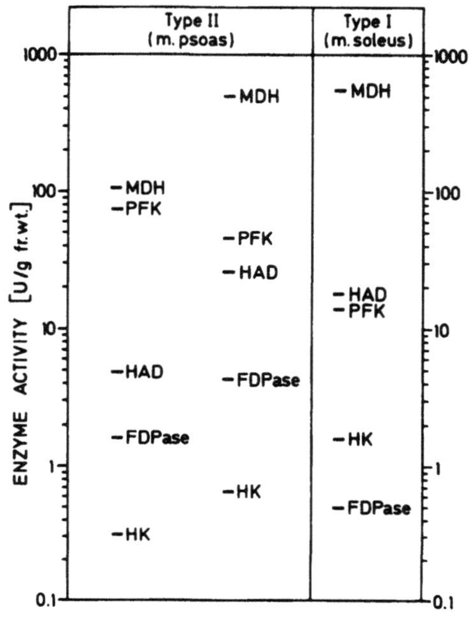

Fig.1 Activities of phosphofructokinase (PFK), hexokinase (HK), fructose-1,6-diphosphatase (FDPase), malate dehydrogenase (MDH) and 3-OH-acyl-CoA dehydrogenase (HAD) in three selected single fibres from rabbit psoas and soleus muscle. The bars one beneath the other represent enzyme activities measured in samples of the same fibre. Fibres were classified histochemically as type I or type II according to their staining for myofibrillar ATPase after acid (pH 4.3 and 4.6) and alkaline (pH 10.3) preincubations.

Fig.2 is a histogram for MDH activity in type II fibres of psoas muscle and type I fibres of soleus muscle. With regard to this enzyme, the two fibre types represent distinct populations, but both display a remarkable scattering. A high percentage of type II fibres falls in the range of low MDH activities between 100 - 300 U/g. Differences exceeding a factor of 10 are found between the extremes of this heterogeneous population. The majority of type I fibres has activities in a range of about 300 - 500 U/g. The range between the extreme values of this population is narrower, having at maximum only fourfold differences.

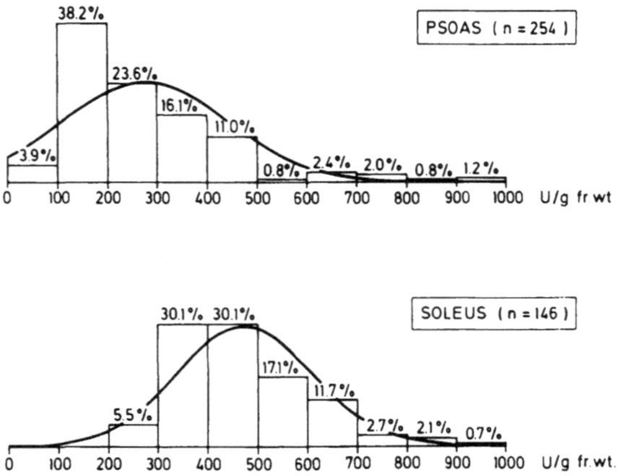

Fig.2 Histogram of malate dehydrogenase activity in fast-twitch psoas and slow-twitch soleus fibres. The histogram was overlayed by a Gaussian curve computed by means of the mean and standard deviation of all values. It is clear that the histogram does not adequately follow this normal distribution.

Fig.3 gives histograms of PFK, HK, FDPase and HAD in fast- and slow-twitch fibres. Each of the enzymes studied, illustrates that fibres belonging to both fast-and slow-twitch groups vary greatly in their quantitative enzymatic make up.

Comparing the two fibre populations, only PFK and FDPase exhibit clearly separate activity distributions. In both enzymes type II fibres are characterized by significantly higher activities than type I, to the extent that activity of FDPase is negligible.

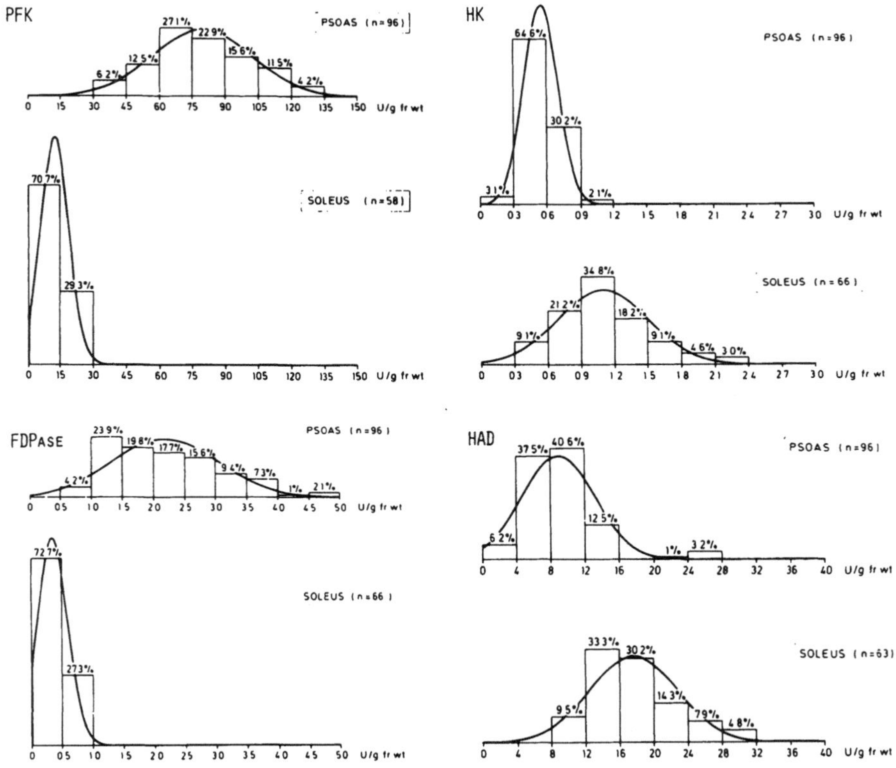

Fig.3 Histograms of phosphofructokinase (PFK), hexokinase (HK), fructose-1,6-diphosphatase (FDPase) and 3-OH-acyl-CoA dehydrogenase (HAD) in fast-twitch psoas and slow-twitch soleus fibres. The histograms were overlayed by computed Gaussian curves. Most of the histograms do not adequately follow this normal distribution.

Metabolic specialization at the level of enzymatic organization may be elucidated by activity ratios of selected enzymes (9,10). A histogram of type I and II fibres for the activity ratio PFK/MDH is given in Fig.4. This ratio may be taken as a qualitative magnitude of metabolic specialization with regard to different glycolytic and aerobic oxidative capacities. It is evident that the two populations vary significantly in this ratio. Moreover, the large scattering indicates that the type II fibre population is not homogeneous. Heterogeneity of the type II fibre population is seen also by the pronounced scattering of absolute PFK and FDPase activities (Fig.3). The similar shapes of the respective histograms of type I fibres might be interpreted as indicating a metabolic coordination. However, neither in type I nor in type IIB fibres could a correlation be detected between these two enzyme activities. Heterogeneity within both fibre populations is thus illustrated by the independent distribution of PFK and FDPase activities.

The histogram for HK activity in Fig.3 may be taken as a further indication of heterogeneity, especially in case of type I fibres. As also shown in Fig.1, HK activity is highest in type I fibres. However, variations by a factor of 8 are found (Fig.3), whereas type II fibres reveal much smaller differences. It has been observed in previous studies (9,10) that a correlation exists between HK activity and key enzymes of the citric acid cycle in various vertebrate and invertebrate muscles. A statistical analysis of the distribution of HK and MDH activities in the same fibres gave correlation coefficients of 0.65 for type I but only of 0.45 for type II fibres. No correlation exists between activity levels of HK and PFK or between HK and FDPase activities within the same fibres.

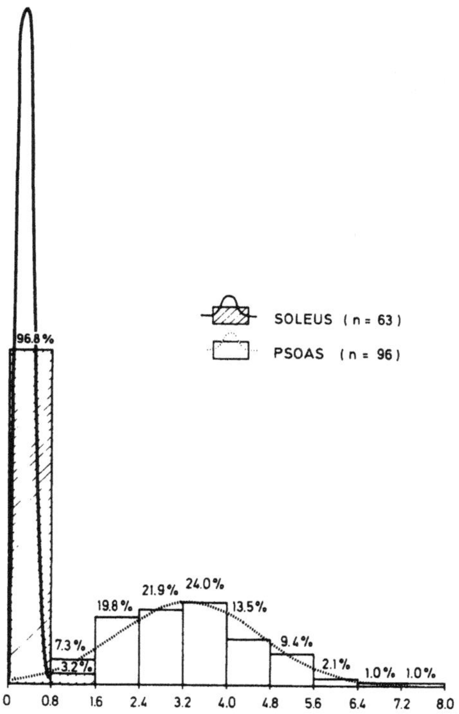

Fig.4 Histogram of relative activities phosphofructokinase (PFK)/malate dehydrogenase (MDH) in fast-twitch psoas and slow-twitch soleus fibres. Activities of PFK had been multiplied by a factor of 10. For further explanation see legend of Fig.2

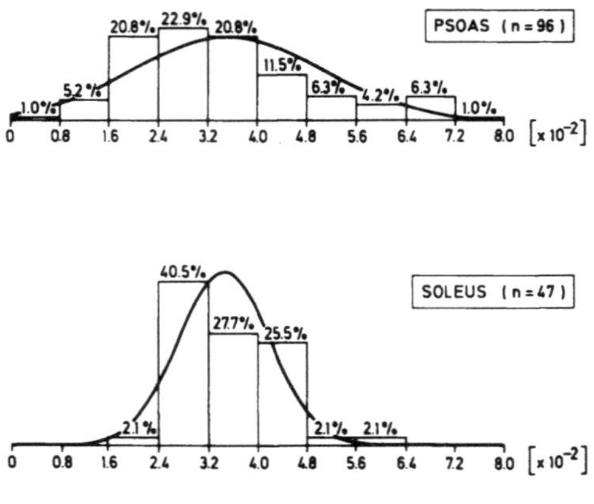

Fig.5 Histogram of relative activities 3-OH-acyl-CoA dehydrogenase (HAD)/malate dehydrogenase (MDH) in fast-twitch psoas and slow-twitch soleus fibres. For further explanation see legend of Fig.2.

Another example of metabolic specialization is represented by the activity of HAD which has been chosen as a reference enzyme of fatty acid oxidation. According to the histograms in Fig.3, the distribution of this enzyme resembles that of MDH in the two fibre populations, and reveals a large scattering. A statistical analysis clearly demonstrates that MDH and HAD activities are closely correlated in type I fibres (r = 0.78), but to a lesser degree in type II fibres (r = 0.45). A comparison of the two fibre populations with regard to the activity ratio HAD/MDH in the same fibres is given in the histograms of Fig.5. The wide range of this activity ratio in type II fibres suggests a pronounced heterogeneity with regard to fatty acid oxidation capacity.

Discussion

While histochemistry of metabolic enzymes provides a useful and easy means for muscle fibre classifications, quantitative data show that with regard to metabolic variability such grouping is an oversimplification. Slow and fast-twitch fibres as defined by their histochemical reaction for myofibrillar myosin ATPase, both represent populations with large scattering of enzyme activities. Although type I and type II fibres may be distinguished with regard to significant differences in absolute activities of some enzymes (e.g. PFK and FDPase), or enzyme activity ratios (e.g. PFK/MDH), there are large variations found within each of the two populations. It appears that a broad continuum of metabolic properties for each population exists rather than distinct metabolic types. This should be taken into consideration when pooled fibres are used for elucidating metabolic properties of these fibre types (1,12). Previous results (2) have shown even greater metabolic variability of fast- and slow-twitch fibres than in the present study, if enzyme activities of the two fibre populations are analysed in different muscles.

Our present findings demonstrate that the enzyme activity patterns of the selected type II fibres in Fig.1 do by no means represent subgroups which might be defined by specific differences in enzyme contents. In the type II fibres studied, the various enzymes and enzyme activity ratios gave no indication of the existence of only two distinct subgroups of type II fibres corresponding to the FG and FOG fibres, as defined by Peter et al. (8). It appears from our findings and also from those of Lowry et al. (5) that these types defined by qualitative histochemistry represent extremes of the type II fibre population.

With regard to MDH activities it should be taken into account that activities of this enzyme measured in the present study represent total activities of both cytosolic and mitochondrial isozymes. Previous studies (13) have indicated that they exist in a constant ratio in "white" and "red" rabbit muscles. If this is valid also for single fibres, this would mean that both cytosolic and mitochondrial activities are heterogeneous.

A further consideration concerns the subgroups of type II fibres in psoas muscle which can be distinguished by staining for myofibrillar myosin ATPase after preincubations at pH 4.3, 4.6 and 10.3. Our present data provide only limited information regarding this point, due to the fact that we measured metabolic enzymes only on IIB fibres. Figs.3 - 5 thus make clear that type IIB fibres cannot be characterized as only "fast-twitch-glycolytic". The respective histograms demonstrate that there is metabolic heterogeneity of this population with regard to enzymes representing anaerobic glucose metabolism (PFK, FDPase) as well as enzymes involved in aerobic substrate oxidation (MDH, HAD). Furthermore, metabolic variability with regard to anaerobic versus aerobic capacities is documented by the large scattering of the ratio between PFK and MDH activities. These findings are in accordance with an independent qualitative study in our laboratory (14, 15). It was shown for two rat muscles that no compatibility exists between the two fibre nomenclatures based on metabolic enzyme profiles and on myofibrillar ATPase staining after preincubation at pH 4.6 for type II fibres. According to this study (14, 15) IIB fibres have a large variability in anaerobic versus aerobic capacities and therefore can be classified as FG and FOG.

It appears thus that the metabolic properties of type II fibres are not related to differences in myosin ATPase pH lability. This point is emphasized by preliminary observations on a few fast-twitch fibres in soleus muscle of a young rabbit. According to myosin ATPase pH lability, these fibres were typed as II A's, II B's and II C's. These different fibre types, however, revealed no significant differences in their activity patterns of energy metabolism. Due to low FDP-ase activity levels, they rather resemble slow-twitch fibres. Most probably, these fibres were in the course of transformation and corresponded to the transition types described by Kugelberg (16). Assuming that differences in pH lability are related to different molecular properties of myosin ATPase, these results suggest that transformation of fibres with regard to metabolic characteristics and to molecular properties of myosin may occur independently. This conclusion is supported by the observation that transformations of metabolic and molecular properties of muscle fibres may be induced at different time courses during chronic stimulation. (17-20). Muscle fibres appear to be capable of transitions between various states. The appearance of a spectrum of fibres rather than of rigid types may be a plausible consequence of their dynamic state.

Acknowledgements

This study was supported by the Deutsche Forschungsgemeinschaft, Sonderforschungsbereich 138 "Biologische Grenzflächen und Spezifität" and by grants from the Bundesinstitut für Sportwissenschaft. Authors are grateful for excellent technical assistance by Ms. M. Emmerich and wish to express their thanks to Dr. Patti Nemeth for reading and discussing the manuscript.

References

1. Essén, B., Jansson, E., Henriksson, J., Taylor, A.W., Saltin, B.: Acta physiol. scand. 95, 153-165 (1975)
2. Spamer, C., Pette, D.: Histochemistry 52, 201-216 (1977)
3. Spamer, C., Pette, D.: Histochemistry 60, 9-19 (1979)
4. Pette, D., Spamer, C.: Diabetes 28, Suppl. 25-29 (1979)

5. Lowry, C.V., Kimmey, J.S., Felder, S., Chi, M.M.-Y, Kaiser, K.K., Passonneau, P.N., Kirk, K.A., Lowry, O.H.: J. Biol. Chem. 253, 8269-8277 (1978)

6. Brooke, M.H., Kaiser, K.K.: Arch. Neurol. 23, 369-379(1970)

7. Lowry, O.H., Passonneau, J.V.: A flexible system of enzymatic analysis. Academic Press, New York and London, 1972

8. Peter, J.B., Barnard, R.J., Edgerton, V.R., Gillespie, C.A., Stempel, K.E.: Biochemistry 11, 2627-2633 (1972)

9. Bass, A., Brdiczka, D., Eyer, P., Hofer,S., Pette, D.: Eur. J. Biochem. 10, 198-206 (1969)

10. Staudte, H.W. Pette, D.: Comp. Biochem. Physiol. 41B, 533-540 (1972)

11. Pette, D., Dölken, G.: Adv. Enzyme Regul. 13, 355-377 (1975)

12. Thorstensson, A., Sjödin, B., Tesch, P., Karlsson, J.: Acta physiol. scand. 99, 225-229 (1977)

13. Pette, D., Bücher, Th.: Hoppe-Seyler's Z. Physiol. Chem. 331, 180-195 (1963)

14. Nemeth, P.M., Hofer, H.W., Pette, D.: Histochemistry 63, 191-201 (1979)

15. Nemeth, P.M., Pette, D.: J. Histochem. Cytochem.: in press

16. Kugelberg, E.: J. Neurol.Sci. 27, 269-289 (1976)

17. Pette, D., Smith, M.E., Staudte, H.W., Vrbová, G.: Pflügers Arch. 338, 257-272 (1973)

18. Sréter, F.A., Gergely, J., Salmons, S., Romanul, F.: Nat. New Biol. 241, 17-19 (1973)

19. Salmons, S., Sréter, F.A.: Nature (London) 263, 30-34(1976)

20. Heilmann, C., Pette, D.: Eur. J. Biochem. 93,437-446 (1979)

HISTOCHEMICAL TYPING OF MUSCLE FIBRES BY MICROPHOTOMETRY

N.C. Spurway
Department of Physiology, University of Glasgow, G12 8QQ, Scotland, U.K.

Histochemistry or Biochemistry?

For the chemical characterisation of muscle fibres, quantitative histochemistry, and single-fibre biochemistry have complementary advantages. Histochemistry allows the study of muscles of all sizes, and all collagen contents; within the muscles, anatomical location can be preserved, and small fibres can be examined as easily as large; more reactions can be applied per specimen, and despite this more fibres can be studied in a given time; and between-fibre variations - or even, with the use of composite blocks, between-muscle variations - can be investigated without being complicated by variation between animals.

On the other hand, though histochemical measurements themselves may be made accurate to the third figure, the information they give about fundamental fibre chemistry can never be more than semi-quantitative. Interfering reagents and structures, which the biochemist would separate from the entity he wished to assay, remain ineluctably present in the histochemical section. In any case, the histochemist is limited in the entities he can demonstrate - be they enzymic activities or bulk cell constituents - by the necessity of producing an insoluble, light-absorbing, final reaction product. (Fluorescent markers, and autoradiography, have so far created only a fairly small number of exceptions to this rule. They will not be discussed further in this paper.)

Problems are not over when an appropriate deposit has been obtained: it has still to be measured. The only conceivable methods at the scale of the individual fibre are those of microphotometry (or microdensitometry). What is measured is the ratio of incident light intensity to the intensity remaining after transmission through the fibre - both at the wavelength of peak absorption. Log_{10} of that ratio is called the 'absorbance' (or optical density) of the field studied. This function is proportional to the amount of the light-absorbing substance, both in a solution (Beer's Law) and in a uniform, solid deposit. (Fig 1a gives an example of a histochemical deposit which reasonably approximates to this condition.) But patterned deposits are more common (e.g. 1b). With these, the effect of unabsorbed light, coming up through the unstained regions, is to swamp the remaining intensity of the light which has passed through stained regions, and - this is the serious bit - to do so proportionately more as the density of the stain increases. Thus the scale of true, chemical density against absorbance is compressed at the upper end. Furthermore, it is not hard to show that the standard methods for circumventing this 'distribution error' cannot be applied to muscle (Spurway, to be published).

Even the way these two sources of error - reaction conditions and optical conditions - interact is unkind to us. The reaction whose deposit we can

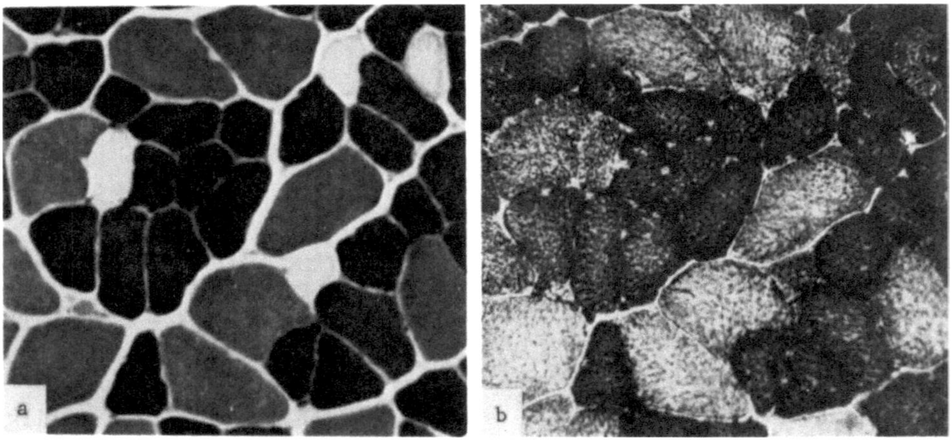

Fig. 1 Serial cryostat sections of mouse gastrocnemius (deep medial part). a Myosin ATPase reaction after formalin fixation and alkaline (pH 10.4) preincubation: three types of fibre are clearly distinguished, and in each type the deposit is reasonably uniform. b Succinate dehydrogenase reaction: a wide but continuous distribution of apparent densities, almost the only common feature being that the deposit is in every instance highly patterned. Light bypassing the deposit through the spaces in this pattern destroys the proportionality (which is well approximated in a) between absorbance and chemical deposit density.

measure best, the myosin ATPase reaction, is the one whose relation to the underlying biological function is least understood; this is because the reaction conditions, chosen to minimise interference from other ATPases, are so unphysiological. The most reliable and well understood reactions are those for enzymes of oxidative metabolism; but these, being associated only with mitochondria, produce the most patterned deposits of all.

The conclusion can only be that, for an accurately quantitative metabolic profile, there is no alternative to single-fibre biochemistry. What the quantification of histochemistry can do is to make the histochemist's traditional interest, fibre typing, objective and rigorous. All the advantages of my opening paragraph suit this role. Interfering reactions seem at least as likely to blur distinctions as to create them; but when they do create them, through an interference occurring in one fibre and not in another, then we have found a real difference even if we take a long time to recognise its nature. As to scale compressions: since some errors are not subject to the compression, the only artefact can be the obscuring of distinctions which are biologically real. Thus if quantitative histochemical properties group in such a way that separate types of fibre can be identified, the case for the real, biological existence of these types is a very strong one indeed.

Experimental Details

Specimens. I shall discuss a completed study of three limb muscles of a mouse, plus the early results of work recently commenced on four muscles of a rabbit. Both animals were cage-bred young adults. Both were killed by stunning. Muscle blocks in both instances were quenched in isopentane at its gelation point. The mouse block was a natural one, of soleus, plantaris and the two parts of gastrocnemius, in normal anatomical relation. The rabbit block was composite: one piece each of soleus, psoas, extensor digitorum longus (EDL) and diaphragm (together with one of masseter, as yet little studied). In both cases, about fifty serial, 10 μm sections were cut at -20°, air dried, and stored dry at -20° for 2-36 hrs.

Reactions. Sections were subjected to one of 9 reactions(mouse) or of 13 (rabbit); of the latter, 12 have been evaluated sufficiently at time of writing to be discussed below. 2-4 sections were put through each reaction with variations of times where appropriate, till at least one good result was obtained for each reaction. Entities demonstrated (in both specimens where not otherwise stated) were:

Lipid (Sudan black stain)
Glycogen (PAS)
Succinate dehydrogenase (hereafter 'SD'; no PMS included)
NADH-tetrazolium reductase (or 'diaphorase': 'NADH-TR')
Malate dehydrogenase ('MD'; rabbit only)
β-hydroxybutyrate dehydrogenase ('HBD'; rabbit only)
Glycogen phosphorylase, a form ('GP_a'; Meijer's method without AMP)
Glycogen phosphorylase, total a plus b ('GP_t'; with AMP; rabbit only)
α-glycerophosphate dehydrogenase ('αGPD'; menadione added)
Myosin ATPase (briefly fixed, then preincubated in Ca; 'BK-mATP')
Ditto (fixed, then preincubated at pH 10.4; 'Alk-mATP')
Ditto (preincubated at pH 4.35; 'Acid-mATP')

The first 9 reactions were performed entirely as in standard texts (e.g. 1) except that none were fixed before incubation and all except the first were mounted anhydrously. The 'brief fixation' carried out at the start of the BK-mATP demonstration consisted of 1min (mouse) and 5min (rabbit) in the cold, fresh 5% formalin fixative of Guth & Samaha (2); these were times sufficient to prevent clumping of the final reaction product (presumably by inhibiting the mechanical interaction of actin and myosin) yet barely to begin the kind of differentiation among fast fibres achieved by the full Alk-mATP procedure. The Ca medium (40mM $CaCl_2$ in 100 mM buffer, pH 9.3) was like one recommended by Brooke & Kaiser, but used by them without any prior fixation; times here were 15min (mouse), 30min (rabbit). The Alk and Acid preincubation procedures were exactly as prescribed by Guth & Samaha, excepting for the times required by the rabbit tissues in the Alk sequence: 15min fixation (5min prescribed, and found satisfactory for mouse) and 45min alkaline (pH 10.4) treatment (15min prescribed, and satisfactory for mouse). (It should be noted also that individual rabbits seemed to vary; another animal, regrettably not the subject of a full series of reactions, gave similar or better results with only 15min fixation, 15min alkali, or again with 45min alkali preceded by no fixation at all. The only consistent rule is that at least one of the two steps must be more protracted for rabbit than for the majority of species with which I am familiar.) Finally, after each of the three pretreatments, mATP activity was demonstrated in the same way - by the 18mM Ca, pH 9.4 procedure of Guth & Samaha, which itself follows closely the long-established method of Padykula & Herman. I depart from Guth & Samaha only in substituting distilled water for the rather viscous buffer-rinse prescribed at step 11 of their sequence.

Sampling. For photometry, regions were chosen in which the same fibres could be confidently identified in each of the 9 or 12 selected sections, and appeared well preserved and demonstrated in each. Within these constraints, areas were not chosen at random: regions of high variation were preferred to those of uniformity, once it appeared that the basic fibre type of each uniform area had been sampled 5-10 times, for the purpose was to study what fibre types occurred - it was not to estimate their overall percentage frequencies of occurrence. In particular the peripheral 1½mm of mouse gastrocnemius was markedly under-sampled for this reason, and in rabbit psoas the only locations considered for study so far have been 3-4 small areas in which some variation of fibre type is seen. Following the application of these criteria, 200 mouse fibres have been fully characterised, but only 66 rabbit fibres have yet been the subjects of reasonably complete measurement.

Photometry. Leitz 'Orthoplan' microscope, with 'Orthomat' camera unit containing no film, is the basic instrument used. Both components are turned on and allowed to stabilize for at least 30 mins before use. Sections are illuminated at constant current (4a) and 3 steps are taken to minimise glare: field diaphragm is maximally closed, condenser diaphragm is set at 1/3 that of the objective, and room lighting is heavily dimmed. Oil immersion objectives have been used in both studies - x40 (mouse), x100 (rabbit). From the objective, light is passed through a filter appropriate to the stain: yellow (K530) for Sudan and all diformazan deposits, green

(S550) for Schiff colorations or blue (CB12) for cobalt sulphide. (It has been established that narrower-band filters than these only marginally increase contrast.) Thence the beam is split, part going to the eyepieces and part to the exposure-control photometer of the Orthomat unit, which is the intensity sensor. Working in the 'detail measurement' mode of the instrument, in which light from a 10μm (x100 objective) or 25μm (x40) circle of the object plane reaches the photometer, a reading indicating light intensity is taken by initiating an 'exposure' (of the non-existent film): exposure time allowed by the Orthomat is inversely proportional to measured intensity. The duration of the 'shutter open' signal is counted by external electronic circuitry - at present these are the timing circuits of a Digital Equipment 'PDP-8' computer. Senistivity adjustment is available in the form of the 'film speed' control. After the mean incident intensity has been measured (as the light transmitted through specimen-free regions of the slide) with the x40 objective 2-5 samples and with the x100 objective 3-10 samples of the light transmitted by each fibre are recorded.

Computation. Mean and standard deviation absorbances are printed, for each fibre and each reaction. When data collection is complete, values for each reaction are normalized by being divided by their own geometric mean. Thus the effects of variation in incubating conditions, producing weaker or stronger overall reactions, are removed - except insofar as they would have altered the relative reaction intensities of two fibres of different types. The same is true of diet-, history-, or even species-variation between animals. The PDP-8 is then programmed to display on orthogonal axes the values of 2 variables for each fibre, and in addition to represent, by choice of symbol, either the muscle from which the fibre came or the range of values of a 3rd variable in which it falls. Regression data, for the variables on the axes, can also be called up. To handle significantly more columns of data at once requires, however, a highspeed digital computer with large memory; cluster analyses, looking for groupings in terms of all data taken together, can then be performed. I have done this for mouse (3), but not yet for rabbit.

Results

Single-reaction histograms & 2-reaction correlations. When histograms displaying the frequencies of occurrence of different absorbances are compared, for reactions traditionally considered equivalent as metabolic markers, they are often found to be rather far from similar (e.g. Fig. 2, top pair). This alerts us to the desirability of regressing such pairs of variables against each other, at least algebraically and where possible graphically. The correlation coefficient (r) for the GP_a/ GPD regression in mouse is 0.61, and values under 0.7 are typical for the glycolysis-associated markers used in the present work - and in much previous muscle histochemistry.

However, in no case explored this far have such mismatches been anything like as severe within a single muscle as when values for the whole block were pooled. The enzymes of oxidative metabolism in the rabbit, although much better matched overall than the 'glycolytic' markers, illustrate this nicely. Within EDL, r for the NADH-TR/SD correlation has the very high value of 0.97, and both MD/SD and HBD/SD show r=0.94. When all muscles of the block are pooled, however, r=0.86 for NADH-TR and 0.77 for HBD (MD looks similar to HBD, but data not yet complete). The statistical explanation is revealed when the constants in the regression equations are considered. Taking NADH-TR as the example (Fig. 3) we find that the slopes of the lines (indicating the contrasts produced by the reactions) are similar in all the muscles, but only for 2 muscles do they start at the origin. For EDL, and even more for soleus, there is an intercept on the NADH-TR axis, indicating residual reaction where SD would show nothing. (Indeed it is striking even to the eye that rabbit soleus is stained more strongly in NADH-TR sections than in SD sections. In mouse, the converse is the case - but mouse soleus contains a far higher percentage of fibres which are not slow.) Thus r is lower for the pooled data than for data from individual muscles simply because it is trying to embrace several lines which, though roughly parallel, have different ordinates. Biochemically, one possible explanation lies in the existence of diaphorases which are not part of the electron transport chain - on this hypothesis, what the histochemistry is showing is that these other diaphorases are particularly active in rabbit soleus.... A similar analysis has been performed for HBD.

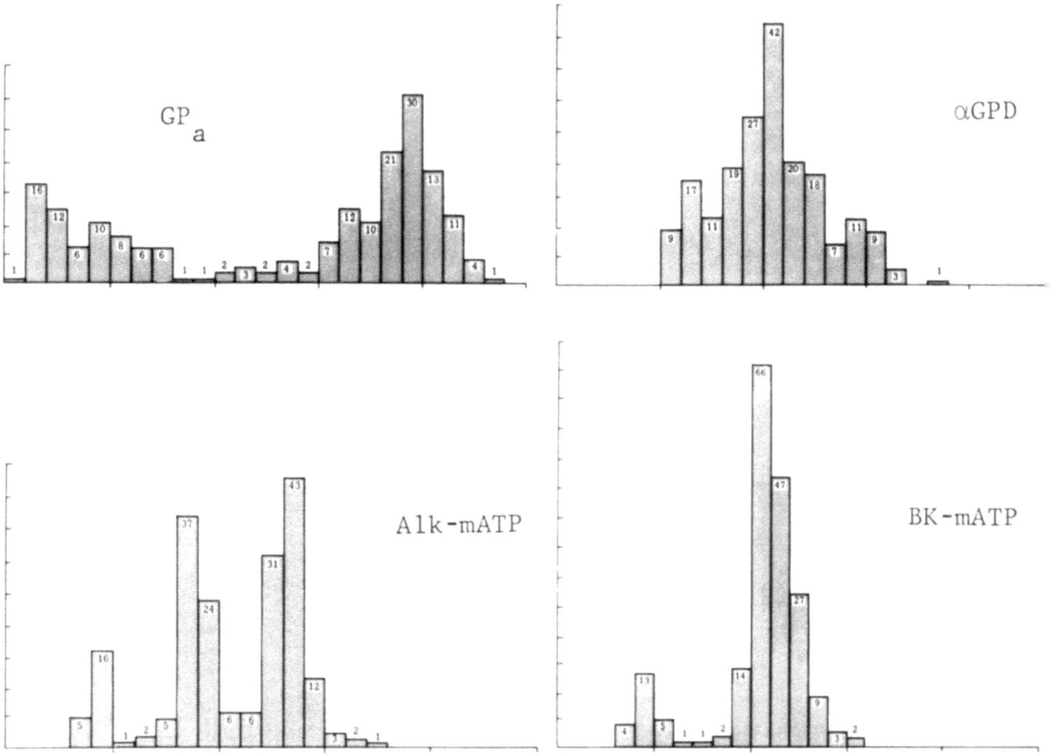

Fig. 2. Mouse data. Distributions of measured absorbances (abscissae) for four individual reactions. Ordinates: number of fibres at each absorbance. Upper row: two reactions, both associated with glycolytic metabolism, correlate imperfectly (r=0.61); GP_a (left) discriminates fibres into "types" more clearly than αGPD. Lower row: both mATP reactions discriminate sharply, but Alk (left) further divides the large peak of BK.

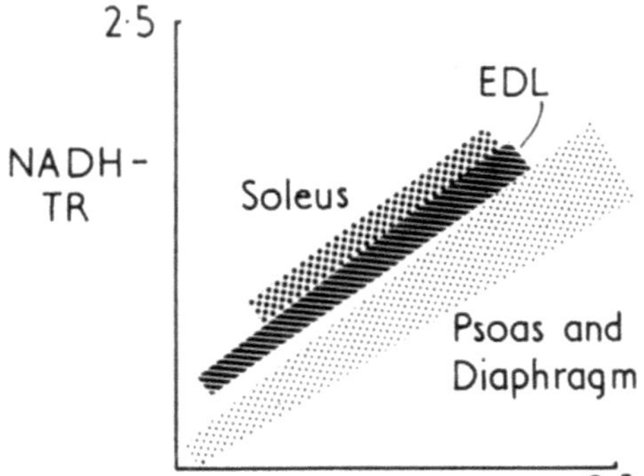

Fig. 3. Rabbit data. Sketch-graph of NADH-TR absorbances plotted against SD absorbances. In all muscles these two oxidative markers vary proportionately; however there is significant background (?cytoplasmic) reaction for NADH-TR in some muscles while not in others.

Again, soleus fibres display activity above the 'zero intercept' trendline. This time, however, they are joined not by EDL but by the more highly oxidative fibres of diaphragm.

When oxidative and 'glycolytic' markers are compared, the textbook concept that these two branches of metabolism vary reciprocally would lead one to expect high negative r. Actually, if GP_a is regressed against SD, r for pooled data, though negative, is small (-0.33 for rabbit); and within the group of fibres reacting strongly in the Alk-mATP sequence (?'FG' fibres - see below) the more strongly glycolytic fibres are also more oxidative (r= +0.68, on present, limited data). One is reminded of similar trends which are seen even across the fibre types in amphibia and fish. However, in those classes of vertebrate, fat metabolism is also positively correlated with glycolytic; the rabbit HBD/GP_a plot suggests a clear negative correlation, though I have yet to calculate r. What the better trained (or less untrained) of these rabbit 'FG' fibres are more probably doing is equipping themselves to metabolize glycogen aerobically.

Clustered plots, & multi-reaction correlations. When 2 variables, which are being plotted orthogonally, each possess strongly multimodal or even discontinuous histograms then 2-dimensional clusters of data points emerge. The mere existence of these clusters commands primary attention, though questions of correlation - particularly within-cluster correlation - may be raised again later. Such clusters are, of course, the graphical representation of 'fibre types'.

Consider the left-hand pair of histograms in Fig. 2. Plotted orthogonally, these variables produce the distribution of data-points seen in Fig. 4. Few would dispute that there are 4 significantly-discrete clusters here, though some might wish to leave more fibres outside the envelopes than I have done, to be regarded as 'intermediates' (and so, perhaps, 'transitionals'). However, we cannot constructively name the clusters until some correlation with oxidative metabolism is also established. Choosing, from plots of SD against GP_a or Alk-mATP, the SD ranges which seem best to reveal distinctions between one cluster of the Fig. 4 plot and another, I then indicate by the symbol used for each fibre the range in which it falls.

The labels, SO, FG, FOG, FO (extending the shorthand which emerged from the work of Edgerton, Peter & colleagues) now place themselves!

Of other possible variables, SD, when itself used as one of the axes, is less successful in giving clear divisions ('moats') between clusters, for its own histogram is far from discontinuous. As alternatives to GP_a, PAS and αGPD both have the merit of showing all SO fibres lower than any FG fibre; however they both lower the FOG cluster sufficiently that separate identity from the FO is lost. Alk-mATP might be replaced by Acid or by BK. In the multi-enzyme series described, Acid did not make the FG/FOG distinction as clear as it is in Fig. 4, though other results indicate that, when time and pH are absolutely optimal, it can do so. BK, on the other hand, shifts the FG cluster into the middle of the FOG and entirely destroys the distinction - as qualitative histochemistry would lead one to expect. The BK and Alk effects may be directly compared in Fig. 2 (bottom).

Formal cluster analysis (3), which can consider together all 9 reactions (plus a 10th measurement: diameter), still finds the 4 clusters indicated

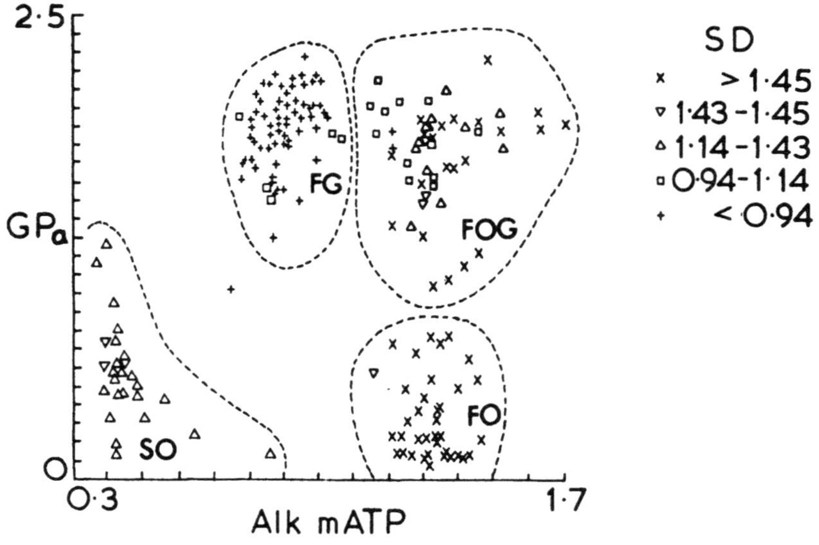

Fig. 4. Mouse data. Absorbance values, for the most discriminative reactions representing two distinct metabolic systems, plotted orthogonally. Points (fibres) fall into four clusters. Values for the third metabolic system (symbols) allow labels of traditional form (FOG etc.) to be added.

in Fig. 4, though it does not if GP_a and Alk data are excluded from consideration. This suggests that the 2 most discriminatory markers are not in opposition to more general trends of the data (or they would be "outvoted", 8:2), but merely introduce discriminations the others fail to make.

Turning to the new rabbit data, we no longer find one reaction whose absorbances, used as ordinates, give sharper and more interpretable clusters than any others. Choosing instead for familiarity, Fig. 5 shows the same 3 variables as Fig. 4. Clusters appear, but how should they be regarded? There is an indisputable SO group (soleus and some diaphragm fibres). Then a few fibres presumably in transition between slow and fast. But should the fast fibres be considered as 2 very elongated, vertical clusters, or would it prove more fruitful to regard them as 2-3 roughly circular clusters, one above another? With due recognition of the smallness of the present sample, I am for the time being firmly disposed toward the first interpretation by 2 facts. Firstly, replacement of GP_a by αGPD (Fig. 6) shifts the psoas and EDL fibres, some up and some down, till those which had formed the apparent cross-over at mid height merely contribute to a slight tilt of both clusters from the vertical. (What happens to the small group of diaphragm fibres is an intriguing but separate point.) The second fact pressing one to think of the columns as the working groups is that, categorised in terms of HBD, all the right hand column are weak, almost all the left hand column, plus all to the left of <u>them</u>, are strong. The exceptions in the left column are just those fibres which are nearest the putative cross-over point in Fig. 5, and which rate slightly less strongly than their fellows for SD activity also. The very few members of the right column which have SD in this, for-them-high range are at the very same GP_a level. Could it be that this cross-over is not a chance geometrical feature but a true transition zone - a level of GP_a activity to which fibres must approximate before making the transit from one column to the other? Obviously far more data are needed before weight can be attached to such a speculation, but it does illustrate another of the kinds of contribution which can be made by quantitative histochemistry.

What remains is to designate the 2 different fast-fibre clusters. No traditional labels, other than those based solely on pretreated mATP reactions,

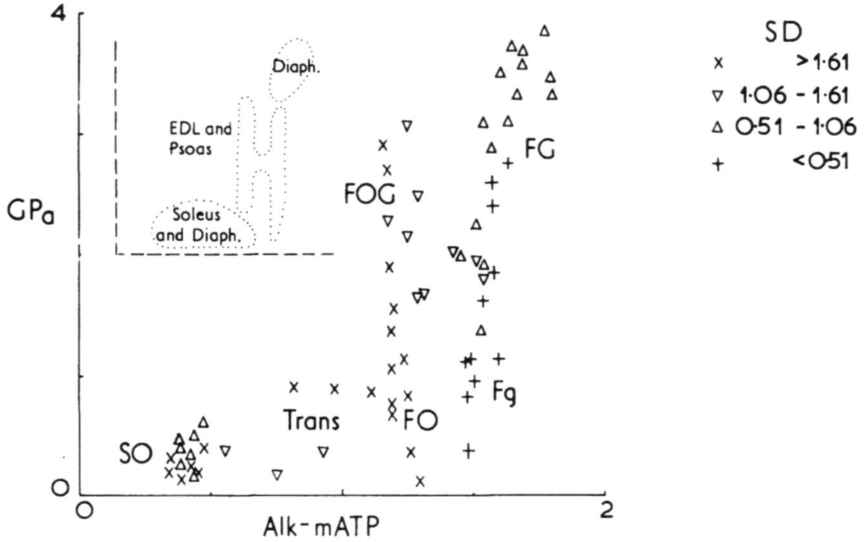

Fig. 5. Rabbit data. Cluster plot, directly equivalent to that for mouse, Fig. 4. This plot, considered alongside Fig. 6, presents fast fibres in two main clusters: one is more oxidative than the other, but each embraces a wide range of glycolytic capacities. "Trans" = transitional (?). "Fg" = fast, modestly glycolytic. Insert indicates site of each fibre.

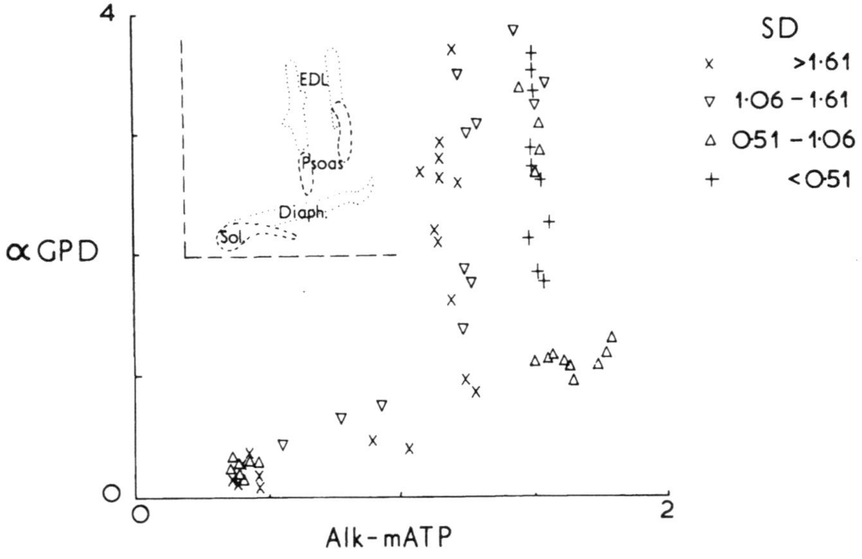

Fig. 6. Rabbit data. As Fig. 5, except for changed ordinate.

can possibly apply to the whole of a cluster elongated, as each of these apparently is, across almost the total range of one variable. Yet there is a difficulty with the mATP reactions as bases for classification: the ambiguity produced by the fact (already well-known, but clearly seen when Figs 4 & 5 are compared) that in rabbit, like other medium and large-sized mammals, the darkest reaction after Alk pretreatment is in the less oxidative fast fibres, whereas in mouse and other small mammals it is in the more oxidative. The left hand cluster can, however, very conveniently be labelled as ranging from FOG to FO, without any implication that a moat exists half-way up. Treating the right hand cluster similarly, I designate the upper end 'FG' - the fibres are also moderately oxidative, but not more so than the most oxidative members of the FG cluster in the mouse. But the lower fibres, though profoundly non-oxidative, are rather weakly glycolytic too - in accordance with the positive correlation between SD and GP_a levels, within this cluster, noted earlier. My tentative coinage 'Fg', expresses this fact. Yet the need to indicate it may be doubted; the GP_a/SD ratio in these fibres is unquestionably high - i.e. they are emphatically glycolytic when they metabolize at all. I therefore wonder if the low absolute level of both enzymes is not an abnormality produced by caging? All this takes no account of the very poor correlation between GP_a and αGPD within the right-hand cluster; but an implication of much of this paper is that one should always specify the markers in terms of which one is classifying.

Comments

1. Validity of correlation figures. The oxidative enzymes all give deposits with essentially the same distribution. Thus both absorbance scales are similarly compressed at the upper ends. Assayed biochemically instead of histochemically each enzyme would need a much longer scale, but there is no reason to think that their correlations would be much altered. The GP_a/SD correlation is the only one discussed which should be viewed with real caution on these grounds; and even GP_a's deposit is mainly interfibrillar, and so networklike.

2. Mechanism of 'Alk' & 'Acid' pretreatments. The importance of pretreated mATP reactions, for discrimination between fast fibres of different types, has been re-emphasized by quantification. Regrettably we still do not know how the pretreatments work. Weeds et al. (4) were the first to show that the more and the less oxidative fast fibres do not differ in the molecular weights of their myosin light chains, but this does not prove that the respective chains are identical in their detailed chemistry. Alternatively, other parts of the molecule may differ.

What seems to me less likely to be fruitful is to look for the differences of myosin lability in differences of protection by the cellular environment. I reject this:

a Because histograms reflecting mitochondrial content show continuous distributions; myosin lability is discontinuous.

b Because then mouse mitochondria would have to be protective, rabbit mitochondria the opposite.

3. Cause of clustering. Mouse and rabbit results differ partly because muscles with a wider range of functions were examined from the rabbit. Within-muscle clustering, like simple within-muscle correlation, is usually tighter than that given by pooled data. This suggests to me that fibre properties cluster, not because of any molecular-biological constraints, but because neuronal firing patterns cluster - and that the details of these pattern-clusters vary with the function of the neurone pool.

I acknowledge with gratitude the technical histochemical assistance of Mrs. Anne Ferrell, the computer-programming of Dr. Victor Moss and the typing of Mrs. Marjory Wright.

References

1. Pearse, A.G.E.: Histochemistry, Churchill-Livingstone, Edinburgh (1972).

2. Guth, L., Samaha, S.J.: Exp. Neurol. 28, 365-367 (1970).

3. Spurway, N.C.: J. Physiol. 280, 39-40P (1978).

4. Weeds, A.G., Hall, R., Spurway, N.C.: Febs Letters 49, 321-324 (1975)

MALATE DEHYDROGENASE HOMOGENEITY OF SINGLE FIBERS OF THE MOTOR UNIT

P.M. Nemeth, D. Pette and G. Vrbová*
Fakultät für Biologie, Universität Konstanz, Konstanz, Germany,
*Department of Anatomy and Embryology, University College London, London, England

Metabolic characteristics of muscle have long been thought to be under the influence of nerve, initially observed with other changes brought about by denervation (for review, 6). That metabolic properties could be, in part, switched by crossing the nerves (2) of muscles of different general metabolic types (1,13), suggested that the influence was a specific property of the innervation. Manipulation of the nerve activity by electrostimulation to a pattern normally supplying slow-twitch muscle has been shown to direct the energy metabolism of the muscle more towards aerobic substrate end oxidation (11). This suggests that the metabolic changes are in some way accountable to the frequency of nerve activity. If so, it should be expected that the muscle fibers composing a motor unit have identical metabolic characteristics. Along with the establishment of a means to detect the fibers of a motor unit (7), histochemical studies have provided qualitative evidence for the homogeneity of the motor unit with regard to certain enzymes of energy metabolism (3,4). With the possibility of making biochemical measurements on single cells (10), it has been shown that histochemical homogeneity of fiber populations does not necessarily indicate quantitative homogeneity (14,15). This study was undertaken to determine whether enzyme activities of fibers of a motor unit are quantitatively identical.

In this laboratory, a host of metabolic enzymes have been analysed in single fibers of muscle, in general and in subpopulations (14, 15). Malate dehydrogenase (MDH) activity has been shown to be highly variable, and was thus considered to be the best enzyme with which to test the existence of metabolic homogeneity.

METHODS

Fibers of EDL muscle of rat belonging to a single motor unit were identified histochemically after physiological depletion of their glycogen stores (7). Depletion was made by electrically stimulating single ventral rootlets of spinal cord levels L7 and S1, while monitoring the mechanical response of the muscle held on a strain gauge. Single rootlets gave an all-or-none contractile response to low voltage electrical pulses. The tension of the initial nearly isometric twitch and of the unfused tetanus to a 40 Hz stimulus were recorded. Stimulation programs consisted of repeated tetanization with 10 pulses of 0.02 msec at 40 Hz/sec, each tetanus lasting 250 msec. Depletion was judged as fatigue of the muscular contractions on the pen recorder of the Digitimer Ltd. DS2 stimulation unit. Motor units showing fatigue sensitivity in the first 15 min were stimulated for 30 min and those showing fatigue resistance were stimulated 1 hour. Immediately following stimulation, the muscle was quickly frozen in melting isopentane (-160°C), and stored below -70°C. A piece of the muscle was processed for the histochemical PAS reaction to insure that one motor unit per muscle was effectively depleted.

For microanalysis, whole cross-sections of the muscle, 4 to 5mm thick, were freeze dried for 48 hours at -40°C at a pressure less than 10^{-3} torr, and stored thereafter under vacuum at -40°C (14).

All fibers, to the extent possible, were microdissected from the area identified by the histological section stained for PAS to contain the depleted motor unit fibers. Small pieces of each dissected fiber were stained with PAS. It was necessary to enhance the contrast between depleted fibers and those non-depleted fibers containing low levels of glycogen. This was accomplished by building-up the existing glycogen molecules by preincubating the fiber pieces in a medium for the enzyme α-glucan-uridine diphosphate glucosyltransferase (12). In 6 muscles effectively depleted, over 3300 single fibers were dissected, catalogued and examined for glycogen depletion. From these, 5 - 13 fibers of the depleted motor unit per muscle for a total of 46 fibers, were found.

Quantitative determinations for MDH (EC 1.1.1.37, MDH) activity were made on 6 - 12 pieces of single fibers of the motor units. Measurements were made concurrently on non-depleted fibers randomly selected from the same muscle. Since only 10 fibers could be analysed per experiment, it was necessary to control for experimental error from day to day. One or two fibers from the previous assay were re-examined with the current assay. The assay mixture contained 70.5 mM triethanolamine-HCl, 7.05 mM EDTA 0.58 mM NADH, 0.75 mg/ml bovine serum albumin, 0.5 mM oxaloacetate, pH 7.6 (14). The production of NAD was measured fluorometrically, and calibrated by enzymatically measured NAD standards. Photometrically measured assay reactions were run in parallel with microassays, using purified MDH enzyme. Activity was shown to be proportional to weight of fiber sample within the range of 4 - 18 ng dry weight and to incubation time.

RESULTS

MDH enzyme activities in single fibers of rat EDL muscle ranged between 20 - 226 U/g fresh weight (Table 1). Enzyme levels of the 6 motor units ranged between 31 - 106 U (Table 1), but differed only slightly in the fibers of the individual motor units. On analysis (Table 2), it is seen that the mean activity

Table 1 MDH activities (U/g fresh wt.) in single fibers of rat EDL muscle

EDL	Motor Unit	Controls
1	35.4 - 38.4	52.7 - 156.2
2	101.8 - 106.8	45.2 - 119.8
3	99.9 - 102.7	35.9 - 120.3
4	47.9 - 52.9	31.2 - 148.7
5	30.7 - 32.3	20.2 - 135.8
6	75.8 - 77.2	46.4 - 225.5

Table 2 Deviations in MDH activities within motor units and within their single fibers

Motor Unit	Number of fibers	Mean activity U/g fresh wt	Mean SD of Motor Unit	% Error	Mean SD of single fibers	% Error
1	13	37.1	0.9	2.4	1.6	4.3
2	6	105.1	1.8	1.7	3.9	3.7
3	5	101.4	1.1	1.0	6.5	6.4
4	8	49.2	2.1	4.2	3.2	6.5
5	7	31.5	0.5	1.6	1.1	3.5
6	7	76.3	0.6	0.7	1.7	2.2

of fibers of the individual motor unit varies by 0.7 to 4.2%. The experimental error, judged by the SD of activities in multiple determinations of the same fiber was 2.2 to 6.5%. Thus, it is evident that the variation in activities among fibers of a motor unit is less than that within pieces of single fibers.

The varying activities of the fibers of the 6 motor units relative to the other fibers of the same muscle, suggest that they might represent specific functional types. Examples of motor units having relatively low (Fig.1) and high (Fig.2) MDH activities in their single fibers are displayed in conjunction with their physiological properties. Mechanical records of (a) initial twitch and (b) unfused tetanus to a stimulus train of 40 Hz show that both motor units are fast-twitch. The speeds are 20 and 23 msec twitch time to peak, and 24 and 26 msec 1/2 time to peak tetanic tension. The third graph of tension (c) is a record of maximal tetanic contractions produced by an intermittent stimulation program at 40 Hz/sec. In figure 1,the tension was reduced by 91% within 15 min, indicating a fatigue sensitive motor unit corresponding to relatively low MDH activity in its fibers. The motor unit illustrated in figure 2, lost only 65% of its tension in 15 min and retained this level for 1 hour. Its resistance to fatigue is accompanied by relatively high MDH activities. The motor units in this study displaying fast-twitch fast-fatigue properties, had a range of 32 - 49 U/g fresh weight MDH activity, while those fast-twitch motor units with fatigue resistance had 101 - 105 U/g fresh weight.

The relationship between enzyme activity level and the fatigue resistance of each fast-twitch motor unit was assessed. An index of fatigue was assigned to each as the fraction of initial tetanic tension remaining after 30 min of continuous stimulation. The enzyme activities were roughly proportional to the fatigue index, such that high MDH activities were assosiated with fatigue resistance (Fig.3).

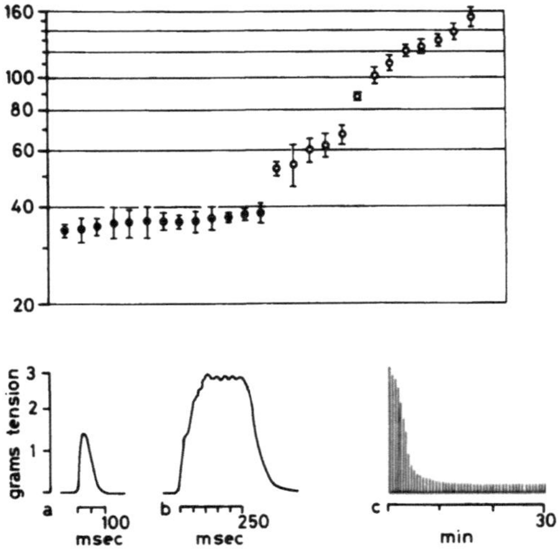

Fig.1 MDH activities in single fibres of a fast-twitch fast-fatigue motor unit (●) and randomly selected fibers (o) of rat EDL muscle. Enzyme activity expressed in U/g fresh wt. ± SD. Mechanical records show a) initial twitch, b) unfused tetanus to 40 Hz stimulus, and c) maximal tetani to repeated stimulation of 10 pulses per sec at a frequency of 40 Hz.

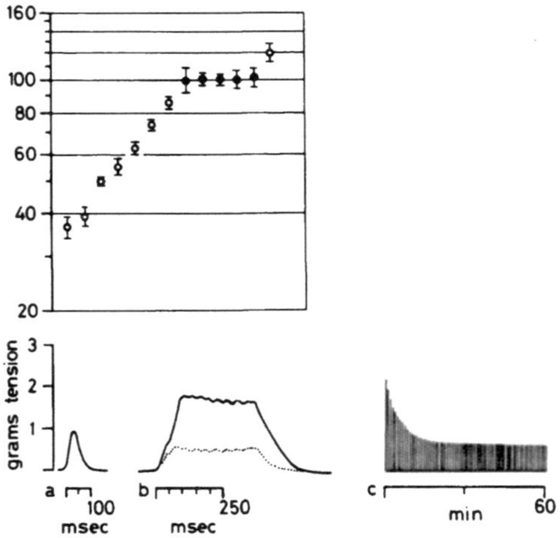

Fig.2 MDH activities in single fibers of a fast-twitch fatigue-resistant motor unit (●) and randomly selected fibers (o) of rat EDL muscle. For further detail, see legend for figure 1.

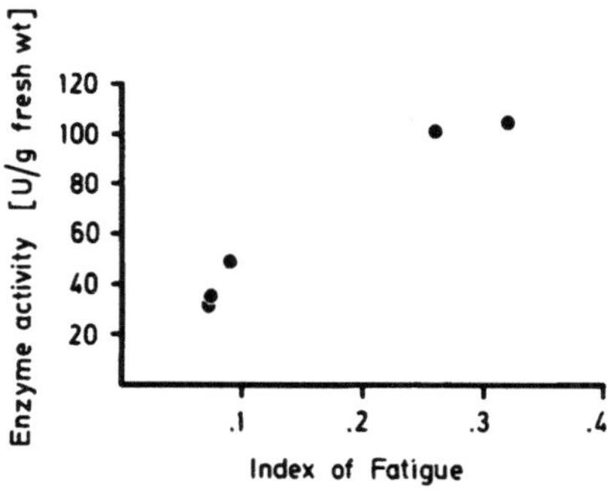

Fig.3 Relationship between MDH enzyme activities in single fibers of fast-twitch motor units and their fatique-resistance, judged as the fraction of initial tetanic tension remaining after 30 min of continuous stimulation (40 Hz/sec) of the motor unit.

DISCUSSION

An evaluation of the individual motor units in this study is in agreement with the findings of Edström and Kugelberg (4) and Burke et al. (3) who showed a direct correlation between fatigue resistance in fast-twitch fibers and, by qualitative assessment, of their aerobic oxidative capacities. Recently, Kugelberg and Lindegren (8) have studied the nature of the fatigue caused by continuous stimulation programs. They reported that the extent to which the reduction of tetanic tension correlates to oxidative capacity suggests a purely muscular type of fatigue. As the present findings show a reasonable correlation between these two properties, the question is opened as to how the extensive chain of events leading to contraction are appropriately matched with the available means of utilizing energy for contraction. Further functional evaluation is beyond the scope of this study as it is limited to a few units studied in great detail biochemically.

Recently, large numbers of single muscle fibers have been evaluated for enzymes of energy metabolism in rabbit (14, 15) and in man (5, 9, 16). It appears that a spectrum of enzyme activities exists throughout a muscle's fiber population. Subgroups, however, have been detected having varying ratios of enzyme activities involved in anaerobic and aerobic metabolism (5, 9, 14, 15, 16). This is in accordance with the present small sample showing two distinct groups with regard to physiological properties having a range of 32 - 49 and 101 - 105 U of MDH activity. These ranges support the evidence against simple fiber grouping.

The present data show a 12-fold range of MDH activity in fast-twitch fibers of EDL muscle. This is in agreement with Spamer and Pette (15) showing 14-fold difference in MDH activity of fast-twitch fibers, distinguished by myosin ATPase. The absolute activities in rat EDL muscle of the present study are lower than those measured by these authors in rabbit soleus and somewhat lower than rabbit psoas. It is clear from their study that this difference in absolute activities is not unlikely since great variations occur even within the same species from muscle to muscle. They found absolute activities within the same range in fast-twitch fibers of soleus muscle and in slow-twitch fibers of psoas muscle.

It is clearly demonstrated to the precision available through microchemistry, that the fibers of a motor unit are homogeneous for MDH activity. The assumption that these fibers are entirely metabolically identical is supported. MDH activity was considered to be the most sensitive enzyme marker, because it shows the greatest variability of the enzymes measured in rabbit psoas and soleus muscles (15). Metabolic heterogeneity in single motor units, the antithesis, would be most likely detected by the measure of this enzyme. Qualitative histochemical evidence for homogeneity in the motor unit fibers for several enzymes of both anaerobic and aerobic glycosis was given by Burke et al. (3) and by Edström and Kugelberg (4). Furthermore, great numbers of enzymes have been measured in single fibers by Lowry et al. (9). There were groups of fibers with profiles of 10 enzymes per fiber showing remarkable similarity in their absolute activities, and great dissimilarity between other groups of the same muscle. The present evidence would suggest that those fibers whose profile were the same, belong to the same motor unit.

From this study it is evident that the control of the muscle fiber's metabolism mechanistically involves the specific neural input. Whether this control is strictly by an electrical phenomenon due to changes in ionic conditions or by electrical activations of a chemical trophic factor of nerve or muscle is not clear. The investigation of the mechanism of neural input as well as other studies dealing with functional and metabolic correlations, require this clarification of the basic assumption of metabolic purity of the motor unit.

Acknowledgements

Authors gratefully acknowledge the advice of Dr. Robert Burke on the stimulation method for glycogen depletion, as well as the skillful assistance of Ms. Margreth Emmerich in microdissection. P. Nemeth wishes to thank the Alexander von Humboldt Foundation for a fellowship. This study was supported by Deutsche Forschungsgemeinschaft, Sonderforschungsbereich 138 "Biologische Grenzflächen und Spezifität."

References

1. Bücher, T., Pette, D.: Verh. Dt.Ges. Inn. Med. 71 Kongress 1965, München, 104-124 (1965).

2. Buller, A.J., Eccles, J.C., Eccles, R.M.: J. Physiol. (Lond.) 150, 417-439 (1960).

3. Burke, R.E., Levine, D.N., Tsairis, P., Zajac F.E.III: J. Physiol. 234, 723-748 (1973).

4. Edström, L., Kugelberg, E.: J. Neurol. Neurosurg. Psychiat. 31, 424-433 (1968).

5. Essén, B., Jansson, E., Henriksson, J., Taylor, A.W., Saltin, B.: Acta Physiol. Scand. 95, 153-165 (1975).

6. Gutmann, E.: The Denervated Muscle, Publishing House of the Czechoslovak Academy of Sciences, Prague (1962).

7. Kugelberg, E., Edström, L.: J. Neurol. Neurosurg. Psychiat. 31, 415-423 (1968).

8. Kugelberg, E., Lindegren, B.: J. Physiol. 288, 285-300 (1979).

9. Lowry, C.V., Kimmey, J.S., Felder, S., Chi, M.M.-Y., Kaiser, K.K., Passonneau, P.N., Kirk, K.A., Lowry, O.H.: J. Biol. Chem. 253, 8269-8277 (1978).

10. Lowry, O.H.,Passonneau, J.V.: A Flexible System of Enzymatic Analysis, Academic Press, New York (1972).

11. Pette, D., Smith, M.E., Staudte, H.W., Vrbová, G.: Pflügers Arch. 338, 257-272 (1973).

12. Pool, C.W. , Donnelaar, Y.E., Griep, P.A.M.: J. Histochem. Cytochem. 26, 742-744 (1978).

13. Romanul, F.C.A., van der Meulen, C.J.: Nature (Lond.) 212, 1369-1370 (1966).

14. Spamer, C., Pette, D.: Histochemistry 52, 201-216 (1977)

15. Spamer, C., Pette, D.: Histochemistry 60, 9-19 (1979)

16. Thorstensson, A., Sjödin, B., Tesch, P., Karlsson, J.: Acta Physiol. Scand. 99, 225-229 (1977).

MYOSIN LIGHT CHAINS, POLYMORPHISM AND FIBRE TYPES IN SKELETAL MUSCLES.

Alan Weeds

MRC Laboratory of Molecular Biology, Hills Road, Cambridge CB2 2QH, England.

Classification of Fibre Types

It is well established that skeletal muscles contain two broad classes of fibres, fast and slow, according to their biological roles. Muscles utilised in maintaining posture contract more slowly than those in rapid voluntary movements, with isometric twitch times about three times greater than fast muscles of the same species (1). These slow muscles operate by a reflex mechanism, their motor units being stimulated in response to stretch of the afferent spindle, as for example in cat soleus muscle. The fibres are small in diameter, rich in mitochondria and have wide Z-lines (See review 2). Their motor units have been classified as type S, and the muscles show high resistance to fatigue, which probably reflects the slow turnover of ATP and its rapid regeneration by mitochondrial oxidative phosphorylation (3). Postural muscles may be slow-twitch, with a propagated action potential, as in mammalian soleus muscle, or slow tonic, i.e. innervated by more than one motoneurone, as in chicken Anterior Latissimus Dorsi (ALD) muscle. In contrast, fast muscles are focally innervated and these fast-twitch fibres may be classified histochemically as fast white or fast red. The former are larger than the latter, have narrower Z-lines and a much lower mitochondrial content (2). Their ATP is provided by anaerobic glycolysis rather than oxidative phosphorylation. They are fast fatiguing and hence used for rapid bursts of activity, in contrast to the fast red fibres (4). A further type of fast fibre, called intermediate, has been classified on the basis of its intermediate mitochondrial content, fatigue resistance and its type of motor unit (4).

Muscles frequently contain a mosaic of different fibre types to give them a wide range of contractile responses in normal movement. Individual motoneurones are activated by different inputs so that different fibre populations can be recruited selectively or together. This makes muscles highly versatile and adaptable to their physiological roles. Thus comparison of particular muscles in different species often reveals a different distribution of fibre types, and within a single species the relative abundance of individual fibre types may vary in response to altered hormone levels or changes in the pattern of stimulus. This has been best demonstrated in experiments involving cross-reinnervation (1,5) or long term artificial stimulation (6,7). Changes in contractile speed have also been observed in response to tenotomy or limb immobilisation or where antagonistic muscles have been denervated and these are all examples of the adaptive response of muscles to changes in functional requirements. They show the remarkable plasticity of muscles in their development and use.

The maximum speed of shortening of a muscle depends on the rate of the cross-bridge cycle of actomyosin interaction and hence on the kinetic cycle of ATP hydrolysis. Barany (8) showed that there was a correlation between maximal shortening speed and myosin ATPase activity, whether this activity was measured in the presence of actin or was the calcium-ATPase of the myosin alone. Histochemical assays for myosin ATPase have shown that fast and slow fibres differ in their staining intensity and these procedures have developed into a standard means of classifying fibre types, particularly when combined with the known lability of fast and slow muscle myosins to acid and alkali respectively (9). Combined with other histochemical stains, a "metabolic profile" may be constructed, which often emphasizes the dynamic nature of fibre types (10). However, one must be cautious in interpreting these histochemical procedures, particularly the myosin ATPase, both because other ATPase enzymes can interfere and because of the differential lability of different myosins to slight variation in the pH of pre-incubation (11). More recently, fluorescently labelled antibodies to fast and slow muscle myosins have been used to demonstrate the presence of particular isoenzymes in different fibre types (12). Antibodies to fast muscle myosin from chicken stained all three types of fast fibre in rat diaphragm muscle but not the slow fibres of soleus muscle, though these showed a positive reaction with antibodies to chicken ALD muscle myosin (12,13). This method has the additional advantage that it can readily detect the presence of more than one type of myosin in a single fibre; it can show whether fibres are promiscuous in the expression of their myosin genotypes. Thus in our attempt to understand the nature of muscle plasticity, we need to know the number of myosin genes and how their expression is controlled both in the early development of the myotube and the following neuronal innervation. We know that the physiological response of a muscle and its metabolic profile are determined both by the pattern of nerve stimulus and its functional requirements, but we would like to know how these factors influence the myosin phenotype. These are some of the questions that will be considered in this book. This paper will review the classification of myosin isoenzymes in different fibre types and the relationship of myosin light and heavy chains to ATPase activity.

Isoenzymes of Myosin

The myosin molecule contains two heavy chains of about 200,000 mol. wt. and two pairs of chemically distinct light chains (14). Because of the size of the heavy chains, amino acid sequence information is at present limited to selected regions (15,16) and the extent of comparative studies with slow and cardiac myosin heavy chains is minimal, though a significant difference has been reported in the 3-methyl histidine content (17). Thus comparative studies of myosin isoenzymes have, until recently been confined to the light chains. Rabbit fast muscle myosin contains two moles of a 18,500 mol. wt. light chain which can largely be removed from myosin by reaction with 5,5'-dithiobis-(2 nitrobenzoic acid) (Nbs_2) without loss of ATPase activity, and these inessential light chains have been termed the Nbs_2 light chains (14). These light chains undergo phosphorylation by a calcium specific light chain kinase and Perry has termed them "P" light chains (18). Because they can partially substitute for the scallop regulatory light chains in restoring calcium sensitivity to desensitised scallop actomyosin (19), they have also been termed "regulatory" light chains, and I will use this term to describe these and equivalent 19,000 mol. wt. light chains from other skeletal myosins. The other class of light chains in rabbit fast-twitch myosin cannot be removed without loss of enzymic activity and appear to be essential for catalytic function. We have termed these the Alkali light chains (14). Two chemically related Alkali light chains, of 20,700 (A1) and 16,500 (A2) mol. wt. are present in single fibres of psoas muscle in non-equal yields, showing that there must be at least two isoenzymes of myosin in these fibres (20). The two Alkali light chains have a common sequence of 141 residues at the C-terminal ends, but there are 5 amino acid substitutions at the N-terminus of Alkali 2, when compared with the corresponding sequence of Alkali 1, and this larger protein contains an additional 41 residues at its N-terminal end (21). This is shown schematically in Fig. 1. Similar chemically related Alkali light chains have been identified in other fast muscle myosins and homologous light chains are also present in cardiac and slow muscle myosins (Fig. 2) (22,23).

The highly basic nature of the N-terminal difference peptide of A1 gives rise to an anomalous mobility on polyacrylamide gel

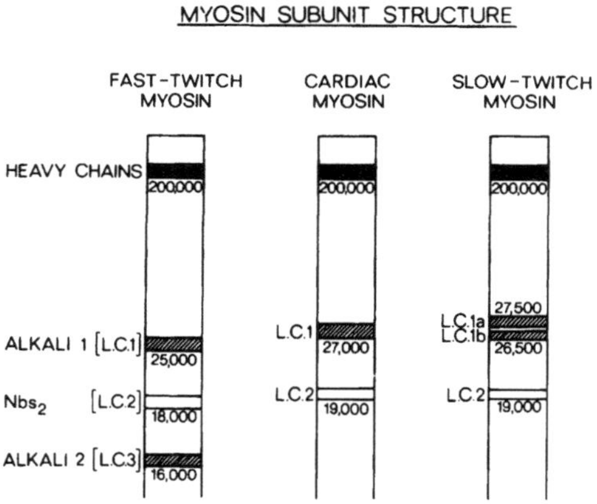

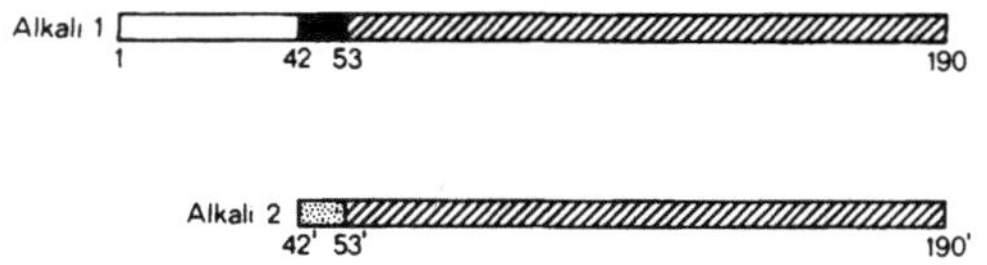

Fig. 1. *Schematic representation of Alkali light chain sequences.* **Sequence identities shown in diagonal hatching (numbered according to the A1 sequence). A1 contains an additional 41 residues and 5 substitutions occur within residues 42-53. (See Ref. 21).**

Fig. 2. *Light chain relationships in different fibre types.* **Distribution of light chains on SDS-polyacrylamide gels. Chemically related Alkali light chains are diagonally shaded, while the regulatory light chains are here denoted LC2.**

electrophoresis in sodium dodecyl sulphate, and these differences in apparent mol. wt. have been used to identify myosin isoenzymes in different muscle types (5,7,23,24). Although this technique has proved valuable, its resolution is very limited and other methods for the rapid identification of myosin isoenzymes are needed. Two such methods are now available:-
1) Gel electrophoresis of native myosins on non-denaturing gels has been used extensively to analyse different muscle types (25,26,27). 2) Whalen et al. have used the O'Farrell technique of two-dimensional gel electrophoresis to characterise the light chain distribution with much improved resolution (28). However, this method does not analyse the heavy chain components and other techniques for fingerprinting these are being developed currently, which combine two-dimensional gel electrophoresis with limited fragmentation of the heavy chains by either CNBr or proteolysis (29 and Whalen, personal communication). Ideally both native and denatured proteins should be examined, since neither technique alone is completely

reliable in giving unequivocal results (27,29). Whilst these methods are valuable for analysing whole muscles, they are less easily applied to single fibres or very small muscles in embryonic systems, and here the immunocytochemical approach described above is particularly useful (12,13). This is an extremely sensitive technique and its reliability depends crucially on the purity both of the antigen used and the antibodies generated. A disadvantage of the method for any quantitative interpretation is the different cross-reactivity between antigens from different species. Antisera to chicken fast and slow muscle myosins were prepared in rabbits, and these antibodies used to analyse myosin isoenzymes in rat muscles. Staining intensity depends not only on the concentration of the myosin isoenzyme, but also on the extent to which it shares common determinants with the chicken antigen. Furthermore, light and heavy chains differ in their antigenic strength and for this reason antibodies to isolated light chains and proteolytic fragments of the heavy chains have been prepared in addition to whole myosin (12,13).

Using these different techniques, we may now investigate the forms of myosin present in different fibre types.
Adult Fast Muscle. Three myosins have been identified by gel electrophoresis of native myosins in both chicken and rabbit muscles (25,27). These appear to separate on the basis of their Alkali light chains, being homodimers containing either A1 or A2 and heterodimers containing both A1 and A2 (30). Differences based on known sequence substitutions in the heavy chains are not resolved by gel electrophoresis and there appears to be no selectivity between light and heavy chains on this basis (31,32). Do different myosin phenotypes occur in fast red and fast white muscles? As yet there is no evidence for this from gel electrophoresis, but this may simply reflect the choice of muscles analysed and the relative insensitivity of the methods used to detect very small quantities of minor components. However, using the immunocytochemical approach,

Gauthier and Lowey have shown differences in the intensity of staining of fast red and fast white fibres when antibodies to subfragment-1 or A2 light chains were used, although these differences did not occur with antibodies to myosin rods or A1 light chains (13). These observations may reflect variable distribution of the antigenic determinants of the light and heavy chain components, but they could indicate the presence of an additional isoenzyme of fast muscle myosin which has not been demonstrated by other means.

Adult Cardiac Muscle. Non-denaturing gels show the presence of three components in rat ventricular muscle and another two in atrial muscle (26). Hypophysectomy and thyroxine replacement produce changes in the distribution of the ventricular myosins, which appear to differ in their ATPase activities. Light chain sequence studies suggest a single Alkali light chain (33), compatible with the single band seen on gel electrophoresis (26), but there are differences in the mobility of the regulatory light chains between atrial and ventricular myosins which may indicate different phenotypes. Gel fingerprinting shows differences in the heavy chains of the two ventricular myosins (34); thus there must be at least two heavy chain phenotypes in this part of the myocardium and possibly additional heavy chain components in the atria.

Adult Slow Muscle. Chicken ALD contains two myosins as identified by non-denaturing gels (25), but gel electrophoresis in sodium dodecyl sulphate suggests only a single form of each light chain (24,30), indicating heavy chain polymorphism in this muscle. The proportion of the minor component is substantially enhanced in embryonic chicken ALD, from which we may deduce that this heavy chain is an embryonic phenotype (35). Rabbit soleus myosin yields a single band on non-denaturing gels (27), with a heavy chain distinct from the fast muscle form. The existence of two different Alkali light chains (23) is not reflected in the electrophoretic mobility of the native myosin. Although these light chains can easily be resolved by two-dimensional gel electrophoresis under denaturing conditions (28), their contribution to the electrophoretic mobility of the native myosin is inadequate to resolve these phenotypes.

Embryonic Muscle. The nature of myosin gene expression in developing muscle is controversial at this time, and will be discussed in detail by others in this book. Three hypotheses will be considered:- 1) Developing fast and slow muscles both expresses only fast muscle myosin genes prior to innervation; slow muscle myosin is synthesized following innervation by slow nerves. This view is supported by studies on tissue culture preparations and isolated muscles from chicken and rat, and the evidence is based on two-dimensional gel electrophoresis of light chains, immunodiffusion and fluorescent antibody staining (36-38). 2) Developing muscles are promiscuous in their gene expression, producing both fast and slow isoenzymes at the earliest stages examined. This view is based on immunocytochemical studies on rat diaphragm and extensor digitorum longus muscles (39), and is consistent with earlier, less rigorous observations (40). 3) The existence of an embryonic form of myosin is evidenced from gel electrophoresis of light chains in developing rat muscles and in tissue culture preparations (28). It is also supported by non-denaturing gels of myosins from 2-day old rabbits (27). These studies suggest a distinct heavy chain phenotype in developing rabbit muscle, which was earlier predicted on the basis of reduced 3-methyl histidine content (17). Combination of non-denaturing gels and light chain analysis suggest the existence of hybrid myosins containing heavy chains from embryonic myosin and light chains of the adult phenotype (27). If substantiated, this will demonstrate a new dimension of polymorphism in myosin phenotypes. Hybrid phenotypes have also been proposed on the basis of the regulatory light chains (41).

On present evidence it is difficult to discriminate between these three different hypotheses. Although very powerful, immunological techniques have their shortcomings. Immunodiffusion is rather insensitive and depends on the production of precipitating complexes, and immunofluorescence may be misleading if, for example, an embryonic myosin contains determinants which cross-react with antibodies to various adult myosin phenotypes. Limitations based on gel electrophoresis of light chains are averted to some extent by two-dimensional fractionation, and non-denaturing gels may be misleading due to protein aggregation or adventitious chemical modification, or additional heterogeneity may arise through partial phosphorylation of the regulatory light chains. Thus only through a combination of different approaches will it be possible to achieve a satisfactory analysis. Two further problems cannot be ignored in evaluating these studies. The first is our lack of precise knowledge of the time of innervation of particular muscles in different developing systems, and the second concerns the effects of multiple innervation which has been demonstrated in newborn mammals (42). Thus correlation of different researches is complicated by ignorance at a physiological level as well as technical difficulty and ambiguity.

Myosin ATPase and Role of Light Chains

Differences in the ATPase activities of myosins from fast and slow muscles have been well documented (8), but the relationship between myosin light chains and catalytic activity remains obscure. Although the regulatory light chains can be removed without significant loss of ATPase activity, it has not proved possible to remove the Alkali light chains without inactivation. However, mild dissociating solvents make the light chains labile (43), and this has facilitated the production of artificial hybrids of subfragment-1 (S-1) to examine the role of light and heavy chain components in different myosins (32).

Our results indicated that the turnover of ATP by S-1 in the absence of actin, whether using Mg.ATP or Ca.ATP or K.ATP as substrate, was determined by the heavy chain component alone: hybrids with fast myosin light chains and slow myosin heavy chains behaved like slow myosins, while those with slow myosin light chains and fast myosin heavy chains behaved like fast phenotypes (32). These results are consistent with much research into the effects of modification of certain heavy chain thiol groups on the ATPase activity (16,44). The "essential" nature of the light chains here may be solely to maintain the native conformation of the active site in the heavy chains. However, in the presence of actin, the ATPase activity was modified in these hybrids to an intermediate level, suggesting that the Alkali light chains play a role in actin.myosin.ATP interaction. Light chains from either cardiac or soleus muscle myosin produced a significant fall in maximum turnover rate (k_{cat}) and markedly reduced the K_m for actin (the apparent dissociation constant in the presence of ATP) in hybrids containing rabbit fast muscle myosin heavy chains. The reverse effect of fast myosin light chains in combination with cardiac heavy chains was more doubtful; although K_m for actin increased in the hybrid, the rise in k_{cat} was minimal. Much more work will be needed to clarify these observations, but they are of interest in shedding light on possible kinetic differences between polymorphic hybrids of myosin in developing muscles (27,41).

More perplexing is the role of different Alkali light chains in myosins from fast-twitch and slow-twitch muscles, which has been described above. Subfragment-1 isoenzymes have been separated on the basis of their Alkali light chain content, which do not appear to differ in their heavy chain components (32). These S-1 isoenzymes have identical ATPase activities in the absence of actin, but in its presence the turnover rate of S-1 (A2) is twice that of S-1(A1) while its affinity for actin in the presence of ATP is much weaker (Fig. 3) (32). These

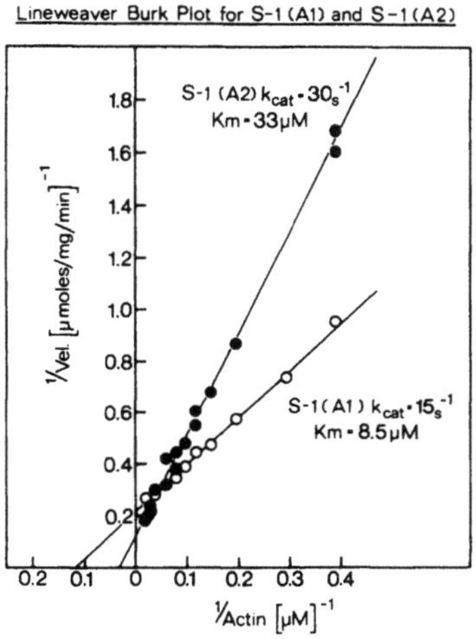

Fig. 3. *Lineweaver-Burk plots of actin-activated ATPase for S-1 isoenzymes.* **The activities were measured in a pH-stat at pH 8.0, 25ºC and the assay mixture contained 6 mM KCl, 1.5 mM $MgCl_2$ and 1.0 mM ATP.**

experiments were carried out at very low ionic strength (6 mM KCl) in order to facilitate reliable extrapolations to k_{cat} with actin concentrations below 100 μM. They showed that the S-1 isoenzymes were capable of hydrolysing ATP at substantially different rates and if these results could be extrapolated to the physiological situation, they would imply that cross-bridges containing different Alkali light chains could undergo their mechanical cycles at different rates. Such a conclusion poses considerable problems since both Alkali light chains are present in the same myofibril (45) and heterodimers of myosin containing both Alkali light chains have been identified as a unique species on non-denaturing gels (30). This apparent problem led us to explore the effects of increasing ionic strength on the actin activated ATPase activity of these S-1 isoenzymes. We found that differences in k_{cat} between S-1(A1) and S-1(A2) were almost eliminated at 26 mM KCl and there was no difference at either 46 mM KCl or 75 mM KCl. Further kinetic analysis indicated that while product release from an actin.S-1.ADP.Pi complex was probably rate limiting in the hydrolysis of ATP by S-1(A2) at either 6 mM or 26 mM KCl and for S-1(A1) at 26 mM KCl; in 6 mM KCl there was a slower step in the ATPase mechanism for S-1(A1) which controlled the turnover of substrate (46). Since this slower process was not dependent on actin concentration (i.e. it was measured at saturating actin concentrations), it probably represented a unimolecular transition occurring before actin reassociation. It therefore resembled the "refractory state" transition discussed earlier (47).

Just as it would be ill-advised to extrapolate the earlier results to the physiological situation, these results too are artificial since they relate to experiments performed on soluble proteolytic fragments of myosin. Recently we have used the hybridisation technique to prepare myosin isoenzymes which contain about 90% A1 or over 75% A2 respectively. Measurement of the ATPase activities of these highly enriched myosin homodimers both at physiological ionic strength and at lower salt concentrations has shown no significant differences in their ATPase activities (Pope, Wagner and Weeds, unpublished work). Thus it appears probable, notwithstanding the earlier studies on S-1, that under physiological conditions the presence of a particular Alkali light chain in fast myosin isoenzymes does not affect the turnover of ATP significantly, but it is premature to conclude that these different phenotypes do not have some specific role in cross-bridge interaction, because all the experiments to date have ignored the highly organised arrangement of these bridges that exists _in vivo_.

The great plasticity of muscle in development, adaptation to requirement and response to disease, depends on fine control of gene expression which is elicited by neuromuscular and metabolic interactions. Just as in a single muscle the recruitment of fibres is orchestrated for particular use, so too there is a great variety of myosin genes whose expression is orchestrated to suit different requirements. Much of the discussion at this meeting will be directed at elaborating the different levels of this genetic control and its integration with other components of muscle that regulate contractile response, including calcium transport and metabolite provision.

References

1. Buller, A.J., Mommaerts, W.F.H.M., Seraydarian, K.: Nature New Biology 233, 31-32 (1971).
2. Close, R.I.: Physiological Reviews 52, 129-197 (1972).
3. Burke, R.E., Levine, D.N., Salcman, M., Tsairis, P.: J. Physiol. 238, 503-514 (1974).
4. Burke, R.E., Levine, D.N., Tsairis, P., Zajac, F.E.: J. Physiol. 234, 723-748 (1973).
5. Weeds, A.G., Trentham, D.R., Kean, C.J.C., Buller, A.J.: Nature 247, 135-139 (1974).
6. Romanul, F.C.A., Sreter, F.A., Salmons, S., Gergely, J.: Exploratory Concepts in Muscular Dystrophy, 2 (edit. by Milhorat), Int. Cong. Series No. 333, 344-348 (1974).
7. Salmons, S., Sreter, F.A.: Nature 263, 30-34 (1976).
8. Barany, M.: J. Gen. Physiol. 50, 197-216 (1967).
9. Guth, L., Samaha, F.J.: Exptl. Neurol. 25, 138-152 (1969).
10. Guth, L., Yellin, H.: Exptl. Neurol. 31, 277-300 (1971).
11. Guth, L.: Exptl. Neurol. 41, 440-450 (1973).
12. Gauthier, G.F., Lowey, S.: J. Cell Biol. 74, 760-779 (1977).
13. Gauthier, G.F., Lowey, S.: J. Cell Biol. 81, 10-25 (1979).
14. Weeds, A.G., Lowey, S.: J. Mol. Biol. 61, 701-725 (1971).
15. Weeds, A.G., Hartley, B.S.: Biochem. J. 107, 531-548 (1968).
16. Elzinga, M., Collins, J.H.: Proc. Natl. Acad. Sci. 74, 4281-4284 (1977).
17. Huszar, G.: Nature New Biology 240, 260-264 (1972).
18. Perrie, W.T., Smillie, L.B., Perry, S.V.: Biochem. J. 135, 151-164 (1973).
19. Kendrick-Jones, J.: Nature 249, 631-634 (1974).
20. Weeds, A.G., Hall, R., Spurway, N.C.: FEBS Lett. 49, 320-324 (1975).
21. Frank, G., Weeds, A.G.: Eur. J. Biochem. 44, 317-334 (1974).

22. Weeds, A.G., Pope, B.: Nature 234, 85-88 (1971).

23. Weeds, A.G.: Eur. J. Biochem. 66, 157-173 (1976).

24. Lowey, S., Risby, D.: Nature 234, 81-85 (1971).

25. Hoh, J.F.Y., McGrath, P.A., White, R.I.: Biochem. J. 157, 87-95 (1976).

26. Hoh, J.F.Y., McGrath, P.A., Hale, P.T.: J. Molec. Cell Cardiol. 10, 1053-1076 (1978).

27. Hoh, J.F.Y., Yeoh, G.P.S.: Nature 280, 321-323 (1979).

28. Whalen, R.G., Butler-Browne, G.S., Gros, F.: J. Mol. Biol. 126, 415-431 (1978).

29. Flink, I.L., Rader, J.H. Morkin, E.: J. Biol. Chem. 254, 3105-3110 (1979).

30. Hoh, J.F.Y.: FEBS Lett. 90, 297-300 (1978).

31. Starr, R., Offer, G.W.: J. Mol. Biol. 81, 17-31 (1973).

32. Wagner, P.D., Weeds, A.G.: J. Mol. Biol. 109, 455-473 (1977).

33. Weeds, A.G.: FEBS Lett. 59, 203-208 (1975).

34. Hoh, J.F.Y., Yeoh, G.P.S., Thomas, M.A.W., Higginbottom, L. FEBS Lett. 97, 330-334 (1979).

35. Hoh, J.F.Y.: FEBS Lett. 98, 267-270 (1979).

36. Rubinstein, N.A., Pepe, F.A., Holtzer, H.: Proc. Natl. Acad. Sci. 74, 4524-4527 (1977).

37. Rubinstein, N.A., Kelly, A.M.: Dev. Biol. 62, 473-485 (1978).

38. Rubinstein, N.A., Holtzer, H.: Nature 280, 323-325 (1979).

39. Gauthier, G.F., Lowey, S., Hobbs, A.W.: Nature 274, 25-29 (1978).

40. Masaki, T., Yoshizaki, C.: J. Biochem. 76, 123-131 (1974).

41. Pette, D., Vrbova, G., Whalen, R.C.: Pflügers Arch. 378, 251-257 (1979).

42. Brown, M.C., Jansen, J.K.S., van Essen, D.: J. Physiol. 261, 387-422 (1976).

43. Dreizen, P., Richards, D.H.: Cold Spring Harbor Symp. Quant. Biol. 37, 29-45 (1972).

44. Reisler, E., Burke, M., Harrington, W.F.: Biochemistry 13, 2014-2022 (1974).

45. Silberstein, L., Lowey, S.: Biophys. J. 21, 45a (1978).

46. Wagner, P.D., Slater, C.A., Pope, B., Weeds, A.G.: Eur. J. Biochem. (In press).

47. Wagner, P.D., Weeds, A.G.: Biochemistry 18, 2260-2266 (1979).

AN IMMUNOLOGICAL APPROACH TO THE ISOLATION OF MYOSIN ISOENZYMES

Susan Lowey

Department of Biochemistry and the Rosenstiel Basic Medical Sciences Research Center, Brandeis University, Waltham, Massachusetts 02254, U.S.A.

Introduction

As the application of immunology to muscle research becomes increasingly popular, it seems appropriate to discuss some of the strengths and weaknesses of the available immunological techniques. Immunofluorescence, in particular, has become widely used to determine the distribution of isoenzymes in individual muscle fibres, both in adult and in developing muscles. Although the main emphasis here will be on the use of antibodies to answer questions about the structure and function of myosin, this discussion should also serve to illustrate some of the problems associated with an immunological approach to cell biology.

Isolation of Myosin Isoenzymes

In order to understand the function of the subunits of myosin, it is important to be able to compare the properties of isoenzymes which differ in the composition of their light chains and/or heavy chains. However, since all myosin molecules are similar in overall shape, size and amino acid composition, the typical fractionation procedures of ion-exchange chromatography and gel filtration are not applicable. Rather, one must employ some type of affinity chromatography to isolate myosin molecules on a preparative scale. We have chosen to use antibodies which are specific for a restricted population of myosin molecules as a means of fractionating myosin isoenzymes.

Adult fast muscle myosin isoenzymes. Antibodies specific for the N-terminal sequence of the alkali 1 or alkali 2 light chain can be used as immunoadsorbents to fractionate myosin into two kinds of homodimers (molecules with the same alkali light chain on each head) (1,2). The protocol used to prepare these immunoadsorbents is outlined in Fig. 1. When myosin from chicken pectoralis is passed through such an antibody affinity column, a fraction rich in alkali 1 or alkali 2 is

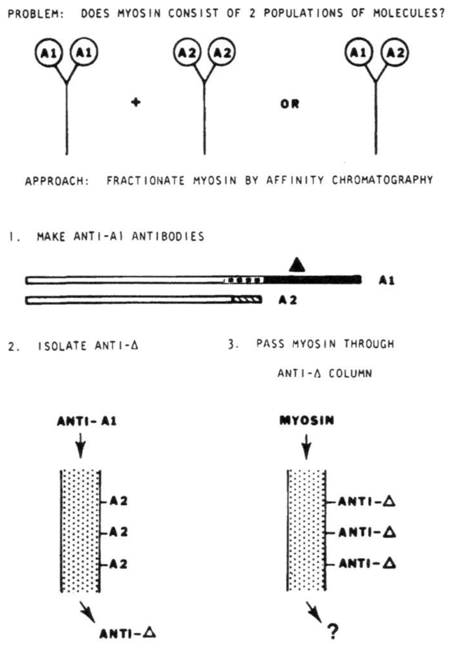

Fig. 1. Schematic representation of the procedure used to separate myosin isoenzymes by affinity chromatography. (For details see Ref. 2.)

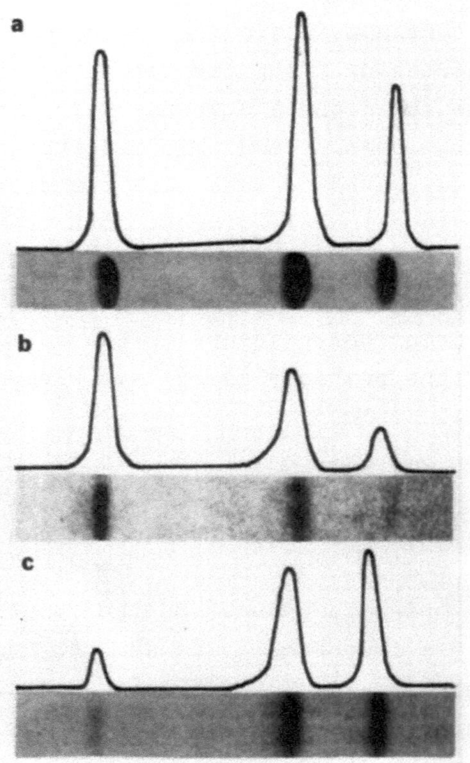

Fig. 2. SDS-polyacrylamide gel electrophoresis of chicken pectoralis myosin (a); A1 myosin (b); and A2 myosin (c). (See Ref. 5.)

obtained in the void volume, depending on the specificity of the immunoadsorbent, Fig. 2. The retained myosin fraction can be eluted only by a denaturant such as guanidine hydrochloride. It is evident that the isolated native homodimers are slightly contaminated by the heterologous light chain; the antibody-bound myosin (not shown) contains an even greater amount of heterologous light chain (1,2). If one assumes that all of the heterologous light chain is derived from a heterodimeric species (molecules containing both alkali 1 and alkali 2), a maximum estimate of about 30% heterodimer is calculated.

It has recently been shown that chicken pectoralis myosin separates into three bands upon electrophoresis in non-denaturing pyrophosphate gels (3). Analysis of the contents of each band suggested that the bands with the slowest and fastest mobility corresponded to A1 and A2 homodimers, respectively; and that the band of intermediate mobility consisted of heterodimers containing both A1 and A2 light chains (4). However, it was not clear to what extent heavy chain heterogeneity might contribute to the observed electrophoretic pattern. Our ability to prepare relatively pure homodimers afforded the opportunity to investigate this question directly. Figure 3 shows non-denaturing gels of native pectoralis myosin and the homodimeric species isolated by immunoadsorption. Both A1 myosin and A2 myosin migrate as single bands, and mixtures of the two do not produce any additional bands. Moreover, when A2 myosin was exchanged with excess A1 light chain under mild dissociating conditions, a myosin species was generated whose mobility was indistinguishable from that of naturally occurring A1 myosin (5). These results confirm the suggestion that the band of intermediate mobility is an alkali light chain heterodimer, and show, moreover, that in chicken pectoralis myosin the alkali light chain is the principal determinant of mobility differences on non-denaturing polyacrylamide gels (5).

Adult slow and fast myosin isoenzymes from a mixed red muscle. The use of immunoadsorbents to fractionate myosin molecules has now been applied to the isolation of "fast" and "slow" myosin isoenzymes from a mixed red muscle of the chicken. Immunoadsorbents were prepared using antibodies specific for "fast" white pectoralis myosin or for "slow" red anterior latissimus dorsi (ALD) myosin, following the general method described in Fig. 1. To test the capability of these columns, we isolated myosin from a circumscribed region of the pectoralis which consists entirely of fibres which are rich in mitochondria (6). Despite being "red" in appearance, only certain fibres react with anti-ALD myosin, while the remainder react with anti-pectoralis myosin (see Fig. 8). Myosin from this mixed red muscle shows the presence of both slow and fast myosin light chains by SDS gel electrophoresis, Fig. 4. Upon application to the immunoadsorbents, the fast component was retained by the anti-pectoralis column, and the slow component was bound by the anti-ALD column, as judged by its light chain pattern.

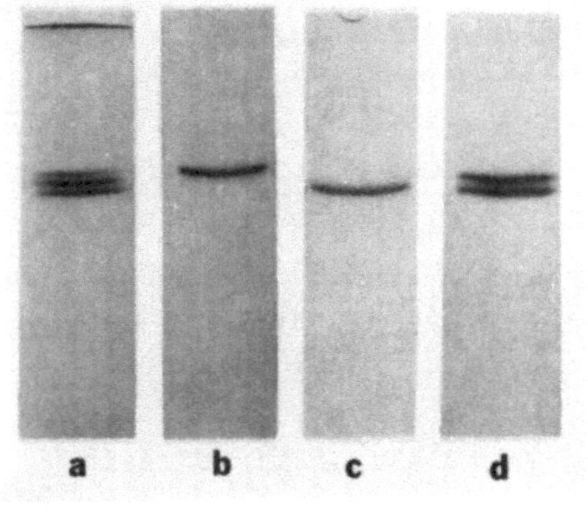

Fig. 3. Native myosin on non-denaturing gels. (a) Myosin control is resolved into three isoenzymes; (b) A1 myosin; (c) A2 myosin; (d) mixture of A1 and A2 homodimers. (see Ref. 5.)

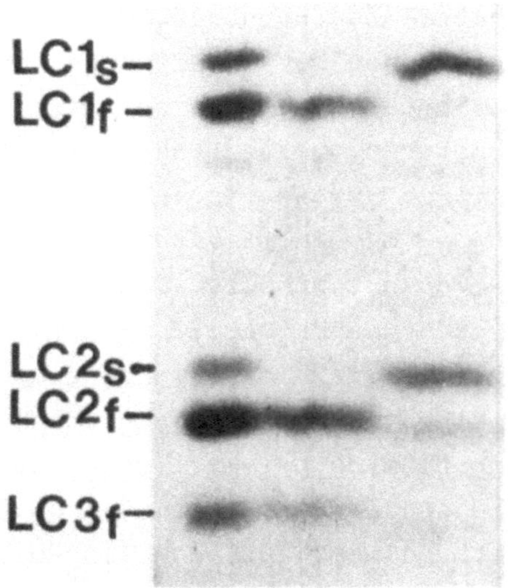

Fig. 4. SDS-polyacrylamide gel electrophoresis of myosin from a mixed "red" region of chicken pectoralis (a); the fast component (b); and the slow component (c).

Peptide mapping of these myosins suggests that the heavy chains are also fractionated into distinct components, although it is still unclear whether the isoenzymes in the mixed muscle are identical to those against which the antibodies were elicited.

Specificity of Antibodies

Only those antibody preparations which possess a high degree of specificity are capable of successfully fractionating myosin isoenzymes. Therefore, all contaminating and cross-reacting antibodies must be removed by prior adsorption of the antisera. Tests for specificity include gel diffusion, radioimmunoassay, and the response of myofibrils or cryostat sections of muscle to fluorescein-labelled antibodies.

Gel diffusion. A single precipitin band in double diffusion is usually taken to mean that the antiserum is specific for a homogeneous protein antigen, Fig. 5 and Fig. 6. Figure 5a shows that alkali 1 and alkali 2 each form a single band which fuses against anti-alkali 1 serum; the spur of partial identity results from some of the anti-alkali 1 antibodies not reacting with alkali 2 and diffusing past its precipitation band (7). These non-cross-reacting antibodies are presumably specific for the 41-residue "difference peptide" unique to alkali 1 (8). In contrast, alkali 1 and DTNB light chain appear to be immunologically distinct since their precipitin lines from the homologous antigen-antibody reactions cross (7), Fig. 5b. It was not realized until later that some cross-reactivity actually existed between alkali 1 and DTNB light chain (9). In the presence of high concentrations of the immunoglobulin fraction, Fig. 5c, or of affinity-purified antibody against DTNB light chain, Fig. 5d, a faint precipitin line forms with alkali 1. Competitive radioimmunoassays confirmed that cross-reactive antibody was definitely present in the anti-DTNB light chain serum, and that the results were unlikely to be an artifact caused by contamination of the immunogen (9).

Antisera against pectoralis myosin and ALD myosin were similarly scrutinized by gel diffusion, Fig. 6. Since neither immunogen is derived from a completely homogeneous muscle (6), antiserum against pectoralis myosin was adsorbed with ALD-myosin, and anti-ALD myosin was adsorbed with pectoralis myosin conjugated to Sepharose. Both antisera showed single precipitin bands when reacted with their homologous antigens, Fig. 6b and c. Even more significant is the single band formed by crude ALD-myosin when reacted with anti-ALD serum,

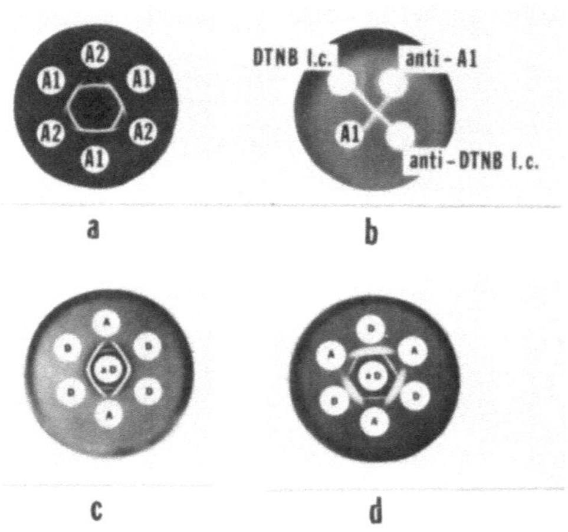

Fig. 5. Immunodiffusion of light chains. (a) A1 and 2 against anti-A1 serum; (b) DTNB and A1 l.c. against homologous antisera (see Ref. 7); (c) A1 (A) and DTNB (D) l.c. against anti-DTNB l.c. IgG; (d) the same antigens as in (c) against affinity-purified anti-DTNB l.c. antibody. (See Ref. 9.)

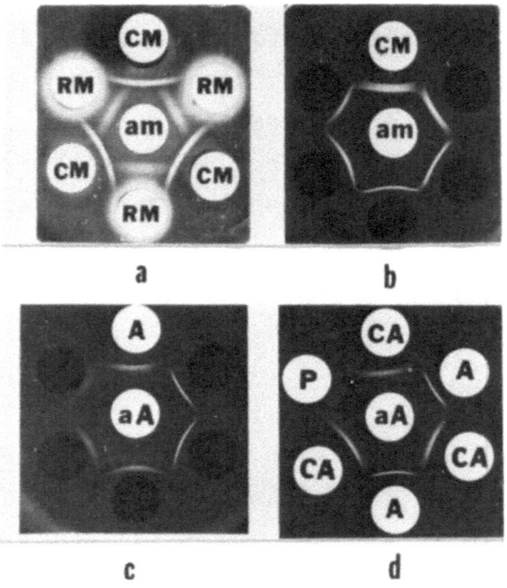

Fig. 6. Immunodiffusion of chicken pectoralis and ALD myosins. (a) affinity purified anti-pectoralis myosin (am) against chicken pectoralis myosin (CM) and rat diaphragm myosin (RM); (b) two-fold serial dilution of chicken pectoralis myosin against homologous antibody (see Ref. 11); (c) two-fold serial dilution of ALD myosin against anti-ALD serum; (d) purified ALD myosin (A), a crude extract of ALD myosin (CA), and chicken pectoralis myosin (P) against anti-ALD serum.

Fig. 6d. This is a more sensitive test for the specificity of an antiserum, since any contaminating antibodies would be expected to form additional precipitin bands upon reacting with the impurities in the antigen. The adsorbed anti-ALD serum, predictably, did not react with pectoralis myosin, Fig. 6d.

Radioimmunoassay. The absence of a precipitin line does not necessarily exclude the possibility that non-precipitating antibody-antigen complexes are present. This is particularly relevant for weak affinity interactions in a heterologous antibody-antigen system, or in the case of few determinants on an antigen. By using a modified Farr assay or a solid phase radioimmunoassay (10), the detection of minute quantities of antibody-antigen complexes becomes possible. The latter assay is particularly convenient and sensitive; the antigen need not

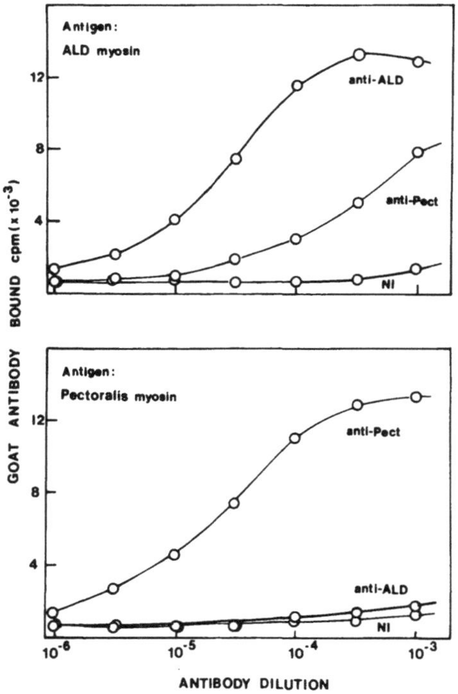

Fig. 7. Solid phase radioimmunoassay. Serial dilutions of non-immune rabbit IgG and rabbit anti-ALD or anti-pectoralis myosin antibody were assayed on antigen-coated plastic wells containing ALD myosin (top) and pectoralis myosin (bottom). ^{125}I-goat anti-rabbit Fc was used at a fixed concentration as the detecting agent.

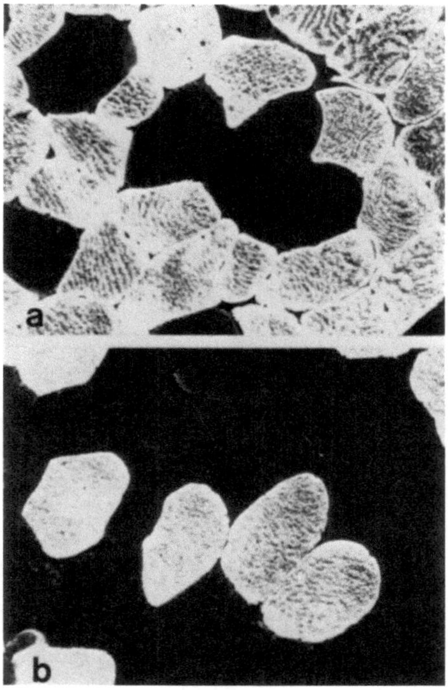

Fig. 8. Indirect immunofluorescence of serial transverse sections of chicken pectoralis "red region." (a) rabbit anti-pectoralis myosin; (b) rabbit anti-ALD myosin. Fluorescein-labelled goat anti-rabbit Fc was used as the detecting agent. (Courtesy of G. F. Gauthier.)

be labelled, and only the goat antibody against the Fc region of rabbit immunoglobulin is radiolabelled. An important advantage of this procedure is that the course of the antibody purification can be conveniently followed. For example, untreated anti-ALD serum reacts strongly with pectoralis myosin; after adsorption the cross-reactivity is markedly reduced, although not eliminated. As shown in Fig. 7 (bottom), it is only after the adsorbed serum is eluted from an ALD-conjugated Sepharose column that the purified antibody displays complete absence of cross-reactivity with pectoralis myosin. A similar immunoadsorption/elution procedure was followed for anti-pectoralis serum. As seen in Fig. 7 (top), most but not all cross-reactivity was eliminated.

Antibodies against the N-terminal sequences of alkali 1 and alkali 2 were routinely tested for specificity by this solid-phase assay. Only those antibodies which showed absolutely no cross-reactivity with the heterologous alkali light chain, or with the DTNB light chain, were capable of fractionating A1 myosin from A2 myosin.

Fluorescent labelling. An alternative test for the detection of soluble antibody-antigen complexes is to react cryostat sections of muscle with fluorescein-labelled antibodies. Figure 8 shows the "red strip" of chicken pectoralis exposed to either anti-pectoralis myosin, Fig. 8a, or anti-ALD myosin, Fig. 8b. The staining in each case is restricted to a certain fibre type, and there is no non-specific staining in the other fibres. Similar results have been obtained with cryostat sections of the "white region" of the pectoralis muscle, which failed to react with anti-ALD myosin (11).

A striking example of what can happen when sera are not rigorously adsorbed for all cross-reacting antibodies is shown by the fluorescent staining patterns in Fig. 9 and Fig. 10. When a given antibody preparation specific for the N-terminal sequence of alkali 1 (anti-Δ1) was reacted with the "red strip," primarily the fast white fibres stained, Fig. 9a; however, this same preparation stained *all* the fibres, both fast and slow, in the rat diaphragm, Fig. 10a. On further examination, this preparation of anti-Δ1 was found to be unable to fractionate the myosin light chain isoenzymes, and by radioimmunoassay showed a small degree of cross-reactivity with the DTNB light chain. After extensive adsorption with DTNB light chain, the anti-Δ1 preparation showed its usual staining pattern of fast fibres in the rat diaphragm (Fig. 10b), and the staining pattern in the "red strip" improved in contrast (Fig. 9b). These experiments suggest that certain determinants in the

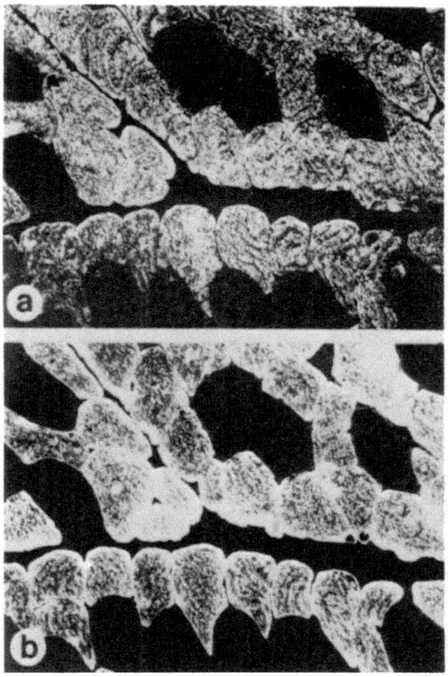

Fig. 9. Indirect immunofluorescence of serial transverse sections of chicken pectoralis "red region." (a) rabbit anti-Δ1 antibody prior to additional adsorption with DTNB l.c.; (b) rabbit anti-Δ1 antibody after extensive adsorption with Sepharose-coupled DTNB l.c. Fluorescein-labelled goat anti-rabbit Fc was used as the detecting agent. (Courtesy G.F. Gauthier.)

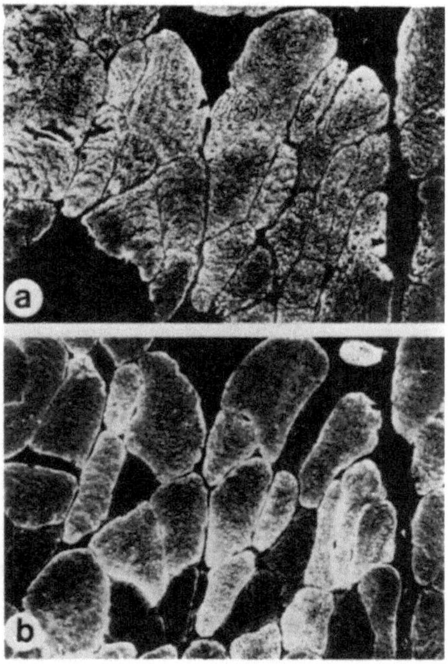

Fig. 10. Indirect immunofluorescence of serial transverse sections of rat diaphragm. (a) rabbit anti-Δ1 antibody prior to adsorption (see Fig. 9a); (b) rabbit anti-Δ1 antibody after extensive adsorption with DTNB l.c. (see Fig. 9b). Fluorescein-labelled goat anti-rabbit Fc was used as the detecting agent. (Courtesy G.F. Gauthier.)

"difference peptide" are shared with the DTNB light chain, and with the light chains of the slow red fibres. The common sequences probably include Pro-Lys-Lys and Ala-Ala-Ala which are located in the N-terminal regions of the alkali 1 and DTNB light chains (12).

Species Cross-Reactivity of Antibodies

The antibodies used in the isolation and localization of the isoenzymes described above were extensively adsorbed with cross-reacting antigens of the same species. These adsorbed antibodies will, nevertheless, cross-react with skeletal myosins from a number of other species. We have shown previously by fluorescence microscopy that antibodies to chicken pectoralis myosin, to S1 and rod, and to both classes of light chains stain the fast fibres of the rat diaphragm; and that antibodies against ALD myosin stain the slow red fibres of rat diaphragm and rat soleus (6,11). These antibodies also selectively stain the fibres of cat muscle (see G. F. Gauthier in this volume). Apart from demonstrations of cross-reactivity by direct and indirect immunofluorescence, myosin isolated from the rat diaphragm reacts with anti-pectoralis myosin in gel diffusion, Fig. 6a, and by quantitative precipitin analysis (11).

Usually an animal is *tolerant* to self proteins; that is, if chicken myosin is injected into a rabbit, the rabbit would be expected to make antibodies against sequences in the chicken molecule which are different from rabbit myosin. The resultant rabbit antibodies would not be expected to recognize rabbit myosin. This pattern, which is common to most antigens which circulate and come in contact with the lymphoid system, is not found for myosin, actin and other sequestered proteins such as the cytochromes c. For example, although all skeletal muscle actins appear to be very similar, or perhaps even identical in sequence (13), antibodies can be raised in rabbits immunized with SDS denatured murine skeletal actin (14). Between rabbit and mouse cytochrome c, there are only two amino acid residue differences, each of which elicits an antibody of the appropriate specificity (15). In addition, however, rabbit cytochrome c elicits a third antibody population in mice, and mouse cytochrome c produces a third antibody in rabbits. As this third determinant is identical in both species, this is clearly an *autoimmune* response (15).

Evidence for autoimmune antibodies in the myosin system was provided by the demonstration that rabbit light chains showed 20-30% cross-reactivity with rabbit antisera specific for both DTNB light chains and alkali light chains, Fig. 11 and Fig. 12 (7). When rabbit light chains were used as inhibitors, even greater cross-reactivity was observed, Figs. 11 and 12, since low affinity interactions could be detected. The "difference

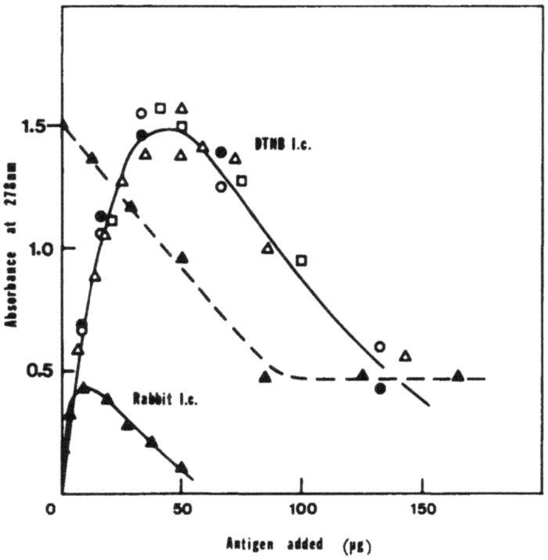

Fig. 11. Quantitative precipitin reaction with rabbit antiserum to chicken DTNB light chain. Total light chains isolated from rabbit myosin were tested both as precipitating antigens (solid line) and as inhibitors of the reaction between antiserum and chicken DTNB light chains (dashed line). (See Ref. 7.)

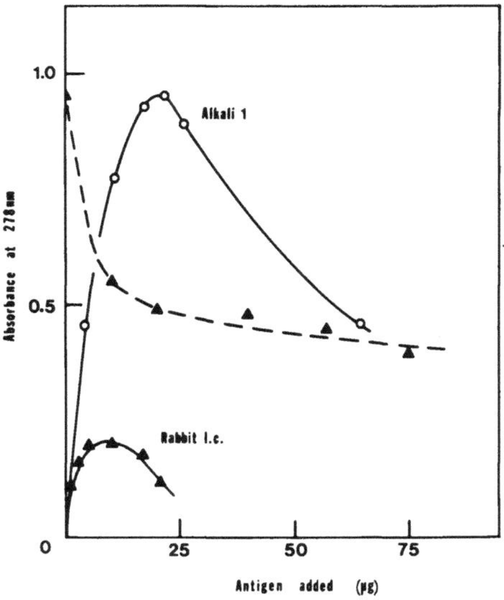

Fig. 12. Cross reactivity between rabbit and chicken alkali light chains. Rabbit total light chains were used as precipitating antigens (solid line) and as inhibitors of the reaction between antiserum and chicken alkali light chains (dashed line). (See Ref. 7.)

peptide" (8) was also effective in inhibiting the reaction between alkali 1 serum and its antigen, irrespective of whether the peptide was derived from chicken or rabbit light chains (7). Moreover, when a preparation of rabbit light chains was isolated from the same rabbit which had produced the anti-DTNB light chain serum, the extent of immunoprecipitation remained unchanged (7).

These data strongly suggest that sera raised against myosin and its components contain a significant population of autoimmune antibodies. This property of the fibrous proteins is extremely useful, since their antibodies can then be used as "probes" to determine the distributions of isoenzymes in other systems (11). Thus in the developing muscles of the rat diaphragm or EDL, immunofluorescence reveals that the myosin isoenzymes at early stages of development share some sequences with adult ALD myosin; however, these experiments cannot distinguish between the presence of an "embryonic myosin" or an adult slow myosin (16).

Failure to detect cross-reactivity between two adult species or between myosins of adult and embryonic tissues does not necessarily mean that no common sequences exist. It could simply mean that the particular serum used is missing the cross-reacting population of antibodies. The specificity of a serum depends on many factors -- the dose of antigen and schedule used for immunization, times of bleeding and differences in the genetic makeup of the animals. Individual animals within a species can respond quite differently to the same determinants on an antigen (6), and it is important to recognize that the antigenicity of a protein also depends on the species in which the response is elicited (15).

Despite these inherent difficulties and ambiguities in an immunological approach, the benefits have far exceeded the problems. The use of antibodies has made it possible for the first time to prepare myosin light chain isoenzymes which can be studied for their enzymatic and structural properties. But in addition to providing a new preparative tool, the availability of antibodies against myosin and its components has provided a highly specific marker for determining the distribution of myosin isoenzymes in adult and developing muscles.

References

1. Holt, J.C., Lowey, S.: Biochemistry 16, 4398-4402 (1977).
2. Lowey, S., Silberstein, L., Gauthier, G.F., Holt, J.C.: In Motility in Cell Function (in press, Academic Press).
3. Hoh, J.F.Y., McGrath, P.A., White, R.I.: Biochem. J. 157, 87-95 (1976).
4. Hoh, J.F.Y.: FEBS Lett. 90, 297-300 (1978).
5. Lowey, S., Benfield, P.A., Silberstein, L., Lang, L.M.: Nature, in press (1979).
6. Gauthier, G.F., Lowey, S.: J. Cell Biol. 74, 760-779 (1977).
7. Holt, J.C., Lowey, S.: Biochemistry 14, 4600-4609 (1975).
8. Frank, G., Weeds, A.G.: Eur. J. Biochem. 44, 317-334 (1974).
9. Silberstein, L., Lowey, S.: Biochemistry 16, 4403-4408 (1977).
10. Klinman, N.R., Pickard, A.R., Sigal, N.H., Gearhart, P.J., Metcalf, E.S., Pierce, S.K.: Ann. Immunol. (Inst. Pasteur) 127c, 489-502 (1976).
11. Gauthier, G.F., Lowey, S.: J. Cell Biol. 81, 10-25 (1979).
12. Collins, J.H.: Nature 259, 699-700 (1976).
13. Vandekerckhove, J., Weber, K.: J. Mol. Biol. 126, 783-802 (1978).
14. Lessard, J.L., Carlton, D., Rein, D.C., Akeson, R.: Anal. Biochem. 94, 140-149 (1979).
15. Urbanski, G.J., Margoliash, E.: J. Immunol. 118, 1170-1180 (1977).
16. Gauthier, G.F., Lowey, S., Hobbs, A.W.: Nature 274, 25-29 (1978).

DISTRIBUTION OF MYOSIN ISOENZYMES IN ADULT AND DEVELOPING MUSCLE FIBERS

Geraldine F. Gauthier
Department of Anatomy, University of Massachusetts Medical School,
55 Lake Avenue North, Worcester, Massachusetts 01605, U.S.A.

Introduction

Mammalian skeletal muscles usually consist of mixtures of structurally different fibers. This fact has been known for more than 100 years, but it was not until the 1950's, when methods of enzyme cytochemistry were developed and applied to the study of muscle, that differences among fibers were easily recongnized. Additional distinguishing characteristices were revealed by electron microscopy (see ref. 5 for review). A number of criteria have been used to identify these fibers, but we have emphasized two which distinguish the fibers on the basis of ultrastructural as well as cytochemical features, namely the form and distribution of mitochondria, and width of the Z line. Small red fibers have abundant mitochondria and wide Z lines. In large white fibers, mitochondria are sparse and Z lines are narrow. Intermediate fibers are moderately rich in mitochondria and thus resemble red fibers, but Z lines are narrow as in white fibers. In addition, by measuring the contractile properties of individual motor units, it has been demonstrated that a particular type of muscle fiber corresponds to a physiologically distinctive motor unit (3, 4). There has been increasing evidence for chemical heterogenety among the myofibrillar proteins of a muscle. For example, there are different isoenzymes of myosin within single muscles, and this heterogenety includes both the heavy chain and the light chains (13, 15, 16, 21, 22). We have been concerned with determining how the isoenzymes are related to the individual fibers which comprise a mixed muscle. In collaboration with S. Lowey, we have used an immunocytochemical approach to localize myosin isoenzymes with respect to the fiber population in various muscles. Specific antibodies have been prepared against myosin from skeletal muscles of the chicken. By taking advantage of the cross-reactivity of myosin in different species (11), we have been able to localize the isoenzymes in fresh frozen sections of

Plasticity of Muscle

skeletal muscles of the rat and cat as well as the chicken. Transverse cryostat sections were exposed to the antibodies, which were prepared in the rabbit. The sections were then reacted with fluorescein-labeled goat antibodies against rabbit immunoglobulin. We have shown that there is a preferential distribution of myosin isoenzymes with respect to individual fibers in a variety of muscles. The pattern of distribution appears to be related to the speed of contraction of the fibers. This relationship and the changes which take place during normal development will be examined in the discussion which follows.

Distribution of Fast and Slow Myosins Among Fibers of a Fast-Twitch Muscle

When transverse sections of rat diaphragm (a fast-twitch muscle) are exposed to labeled antibody specific for white ("fast") myosin from the chicken pectoralis, a differential response is evident among the fibers, and those fibers which react with the antibodies appear bright with the fluorescence microscope (6). Fiber types can be identified by comparison with serial sections in which the mitochondrial enzyme, succinic dehydrogenase, has been localized (fig. 1a). All white, intermediate and about 40% of the red fibers react with the antibodies (fig. 1b). In addition, an alkali-stable (myosin) ATPase has the same pattern of distribution among fibers (fig. 1c), which is consistent with the existence of two categories of red fibers. Because myosin ATPase activity is directly related to speed of contraction (1), we have designated red fibers which have a positive or negative response to this antimyosin as "fast" or "slow" red fibers, respectively.

The myosin of the fast red fiber and that of the white fiber are sufficiently similar that they have antigenic determinants in common. In order to determine whether this homology is limited to a particular region of the myosin molecule, we examined the response to antibodies directed against the proteolytic subfragments ("head" (S1) and "rod") of myosin and also the low molecular-weight subunits (alkali- and DTNB- light chains) of white myosin. Antibodies capable of distinguishing between alkali 1 or alkali 2 light chains were also used to localize these isoenzymes (see S. Lowey, this volume). The same fibers which had reacted with antibodies against whole white myosin also reacted with antibodies against the proteolytic subfragments and light chains (8). White, intermediate, and the same red fibers react with all the antibody preparations. This confirms our conclusion

Figure 1 Rat diaphragm, serial transverse sections. (a) succinic dehydrogenase. (b) anti-pectoralis (white) myosin. (c) alkali-stable ATPase. Fibers are identified by their mitochondrial content (a), which is highest in small red fibers (R), lowest in large white fibers (W), and moderate in intermediate fibers (I). All white (W) and intermediate (I), and about 40% of the red fibers (fast red fibers, black R) react with anti-white myosin (b) and therefore appear bright with the fluorescence microscope. The same fibers have a high level of ATPase activity (c). From Gauthier and Lowey, 1979.

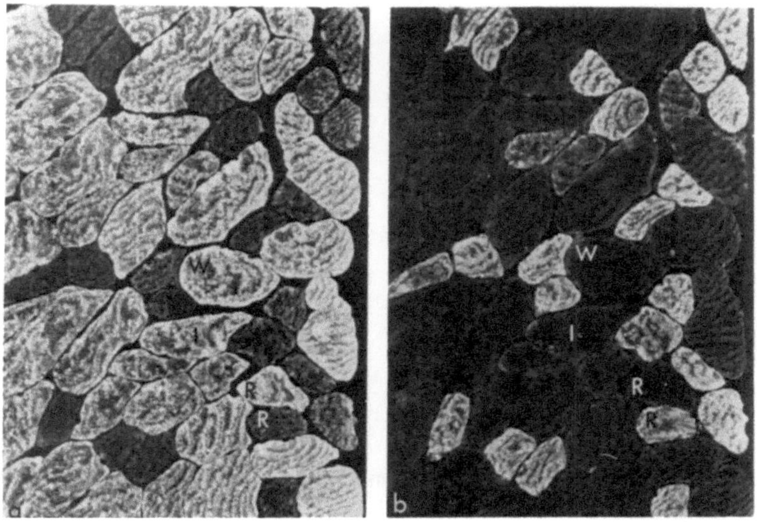

Figure 2 Rat diaphragm, serial transverse sections. (a) anti-pectoralis myosin. (b) anti-ALD myosin. The pattern of response to anti-ALD myosin (b) is reciprocal to that seen with anti-pectoralis myosin (a). It reacts strongly with slow red fibers (black R), but not with fast red fibers (white R) or white (W) or intermediate fibers (I). An occasional fast red fiber reacts with anti-ALD myosin (upper center, Fig. 2b). From Gauthier and Lowey, 1979.

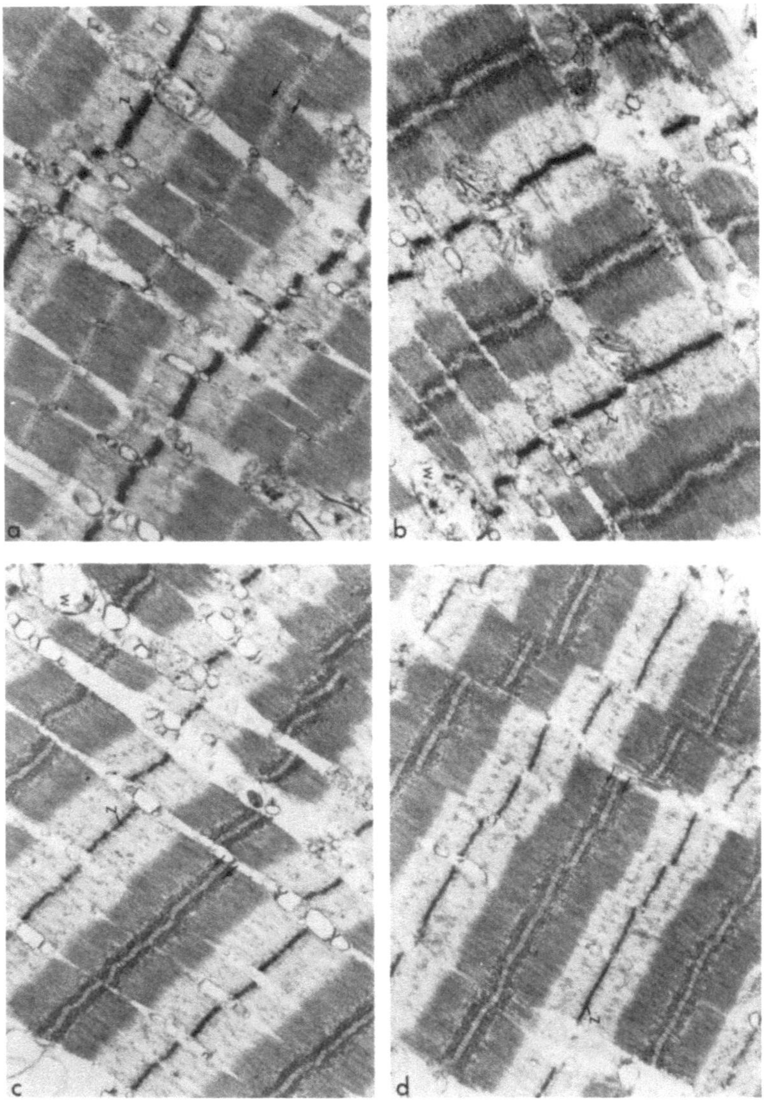

Figure 3 Red (a,b), intermediate (c), and white (d) fibers exposed to anti-white myosin. In both red fibers (a,b) Z lines (Z) are wide, and membranous remnants of large mitochondria (M) are present between myofibrils. There is no evidence of binding of antibody with myofibrils of the "slow" red fiber (a), but in the "fast" red fiber (b), a distinct electron-opaque band (arrows) is apparent on either side of the unstained bridge-free zone. Intermediate (c) and white fibers (d) have narrow Z lines (Z) and both react with anti-white myosin. A circumscribed electron-opaque band (arrows) is present on either side of the bridge-free zone. From Gauthier, 1979.

that the myosins of the reactive (fast) fibers in the rat diaphragm are sufficiently similar to share antigenic determinants. The homology, furthermore, is not confined to a limited region of the myosin molecule, but includes the head and rod portions and all classes of light chains. In addition, the red fibers which did not react with antibodies against pectoralis myosin did react strongly with antibodies to red myosin from the anterior latissimus dorsi (ALD) of the chicken, a slow red muscle. The pattern of response to this antibody is reciprocal to that observed with anti-white myosin (figs. 2a, b). It reacts with slow red fibers but not with fast red or with white or intermediate fibers. It is concluded that there are two classes of red fibers, one of which has a myosin which is similar to white myosin.

Red fibers, identified by abundant mitochondria can be differentiated by their myosins. However, because there is also a difference in Z-line width among the fibers of the same muscle, the possibility existed that the two red fibers, which are similar in appearance, might differ in dimension of the Z line as well as in the type of myosin present. It was therefore important to establish whether both fast and slow red fibers have wide Z lines. We examined, with the electron microscope, fibers which had been exposed to antimyosin (9). To label the fibers, 50 um transverse sections of glycerinated muscle were reacted with anti-white (fast) myosin, using an indirect procedure similar to that used for light microscopy, but no exogenous label was necessary. In fibers which react with the antibody, two elctron-opaque bands are evident in the H-band region, on either side of the bridge-free zone. Thus reactive fibers are readily distinguished from unreactive fibers (figs. 3a,b). All fibers with narrow Z lines, namely white and intermediate fibers, react with anti-fast myosin (figs. 3c and d). Fibers with abundant mitochondria and wide Z lines can be either reactive (fig. 3b) or unreactive (fig. 3a). That is, both fast and slow red fibers have wide Z lines and thus can be distinguished from white and intermediate fibers. We have also shown that the negative red fiber, but not the positive red fiber reacts with anti-slow myosin from the ALD (Gauthier, unpublished observations).

On the basis of combined immunochemical and ultrastructural characteristics, four types of fibers can be recognized in the rat diaphragm. We have suggested that these four fibers correspond to four types of motor units which Burke and his associates have described in fast-twitch muscles of the cat (Table 1). This correlation is based on the fact that succinic dehydrogenase and ATPase activity have been localized by us in muscles of the rat and by Burke and his associates in the cat. These enzymic activities can therefore be used as a basis for comparison. Additional justification for this correlation is is provided by recent experiments with skeletal muscles of the cat. We have examined, using the same immunocytochemical procedures, cat muscles which are equivalent to those used in the physiological analyses. We have found that four types of fibers can be recognized by the same criteria applied to the rat, and response to anti-myosin is, for the most part, comparable to that observed in the rat.

Evidence for a Distinctive Myosin in the Fast Red Fiber

Despite the similarities which are evident among fast fibers, we have not implied that the myosins of these fibers are identical. In fact, our observations indicate that the myosin of the fast red fiber is different from that of other fast fibers. Labeled antibodies to S1 do not stain the fast red fiber as strongly as they do white or intermediate fibers. (8). Non-uniform staining is also observed with antibodies specific for the alkali 2 light chain (anti - Δ 2); the fast red fiber again shows weaker fluorescence than do the other fast fibers of the rat diaphragm (figs. 4 a, b). These results could indicate a variable distribution of myosin isoenzymes according to their light chain composition. Alternatively there may exist a third type of myosin isoenzyme which is localized in the fast red fiber. Our observations indicate that the myosins within the population of fast fibers of the cat gastrocnemius and flexor digitorum (FDL) are also not identical (Gauthier, unpublished observations). The level of response to antibody specific for the alkali 1 light chain (anti-Δ1) is particularly high in one type of fiber, which is comparable to the fast red fiber of the rat diaphragm (fig. 5a), and the response to anti -Δ2 is somewhat less intense than in other fast fibers (fig. 5b), as in the rat. The presence of a distinctive type of myosin in this fiber would be consistent with the physiological characteristics of the corresponding motor unit. The FR unit (see Table 1) has a somewhat longer contraction time than the other fast units of the same muscle (3), which suggests that it represents a separate category based on speed of contraction. The corresponding fast red fiber represents a separate category having a myosin which is different from that of other fast fibers.

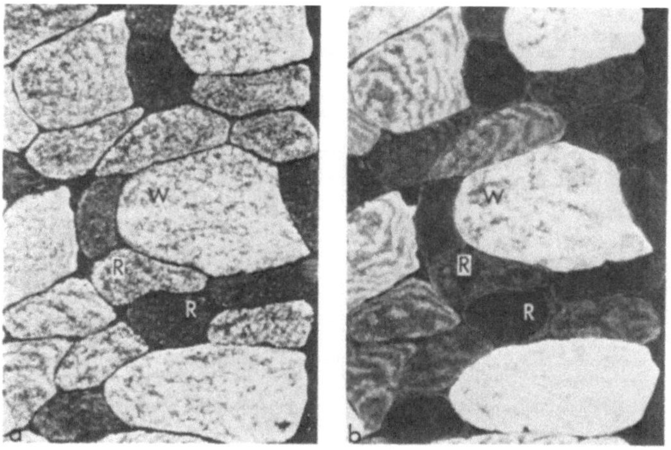

Figure 4 Rat diaphram, serial transverse sections. (a) anti-△1. (b) anti-△2. Both antibodies react with the same fibers (W, I, black R) that react with antibodies against whole white myosin. However, the level of response to anti-△2 is lower in fast red fibers (black R) than in other fast fibers (W, I).

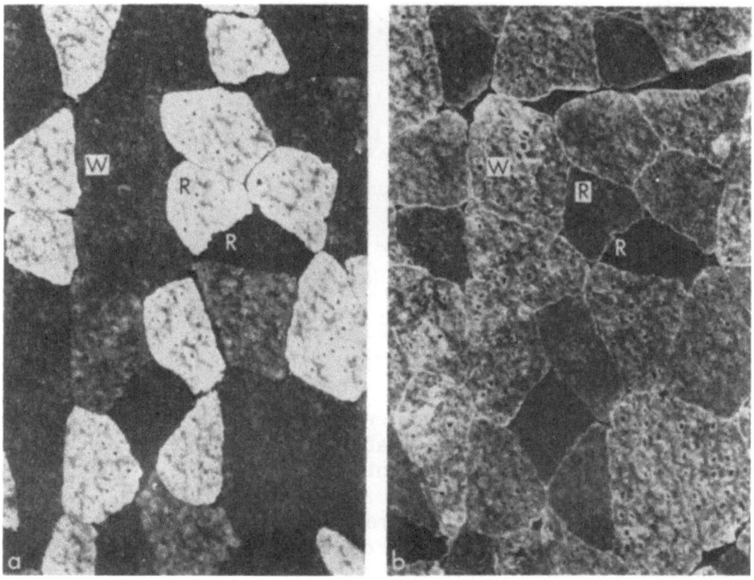

Figure 5 Cat FDL, serial transverse sections. (a) anti-△1. (b) anti-△2. The response of most fast fibers (w) is weak; compare with unreactive fibers (white R). However, the level of response to anti-△1 (a) is more intense in one type of red fiber (black R) than in other fast fibers. Response to anti-△2 (b) is less intense in this fiber (black R) than in other fast fibers.

TABLE 1

ULTRASTRUCTURAL AND CYTOCHEMICAL CHARACTERISTICS OF FIBER TYPES IN RELATION TO PHYSIOLOGICAL PROPERTIES OF MOTOR UNITS

FIBER TYPE[1]	MITO-CHONDRIA (SDH)	Z LINE	ATPase (alkali-stable)	ANTI-WHITE MYOSIN (Pectoralis)	ANTI-RED MYOSIN (ALD)	MOTOR UNIT[2]	SPEED OF CONTRACTION	FATIGUE RESISTANCE
WHITE	Low	Narrow	+	+	-	FF	Fast	Low
INTER-MEDIATE	Int.	Narrow	+	+	-	F(int)	Fast	Int.
RED (fast)	High	Wide	+	+	-	FR	Fast	Int. to High
RED (slow)	High	Wide	-	-	+	S	Slow	Very High

1 Based on ultrastructural and cytochemical characteristics of muscle fibers in the rat diaphragm and semitendinosis (Gauthier, 1969, 1971, 1979. Gauthier and Lowey, 1977).

2 Data represent individual muscle units from the cat gastrocnemius (Burke, et al., 1973; Burke, 1975).

From Gauthier and Lowey, 1979.

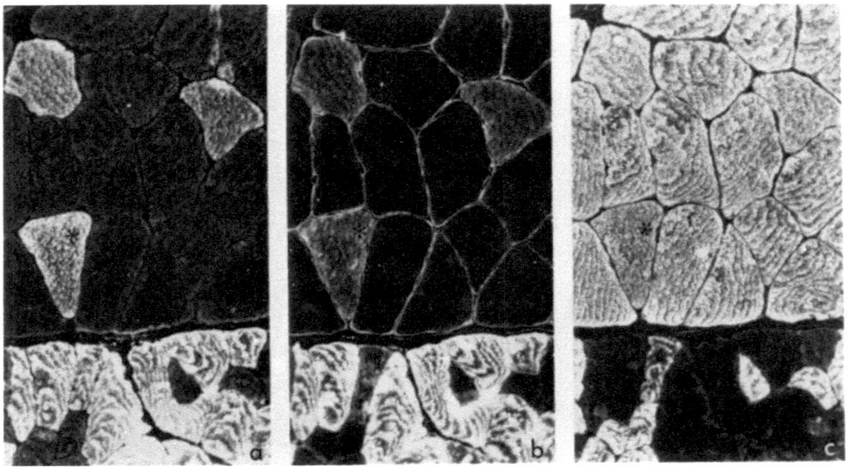

Figure 6 Rat soleus, serial transverse sections. Soleus is sectioned simultaneously with rat diaphragm so that the two muscles can be compared directly under identical conditions. (a) anti-Δ1. (b) anti-Δ2. (c) anti-ALD. A minority of the fibers react with antibodies against the two fast light chains (a, b). The level of fluorescence in soleus fibers reacting with anti-Δ2 (b) is less intense than in white or intermediate fibers of the diaphragm; it is comparable to that of the fast red fibers of the diaphragm. Antibody against slow myosin (c) reacts with the same fibers that reacted with antibodies against fast myosin (a, b) as well as with the majority of the fibers, which did not react with anti-fast myosin. From Gauthier and Lowey, 1979.

Pattern of Distribution of Myosins in a Slow-Twitch Muscle

The fibers of a fast-twitch muscle normally have a reciprocal response to antibodies against slow and fast myosin; that is, they contain either slow or fast isoenzymes. Both types coexist, however, in individual fibers during development (7) or following chronic stimulation (20). In the rat soleus (a slow-twitch muscle), the majority of the fibers react only with antibody against myosin from the slow ALD muscle. A minority (about 25%) react with antibodies to pectoralis (figs. 6a, b) as well as ALD myosin (fig 6c). This property is not shared by any of the fibers of the rat diaphragm, with the possible exception of an occasional fast red fiber, which my exhibit a low level of response to anti-ALD myosin. This indicates that both fast and slow myosin can coexist within the same fiber of a normal adult muscle.

Myosin Isoenzyme in Developing Muscle Fibers

During the early development of skeletal muscles, speed of contraction tends to be slow, even in those muscles which contract rapidly in the adult animal, and this includes the rat diaphragm. Using the procedures of immunocytochemistry, we have demonstrated that significant amounts of myosin with slow as well as fast characteristices are present in developing fibers of this muscle (7). Both are present, moreover, in the same fiber, and this coincides with the presence of polyneuronal innervation. At 19 days gestation, all fibers react strongly with antibody against slow myosin (anti-ALD; fig. 7c) as well as the alkali 1 light chain of fast myosin (anti -Δ1; fig. 7a). Response to antibody specific to the alkali 2 light chain (anti -Δ2; fig. 7b) is weak. During the first postnatal week, when speed of contraction of the diaphragm is still slow (12), myosin isoenzymes have begun to be segregated into different populations of fibers, but the overall levels of staining remain the same. The response to anti-slow myosin is still greater than in the adult muscle. By 21 days, when speed of contraction has increased, the response to anti-slow myosin has decreased to adult levels (fig. 8c). Only a minority of the fibers now react with this antibody, whereas a majority of the fibers react strongly with antibodies against both light chains of fast myosin (figs. 8a,b). Fast and slow myosins are therefore distributed among different populations of fibers and, in the same muscle, the fibers are now innervated by single axons (2).

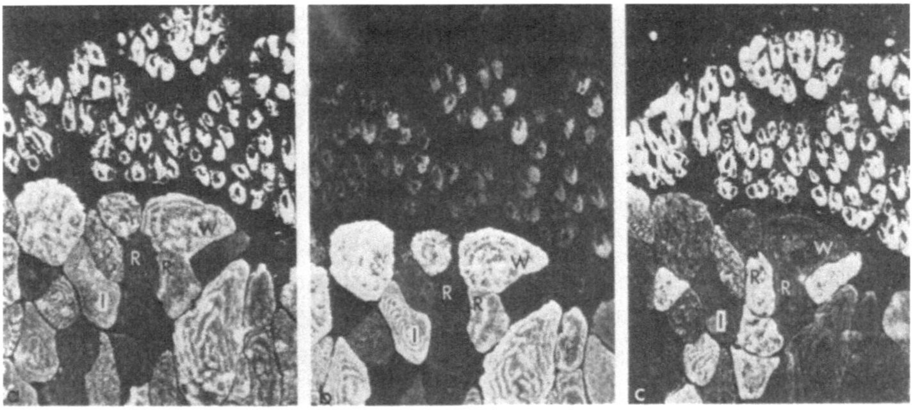

Figure 7 Rat diaphragm, serial transverse sections. 19 days gestation. Antibodies against a, A1 myosin (anti-△1); b, A2 myosin (anti-△2); c, slow myosin (anti-ALD). Developing muscle (upper) was sectioned simultaneously with a specimen of adult diaphragm (lower) so that differentiating fibers can be compared directly with adult fibers of equal thickness under identical conditions. Although intesity of fluorescence fades during exposure to ultraviolet light, the adult muscle provides a reference against which the relative intensity at each stage of development can be compared. All fibers in the developing muscle (upper) react strongly (greater than the adult) with both anti-△1 (a) and anti-ALD (c) and weakly (less than the adult) with anti-△2 (b). From Gauthier, Lowey and Hobbs, 1978.

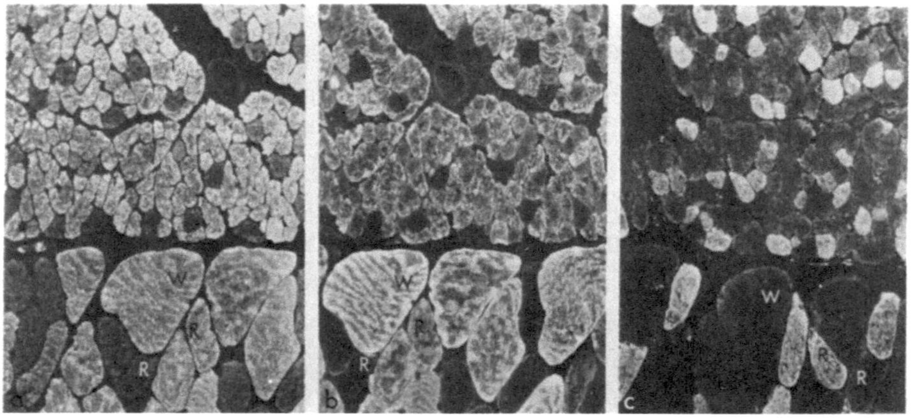

Figure 8 Rat diaphragm, serial transverse sections. Day 21 postnatal. A differential response to anti-△2 is evident (b). The positive response to anti-△1 in the majority of the fibers is now equivalent to the adult (a), and the majority of the fibers fail to react with anti-ALD (c). The overall response to all three antibodies is comparable to the adult muscle (lower) in each case, and the pattern of response to anti-ALD (c) is reciprocal to that seen with antibodies against the light chains of fast myosin (a,b). The section in (c) is not serial to those shown in (a) and (b) in this illustration. From Gauthier, Lowey, and Hobbs, 1979.

The abundance of slow myosin during the development of a fast muscle is not unique to the diaphragm, nor is the coexistence of slow and fast myosin in the same fiber. This is also evident in the extensor digitorum longus (EDL) of the rat (7, 10), which consists, in the adult, almost entirely of fibers with fast myosin. At 19 days gestation, all fibers react strongly with anti-Δ1 as well as anti-ALD myosin. The results suggest that synthesis of slow myosin or of an embryonic myosin with characteristics of slow myosin is "switched off" in most of the fibers of fast-twitch muscles during the normal course of differentiation. In a typical slow muscle, the soleus, slow myosin predominates in the adult. However, we have observed that, as in the fast muscle, both slow and fast myosin are present in all fibers during development (10). Therefore, synthesis of fast myosin appears to be terminated in the majority of the fibers as this slow-twitch muscle differentiates (Gauthier, unpublished observations; see also ref. 18).

Although we have demonstrated abundant slow myosin in developing fast mammalian muscles using immunocytochemistry, this type of myosin is not evident in avian muscles analyzed by gel electrophoresis (17, 19). This apparent discrepancy could be related to difference in species. It is also possible that a myosin with immunological properties resembling slow myosin and electrophoretic characteristics resembling fast myosin is responsible for the conflicting data. We have therefore analyzed developing muscles of the rat using two-dimensional O'Farrell gels. The results show that at postnatal day 1, there is very little light chain typical of adult slow myosin in the diaphragm and EDL, and this is consistent with previous reports. There is, however, a small amount of light chain which has the same molecular weight as the alkali 1 light chain of adult fast myosin, but which has a slightly lower isoelectric point. The existence of such an "embryonic" light chain has been reported by Whalen and his associates (23). Secondly, we have examined selected muscles of the chicken by immunocytochemical procedures (10), and we have observed that, as in the rat, slow myosin is present along with fast myosin in all the fibers of slow (ALD) as well as fast (posterior latissimus dorsi (PLD) and pectoralis) muscles during early development (figs. 9a,b). The observation of slow myosin in developing muscle is therefore not apparently related to species, and this is consistent with the principle that speed of contraction is related to the type of myosin present, a principle which would most likely hold for all species.

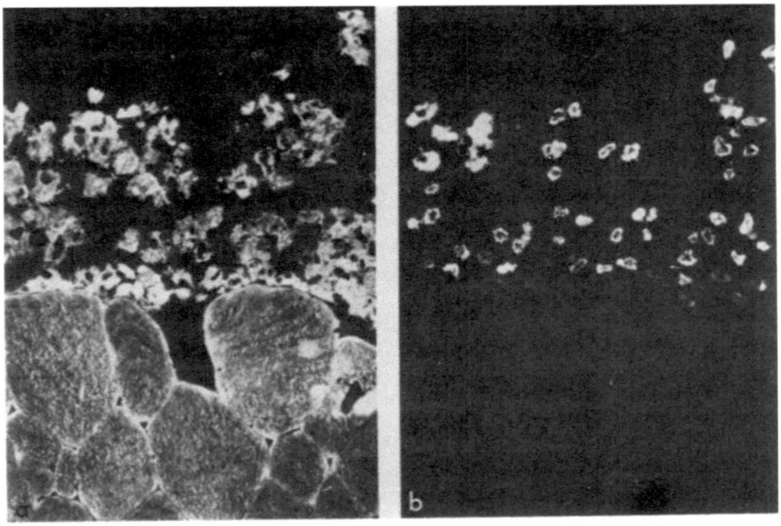

Figure 9 Chicken PLD. 11-day embryo (upper) sectioned with adult PLD (lower). (a) anti-Δ1. (b) anti-ALD. In the adult muscle (lower), all fibers react with anti-Δ1, but none react with anti-ALD. All fibers in the developing muscle (upper) react with antibodies against both "fast" (a) and "slow" (b) myosin.

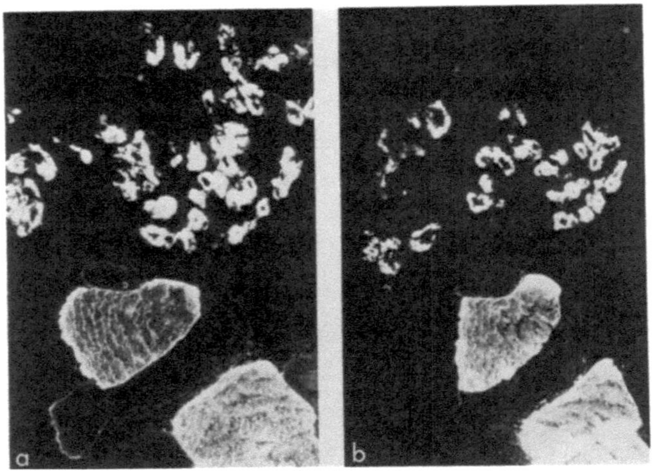

Figure 10. Chicken PLD. 10-day embryo (upper) sectioned with a region of the adult pectoralis (lower) which has a mixed population of fibers. (a) anti-ALD. (b) anti-ALD adsorbed with total ALD light chains. Both reactive and unreactive fibers are present in the adult muscle (lower). In the developing muscle (upper), all fibers react with antibodies against "slow" myosin before (a) and also after (b) adsorption, that is, with antibodies against the heavy chain of myosin.

These results are consistent with the existence of a distinctive "embryonic" myosin during development. The possibility exists that the isoenzyme which reacts with antibody against slow myosin has a light chain which is characteristic of the adult alkali 1 light chain or the newly-described embryonic light chain, and that this light chain is associated, during development, with a heavy chain more characteristic of slow myosin. There is chemical evidence that the heavy chain of myosin is distinctive in the newborn rabbit (14). To investigate the possibility that the heavy chain is responsible for staining with anti-ALD myosin, we have adsorbed anti-ALD myosin with total ALD light chains. The resulting antibody preparation would be expected to react only with the heavy chain of myosin. This adsobed antibody continues to react with developing fibers in two fast muscles (PLD and pectoralis) of the chicken embryo (figs. 10a,b), and therefore it is concluded that the response reflects a property of the heavy chain. This suggests the existence of a myosin which has a slow or "embryonic" type of heavy chain but a light chain similar to the alkali 1 light chain of fast myosin.

Conclusions

Adult skeletal muscles consist of mixed populations of fibers, and each fiber can be recognized by immunological differences in the myosin present. In early development, before the adult pattern of contractile activity is attained, these differences are not apparent. "Fast" and "slow" myosin coexist in all the fibers of both fast and slow muscles at a time when speed of contraction is slow and the fibers are polyneuronally innervated. As the contractile properties differentiate, and the fibers become innervated by single motoneurones, different isoenzymes of myosin become segregated into different populations of fibers, giving rise to the typical adult pattern of fiber types.

References

1. Bárány, M.: J. Gen. Physiol. 50, 197-218 (1967).
2. Bennett, M.R., Pettigrew, A.G.: J. Physiol. 241, 515-545 (1974).
3. Burke, R.E., Levine, D.N., Tsairis, P., Zajac, F.E., III: J. Physiol. 234, 723-748 (1973).

4. Edström, L., Kugelberg, E.: J. Neurol. Neurosurg. Psychiat. 31, 424-433 (1968).
5. Gauthier, G.F.: In: The Contractility of Muscle Cells and Related Processes pp. 131-150. (R.J. Podolsky, editor). Prentice-Hall, Englewood Cliffs, N.J. (1971)
6. Gauthier, G.F., Lowey, S.: J. Cell Biol. 74, 760-779 (1977).
7. Gauthier, G.F. Lowey, S., Hobbs, A.W.: Nature 274, 25-29 (1978).
8. Gauthier, G.F., Lowey, S.: J. Cell Biol. 81, 10-25 (1979)
9. Gauthier, G.F.: J. Cell Biol. 82, 391-400 (1979)
10. Gauthier, G.F., Benfield, P.A., Lowey, S., Hobbs, A.W.: Manuscript in preparation (1979).
11. Gröschel-Stewart, U., Schreiber, J., Mahlmeister, C., Weber, K.: Histochemistry 46, 229-236 (1976).
12. Gutmann, E., Melichna, J.: Physiol. bohemoslov. 21, 1-7 (1972).
13. Hoh, J.F.Y., McGrath, P.A., White, R.I.: Biochem. J. 157, 87-95 (1976).
14. Huszar, G.: Nature New Biol. 240, 260-264 (1972)
15. Jean, D.H., Albers, R.W., Guth, L, Aron, H.J.: Exp. Neurol. 49, 750-757 (1975).
16. Lowey, S., Risby, D.: Nature 234, 81-85 (1971)
17. Pette, D., Vrbová, G., Whalen, R.C.: Pflügers Arch. 378, 251-257 (1979).
18. Rubinstein, N.A., Kelly, A.M.: Dev. Biol. 62, 473-485 (1978).
19. Rubinstein, N.A., Pepe, F.A., Holtzer, H.: Proc. Nat. Acad. Sci. 74, 4524-4527 (1977).
20. Rubinstein, N., Mabuchi, K., Pepe, F., Salmons, S., Gergely, J, Sreter, F.: J. Cell Biol. 79, 252-261 (1978).
21. Sarkar, S., Sreter, F.A., Gergely, J.: Proc. Nat. Acad. Sci. 68, 946-950 (1971).
22. Weeds, A.G., Pope, B.: Nature 234, 85-88 (1971).
23. Whalen, R.G., Butler-Browne, G.S., Gros, F.: J. Mol. Biol. 126, 415-431 (1978).

FIBRE POPULATIONS IN RABBIT SKELETAL MUSCLES FROM BIRTH TO OLD AGE

E. Jenny, H. Weber, H. Lutz and R. Billeter
Institut für Pharmakologie und Biochemie, Universität Zürich, CH-8057 Zürich, Switzerland

Introduction

A confusing number of systems has been used to designate fibre types in adult vertebrate skeletal muscles (1). A widely accepted scheme combining the reaction of the myofibrillar ATPase (2) with physiological and metabolic properties (3, 4) classifies fibres into type I (slow twitch, oxidative; motor unit S, slow contracting, fatigue resistant), type IIA (fast twitch, oxidative, glycolytic; motor unit FR, fast contracting, fatigue resistant) and type IIB (fast twitch, glycolytic; motor unit FF, fast contracting, fast fatigue). A fourth type, IIC, has been described by Brooke and Kaiser (5) and called an undifferentiated fibre type. These classes have also been correlated with ultrastructural properties (6). In addition, fibre analysis has been extended to include structural proteins like actin, myosin and the tropomyosin-troponin-complex (7). About 22 structural genes are involved in their synthesis, but only 8 of them need to be expressed to produce a single variant of each protein (8). It still is a matter of dispute how many of them are actually synthesized within individual muscle fibres. From the information currently available, it seems that the existence of polymorphic protein forms within a single cell is more pronounced during development than in adult age.

Until now, most efforts have been directed at resolving the

Plasticity of Muscle

myosin isoenzyme compositions prevailing in the various fibre types. However, exceptional difficulties have been encountered as each of the five or even more isoenzymes of myosin (9, 10) is the product of at least three structural genes, and some evidence even suggests the existence of hybrids combining light (LC) and heavy (HC) chains of fast (FM) and slow (SM) myosins (9, 11, 12).

In the experiments reported here we attempted to establish the size of the fibre populations of a fast (m. tibialis anterior) and a slow (m. soleus) rabbit muscle from birth to old age and to analyze the myosin composition of the individual fibre types using novel and refined immunological methods. The purpose of our work was to obtain a basis for studying changes possibly occurring during physical activity (13) and to confirm results obtained earlier showing that in old age the slow type rabbit soleus muscle contains appreciable amounts of FM (14).

Materials and Methods

1) Animals. New Zealand white rabbits were used for all experiments, except in the instance of two old rabbits, which were of undefined breed.

2) Antigens and antisera. Antisera against FM (anti-FM) and SM (anti-SM) were induced in guinea pigs of high responder strains (Himalayan spotted white) as described elsewhere (15) using a low dose, long term procedure (16) resulting in non precipitating antibodies of high specificity and low heterogeneity (17). Antibodies against the LC of heart myosin (anti-heart-LC) were prepared with the same methods. Anti-heart-LC was found to be an excellent probe for the LC of SM giving only a negligible cross-reaction with the LC of FM (18). The specificity of all antisera was assessed using the GEDELISA technique (15).

3) Immunohistochemistry for myosin types. Myosin types in muscle fibres were visualized using a protein A peroxidase technique (19). The details of the procedure are described elsewhere (13).

4) Enzyme histochemistry for fibre typing. Alternating sections were stained for myofibrillar ATPases after preincubation at pH 4.35 (5 min, room temperature), pH 4.6 (5 min, room temperature) and pH 10.8 (15 min, 37°C) (2).

Results and Discussion

1) Fibre populations in the m. tibialis anterior. The size of the fibre populations from birth to old age in this fast muscle and the reaction of the fibre types with the different antisera are summarized in Table 1. In newborn animals the fibres were of types I, IIB and IIC. With regard to the difficulties of fibre typing at this period (20), however, we consider these classifications as tentative. At the age of five months, mainly fibres of types I, IIB and IIA were found. Thereafter, the size of these populations does not change significantly. Consistent with these results, Farkas-Bargeton et al. (21) have described the appearance of fibre types in developing human muscles in the sequence IIC, I, IIB and IIA. The staining reaction of the fibres at birth with anti-FM and anti-heart-LC is shown in Fig. 1. At this age, the largest fibres (type I), situated in or close to the center of the fibre bundles, react with anti-heart-LC, the fibres of slightly smaller size (type IIB) with anti-FM and the very small fibres (type IIC) with both antisera. Type I and IIC fibres also stain with anti-SM (not shown). The positive reaction of the so called undifferentiated (5) type IIC fibres with all three antisera could be explained by the presence of embryonic (10)

Table 1 - The size of the fibre populations (%) in the m. tibialis anterior of rabbits (n = 5) from birth to old age. Classification into types I, IIA, IIB and IIC according to Brooke and Kaiser (2, 5). Reaction of the fibres (protein A immunoperoxidase technique) (19) with antisera induced against fast myosin (anti-FM), slow myosin (anti-SM) and the light chains of heart myosin (anti-heart-LC).

age	anti-FM	anti-hLC anti-SM	I	IIA	IIB	IIC
birth	94				42	52
		58	6			52
5-7 mo	94			38	56	<1
		44	6	38		<1
12-18 mo	94			30	64	<1
		36	6	30		<1
4-7 y	96			34	62	<1
		38	4	34		<1

or fetal (22) LC of myosins having antigenic determinants in common with the LC of the adult myosins. This hypothesis is based on analogy, since in the calf, Brevet and Whalen (23) demonstrated the presence of some of the typical peptides of HC from adult FM and SM in the tryptic hydrolyzate of HC from fetal myosins. On the other hand, the coexistence of FM and SM within these fibres would also explain the facts. In adult muscles, type I fibres react with anti-SM and anti-heart-LC, type IIB fibres with anti-FM and type IIA and IIC fibres with all three antisera. The reaction of the fibres with anti-FM and anti-SM is demonstrated in Fig. 2. With both antisera, clearly three populations of fibres emerge, one of them staining with anti-FM and anti-SM. To substantiate this result, a computer assisted analysis of the photometrically measured

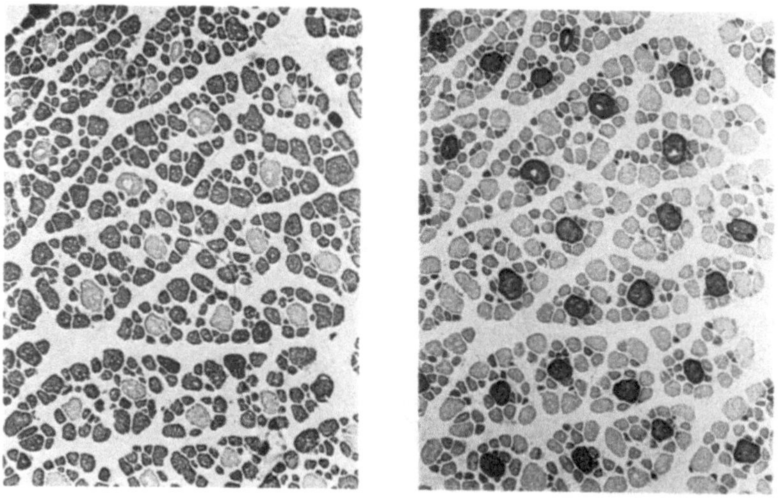

Fig. 1 - Reaction of the fibres in the m. tibialis anterior of rabbits at birth with sera induced against fast myosin (left) and the light chains of heart myosin (right). Protein A immunoperoxidase technique (19).

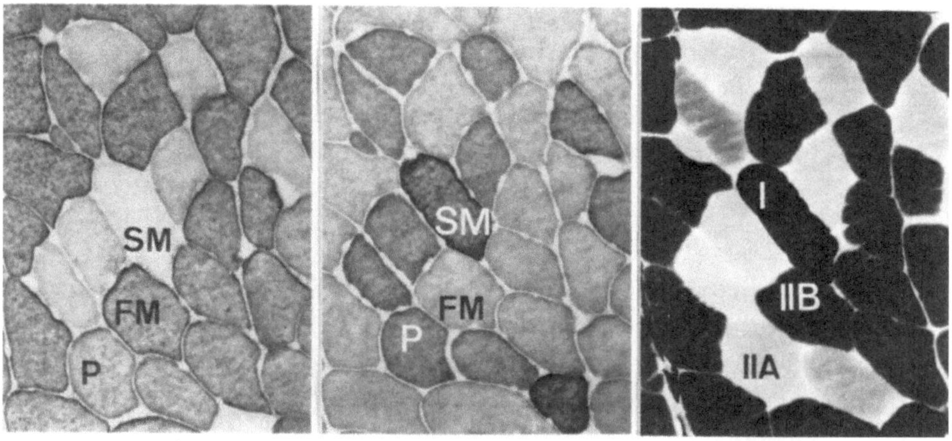

Fig. 2 - Fibres in the adult m. tibialis anterior of rabbits. Reaction (19) with antiserum induced against rabbit fast (left) and slow myosin (middle). Myofibrillar ATPase staining pH 4.6 (2, 5) right. P, promiscuous fibre containing antigenic determinants for fast and slow myosin.

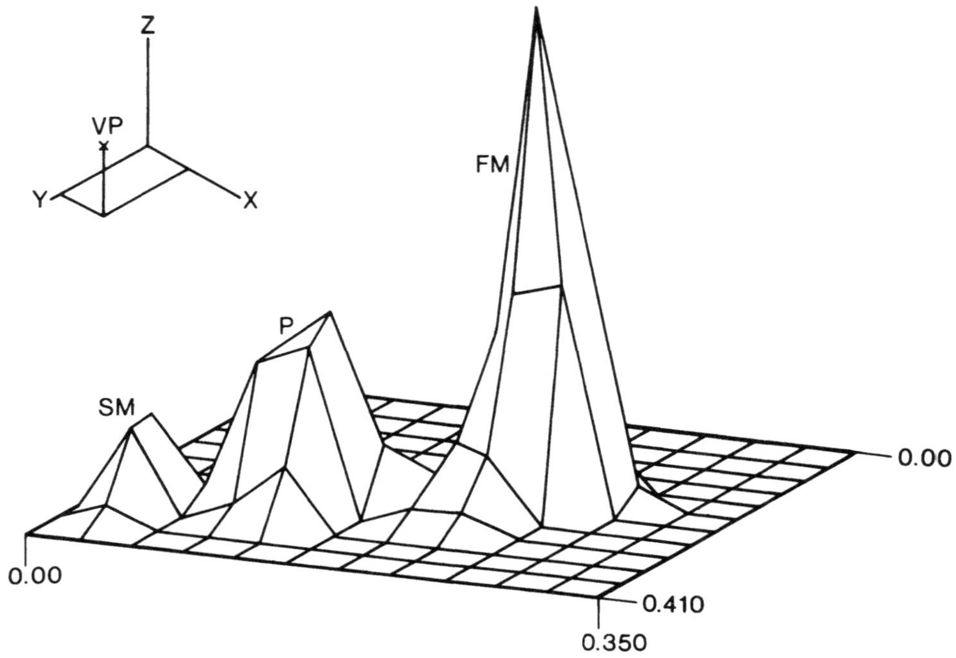

Fig. 3 - Computer assisted analysis of the staining intensities of 132 randomly chosen fibres reacted with antisera induced against rabbit fast and slow myosin. The immune complexes formed were visualized using the protein A immunoperoxidase technique (19).

x, absorbance after reaction with anti-FM serum

y, absorbance after reaction with anti-SM serum

z, relative frequency of absorbance values

VP, computer adjusted viewpoint for optimal three dimensional impression

P, promiscuous fibres containing antigenic determinants of FM and SM.

staining intensities of 132 randomly chosen fibres reacted with anti-SM and anti-FM was performed. Three clusters are visible, one of them (P, for promiscuous) comprising fibres giving an intermediate reaction with both antisera (Fig. 3). Fibres staining both with anti-FM and anti-SM have also been discovered in diaphragm and m. soleus of adult rats (6, 24). These findings are in favour of the assumption that SM and FM coexist within these fibres. However, in histochemically identified single fast fibres of rabbit psoas muscles, Weeds et al. (25) and Pette and Schnez (26) could only demonstrate the presence of LC of FM using microgel-electrophoresis, and Rubinstein et al. (27) found no fibres in the m. peroneus longus of rabbits staining with both anti-SM and anti-FM using a direct immunofluorescence technique. The reasons for these discrepancies could simply be of technical nature. Highly sensitive indirect immunological methods should detect smaller amounts of one myosin species in a mixture of two than microgel-electrophoresis of the myofibrillar proteins of single fibres (25, 26) and should be superior to direct immunofluorescence (27), especially in the case of low titering antibodies. However, the presence of hitherto unknown myosin species has also to be considered, e.g. hybrids consisting of components of FM and SM. Such "pink" myosins have been proposed by several authors (9, 11, 12), but their existence has still to be demonstrated. A different myosin in type IIA as opposed to type IIB fibres could explain the well known differences in the pH-sensitivity (2, 5) and sulfhydryl dependence (28) of the myofibrillar ATPase.

2) Fibre populations in the m. soleus. In newborn animals more than half of the fibres of this prospective slow muscle react strongly with anti-FM and all the fibres with anti-SM, albeit in slightly varying intensities (Fig. 4). After preincubation at pH 10.8 myofibrillar ATPase was only detected in fibres reacting with anti-FM, but after preincubation at pH 4.6 and 4.35 all fibres stained for this enzyme. With due reservation (20), fibres reacting with both antisera classify as type IIC and those reacting only with anti-SM as type I. These results are in agreement with the assumption that in presumptive slow muscles myosin of the FM type (fetal or adult) is synthesized

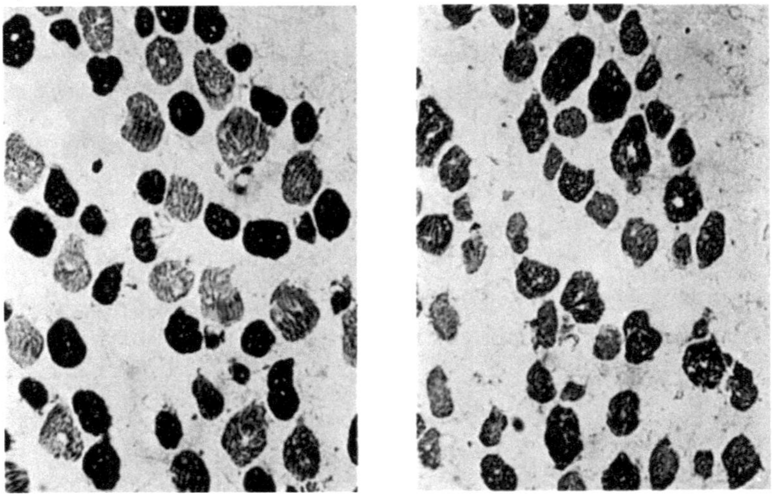

Fig. 4 - Reaction of the fibres in the m. soleus of rabbits at birth with sera induced against rabbit fast (left) and slow myosin (right). Protein A immunoperoxidase technique (19).

first but is later replaced by SM (9, 11, 12, 23, 29). After birth, the number of the FM-containing fibres diminishes and at the age of 12 - 18 months all fibres contain SM only (Fig. 5) thus classifying as type I.

In old age up to 50% of the fibres have resumed the synthesis of FM. Their myofibrillar ATPase is not inactivated after preincubation at pH 10.8 (Fig. 6) and polyacrylamide-gel-electrophoresis reveals the presence of LC of FM and SM in old soleus muscles (Fig. 7). Using indirect immunofluorescence, an abundance of FM-containing fibres in the m. solei of old rabbits has already been demonstrated (14). The fibres reacting with anti-FM still stain with anti-SM and anti-heart-LC. Therefore, they are promiscuous for FM and SM. According to their myofibrillar ATPase staining they classify as type IIA and IIC. These fibres have a somewhat edgy form (Fig. 6), possibly indicating incipient atrophy. The reappearance of FM may then be indicative of a loss of neuronal control. In support of this

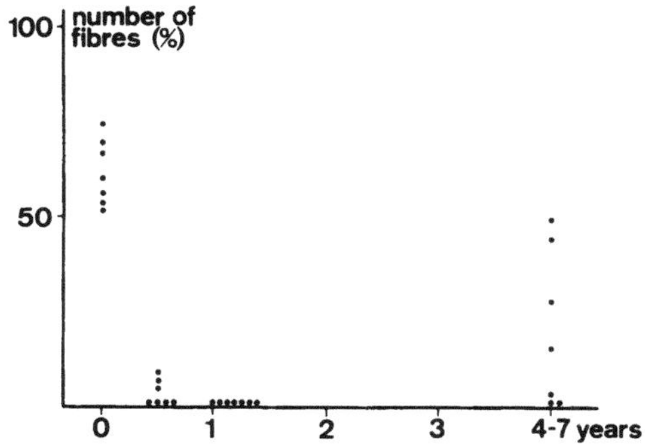

Fig. 5 - Number of fibres in the m. soleus of rabbits of different age reacting with an antiserum induced against rabbit fast myosin.

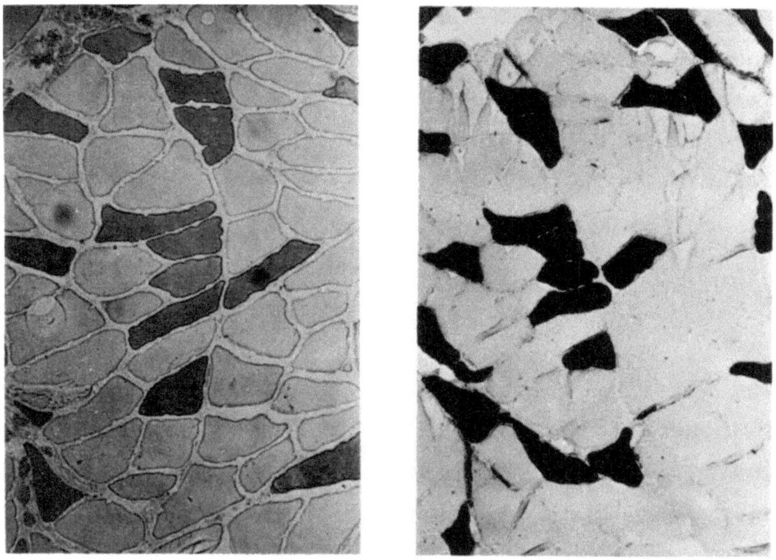

Fig. 6 - Fibres in the soleus muscle of an old rabbit (7 years) Reaction with an antiserum induced against rabbit fast myosin (protein A immunoperoxidase technique (19)) left, and staining for myofibrillar ATPase (2, 5) after preincubation at pH 10.8, right.

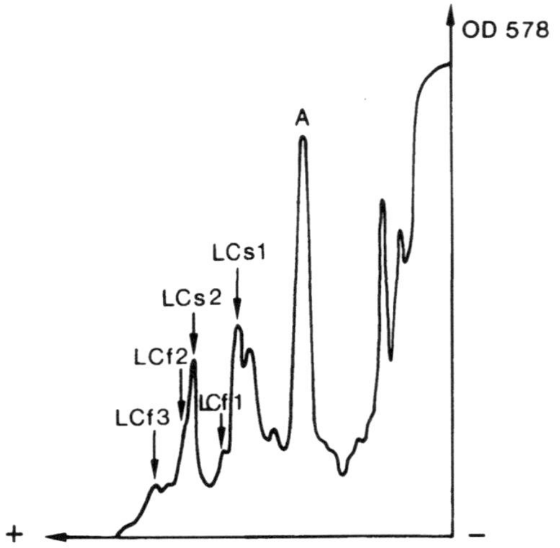

Fig. 7 - SDS-gel-electrophoresis of dissociated myofibrils from a rabbit (7 years) soleus muscle, containing about 30% anti-FM positive fibres.

it has been observed that after neurectomy FM is preferentially synthesized in slow muscles (30, 31). However, it remains to be elucidated if this FM species is of the fetal or of the adult type (22).

Summary

By combining enzyme histochemistry for fibre typing with immunocytochemistry for slow (SM) and fast (FM) myosin and for the light chains (LC) of heart myosin a correlation between fibre type and myosin type was sought in a fast (m. tibialis anterior) and a slow (m. soleus) rabbit skeletal muscle. In addition, the size of the fibre populations was monitored from birth to old age.

In the m. tibialis anterior of newborn animals the fibres were of types I, IIC and IIB. At the age of five months the fibres were of types I, IIA, IIB and very rarely of type IIC. The size of the fibre populations then remained stable throughout life. Type I fibres reacted with antisera against SM and heart-LC, type IIB fibres with antisera against FM and type IIC and IIA fibres with all three antisera. The presence of antigenic determinants for adult FM, SM and heart-LC in type IIC and IIA fibres is indicative for a special myosin or a combination of different myosin species within these cells.

In the m. soleus of newborn animals more than half of the fibres preferentially stained with anti-FM sera whereas the rest predominantly reacted with anti-SM sera. Thereafter, the FM-containing fibres diminished and at the age of 12 - 18 months practically all the fibres were of type I, SM being the only detectable myosin species. In old age, up to 50% of the fibres again reacted with anti-FM sera, while continuing to stain with anti-SM sera. Their myofibrillar ATPase was not inactivated after preincubation at pH 10.8, and the LC of FM were detectable by polyacrylamide-gel-electrophoresis. This probably indicates that these fibres are promiscuous for FM and SM.

Acknowledgments

We thank Dr. C. Kuenzle for reading the manuscript, Mr. W. Roemer for technical assistance and Dr. H. Fey for a generous gift of protein A peroxidase.

References

1. Close, R.I.: Physiol. Rev. 52, 129-197 (1972)
2. Brooke, M.H., Kaiser, K.K.: Arch. Neurol. 23, 369-379 (1970)
3. Peter, J.B., Barnard, R.J., Edgerton, V.R., Gillespie, C.A., Stempel, K.E.: Biochemistry 11, 2627-2633 (1972)
4. Burke, R.E., Levine, D.N., Tsairis, R., Zajac, F.E.: J. Physiol. 234, 723-748 (1973)

5. Brooke, M.H., Kaiser, K.K.: Ann. New York Acad. Sci. USA 22, 121-144 (1974)

6. Gauthier, G.F., Lowey, S.: J. Cell Biol. 81, 10-25 (1979)

7. Dhoot, G.K., Perry, S.V.: Nature 278, 714-717 (1979)

8. Perry, S.V.: in Exploratory Concepts in Muscular Dystrophy II (ed. Milhorat, A.T.), 319-328 (Excerpta medica, Amsterdam, 1974)

9. Hoh, J.F.Y.: FEBS Lett. 98, 267-269 (1979)

10. Whalen, R.G., Butler-Browne, G.S., Gros, F.: J. Mol. Biol. 126, 415-431 (1978)

11. Rubinstein, N.A., Pepe, F.A., Holtzer, H.: Proc. Natl. Acad. Sci. USA 74, 4524-4527 (1977)

12. Pette, D., Vrbovà, G., Whalen, R.G.: Pflügers Arch. 378, 251-257 (1979)

13. Billeter, R., Weber, H., Howald, H., Eppenberger, H.M., Jenny, E.: Submitted for publication

14. Lutz, H., Ermini, M., Jenny, E., Joris, F.: Akt. Gerontol. 8, 667-673 (1978)

15. Lutz, H., Ermini, M., Jenny, E., Bruggmann, S., Joris, F., Weber, E.: Histochemistry 57, 223-235 (1978)

16. Portmann, A.J., Levison, S.A., Dandliker, W.B.: Immunochemistry 12, 461-466 (1975)

17. Roitt, I.: Essential Immunology, 2nd ed., Oxford/London, Blackwell Scientific Publications (1974)

18. Weber, H., Schaub, M.C., Römer, W., Lutz, H., Jenny, E.: In preparation

19. Celio, M.R., Lutz, H., Binz, H., Fey, H.: J. Histochem. Cytochem. 27, 691-693 (1979)

20. Guth, L., Samaha, F.J.: Exptl. Neurol. 34, 465-475 (1972)

21. Farkas-Bargeton, E., Diebler, M.F., Arsénio-Nunes, M.L., Wehrle, R., Rosenberg, B.: J. Neurol. Sciences 31, 245-259 (1977)

22. Hoh, J.F.Y., Yeoh, G.P.S.: Nature 280, 321-323 (1979)

23. Brevet, A., Whalen, R.G.: Biochimie 60, 459-466 (1978)

24. Gauthier, G.F., Lowey, S., Hobbs, A.W.: Nature 274, 25-29 (1978)

25. Weeds, A.G., Hall, R., Spurway, N.C.S.: FEBS Lett. 49, 320-324 (1974)

26. Pette, D., Schnez, U.: Histochemistry 54, 97-107 (1977)

27. Rubinstein, N., Mabuchi, K., Pepe, F., Salmons, S., Gergely, J., Sréter, F.: J. Cell Biol. 79, 252-261 (1978)

28. Brooke, M.H., Kaiser, K.K.: J. Histochem. Cytochem. 18, 670-672 (1969)

29. Pelloni-Müller, G., Ermini, M., Jenny, E.: FEBS Lett. 67, 68-74 (1976)

30. Rubinstein, N.A., Kelly, A.M.: Developmental Biol. 62, 473-485 (1978)

31. Carraro, U., Catani, C., Biral, D.: Exptl. Neurol. 63, 468-475 (1979)

ADAPTIVE FIBRE AND MOTOR UNIT TRANSFORMATION IN RAT SOLEUS DURING GROWTH

E. Kugelberg

Department of Neurology, Karolinska Sjukhuset,
S-104 01 Stockholm 60, Sweden

There is ample evidence that muscle fibres respond to endurance training with increase in oxidative enzymes and capillary supply. So far is known this is a quantitative change. A change in speed of contraction of fibres from fast twitch to slow twitch or the reverse, involves on the other hand qualitative changes in contractile proteins and is less easily obtained under normal conditions.

However, slow muscles undergo a moderate slowing from about 5 weeks postnatally as observed in the cat (1), rat (2) and rabbit (3). Moreover, type I fibres increase from early age (4). I have studied these changes in some detail. Only a short account of the work will be given here, since it has been published in full (5).

In rat soleus there is between the 5th and 34th postnatal week a relative decrease in type II fibres and an increase in type I fibres. It is a continuous process which proceeds with increasing weight of the animal during a large part of its life-span (fig. 1). Since the number of fibres in the muscle remains constant during growth and since selective atrophy of type II fibres seems not to occur, the type II fibres are likely transformed into type I.

The alteration in ATPase activity of fibres is parallelled by a relative decrease of fast contracting units and a relative increase of slow units (fig. 2). The increased contraction time of whole muscle earlier observed during growth is thus mainly caused by a shift in unit composition of the muscle rather than an over all slowing of motor units.

Plasticity of Muscle

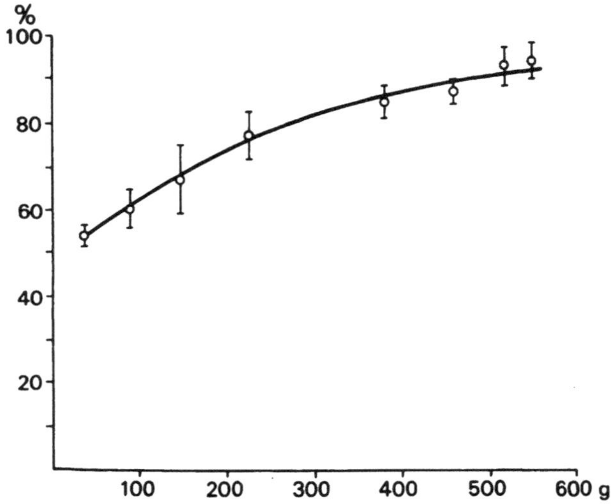

Fig. 1. Percentage type I fibres in rat soleus in relation to body weight. (Mean ± SD of observations in 5 animals). Reproduced by permission of J.Neurol. Sci. 1976

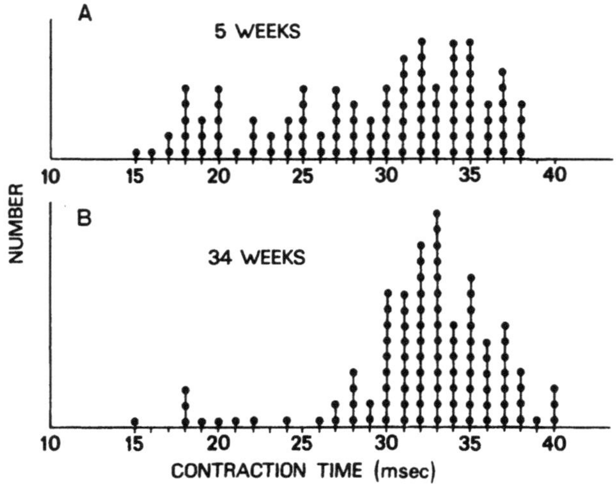

Fig. 2. Distribution of isometric contraction times of 100 rat soleus motor units stimulated through ventral root filaments at 5 and 34 weeks of age. Reproduced by permission of J.Neurol. Sci. 1976.

The question arises, where are the fast motor units gone? Have their motoneurones withdrawn from innervation and are the muscle fibres taken over by slow units e. g. during elimination of polyneuronal innervation of fibres or by newly formed sprouts? Or are the fast motor units transformed in their entirety to slow units? In other words is it the innervation field which is reshaped during growth by competition between motoneurones, or is it primarily the motoneurone itself which is differentiated with all its muscle fibres.

In the rat polyneuronal innervation is extensive at birth but decreases thereafter within the second to fourth postnatal week to one axon for each muscle fibre (6-7) with no change in number of motoneurones innervating the muscle (8). At 5 weeks and beyond elimination of multiple innervation would hardly be expected to play any major role in reshaping the motor units.

Furthermore, if the slow motor units present at 5 weeks had adopted the fast fibres, the slow units would at 34 weeks increase in number of fibres by about 25%. Estimation of unit size by way of mapping with the glycogen depletion method did not show any increase, admitted that unit mapping in small 5 week rats is technically difficult owing to disturbing fasciculations in connection with dissection of the motor root.

If on the other hand fast motor units are transformed as a whole to slow units, one would find motor units in all intermediate stages between fast and slow units. This is actually the case.

The units with contraction times of less than 21 msec. are fast twitch type II fibre motor units composed of fibres with alkali-stable (pH 9.4) and acid-labile (pH 4.3) ATPase. Units with twitch time to peak over 26 msec. are composed of type I fibres showing the reverse pH sensitivity of ATPase.

Between the type II and type I fibre units are intercalated units with intermediate contraction times of 22-25 msec. composed of fibres containing a mixture of alkali-stable and acid-stable ATPase in various proportions. The relative intensity of alkali- and acid-stable ATPase is almost identical for the majority of fibres in the units, but varies among different units so that all stages between type II and type I fibre units are represented.

Transitional fibres are, for reasons not well understood, smaller than type II and type I fibres and the intensity of succinate dehydrogenase activity, SDH, is intermediate between type II, where it is high and type I, where it is moderate. Interestingly the fastest type I fibre motor units have significantly smaller fibres and higher SDH activity than the slower type I motor units (fig. 3). It is suggesting they represent transitional units at the first stage after transformation to type I.

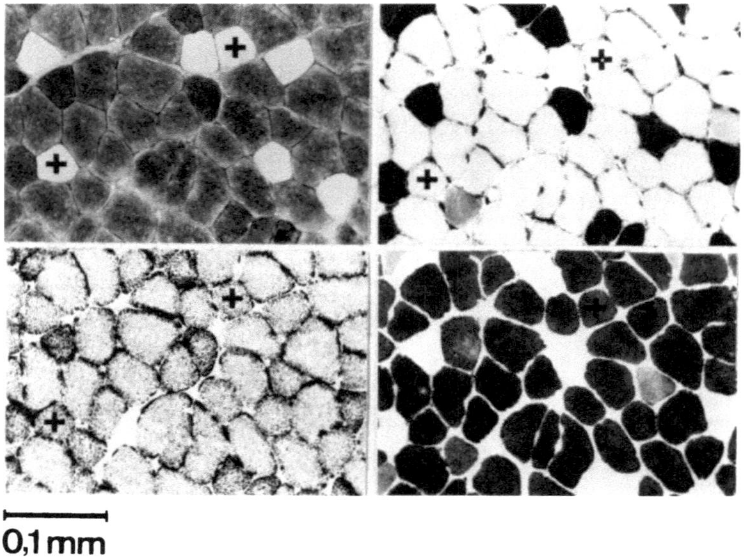

Fig. 3. Serial sections showing part of a type I motor unit of 28 msec. contraction time. The fibres identified by glycogen depletion in the PAS stained section (top left) have all similar size, intensity of SDH activity (bottom left), ATPase activity at pH 9.4 after formaldehyde fixation (top right) and ATPase activity at pH 4.35 (bottom right).

Note the fibres have similar size and SDH activity as the three transitional fibres identified in the top right section incubated for ATPase at pH 9.4, but smaller fibres and higher SDH activity than the majority of type I fibres, which belong to motor units of longer contraction times.

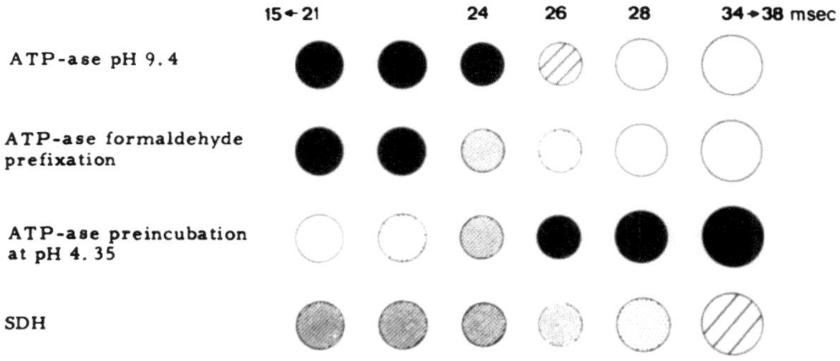

Fig. 4. Schematic presentation of the relationship between contraction time, ATPase reactions, SDH reaction and fibre area in motor units of rats aged 5 weeks. Reproduced by permission of J. Neurol. Sci. 1976.

Thus all tested markers for soleus unit types shift gradually from one into the other, including contraction time, intensity of fast and slow fibre ATPase, SDH reaction and fibre size (fig. 4). It is a picture consistent with a gradual transformation of fast twitch type II units as well as transitional units to slow twitch type I units.

If it is accepted that the muscle part of the unit is transformed as a whole, the motoneurone itself must have changed in a number of ways. The motoneurones supplying slow muscle fibres are for example characterized not only by lower discharge frequencies, but also by longer after-hyperpolarization following the discharge of an impulse (9), lower axonal conduction velocity (10) and morphologically different nerve terminals (11). A plastic change of motoneurones from fast to slow seems not to be known. However, so far as present observations go, the transformation of soleus fibres during growth may actually be produced by such motoneurone transformation.

Tentatively the events leading to differentiation of soleus muscle fibres during the studied period of growth includes a prolongation of after-hyperpolarization and a lowered discharge frequency of motoneurones supplying type II and transitional fibres. Associated with the altered frequency characteristics of the motoneurone, genes coding for fast fibre contractile proteins will progressively be repressed and synthesis of slow fibre proteins promoted until the motor unit is fully transformed to type I. The transitional units which share contractile and histochemical properties of both fast and slow twitch fibre units presumably occur when the motoneurone discharges at a frequency range between the range for type II and type I motoneurones. At this stage genes coding for both specific fast and specific slow fibre proteins are probably simultaneously turned on in degrees varying inversely.

Transformation of soleus fibres during growth is clearly an adaptive process. With increase in leg length the demand for fast fibres will diminish and with increase in body weight, as in a postural muscle like soleus, the demand for slow fibres will increase. It is at least not entirely preprogrammed, since a speed-up in the process has recently been demonstrated if the rat is placed for some time in a centrifuge (12). The reverse results will probably be obtained if the rat is placed in a satellite.

This work was supported by a grant from the Swedish Medical Research Council.

References

1. Buller, A.J., Eccles, J.C., Eccles, R.M.: J. Physiol. (Lond.) 150, 399-416 (1960).
2. Close, R.: J. Physiol. (Lond.) 173, 74-95 (1964).
3. Gutmann, E., Melichna, J., Syrový, I.: Physiol. bohemoslov. 23, 19-27 (1974).
4. Karpati, G., Engel, W.K.: Arch. Neurol. (Chic.) 17, 542-545 (1967).
5. Kugelberg, E.: J. neurol. Sci. 27, 269-289 (1976).
6. Redfern, P.A.: J. Physiol. (Lond.) 209, 701-709 (1970).
7. Bennett, M.R., Pettigrew, A.G.: J. Physiol. 241, 515-545(1974)

8. Brown, M.C., Jansen, J.K.S., Van Essen, D.: J. Physiol. 261, 387-422 (1976).

9. Eccles, J.C., Eccles, R.M., Lundberg, A.: J. Physiol. (Lond.) 142, 275-291 (1958).

10. Appelberg, B., Emonet-Dénand, F.: J. Neurophysiol. 30, 154-160 (1967).

11. Duchen, L.W.: J. neurol. Sci. 14, 37-45 (1971).

12. Martin, W.D.: Aviat. Space Environ. Med. 49 (6), 792-797 (1978).

PLASTICITY OF THE MYOFIBER-SATELLITE CELL COMPLEX IN CULTURE

Richard Bischoff
Department of Anatomy and Neurobiology, Washington University School of Medicine, St. Louis, Missouri U.S.A.

Introduction

Mature skeletal muscle responds to various types of trauma such as physical and chemical injury, denervation and tenotomy by initiating a series of repair and regenerative processes. The first step, common to most types of trauma, involves an increase in the frequency of mononucleated myogenic cells within the basement lamina. Most investigators agree that these cells are responsible for the formation of new muscle fibers during any subsequent regeneration that occurs (1). The origin on these myogenic cells however, remains controversial.

Ultrastructural studies of injured or denervated muscle have suggested that mature fibers are capable of forming new mononucleated cells by formation of plasmalemma around a myonucleus and adjacent sarcoplasm (2-7). These new cells would then be identified as satellite cells, or mononucleated cells situated beneath the basement lamina of mature fibers (8). Other evidence, primarily from studies in which cell populations are labeled with ^{3}H-thymidine (TdR) before injury, supports the view that new myoblasts proliferate from preexisting satellite cells; myonuclei do not contribute (9-12).

Cellular events following injury *in vivo* cannot be observed continuously and various adventitial cells obscure the identification of myoblasts. To avoid these problems I developed a tissue culture approach to study regeneration (13-15).

Direct observation of single fiber segments from adult rat muscle *in vitro* showed that the new myotubes which form are derived entirely from the progeny of preexisting satellite cells (15). This has been confirmed with quail muscle (16). Since the fibers used in these studies are irreversibly damaged as a result of mechanical dissection, the response of the myofiber itself cannot be studied.

The present experiments are designed to obtain *intact* cultured fibers in which both the satellite cells and myonuclei survive isolation. These preparations will be used for studying interactions between satellite cells, myofibers and motor neurons. Preliminary results confirm that satellite cells are reserve myogenic stem cells and can form new myotubes even in association with an intact fiber. In addition, mature fibers are capable of morphological dedifferentiation to the myotube stage but do not give rise to mononucleated cells.

Results

Preparation of fibers

The technique for obtaining dissociated fibers was similar to that described by Bekoff and Betz (17). Briefly, the flexor digitorum brevis muscle was removed from the hind feet of 150-200 g rats and gently cleaned of as much epimysial connective tissue and vessels as possible. The muscles were then incubated for 1-2 h with continuous slow stirring in 0.2% collagenase in Earle's BSS. After washing, the individual fibers were liberated by gentle trituration through a wide-mouth pipet. The fibers were separated from non-muscle cells and debris by layering the suspension on a 7-12-20% Ficoll gradient and collecting the fibers from the 12-20% interface after centrifugation (30 g, 5 min). Alternatively, fibers were purified by repeated 1 g sedimentation until no mononucleated cells remained. The fibers were attached to coverslips in Petri dishes with fibrin clots made from fresh rooster plasma and thrombin. The culture medium was Eagle's MEM with 10% horse serum and 5% chick embryo extract (EE).

Characteristics of isolated fibers

Muscles from one rat yielded from several hundred to about 1000 fibers, depending upon the age of the rat, batch of enzyme and method of purification. The individual fibers are quite uniform in size and appear to be morphologically intact. Many of the physiological properties of these fibers are approximately normal during the early culture period, before denervation changes are expressed (17,18). During the first few hours of culture many fibers become slightly coiled as a result of slow contraction. Aside from this initial shortening, the fibers are quiescent during the first several days in culture. Later, however, many fibers may show rhythmic twitching. This is especially frequent in cultures that contain contaminating non-muscle cells or cultures to which spinal cord explants have been added.

Attached to most of the isolated fibers are a few mononucleated cells that are tentatively identified as satellite cells. The cells can be visualized in fixed and stained fibers (Fig. 1) or in living fibers after exposure to fluorescein isothiocyanate (FITC) labeled lectins such as lens culinaris lectin or concanavalin A (Fig. 2). In addition to pronounced

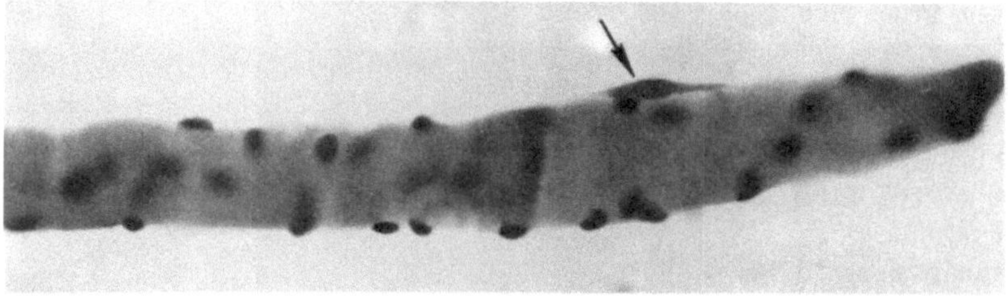

Fig. 1 Intact myofiber fixed after 12 h of culture and stained with hematoxylin. A single spindle-shaped cell (arrow) is attached to one side of the fiber. All other nuclei in the fiber are myonuclei. X325

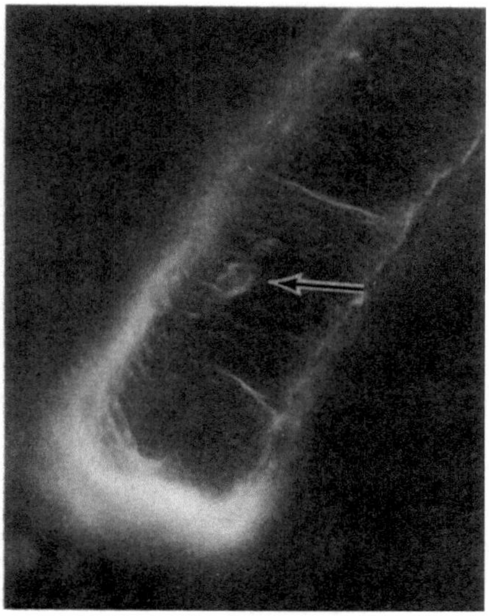

Fig. 2 Fluorescent image of fiber stained with FITC-lentil lectin immediately after being placed in culture. A small spindle-shaped cell (arrow) is visible on the surface of the fiber because of increased lectin binding. Note the very strong fluorescence at the end of the fiber. The staining can be blocked with α-methyl mannoside. X590

staining of the satellite cells, the lectins bind strongly to the ends of the fibers (Fig. 2). This may reflect the folding of the sarcolemma at the myotendinous junction (19) or indicate an increased concentration of glyco-substances involved in mediating attachment of the sarcolemma to the tendon collagen.

Motor end plates can be readily identified in the living fibers by staining with FITC-α-bungarotoxin (Fig. 3). In addition, a moderate amount of acetylcholinesterase (AChE) activity is retained at the myoneural junction (Fig. 4).

Response of the fibers to explantation

There are no gross degenerative processes during subsequent culture of

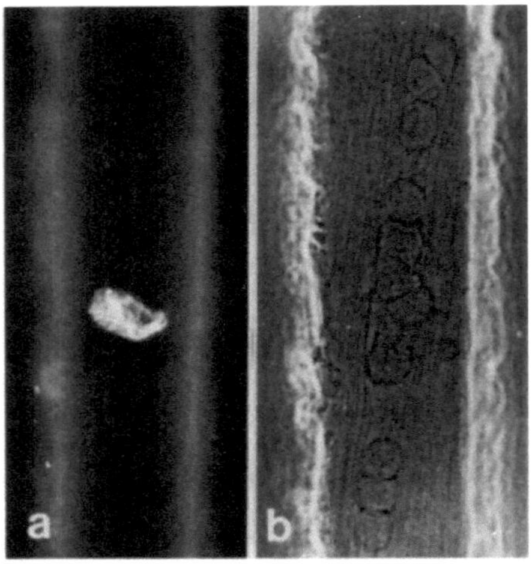

Fig. 3 (a) fluorescent and (b) phase contrast image of living fiber stained at 0-time with FITC-α-bungarotoxin. The acetylcholine receptors of the postsynaptic membrane are located near the middle of the fiber over a cluster of myonuclei. All fibers have a single synapse. X750

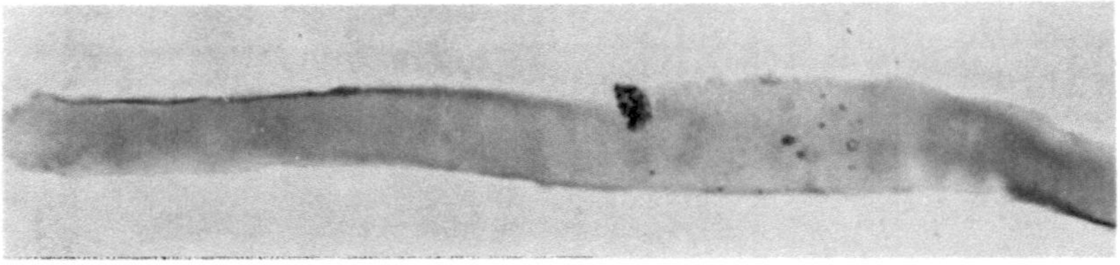

Fig. 4 Fiber stained for cholinesterase (30) during the first day in culture. X300

single intact fibers such as loss of cytoplasm or pyknosis of nuclei. Both the myonuclei and satellite cells survive. The most obvious change in the fibers during the first few days in vitro is the appearance of a proliferating population of mononucleated cells (Fig. 5). These cells remain adherent to the fiber and are probably situated beneath its basement lamina. In earlier studies I showed that the basement lamina acts as a physical restraint to prevent satellite cell progeny from leaving degenerating fiber segments and migrating into the fibrin clot (14,15). In addition to holding the fibers in place, the clot acts as a matrix into which any endomysial connective tissue cells that might be retained on the external surface of the basement lamina can migrate and thus be detected.

After about a week in culture new myotubes appear in association with the original fiber. In living preparations the myotubes usually become evident as outgrowths from the ends of the fibers. In fixed and stained cultures some myotubes can be observed to parallel the shaft of the mature fiber (Fig. 6). The myotubes elongate rapidly and often contract spontaneously.

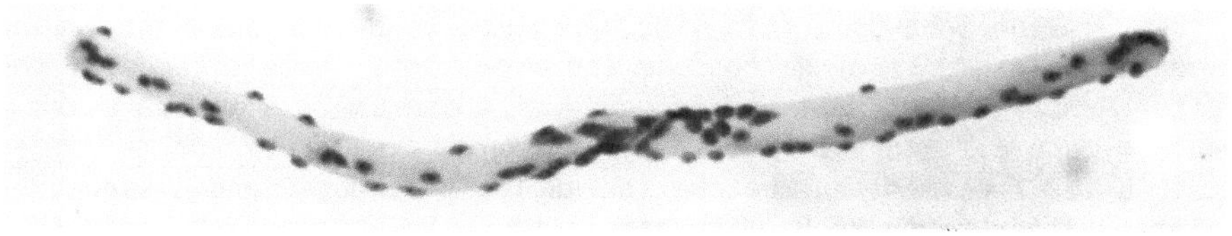

Fig. 5 Fiber fixed and stained after 2d in culture showing a small group of mononucleated myoblasts with basophilic cytoplasm near the center of the fiber. Some of these cells are in mitosis. The myonuclei continue to appear normal and show no evidence of degeneration. X200

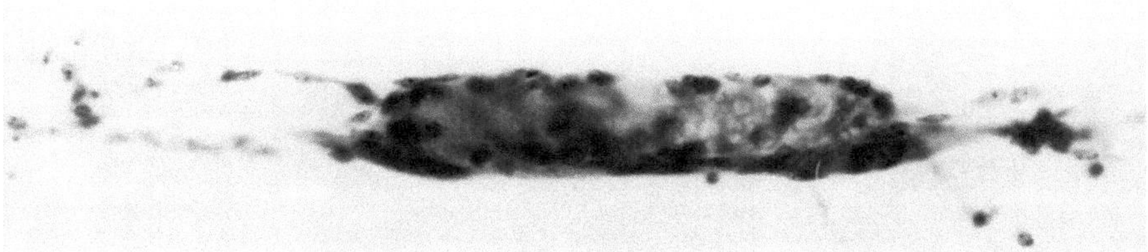

Fig. 6 Fiber fixed after 5d in culture showing a multinucleated myotube with basophilic cytoplasm running along the lower edge of the fiber. The mature fiber has become vacuolated but the myonuclei are normal. This fiber was contracting before fixation. X280

The regeneration of new myotubes is surprising in view of the apparently normal viability and maintenance of the mature fiber. To study these regenerative processes in more detail and to characterize separately the behavior of the satellite cells and the myofiber, techniques were developed that would allow each cell type to be analyzed separately.

Behavior of satellite cells alone

Muscle regeneration can be induced experimentally with the myotoxic anesthetic Marcaine. Local injection of Marcaine into muscle (20) or exposure of cultured muscle to the drug (21) results in rapid degeneration of muscle fibers while sparing satellite cells and mononucleated myoblasts. These cells subsequently initiate regeneration. A range of Marcaine concentrations from 1-10 mM was tested with freshly isolated single fibers _in vitro_ by adding the drug to the culture medium for from 15 min to 3 h. Exposure to 7.5 mM Marcaine for 15 min was found to result in degeneration of virtually all fibers without killing satellite cells or reducing their myogenic activity. Within several minutes after adding Marcaine the fibers develop surface blebs and almost immediately thereafter undergo hypercontraction to about a quarter of their original length. This contraction is not altered by applying the Marcaine in Ca^{++}-free medium with 0.5 mM EGTA. After contraction, the space occupied by the fiber is outlined by a thin line, presumed to be the basement lamina, and contains fragments of sarcoplasm and myonuclei. Satellite cells round up after Marcaine treatment and, because of their small size and scant cytoplasm, are difficult to identify. The myonuclei appear normal immediately after treatment but become increasingly pyknotic and disappear completely after about 24 h. The satellite cells enlarge and begin to proliferate during the first day. Degenerating fibers examined after that time contain a small number of satellite cell progeny. Sequential photos of the same fiber at various times after Marcaine treatment are shown in Fig. 7. The myoblasts later fuse and form one or two new multinucleated myotubes in the space originally occupied by the mature fiber (Fig. 8).

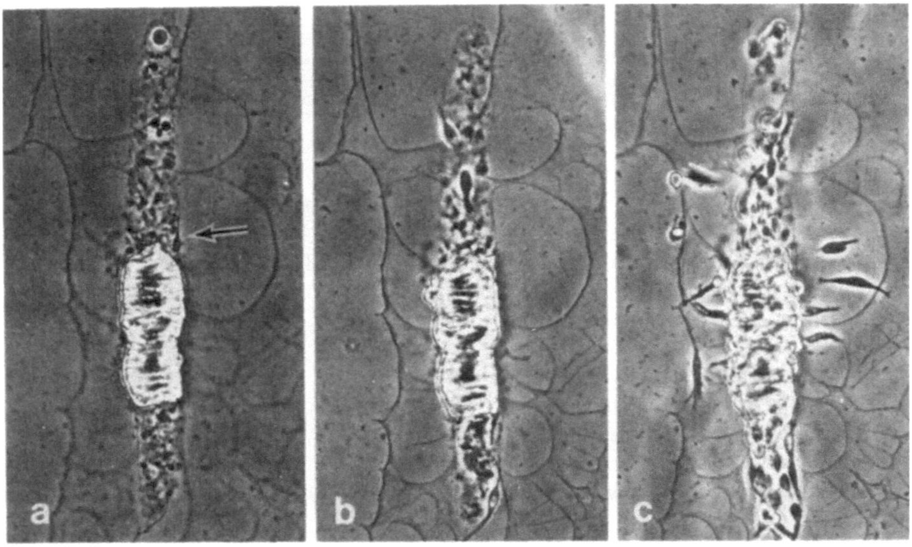

Fig. 7 Phase contrast micrograph of the same fiber at various times in culture. The fiber was exposed to 10 mM Marcaine for 15 min after 12 h in vitro. (a) Immediately after exposure. A single myoblast (arrow) is visible at the end of the contracted fiber. (b) After 2d in vitro several myoblasts are present. (c) At 3.5d a large population of myoblasts surrounds the fiber and is beginning to migrate into the surrounding fibrin clot. X280

Fig. 8 A fiber killed with 7.5 mM Marcaine immediately after explantation and fixed after 6d in culture. Several myotubes have regenerated near the original fiber. The degenerating fiber (asterisk) lacks myonuclei or other identifiable structures. X190

In other experiments isolated fibers were killed by cutting with a razor blade fragment held in a micromanipulator. The fibers immediately contract when cut and undergo essentially the same sequence of events as do the Marcaine-treated fibers. In the cut fibers, however, the basement lamina appears to remain attached to the plasmalemma during contraction.

These studies designed to kill myonuclei show that the mature fibers, as isolated, contain satellite cells that survive injury, proliferate and form new myotubes. What is the mitogenic signal for the satellite cells and how rapidly do they proliferate? Incorporation of ^{3}H-thymidine (TdR) by satellite cells associated with intact single fibers was used to measure proliferation kinetics. Radioautographs of cultures exposed to ^{3}H-TdR for 15 min at various times during the first 3 days in vitro show that the number of labeled cells per fiber increases logarithmically with a doubling time of 13 h. If the duration of the S phase is constant, 13 h is probably a good estimate of the generation time of the satellite cell progeny. Silver grains were found only over the nuclei of mononucleated cells; myonuclei never contained label.

Satellite cells in these preparations begin to proliferate despite the continued viability of the associated mature fiber. Because of this, it is clear that the signal for satellite cell proliferation involves more than just the degeneration of the myofiber. Previous studies in vivo have also shown that slight compression (7) or denervation (11,22) can activate satellite cells without gross damage to the myofiber. In order to investigate the nature of the mitogenic signal using cultured fibers it will be necessary to obtain myofiber-satellite cell units in which the satellite cells remain quiescent. As a first approach, fibers were cultured in medium without EE since EE contains substances that stimulate proliferation of myogenic cells (24). Fibers cultured for 2 d in medium without EE contained only 25% the number of cells labeled with ^{3}H-TdR when compared with fibers from the same muscle cultured with EE. We are currently attempting to reduce the satellite cell response still further by innervating the fibers in culture. When satellite cells remain inactive in culture, we will be able to study the nature of endogenous mitogenic signals after various types of trauma to the myofiber.

Behavior of the myofiber alone

Although it is clear from a variety of experiments that preexisting satellite cells play a major role in muscle repair and regeneration following injury, the myofiber has been implicated as a possible source of new satellite cells (2-7). Also, sarcoplasmic buds, or immature, myotube-like structures have been reported to sprout from, but remain connected with, the mature fiber after injury (25). To examine the response of the myofiber alone to explantation, the satellite cells were selectively killed or removed from the fibers before culture. Cytosine arabinoside (ara C) is a purine analog that inhibits DNA synthesis and eventually leads to the killing of cells in S (26). Myofibers exposed to 10^{-5}M ara C for 2-3 days are completely free of mononucleated cells (Fig. 9). When continued in vitro the fibers show a slow progressive disorganization of myofibrils as reflected initially by loss of sarcomere alignment and later by vacuolation of the sarcoplasm. After about a week in culture small pseudopodial outgrowths appear at both ends of the fiber (Fig. 10). The sprouts elongate rapidly, fill with nuclei that migrate from the mature fiber, and come to resemble embryonic myotubes (Fig. 11). The mature fiber itself diminishes in size and eventually becomes undetectable. To what extent the morphological dedifferentiation involves reorganization of existing components or new protein synthesis is not known. Radioautographs of fibers after exposure to ^{3}H-leucine show that both the mature fiber and the outgrowths are synthesizing proteins.

In addition to these structural changes, there is a reorganization of cell surface properties in cultured fibers. Extrajunctional acetylcholine receptors appear (17,18) and acetylcholinesterase activity becomes uniformly distributed along the fiber (Fig. 12).

The behavior of myofibers alone was also studied by enzymatically removing satellite cells from the fibers before explantation. Fibers prepared with the standard method using collagenase were then exposed to 0.01% trypsin for 15 min to digest the basement lamina and liberate satellite cells (13). The fibers were purified from the suspended satellite cells and cultured as before. When cultured in normal medium, these fibers did not bear, nor subsequently give rise to any mononucleated cells. The dedifferentiation of the fibers was similar to that described for the ara C-treated fibers.

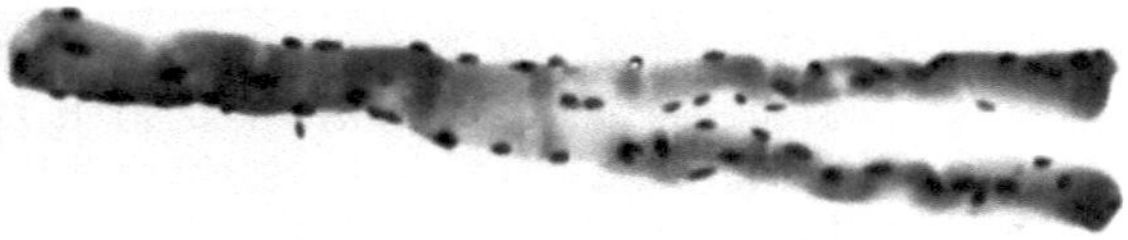

Fig. 9 Branched fiber from a culture exposed to 10^{-5} M ara C for the first 3 days and fixed after 4d. There are no mononucleated cells visible although the myonuclei appear normal. X160

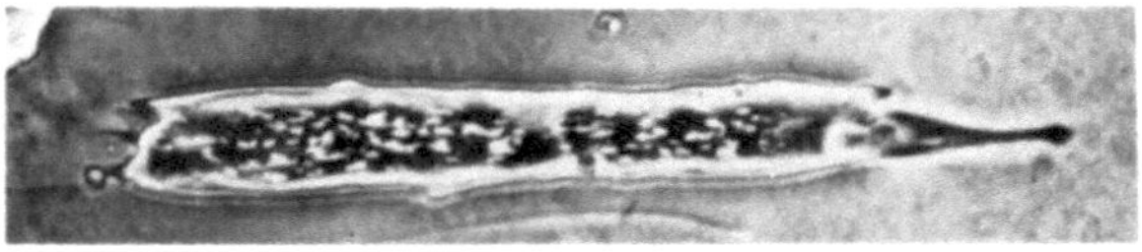

Fig. 10 Phase contrast photo of living fiber at 6d in vitro after killing satellite cells with ara C. Note the vacuolation of the fiber and small outgrowths at either end. The fiber was contracting. X220

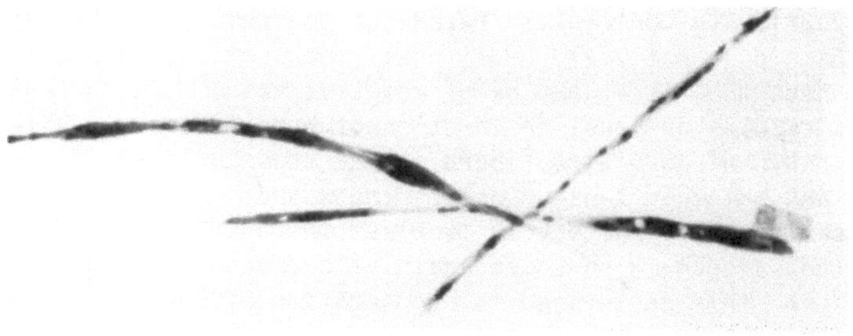

Fig. 11 Several fibers fixed after 9d in vitro after killing satellite cells with ara C. The fibers now resemble immature myotubes. X120

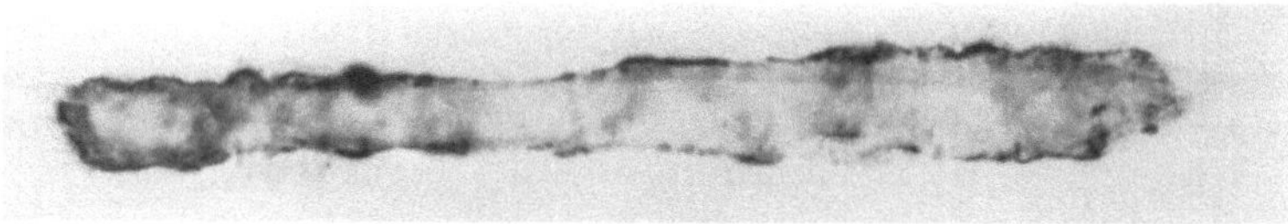

Fig. 12 Fiber depleted of satellite cells with ara C and stained for cholinesterase (30) after 7d in culture. Enzyme activity is distributed over the entire surface of the fiber but is absent from the pseudopodial sprouts at either end. Compare with Fig. 4. X250

Discussion

Several conclusions can be drawn from these preliminary studies of myofiber-satellite cell complexes. First, the in vitro behavior of isolated satellite cells compared with isolated myofibers shows that satellite cells are myogenic stem cells and are the only cell type within the basement lamina capable of giving rise to new myoblasts. Killing myonuclei by exposure to Marcaine or by cutting the fiber does not prevent the appearance of a new population of myogenic cells from surviving satellite cells. Conversely, if the satellite cells are killed with ara C or enzymatically removed from the fibers, no new mononucleated cells appear. Fibers without satellite cells do not form myoblasts.

Second, satellite cells are stimulated to enter the cell cycle as a result of explantation of the fibers. The cells begin to synthesize DNA during the first day in vitro and increase exponentially for at least several days with a doubling time of 13 h. The nature of the stimulus for division has not been determined but it may involve denervation or mitogenic substances in the medium.

Finally, fully mature myofibers are capable of morphological dedifferentiation after explantation. The extent to which this process occurs in vivo following trauma is not known. There have been several reports, however, of nucleated sprouts arising from surviving mature fibers following injury (25,27,28). In contrast to regeneration from satellite cells, this type of repair does not yield additional fibers or increase the number of myonuclei; nevertheless it may be of adaptive value. In the case of an injury that results in local destruction of both fibers and satellite cells, outgrowth of surviving fibers at the edge could rapidly restore continuity and maintain tension across the wound. Also, since mature fibers appear to be refractory to myoblast fusion (29), dedifferentiation may alter the sarcolemma so as to again allow fusion. If new myoblasts were available, this could lead to augmentation of nuclear number in the surviving fibers.

Acknowledgements

I would like to thank Ms. Debbie Grossman for expert technical assistance. This work was supported by the Muscular Dystrophy Association of America.

References

1. Carlson, B.M.: Am. J. Anat. 137, 119 (1973).
2. Hay, E.D.: Dev. Biol. 1, 555 (1959).
3. Hess, A. and Rosner, S.: Am. J. Anat. 129, 21 (1970).
4. Lee, J.C.: Exp. Neurol. 12, 123 (1965).
5. Reznik, M.: Lab. Invest. 20, 353 (1969).
6. Reznik, M.: Differentiation 7, 65 (1976).
7. Teräväinen, H.: Z. Zellforsch. 103, 320 (1970).
8. Mauro, A.: J. Biophys. Biochem. Cytol. 9, 493 (1961).
9. Hsu, L., Trupin, G.L. and Roisen, F.J.: In: Muscle Regeneration. A. Mauro, editor. p 115. Raven Press, N.Y. 1979.
10. Moss, F.P. and LeBlond, C.P.: Anat. Rec. 170, 421 (1971).
11. Ontell, M.: Cell Tissue Res. 160, 345 (1975).
12. Snow, M.H.: Anat. Rec. 188, 201 (1977).
13. Bischoff, R.: Anat. Rec. 180, 645 (1974).
14. Bischoff, R.: Anat. Rec. 182, 215 (1975).
15. Bischoff, R.: In: Muscle Regeneration. A. Mauro, editor. p 13. Raven Press, N.Y. 1979.
16. Konigsberg, U.R., Lipton, B.H. and Konigsberg, I.R.: Dev. Biol. 45, 260 (1975).
17. Bekoff, A. and Betz, W.J.: J. Physiol. 271, 25 (1977).
18. Bekoff, A. and Betz, W.J.: J. Physiol. 271, 537 (1977).
19. Maruyama, K. and Shimada, Y.: Tissue and Cell 10, 741 (1978).
20. Hall-Craggs, E.C.B.: Exp. Neurol. 43, 349 (1974).
21. Schultz, E. and Lipton, B.H.: Anat. Rec. 191, 351 (1978).
22. Ontell, M.: Anat. Rec. 178, 211 (1974).
23. Schultz, E.: Anat. Rec. 190, 299 (1978).
24. Slater, C.R.: Dev. Biol. 50, 264 (1976).
25. Shafiq, S.A.: In: Regeneration of Striated Muscle and Myogenesis.

A. Mauro, editor. p 127. Excerpta Medica, Amsterdam. 1970.
26. Kihlman, B.A.: Action of Chemicals on Dividing Cells. Prentice-Hall, N.J. 1966.
27. Clark, W.E. Le Gros: J. Anat. 80, 24 (1946).
28. Sloper, J.C. and Pegrum, G.D.: J. Path. Bact. 93, 47 (1967).
29. Bischoff, R. and Holtzer, H.: J. Cell Biol. 41, 188 (1969).
30. Karnovsky, M.J. and Roots, L.: J. Histochem. 12, 219 (1964).

Section II. Development and Growth

DIFFERENCES IN DIFFERENTIATION PROGRAMS BETWEEN PRESUMPTIVE MYOBLASTS AND THEIR DAUGHTERS, THE DEFINITIVE MYOBLASTS AND MYOTUBES

H. Holtzer, J. Croop, Y. Toyama, G. Bennett, S. Fellini, C. West
Department of Anatomy, Medical School, University of Pennsylvania, PA.

Introduction

Most papers presented at this Symposium focus on altering the rates of synthesis of muscle-specific molecules in relatively mature, postmitotic myotubes *in vitro*, or innervated, or cross-innervated postmitotic mature muscles *in vivo*. Such studies reveal the "quantitative physiological variations" permissible within the confines of a given, definitive muscle's homeostatic memchanisms. In contrast, our contribution focuses on instances of "qualitative epigenetic variations" in the kinds of molecules synthesized by cells in different compartments of the myogenic lineage. Shifts in the coordinated sets of molecules synthesized in different types of cells in the myogenic lineage occur prior to, and in the total absence of, neural input. Our experiments center on the epigenetic mechanisms that establish the limits of the various physiological options a given kind of myogenic cell inherits from its mother cell. These qualitative changes, some irreversible, some reversible, occur between generations of precursor cells, and within the life-span of one cell in the myogenic lineage.

A. Background Experiments Relating to Early Myogenesis

Though my colleagues and I first demonstrated fusion of myoblasts over 20 years ago (1,2), we always maintained that fusion is only one of a coordinated cluster of options available to the postmitotic, mononucleated myoblast. Fusion does not trigger, it is but one of the consequences of, the differentiation program of the myoblast. Withdrawal from the cell cycle is the event that coordinately primes the cell surface for fusion (3,4),

Plasticity of Muscle

as well as initiating the synthesis of the muscle-specific myosin heavy and light chains (5-8), tropomyosins (3,4) and probably a-actin (9-11) and the troponins (12). Withdrawal from the cell cycle is not an option of normal replicating presumptive myoblasts, but an endogenous option unique to both its daughters, the postmitotic, mononucleated myoblasts.

The following experiments led to these conclusions: (1) When the entire 3-day embryo is stained with fluorescein-labelled antibodies to skeletal myosin, only the postmitotic, mononucleated myoblasts in the developing myotomes bound the skeletal-specific antimyosin. The replicating presumptive myoblasts in the somites, very early limb bud cells (see below), lateral plate or unsegmented trunk mesoderm, did not bind the antimyosin. Neither in vivo nor in vitro did these postmitotic, mononucleated myoblasts incorporate ^{3}H-TdR into their DNA (5,13); (2) We observed under the phase microscope two living postmitotic myoblasts, growing at low density, extend pseudopodial processes, adhere to one another, and fuse. The event of fusion required between 30 min and 1 hour and was monitored by intracellular particles moving from one cell to the next (1); (3) Replicating presumptive myoblasts were allowed to incorporate ^{3}H-TdR and mixed with unlabelled replicating presumptive myoblasts. The myotubes that formed contained a random mixture of labelled and unlabelled nuclei. Under no conditions in vivo or in vitro did normal or viral-infected nuclei within a myotube duplicate their DNA and undergo mitosis (3,4,14,15). Other experiments (16) demonstrated that only myoblasts in G1 had a cell surface that permitted fusion.

With respect to the synthesis of muscle-unique molecules, three, far-reaching conclusions were deduced from these experiments: (1) The differentiation program of the replicating presumptive myoblasts were qualitatively more related to fibroblasts or presumptive chondroblasts, than to their daughters, the postmitotic myoblasts; (2) Fusion, though obviously a challenging problem involving cell-cell recognition, aggregation, and membrane-membrane interactions unique to postmitotic myoblasts (3,4,6), nevertheless does not trigger either the transcription or translation of the unique mRNAs for synthesizing the muscle-specific contractile proteins; and (3) One of the key events in myogenesis is understanding how, when a presumptive myoblast replicates its DNA and histones and assembles its chromosomes during S and G2, that then and there that "mother" cell transmits to its future daughters via a "quantal cell cycle" the option to (a) withdraw from cycling, (b) transcribe those structural genes that are only available for transcription in the definitive, mononucleated myoblast, and (c) assemble a cell surface primed with molecules permitting fusion.

Each of the above conclusions have been both disputed and have been confirmed and extended by other investigators.

Replicating Presumptive Myoblasts, Postmitotic Myoblasts, and the Minor Role of Fusion

Many investigators have argued; (1) that fusion only occurs when cells become confluent; (2) that only after fusion are the muscle-specific proteins synthesized; and (3) that pending fusion any myogenic cell can re-enter the cell cycle (19-30).

More recently we have designed another series of experiments demonstrating that density is irrelevant to fusion and that only cells already postmitotic have the option to fuse. Appropriate doses of Cytochalsin-B permit separating postmitotic myoblasts from myotubes, presumptive myoblasts and fibroblasts (31-34). These postmitotic myoblasts were plated at densities that ranged from 2×10^3 to 2×10^4 cells per 35mm petri dish. Fifty to sixty percent of these cells fused within 24 hours. The addition of conditioned medium (28) neither increased the rate nor numbers of cells that fused. The cells that did not fuse at these low densities did not re-enter the cell cycle as judged by the incorporation of ^{3}H-TdR. Over 95% of the cells that did not fuse bound antibodies to skeletal-specific myosin heavy chains and to LMM. Similar results have been reported using postmitotic myoblasts blocked from fusing by EGTA (35). That myogenic cells in complete growth medium pull out of the cell cycle as a precondition to fusing, and initiate the synthesis of muscle-specific proteins, is a conclusion recently confirmed and extended by many investigators (36-40).

If myogenic cells fuse only in G1 (16), does this mean that <u>all</u> daughters of all presumptive myoblasts that had acquired the option to fuse were obligated to withdraw from the cell cycle, or if failing to find a cell to fuse with, would transitorily loose their capacity to fuse and recycle. In brief, we asked (6) whether the option to fuse might be a decision each presumptive myoblast made sometime in G1 and that the capacity to fuse "waxes and wanes" with phases of the cell cycle? This view would introduce a stoachastic element, but more importantly support the notion that irreversible withdrawal from the cell cycle is a consequence of, not a precondition to, fusion, and that until cells actually fused they could be induced to re-enter the cell cycle.

Konigsberg and co-workers (22,23) using a conditioned medium depleted in glucose and one that depressed cycling claimed that their pre-labelled ^{3}H-TdR cells fused within 24 hours. From this we concluded that postmitotic cells fused in G1 in less than 24 hours (16). Nadal-Ginard (41) using the abnormal L_6E_9 line of rat cells concluded that (1) after every cell division L_6E_9 cells have the option to fuse or to proliferate, but that (2) "myoblasts withdraw from the cell cycle prior to and not as a consequence of fusion." These experiments pre-suppose that the L_6E_9 are all "undifferentiated" and all equivalent. More likely they consist of a mixture of presumptive myoblasts, myoblasts, and what we termed "presumptive myoblasts - fibroblasts" (18,42). The different cloning conditions used with these abnormal cells could have selected for one of the above cell types. The likelihood for this is suggested by a series of experiments using normal rat primary myogenic cells. Bronk (43) using normal rat primary cultures concurs with the second observation above, but concluded that cells that have the option to replicate do not have the option to fuse or to synthesize skeletal-specific proteins. We suggest that the abnormality of the L_6E_9 line is due to a lesion in this particular decision-making control.

Kligman and Nameroff (44) isolated single myogenic cells and allowed them to cycle once. The daughters were stained with antibody against the MM isozyme of CPK. They found either both daughters were M-CPK$^+$ or M-CPK$^-$. No mixed pairs were observed, and conditioned medium did not affect the results. M-CPK$^-$ pairs both incorporated ^{3}H-TdR, the M-CPK$^+$ pairs did not incorporate ^{3}H-TdR. They concluded that the myogenic lineage ends in a symmetrical division, and that the decision to replicate or become postmitotic is endogenously programmed. Croop and Holtzer (45) have confirmed their observations using antibody against skeletal myosin. These results would be consistent with the myogenic lineage outlined in Holtzer (18,42).

Options of Early Limb Bud Myoblasts

Our claim (3-6,31-34,43) that postmitotic, mononucleated myoblasts synthesize skeletal-specific myosin monomers and organize them into thick filaments and that such myoblasts assemble normal-sized sarcomeres of interdigitating thick and thin filaments has been confirmed by Emerson and coworkers (29,30). In spite of these results, Emerson (29,30) and Koingsberg (22-25) maintain that myogenesis in somites differs from myogenesis in leg or breast muscles. They contend only in the former is there a stable population of postmitotic, mononucleated myoblasts, whereas such myoblast do not emerge in normal leg or breast muscle. They claim that postmitotic myoblasts can be induced in non-somitic muscles only by "drugs". Yaffe and co-workers (19-21) and Gros and co-workers (26,27) have also claimed that the mRNAs for the definitive skeletal-specific proteins though transcribed in presumptive myoblasts are not translated until after fusion into myotubes.

To re-investigate these two issues, we cultured stage 23-25 limb buds. The trypsinized limb bud cells were plated at 5 X 10^5 cells/ per dish. As illustrated in Figs. 1-6 sizeable numbers of postmitotic myoblasts were distributed throughout these cultures by day 3. In contrast to presumptive myoblasts or fibroblasts these mononucleated, postmitotic myoblasts synthesize the muscle-specific; (1) fast myosin heavy chains, (2) LMM, and (3) tropomyosins (5-8,46). The number of mononucleated myoblasts increases between 24 hours and 5 days. In the latter cultures there were areas with myotubes indistinguishable from cultures established with 10-day limb or breast presumptive myoblasts. However the majority remained as mono-, bi- or oligonucleated myotubes. Konigsberg and co-workers (22-25) have claimed they can distinguish fibroblasts from replicating myoblasts or myoblasts;

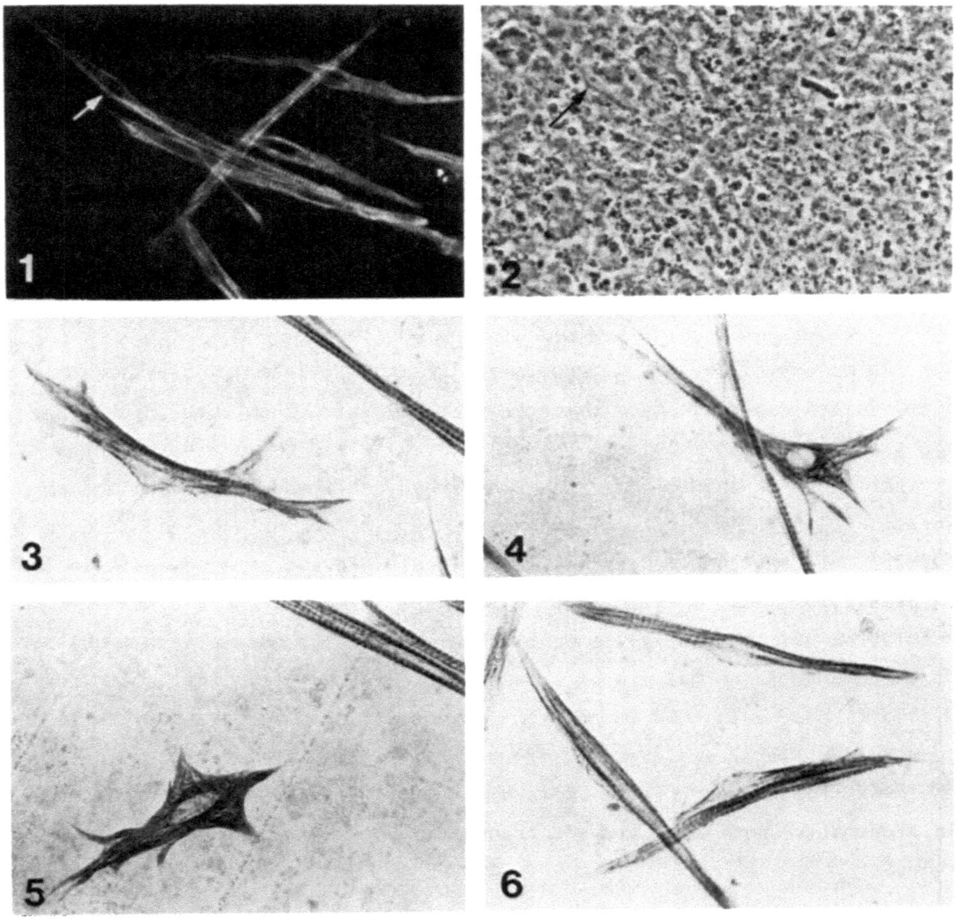

Stage 23-24 dissociated limb buds reared in culture and stained with antibodies to skeletal-specific LMM. Figs. 1 and 2 of the same field viewed under the fluorescent and phase microscopes, respectively. Note total absence of staining of presumptive myoblasts and fibroblasts. Figs.3-6 visualized using the immunoperoxidase technique. Again note that presumptive myoblasts and fibroblasts that are present in the microscopic field do not bind the anti-LMM. Note particulary the "fibroblastic" morphology of the myoblasts in figs. 3,4, and 5.

that the latter two are invariable bipolar. As shown in Figs. 3,4 and 5 such criteria would have lead to classifying erroneously many postmitotic myoblasts as fibroblasts. In populations of rapidly dividing presumptive myoblasts, chondroblasts, and fibroblasts from early somites, or early limb buds, some daughters of the presumptive myoblasts withdraw from the cell cycle, and as mononucleated cells coordinately initiate (47) the synthesis and assemble muscle-specific contractile proteins into definitive sarcomeres. These findings confirm the cloning experiments of Dienstman, et al. (18,42,48,49) that early limb bud cell's are not "multipotential", but parcelled into myogenic-fibrogenic, or chondrogenic-fibrogenic lineages (50,51).

Changes in Kind and Distribution of the 100 Å Filaments in Myogenic Cells

Presumptive myoblasts, myoblasts and myotubes display large numbers of 100 A filaments (4,6,34,52,53). These intermediate-sized filaments (IFs) tend to form side-by-side aggregations in colcimide or vinblastine (6,32,33,52-55). If presumptive myoblasts or fibroblasts are exposed to Cytochalsin-B and colcimide 30% of the cell's total volume can be occupied by these immense cables of 100 A filaments. The IFs in presumptive myoblasts, fibroblasts, chondroblasts, and pigment cells are composed of a protein with a MW of 58K (54,55,56). Franke, et al. (57,58) have termed this protein "vimentin". In fibroblasts and presumptive myoblasts the quantity of the 58K protein varies up to 10% of the cell's total protein; in cells grown at high density the 58K/actin ratio approximates 2.0. Following a quantal cell cycle, the postmitotic daughters when no older than 10-15 hours, initiate the synthesis of another IF protein with a MW of 55K (55, 56,59,60). In immature myotubes the IFs course parallel to the striated myofibrils. Whether each IF consists of either monomers of 58K, or 55K, or whether they are hybids of the two, has not been determined. As the myotube matures between days 3 and 6 and the complement of striated myofibrils increases, the numbers of IFs diminish. Associated with the loss of recognizable IFs, the 58K protein is no longer detectable using antibodies (60). Paradoxically in 7-9 day myotubes the 55K protein formerly associated with the longitudinally-oriented IFs, now is exclusively associated with the I-Z band regions (60). Owing to the limitations of the fluorescent microscope we are not certain whether the distribution of our 55K protein corresponds to that described by Lazardies and co-workers for "desmin" (61,62).

Effects of Tumorigenic Agents on Myogenesis

We have been examining the inverse relationship between the expression of the differentiation program of myogenic cells and tumorigenic agents. To this end we have analyzed the effects of a temperature-sensitive mutant of the Rous sarcoma virus (ts-RSV) and of the tumor promoter, phorobol-12-myristate-13 acetate on cells in the penultimate and ultimate compartments of the myogenic lineage (15,18,63,64).

Postmitotic myoblasts or myotubes cannot be transformed by ts-RSV. Whether this is due to failure of the virus to enter the postmitotic cell, to form a provirus, or to be integrated as viral DNA into the host's nucleus, are yet unknown. Replicating presumptive myoblasts, however, are readily transformed with ts-RSV (15,18,65,66). At permissive temperature (37°) infected presumptive myoblasts replicate, do not initiate the synthesis of muscle-specific proteins and do not fuse. Biochemically, they cannot be distinguished as yet from ts-transformed chondroblasts, fibroblasts, or retinal pigment cells (67-69). If transformed myogenic cells are shifted to non-permissive temperature (41°) within 24 hours many of the infected cells irreversibly withdraw from the cell cycle, initiate the synthesis of muscle-specific proteins, fuse and contract spontaneously (15,18). Presumably in the absence of active pp60 SRC protein kinase (70,71), the presumptive myoblasts are blocked from undergoing the quantal cell cycle that would generate postmitotic daughter myoblasts. Of considerable theoretical interest is that for many generations these transformed cells transmit their status as covert myogenic, or chondrogenic, or pigment cells, respectively. The virus does not render these cells "more embryonic" or in any detectable manner alter their limited options for differentiation. These transformed cells are not "undifferentiated" or "multipotential" anymore than the L_6E_9, or Friend erythroleukemic cells or embryonal carcinoma cells are "undifferentiated" or "multipotential" (17, 18,42).

When viral-infected myotubes formed at non-permissive temperature are dropped to permissive temperature 100% of the postmitotic myoblasts and/or myotubes vacuolate and lyse. The presumptive myoblasts or fibroblasts in the same culture continue to replicate. When viral-infected chondroblasts or pigment cells are dropped to permissive temperature though they cease to synthesize their "luxury" molecules, they nevertheless continue to replicate. The death of the infected postmitotic muscle at permissive temperature may be due to conflicting signals. The integrated virus might signal for continuous DNA synthesis and cell replication - the sine qua non of a tumor cell - but the differentiation program of the myotube calls for inhibition of DNA synthesis. These conflicting demands on the synthetic machinery of the myotube, we propose, leads to cell death (15,18).

Further evidence for an inverse linkage between differentiation of myogenic cells and tumorigenic agents derives from experiments using the tumor promoter, (PMA). Added to normal myogenic cultures PMA promptly blocks fusion (63). The cells continue to replicate but their morphology changes - the cells display long fine processes and multilayer. If transferred to normal medium within hours many of the cells re-acquire their normal morphology and fuse to form normal myotubes. The tumor promoter may act directly on the cell surface, for if added to a population of pure, postmitotic myoblasts it immediately blocks fusion. These experiments, in conjunction with the effect of PMA on the surface of chondroblasts (67,68, 72,73) and pigment cells (74) suggest that the tumor promoter may act on the events of aggregation leading to, or on the event of, fusion itself (3,4,6,18,75-77).

PMA has an exceedingly striking, differential effect on postmitotic myotubes (78). Day 5 myotubes are rich in striated myofibrils and elements of the SR and T-system. The myotubes contract spontaneously. They display sizeable numbers of longitudinally-oriented IFs that bind antibody to the 58K and 55K proteins. When PMA is added to such cultures the spontaneous contractions cease. Within the next 24-48 hours the existing striated myofibrils are gradually dismantled. By 72 hours virtually all

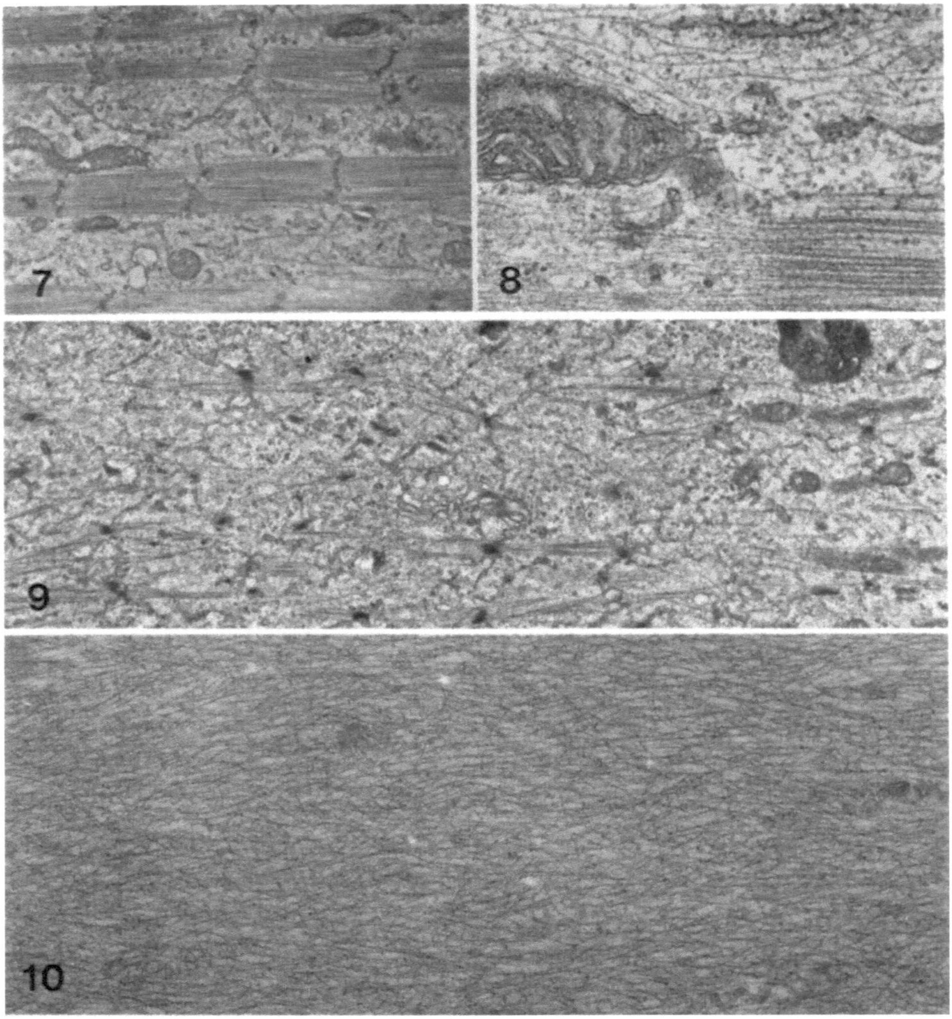

Fig. 7: Low power micrograph of a day 5 cultured myotube (X 11,000). Note the normal, full size, tandemly arranged sarcomeres of interdigitating thick and thin filaments. Fig. 8: High power micrograph illustrating the typical density and longitudinally-oriented 10nm filaments between the striated myofibrils (X 42,000). Fig.9: A day 7 myotube that had been treated with PMA for the previous 48 hours. Note the dismantling of the striated myofibrils. Scattered clusters of interdigitating thick and thin filaments are still present. Note the atypical Z-bands (X 17,000). Fig. 10: A day 8 myotube that had been treated with PMA for the previous 72 hours. The striated myofibrils that had been present in day 5 myotubes have been totally dismantled. Note the extraordinary density of the 10nm filaments (X34,000).

thick and thin filaments have disappeared. The interior of the myotube is occupied by a skein of randomly organized IFs, in which are suspended large numbers of normal mitochondria, SR, and lysosomes. The microfilaments subtending the plasmalemma - consisting presumably of b- and g-actins and other "constitutive" contractile proteins - are not affected by PMA. Figures 7-10 illustrate this sequence of changes. Most remarkable about this response to PMA is the rapidity of its reversal. Twenty-four hours after removing the PMA these myotubes display their typical complement of striated myofibrils, the number of IFs have dwindled, and eventually the 55K protein is confined to the I-Z regions.

PMA not only dismantales the formed striated myofibrils, it inhibits the synthesis and enchances the rate of degradation of many of the muscle-specific contractile proteins. After 24 hours in PMA the synthesis and/or accumulation of a-actin, the two fast myosin light chains and the a- and b-tropomyosins are between 40 and 50% of controls. The b- and g-actins are inhibited less than 15% whereas the synthesis of the IFs 58K protein is stimulated 40%. Much remains to be done regarding rates of synthesis and rates of degradation of the muscle muscle-specific "luxury" molecules vs the "household" molecules in this system. Nevertheless it is already clear that PMA differentially depolymerizes formed myofibrils and blocks the synthesis and enhances the rate of degradation of many proteins unique to the differentiation program of the definitive muscle cell. In contrast, it has little obvious effects o general protein synthesis. This discrimination between the a-actins and the b- and g-actins, as well as the "constitutive" vs the "muscle-specific" fast myosin heavy and light chains is particularly notable.

References

1. Holtzer, H., Abbott, J., and Lash, J.: Anat. Rec. 131, 567-568 (1957).
2. Lash, J., Holtzer, H., Swift, H.: Anat. Rec. 128, 679-693 (1957).
3. Okazaki, K. and Holtzer, H.: J. Histo. & Cytochem. 13, 726-739 (1966).
4. Holtzer, H. and Sanger, J.: Research in Muscle Development and the Muscle Spindle (eds. B. Banker and R. Przybylski) Exerpta Medica (1972).

5. Holtzer, H., Marshall, J. and Finck, H.: J. Cell Biol. 3, 705-723 (1957).

6. Holtzer, H., Sanger, J., Ishikawa, H., and Strahs, K.: Cold Spring Harbor Quant. Biol. 37, 549-565 (1972).

7. Rubinstein, N. and Holtzer, H.: Nature, 280, 323-325 (1979).

8. Chi, J., Fellini, S. and Holtzer, H.: PNAS, 72, 4999-5003 (1975).

9. Elinga, M., Maron, B. and Adelstein, R.: Science, 191, 94-95 (1976).

10. Garrels, J. and Gibson, W.: Cell, 9, 793-805 (1976).

11. Vanderkerckhove, J., Weber, K.: J. Mol. Biol. 126, 783-802 (1978).

12. Dhoot, G., Perry, S.: Nature, 278, 714-718 (1979).

13. Okazaki, K. and Holtzer, H.: PNAS, 56, 1484-1499 (1965)

14. Stockdale, F. and Holtzer, H.: Exp. Cell Res. 24, 508-520 (1961).

15. Holtzer, H., Biehl, J., Yeoh, G., Meganathan, R. and Kaji, A.: PNAS, 72, 4051-4055 (1975).

16. Bischoff, R. and Holtzer, H.: J. Cell Biol. 36, 111-127 (1967).

17. Holtzer, H., Weintraub, H. and Mayne, R.: Current Topics in Developmental Biology (eds. Moscona and Monroy) Academic Press (1972).

18. Holtzer, H.: Cell Lineages, Stem Cells and the Quantal Cell Cycle Concept (eds. Lord, Potten and Cole) Cambridge University Press (1978).

19. Yaffe, D.: Exp. Cell Res. 66, 33-41 (1971).

20. Shainberg, G., Yagil, D. and Yaffe, D.: Dev. Biol. 25, 1-16

21. Yaffe, D. and Dym, H.: Cold Spring Harbor Symp. on Quant. Biol. 38, 543-547 (1972).

22. Konigsberg, I.: Dev. Biol. 26, 133-145 (1971).

23. Buckley, P. and Konigsberg, I.: Dev. Biol. 37, 193-205 (1974).

24. Konigsberg, I., Sollmann, P. and Mixyer, L.: Dev. Biol. 63, 11-26 (1978).

25. Konigsberg, I.: Pathogenesis of Human Muscular Dystrophies (ed. Rowland) Excerpta Medica (1978).

26. Buckingham, M., Caput, D., Cohen, A., Whalen, R. and Gros, F.: PNAS, 71, 1466-1470 (1974.

27. Merlie, J., Buckingham, E., and Whalen, G.: Current Topics in Developmental Biology (eds. Moscona and Monroy) Academic Press (1977).

28. Doering, J. and Fischman, D.: Dev. Biol. 36, 225-235 (1974).

29. Emerson, C. and Beckener, S.: J. Mol. Biol. 93: 431-477 (1975).

30. Emerson, C.: Pathogenesis of Human Muscular Dystrophies (ed. Rowland) Excerpta Medica (1978).

31. Holtzer, H., Rubinstein, N., Dienstman, S. and Somylo, A.: Biochimie, 56, 1575-1580 (1974).

32. Croop, J. and Holtzer, H.: J. Cell Biol. 65: 271-285 (1975).

33. Holtzer, H., Croop, J., Dienstman, S., Ishikawa, H. and Somylo, A.: PNAS, 72, 513-517 (1975).

34. Holtzer, H., Strahs, K., Biehl, J. and Ishikawa, H.: Science, 188, 943-945 (1975).

35. Yeoh, G., and Holtzer, H.: Exp. Cell Res., 104, 63-73 (1977).

36. Keller, J. and Nameroff, M.: Differentiation, 2, 19-23 (1974).

37. Moss, P. and Strohman, R.: Dev. Biol. 48: 431-441 (1975).

38. Turner, D., Gmur, R., Burckhardt, E. and Eppenberger, H.: Dev. Biol. 48, 258-283.

39. Turner, D.: Differentiation, 10, 81-93 (1978).

40. Trotter, J. and Nameroff, M.: Dev. Biol. 49, 548-555 (1976).

41. Nadal-Ginard, B.: Cell, 15, 855-864 (1978).

42. Holtzer, H., Rubinstein, N., Fellini, S., Yeoh, G., Chi, J. and Okayama, M.: Quart. Rev. Biophys., 8, 523-557 (1975).

43. Brunk, C.: Differentiation, 14, 95-99 (1979).

44. Kligman, T. and Nameroff, M.: Exp. Cell Res.

45. Croop, J. and Holtzer, H. (in press).

46. Fellini, S. and Holtzer, H.: Differentiation 6, 71-74 (1976).

47. Rubinstein, N. and Holtzer, H.: Exp. Cell Res. 97, 387-393 (1976).

48. Dienstman, S., Biehl, J., Holtzer, S. and Holtzer, H.: Dev. Biol. 39, 83-94 (1974).

49. Dienstman, S. and Holtzer, H.: Cell Cycle and Cell Differentiation (eds. Reinert and Holtzer) Springer-Verlag, Heidelberg (1975).

50. Chevallier, A. Kiney, M., and Mauger, A.: J. Embryol. exp. Morph.: 41: 245-258 (1977).

51. Christ, B., Jacob, H. and Jacob, M.: Anat. Embryol.: 150, 171-186 (1977).

52. Ishikawa, H., Bischoff, R. and Holtzer, H.: J. Cell Biol. 38: 538-555 (1968).

53. Holtzer, H., Fellini, S., Rubinstein, N. and Strahs, K.: Cell Motility (ed. Goldman, Pollard and Rosenbaum). Cold Spring Harbor Press (1976).

54. Bennett, G., Fellini, S., Croop, J., Otto, J., Bryan, J. and Holtzer, H.: PNAS, 75, 4364-4368 (1978).

55. Fellini, S., Bennett, G., Toyama, Y. and Holtzer, H.: Differentiation 12, 59-69 (2978).

56. Bennett, G., Fellini, S., and Holtzer, H.: Differentiation, 12, 71-82 (1978).

57. Franke, W., Schmid, E., Osborn, M., and Weber, K.: PNAS, 75, 5034-5036 (1978).

58. Franke, W., Schmid, E., Osborn, M. and Weber, K.: J. Cell Biol. 81, 570-580 (1979).

59. Fellini, S., Bennett, G., and Holtzer, H.: Am. J. Anat. 153, 451-457 (1978).

60. Bennett, G., Fellini, S., Toyama, Y., and Holtzer, H.: J. Cell Biol. 82: 577-584 (1979).

61. Lazarides, E. and Hubbard, B.: PNAS, 73, 4344-4348 (1976).

62. Lazarides, E. and Balzer, D.: Cell, 14, 429-438 (1978).

63. Cohen, R., Pacifici, M., Rubinstein, N., Biehl, J., and Holtzer, H.: Nature, 266, 538-540 (1977).

64. Toyama, Y., West, C. and Holtzer, H.: Am. J. Anat. Sept. issue (1979).

65. Fizman, M.: Cell Differentiation, 7, 89-101 (1978).

66. Miskin, R., Easton, T. and Reich, E.: Cell, 15, 1301-1312 (1978).

67. Ruby, K., Boettiger, D., Pacifici, M. and Holtzer, H.: Am. J. Anat. 147, 401-405 (1976).

68. Pacifici, M., Boettiger, D., Roby, K. and Holtzer, H.: Cell, 11, 891-899 (1977).

69. Boettiger, D., Roby, K., Brumbaugh, J. and Holtzer, H.: Cell, 11, 881-890 (1977).

70. Erickson, E., Collett, M. and Erickson, R.: Nature, 274, 919-921 (1977).

71. Collett, M., Erickson, R.: PNAS, 75, 2021-2024 (1978).

72. Pacifici, M. and Holtzer, H.: Am. J. Anat. 150: 207-212 (1978).

73. Lowe, M., Pacifici, M., and Holtzer, H.: Cancer Res. 38, 2350-2356 (1978).

74. Payette, R., Biehl, J., Toyama, Y. and Holtzer, H.: in press.

75. Nameroff, M., and Munar, E.: Dev. Biol. 49: 288-293 (1978).

76. Knudsen, K. and Horowitz, A.: Dev. Biol. 58: 328-338 (1977).

77. Horowitz, A., Wight, A., Ludwig, P. and Cornell, R.: J. Cell Biol. 77, 334-357 (1978).

78. Toyama, Y., West, C. and Holtzer, H.: Am. J. Anat. Sept. issue (1979).

THE SEQUENTIAL APPEARANCE OF FAST AND SLOW MYOSINS DURING MYOGENESIS

N.A. Rubinstein and A.M. Kelly
Department of Anatomy, School of Medicine, and Laboratory of Pathology, School of Veterinary Medicine, University of Pennsylvania, Philadelphia, Pennsylvania 19104

Introduction

Although many properties have been used to define distinct fiber types, the use of antibodies to the different polymorphic forms of the contractile proteins has become popular. For example, in adult animals, two antigenically - as well as biochemically - distinct classes of myosin exist: fast and slow (1). Each class, moreover, can be further divided into two [slow] or three [fast] isoenzymes (2) with different light chain complements and probably different heavy chains (3). In general, each normal adult fiber contains one class of myosin (1, 4-7), although the proportions of the different isoenzymes of that class may differ among fibers (3,6). Exceptions to the "one myosin class per fiber" rule do exist, however. In the rat soleus, many fibers containing fast myosin also contain a larger amount of slow myosin, although most fibers appear to contain either one or the other (6). Also, in the rabbit EDL, many fibers with predominantly fast myosin also contain a small amount of slow myosin (Rubinstein, unpublished observation). How this diversity among fiber types is generated during embryogenesis is the subject of this paper. In particular, how do different fibers come to synthesize different myosins?

During myogenesis, immature muscle fibers (myotubes) are formed by fusion of mononucleated myoblasts (8,9). Prior to

Plasticity of Muscle

innervation, both the post-mitotic myoblasts and the immature myotubes begin synthesizing myosin. To generate fibers with different myosin types, these myotubes could initially synthesize either fast or slow myosin and continue synthesizing this myosin through adult life. In the adult, however, muscle fibers are labile and capable of switching from the synthesis of one myosin type to the synthesis of another (4,10). As experiments with cross innervation and chronic stimulation have strongly suggested, the decision to produce a particular myosin isoenzyme is a result of the frequency with which the fiber is being stimulated (11-13). Since these earliest myotubes are not innervated and unless these young cells are responding to cues not operative in the adult, it is unlikely that diversity is generated at the time of myotube formation. As most recent papers have concluded, then, all myotubes must initially synthesize the same myosin[s]. What has not yet been decided, however, is whether all myotubes initially synthesize [a] slow myosin, [b] fast myosin (14-18), [c] both fast and slow myosins (19,20), or [d] an embryonic myosin distinct from both adult myosins (21,22).

We have investigated the initial myosin synthesized by newly formed myotubes both during embryogenesis and during regeneration. Using immunohistochemical localization of myosin types in individual fibers plus two dimensional electrophoresis of myosin light chains, we conclude that all embryonic fibers initially synthesize fast myosin, while the synthesis of slow myosin is initiated in fibers already containing fast myosin and only in fibers destined to be slow in the adult animal.

Results

Attention was focused on the rat extensor digitorum longus (EDL, a predominantly fast muscle in the adult) and soleus (predominantly slow muscle in the adult), although a number

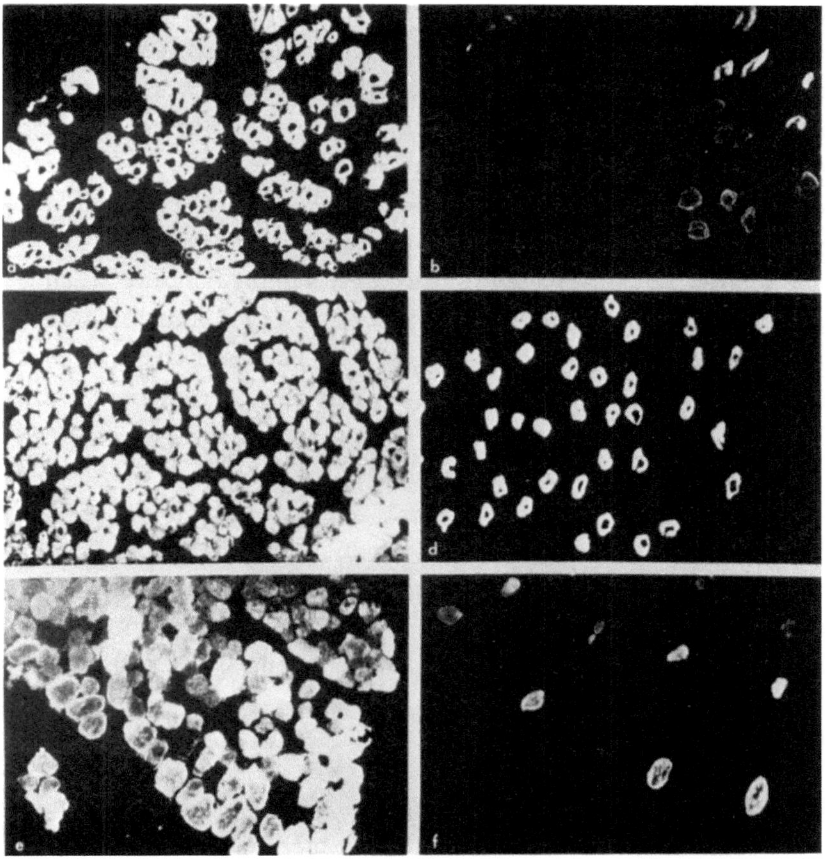

Figure 1. Staining of rat EDL muscles at various ages with fluorescent antimyosin antibodies. Affinity purified antibodies were prepared from IgG by two rounds of absorptions against myosin bound to agarose. First, IgG was absorbed to the homologous myosin [e.g. antifast IgG to fast myosin] to remove non-myosin IgG. The absorbed antibody was recovered and reabsorbed against the heterologous myosin [e.g. antifast myosin IgG absorbed against slow myosin] to remove cross reacting antibodies. The final specific antifast or antislow myosin antibody showed specificity on immunodiffusion gels identical to those shown in a previous paper (4). Six micron frozen sections of EDL muscles at various ages were stained with fluorescent antimyosin as described by Gauthier and Lowey (5). [a] EDL at 18 days *in utero* stained with antifast myosin antibody (AF); [b] serial section of [a] stained with antislow myosin antibody (AS); [c] EDL 2 days after birth stained with AF; [d] serial section to [c] stained with AS; [e] EDL 14 days after birth stained with AF; [f] serial section to [e] stained with AS.

of other muscles were also examined. Cross sections of rat legs from 18 day fetuses were stained with fluorescein labelled antibodies specific to adult fast and adult slow muscle myosins. Individual muscles were located and examined under epifluorescence with a Zeiss Photomicroscope II. In the entire cross section of the rat leg, every fiber reacted uniformly and intensely with antifast myosin antibody [AF], while a subset of fibers additionally stained light with antislow myosin antibody [AS]. Thus, every fiber of both the EDL (fig. 1a) and the soleus (fig. 2a) reacted with AF. In the EDL, however, only a few scattered fibers stained with AS (fig. 1b), while most of the soleus fibers also reacted lightly with AS (fig. 2b) as well as intensely with AF. Each fiber shows intense subsarcolemmal staining with a non-stained central region. This pattern is probably due to the central position of nuclei and other organelles in these immature myotubes plus the subsarcolemmal site of sarcomere formation. Variable numbers of fibers in different muscles reacted with AS. The percentage of AS positive fibers ranged from more than 90% in the soleus to 0% in some parts of the gastrocnemius.

Two days after birth, the pattern has not changed radically. Once again, each fiber in the rat leg reacted with AF, while some cells also reacted with AS. Staining with AF, however, was no longer uniform in intensity from fiber to fiber. In the EDL, several levels of staining were seen. Those which stained less intensely with AF were the fibers which reacted most strongly with AS (fig. 1c,1d). In most fibers, the soleus showed less intense staining with AF and a greatly increased intensity with AS. However, a number of small AS-negative fibers were strongly stained with AF (fig. 2c,2d).

Fourteen days after birth, this pattern progressed. The EDL now contained a large proportion of AF positive fibers (fig. 1e) and a small percentage of AS positive fibers (fig. 1f). The AS positive fibers, moreover, failed to react with AF.

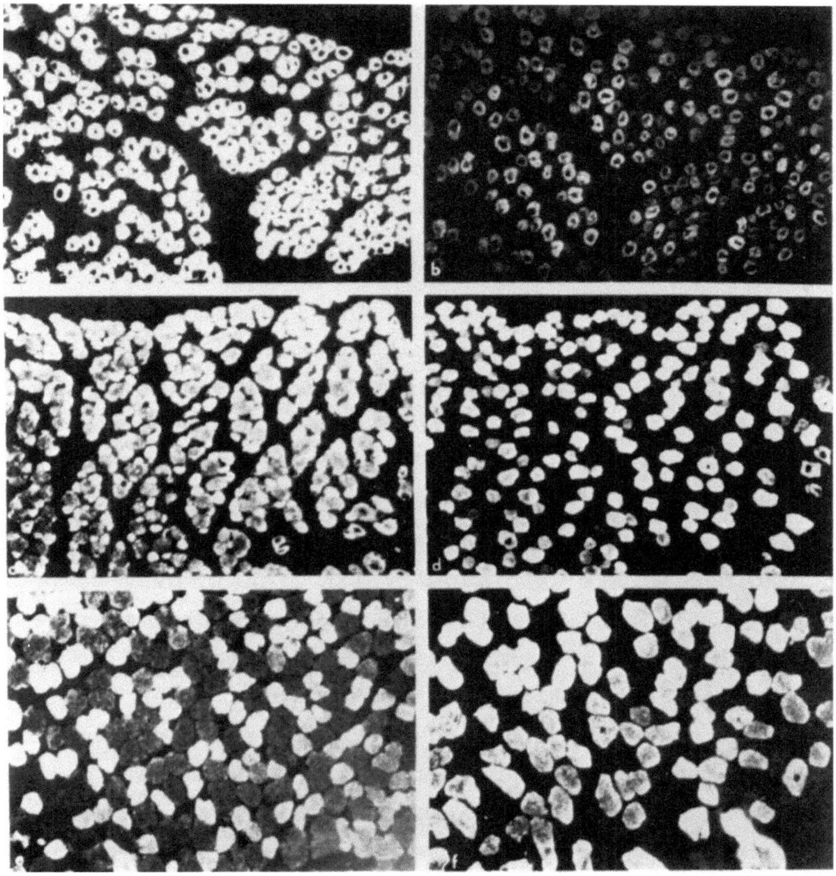

Figure 2. Staining of rat soleus muscles at various ages with fluorescent antimyosin antibodies. Antibodies and frozen sections were prepared as described in figure 1. [a] soleus from 18 days *in utero* stained with AF; [b] serial section to [a] stained with AS; [c] soleus 2 days after birth stained with AF; [d] serial section to [c] stained with AS; [e] soleus 14 days after birth stained with AF; [f] serial section to [e] stained with AS.

The soleus showed a less striking contrast with two levels of AF staining: intense AF staining correlating with AS negative fibers, and light AF staining correlating with AS positive fibers (fig. 2e,2f). By 14 days after birth, then, both the EDL and soleus have clearly approached the definitive pattern of myosin distribution seen in the adult animal.

Although analyses of myosin light chains from whole muscles cannot reveal the myosin composition of individual fibers, they do confirm the relative antibody staining intensities seen with the immunohistochemistry. At all stages of development, the EDL contains predominantly fast myosin light chains. At 18 days gestation, only LC_{1f} and LC_{2f} can be seen (fig. 3a); negligible amounts of LC_{3f} are found, as others have frequently reported for fetal muscles (16, 17, 23). Because of the light loading of this gel, slow myosin light chains cannot be seen in this figure. The original gel, however, did reveal traces of slow myosin light chains, LC_{1s} and LC_{2s}. An additional band, slightly more acidic and with a lower mobility than LC_{1f}, can be seen abutting the top right hand side of LC_{1f}. This band has been called an embryonic light chain by Whalen et al. (22); its slightly different position on our gels is due to different electrophoresis conditions. This band is seen at all stages of EDL development in the rat (fig. 3b,3c) and in embryonic and adult chickens, cats, and rabbits (18; Kelly and Rubinstein, unpublished observation).

The soleus shows a progression from predominantly fast light chains in utero (fig. 3d) to almost exclusively slow light chains as development proceeds (fig. 4e,4f).

From the antibody staining data and the two dimensional gels, one can conclude that most fibers in the EDL and some fibers in the soleus initially contain only fast myosin. A small number of fibers in the EDL and more fibers in the soleus do contain both myosins (although significantly less slow myosin than fast myosin) at 18 days in utero, the earliest stage

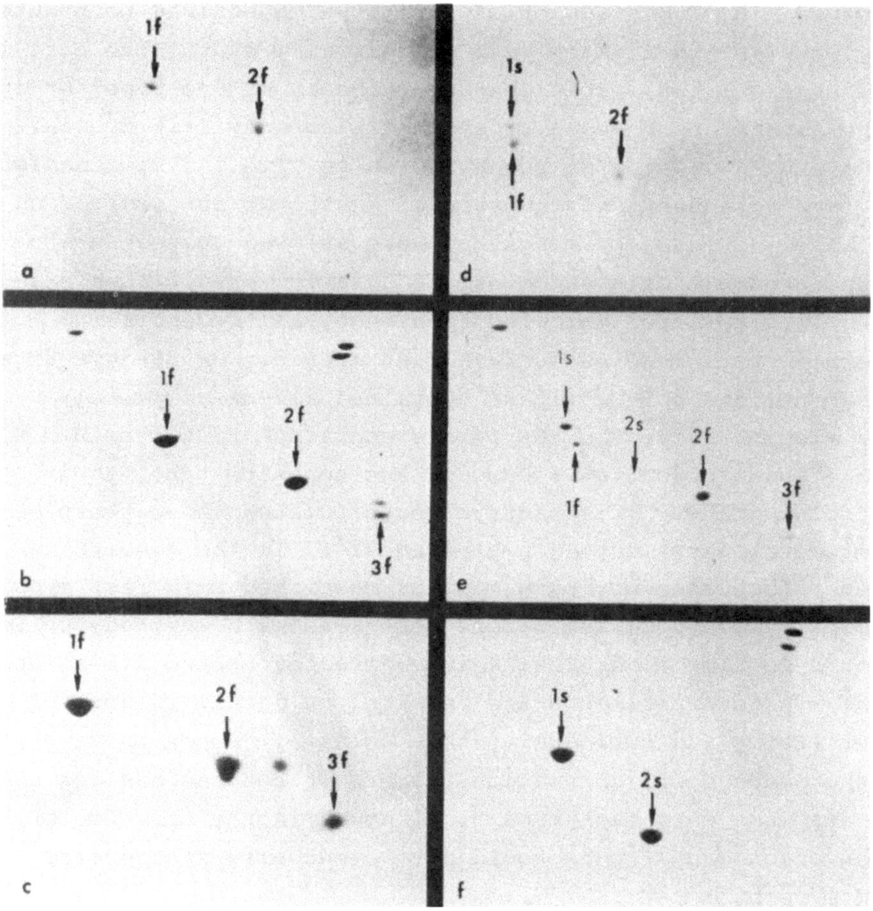

Figure 3. Two dimensional electrophoresis of EDL and soleus myosin light chains at different ages. Myosin or actomyosin was prepared as described previously (16) and electrophoresed on two dimensional gels (25). EDL myosin light chains: [a] 18 days *in utero*, [b] 2 days after birth, [c] adult. Soleus myosin light chains: [d] 18 days *in utero*, [e] 2 days after birth, [f] adult. Fast myosin light chains are LC_{1f}, LC_{2f}, and LC_{3f}. Slow myosin light chains are LC_{1s} and LC_{2s}.

examined. Did this subset of fibers which stained both intensely with AF and lightly with AS initially synthesize both myosins and later stabilize the synthesis of one type? Or did these fibers, too, begin by synthesizing only fast myosin? We have used studies of regeneration in vivo and myogenesis in vitro to answer this question. Adult rat EDL and soleus muscles, destroyed by marcaine, were allowed to regenerate in the presence or absence of the sciatic nerve. When examined 12 days after marcaine treatment, all regenerating fibers in both muscles contain fast myosin. By 30 days after treatment, the soleus fibers contained almost completely slow myosin, while the EDL fibers contained fast myosin. Each muscle contained a few fibers of the opposite type, as in control muscles. This pattern recapitulated the pattern of fiber development during embryogenesis. In the denervated regenerates, however, both muscles contained only fast myosin. No slow myosin positive fibers were seen with antibody staining and no slow light chains were detected on two dimensional gels. These experiments are reported in detail in another paper (Kelly and Rubinstein, this volume). Hence, in vivo in the absence of innervation, fibers of both soleus and EDL muscles make only fast myosin, although in the presence of innervation many fibers would have eventually synthesized slow myosin.

Similar results have recently been reported in avian fast and slow muscles in tissue culture, grown in the absence of motoneurons (18). In tissue culture, both the ALD (a slow muscle) and the pectoralis (a fast muscle) reacted exclusively with AF and contained only fast myosin light chains. Similarly, rat embryo leg muscle in culture also contained exclusively fast myosin light chains and failed to stain with AS (not shown). Finally, post-mitotic, mononucleated myoblasts cultured from 3 day chick embryo limb buds synthesize only fast myosin. These cells had not been innervated either in vivo or in vitro (Croop, Holtzer, and Rubinstein, unpublished).

To confirm that those fibers synthesizing exclusively fast myosin in tissue culture are also capable of synthesizing slow myosin, we stimulated muscle in tissue culture with exogenous electrodes. Cultures were stimulated at 3 Hz for 24 days in normal medium supplemented with 3 ug/ml cytosine arabinoside. Cultures were washed, extracted with high salt, and the extract subjected to two dimensional electrophoresis. Fig. 4a shows the light chains of control cultures. LC_{1f} and LC_{2f} as well as fibroblast light chains can be seen. No slow light chains are evident. In the stimulated culture, however, both LC_{1s} and LC_{2s} can be seen in addition to the fast light chains (fig. 4b). In most experiments, LC_{2s} is the predominant slow light chain synthesized and in some cases is the only slow light chain seen.

Discussion

Immunohistochemical staining of development fast and slow fibers with antibodies specific to adult fast and slow myosins reveals that all immature fibers contain fast myosin, while a subset also contain slow myosin. With development, most of those fibers which contain slow myosin lose their fast myosin. In this way, the adult pattern of one myosin class per fiber (with certain exceptions) is attained. Those fibers which contain both predominantly fast myosin and a small amount of slow myosin at 18 days _in utero_ probably initially also synthesized fast myosin exclusively, since regenerating fibers without innervation fail to initiate slow myosin synthesis; and muscle fibers in tissue culture without motoneurons also fail to synthesize slow myosin. We have previously shown (17) that rat hind limb muscles denervated at birth fail to increase their slow myosin content.

That these exclusively fast myosin producing fibers are the ones capable of producing slow myosin under appropriate con-

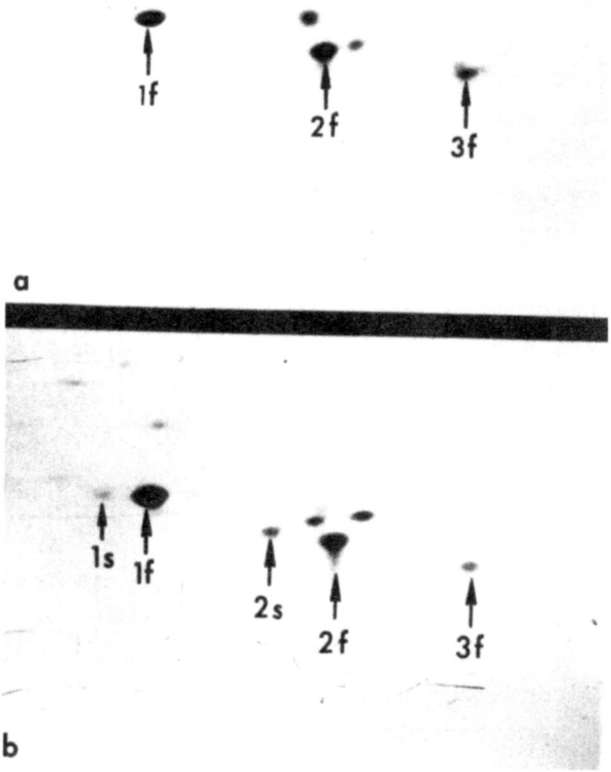

Figure 4. Myosin light chains after chronic stimulation of chicken breast muscle in tissue culture. Chicken breast muscle cultures prepared from 11 day chick embryos were grown for 5 days; at this time, cultures contained numerous cross-striated, multinucleated myotubes. Platinum electrodes were placed in the culture dishes and the cultures stimulated at 3 Hz continuously for an additional 24 days. Medium contained 3 ug/ml cytosine arabinoside to reduce fibroblast proliferation which is greatly stimulated by the electrical field. Cultures were fed every other day; control cultures were treated similarly, except no current was applied to the electrodes. [a] light chains from control culture. [b] light chains from stimulated culture. The peptides above LC_{2f} and LC_{3f} are fibroblast light chains. Fast light chains are LC_{1f}, LC_{2f}, and LC_{3f}. Slow light chains are LC_{1s} and LC_{2s}. The stimulated cultures contain a small amount of 1s and a larger proportion of 2s. Control cultures contain neither 1s nor 2s.

ditions has been shown by stimulating muscle in tissue culture in the presence of cell cylcle inhibitors. These inhibitors prevent the emergence of new myotubes and ensure that the myosin switch is occurring within pre-existing fibers.

Changes of contractile protein isoenzymes within individual pre-existing fibers, then, occur under normal as well as experimental conditions. Experimentally, slow myosin can be induced in a fiber already containing fast myosin by chronic stimulation of the fiber's intact nerve (4,10) and probably also in fast fibers re-innervated by slow motoneurons. The opposite type of change, from slow to fast myosin, also occurs in cross innervation and within denervated slow muscles (Kelly and Rubinstein, unpublished observation). Finally, during normal development, as we have shown, fibers containing fast myosin begin synthesizing slow myosin and eventually lose their fast myosin complement. A similar type of change in troponin components also occurs during development (24). The cue for all these myosin switches could be either the different frequencies of stimulation imposed by fast and slow motoneurons or neurotrophic factors passed from neuron to muscle fiber. In tissue culture, no neurons are present and the frequency of stimulation is exogenously imposed by platinum electrodes. Hence, a frequency of stimulation is sufficient to initiate a myosin switch. The change, however, is not complete; and whether this failure is due to imperfectly defined stimulating conditions or to some non-frequency related motoneuronal contribution cannot yet be answered.

Similarly, we cannot yet attribute the synthesis of fast versus slow myosin during myogenesis to a frequency or non-frequency related event. During early development, only certain fibers are initially innervated by the first axons to arrive, the "pioneering" axons. These fibers, termed "primary fibers" (26), usually become slow fibers. The adjacent, more immature fibers clustered around the primary fiber do not have well-formed neuromuscular junctions at first, but do have junctions with the primary fiber. Hence, contractile and electrical impulses should spread from primary to secondary fiber and both should be identical with regard to myosin type if frequency alone is the determinant. However, the primary and secondary fibers in fast muscles are not identical with regard to myosin type: the primary fibers stain intensely for both fast and slow myosins, while the secondary fibers stain only for slow myosin (Rubinstein and Kelly, unpublished observation). Hence, the determinants of myosin type during myogenesis remain to be fully understood.

We do understand, however, the sequence of myosin types during myogenesis as described in this paper. All embryonic muscle fibers initially synthesize fast myosin, while the synthesis of slow myosin is initiated in fibers already containing fast myosin and only in fibers destined to be slow in the adult animal.

We would like to thank Mrs. Z. Paltzmann and Ms. Jean Collenberg for expert technical assistance. Research for this publication was supported by NIH Grant No. 15835 to the Pennsylvania Muscle Institute, by NIH Grant No. 14332, and by a grant from the Muscular Dystrophy Association of America.

References

1. Arndt, I., Pepe, F.: J. Histochem. Cytochem. 23, 159-168 (1975).
2. Hoh, J.F.Y., McGrath, P., White, R.: Biochem. J. 157, 87-95 (1976).
3. Hoh, J.F.Y.: FEBS Lett. 90, 297-300 (1978).
4. Rubinstein, N., Mabuchi, K., Pepe, F., Salmons, S., Gergely, J., Sreter, F.: J. Cell Biol. 79, 252-261 (1978).
5. Gauthier, G., Lowey, S.: J. Cell Biol. 74, 760-779 (1977).

6. Gauthier, G., Lowey, S.: J. Cell Biol. 81, 10-25 (1979).

7. Pette, D., Schnez, U.: FEBS Lett. 83, 128-130 (1977).

8. Lash, J., Holtzer, H., Swift, H.: Anat. Rec. 128, 679-693 (1957).

9. Okazaki, K., Holtzer, H.: J. Histochem. Cytochem. 13, 726-739 (1966).

10. Pette, D., Schnez, U.: Histochemistry 54, 97-107 (1977).

11. Sreter, F., Gergely, J., Salmons, S., Romanul, F.: Nature 241, 17-18 (1973).

12. Weeds, A., Trentham, D., Kean, C., Buller, A.: Nature 247 135-139 (1974).

13. Salmons, S., Sreter, F.: Nature 263, 30-34 (1976).

14. Pelloni-Müller, G., Ermini, M., Jenny, E.: FEBS Lett 67, 68-74 (1976).

15. Syrovy, I., Gutmann, E.: Pflügers Arch. 369, 85-89 (1977).

16. Rubinstein, N., Pepe, F., Holtzer, H.: Proc. Nat. Acad. Sci. (USA) 74, 4524-4527 (1977).

17. Rubinstein, N., Kelly, A.: Develop. Biol. 62, 473-485 (1978).

18. Rubinstein, N., Holtzer, H.: Nature 280, 323-325 (1979).

19. Masaki, T., Yoshizaki, C.: J. Biochem. 76, 123-131 (1974).

20. Gauthier, G., Lowey, S., Hobbs, A.: Nature 274, 25-29 (1978).

21. Sreter, F., Balint, M., Gergely, J.: Develop. Biol. 46, 317-325 (1975).

22. Whalen, R., Butler-Browne, G., Gros, F.: J. Molec. Biol. 126, 415-431 (1978).

23. Dow, J., Stracher, A.: Proc. Nat Acad. Sci. (USA) 68, 1107-1110 (1971).

24. Dhoot, G.K. Perry, S.V.: Nature 278, 714-718 (1979).

25. O'Farrell, P.: J. Biol. Chem. 259, 4007-4021 (1975).

26. Kelly, A., Schotland, D.: Research in Muscle Development and the Muscle Spindle, 32-48 (1971).

PATTERNS OF MYOSIN SYNTHESIS IN REGENERATING NORMAL AND DENERVATED MUSCLES OF THE RAT

A.M. Kelly and N.A. Rubinstein

Departments of Pathology and Anatomy, School of Veterinary Medicine and Medicine, University of Pennsylvania, Philadelphia, PA

In the previous study (Rubinstein & Kelly 1979), we have examined the characteristics of myosin synthesis in differentiating fast and slow muscles of the rat. The present study uses immunocytochemical localization of myosin types in individual fibers and two dimensional polyacrylamide gel electrophoresis of myosin light chains to investigate characteristics of regenerating muscle after experimental injury.

In the first set of experiments, attempts were made to achieve a homogenous population of new, regenerating myofibers by completely destroying either the soleus or extensor digitorum longus (E.D.L.) muscles of mature rats. Necrosis was obtained with a combination of ischaemia and marcaine injections after the procedure of Hall-Craggs (1978). Pertinent questions these experiments sought to answer are as follows:
(a) are the patterns of myosin synthesis in regenerating muscles of mature animals comparable to those of the same muscle during differentiation?
(b) as new muscle cells will regenerate in the absence of a nerve supply (Mong, 1975; Carlson & Gutmann, 1976), what form of myosin occurs in such cells never having been innervated? This question is particularly relevant to the regenerating soleus for during normal differentiation primordia of this muscle so far investigated contain both fast and slow muscle myosin. Hence, it has not been shown whether differentiating slow fibers in vivo are endogenously programmed to synthesize

both fast and slow myosins or whether synthesis of slow myosin is imposed by the nervous system at a very early stage of development.

The second set of experiments examines the types of myosin within innervated myofibers of soleus muscle which are undergoing repair after a local crush injury. Innervation of muscle is not compromised in these experiments. We have therefore asked the question, does the locally regenerating region of a slow fiber initially contain fast myosin and then mature to a program of slow myosin synthesis; or does the influence of the nervous system dictate that slow myosin is produced from the inception of regeneration?

Materials and Methods

Animals used in these experiments were approximately 200 gm. female Wistar rats.

Necrosis of either the soleus or E.D.L. was achieved essentially following the procedure of Hall-Craggs (1978). Animals were anaesthatized with Halothane (Halocarbon Labs. Inc.) and the muscle exposed. Myotendinous junctions were crushed using haemostats in order or disrupt the vascular supply to the muscle via this route. Peripheral nerve innervating the muscle together with the blood supply was severed close to the muscle belly and carefully reflected. The muscle was then isolated within a sheath made of parafilm and repeatedly injected with marcaine. The parafilm sheath was also filled with the drug and the muscle was bathed in this for a least 15 minutes. In animals in which the regenerate was to be re-innervated, the severed nerve stump was inserted into the damaged muscle belly with fine forceps and was loosely held in place by suturing with 7-0 Vicryl (Ethicon Inc. Summerville, NJ). Denervation was achieved by removing a 2-4mm segment of the sciatic nerve in the proximal thigh.

After the surgery the animals were injected subcutaneously with

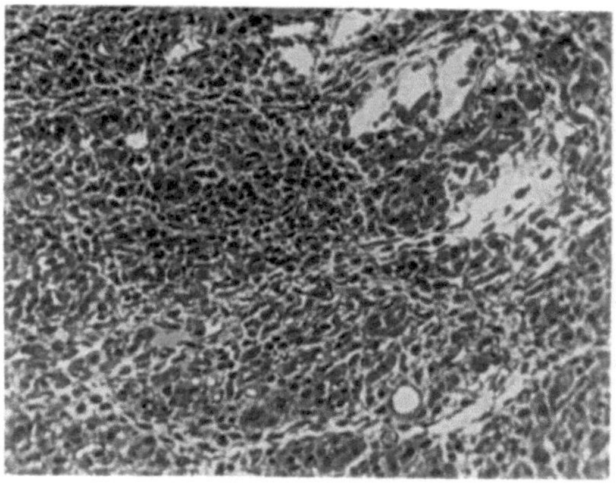

Figure 1 Regenerating soleus 3 days after injury. The normal architecture has been destroyed and is replaced by mononuclear cells many of which are presumptive myoblasts. H & E stain x 140.

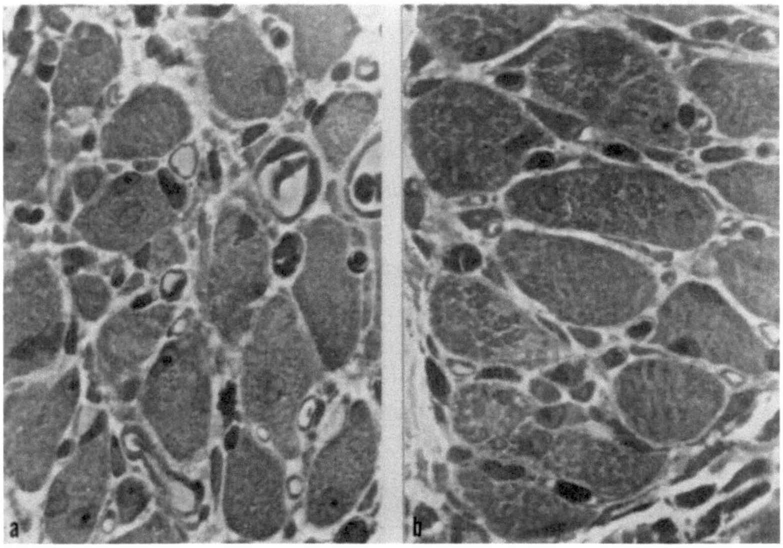

Figure 2 Regenerating soleus 12 days after injury; 2a is from an innervated limb and 2b from a denervated limb. They are morphologically very similar. H & E stain x 300.

150,000 units of Flo-cillin (Bristol Labs).

Local crush injury was inflicted by tightening a loop of 6-0 Mersilene (Ethicon, Inc.) around the distal belly of the soleus so as to produce a complete transverse lesion. Care was taken to avoid any interference with innervation of the muscle during this procedure.

Patterns of myosin light chains were studied using acto-myosin preparations and the two-dimensional polyacrylamide gel electrophoresis system of O'Farrell (1975). The preparation of affinity purified antimyosins specific to adult fast and slow myosins was as described by Rubinstein et al 1978. Cryostat sections were cut at 6_u from muscle which had been frozen in isopentone cooled to 150° F by liquid nitrogen. Sections to be used for immunofluorescence were mounted on coverslips coated with EGTA according to the method of Pearson and Saberra (1974). Sections were stained with the fluorescein labelled antimyosins and individual cells were located and examined under epifluorescence with a Zeiss Photomicroscope II. Material to be examined with phase optics was embedded in JB 4 plastic (Polysciences, Warrington, PA) sectioned at 2_u and stained with hematoxylin and eosin (H & E).

(a) Regeneration by new fiber formation in the innervated and denervated soleus and E.D.L. muscles.

Figure 1 is from a soleus muscle, 3 days after surgery. The normal muscle architecture has been destroyed and replaced by mononuclear cells. Many of the cells are undoubtly myoblasts derived from satellite cells of the pre-existent muscle.

Figure 2 is of regenerating muscle, 12 days after injury; Figure 2a is from an innervated limb and Figure 2b is from a denervated limb. As has been observed in previous studies, it is not easy to distinguish the two, probably because in both systems, neuromuscular relations are initially interrupted and have not ye been re-established in the innervated limb. Functionally, both muscles failed

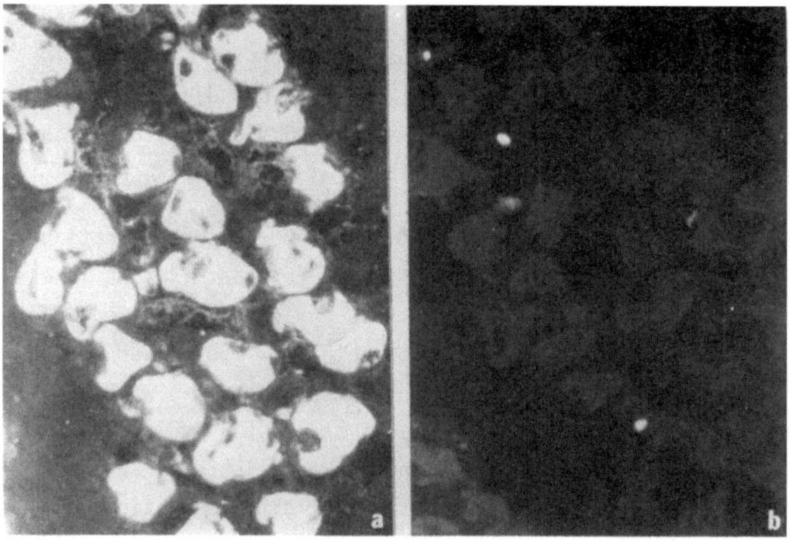

Figure 3 Regenerating muscle from an innervated limb, 12 days after injury. All fibers stain with antifast myosin (3a) and do not stain with anti slow myosin (3b). x 300.

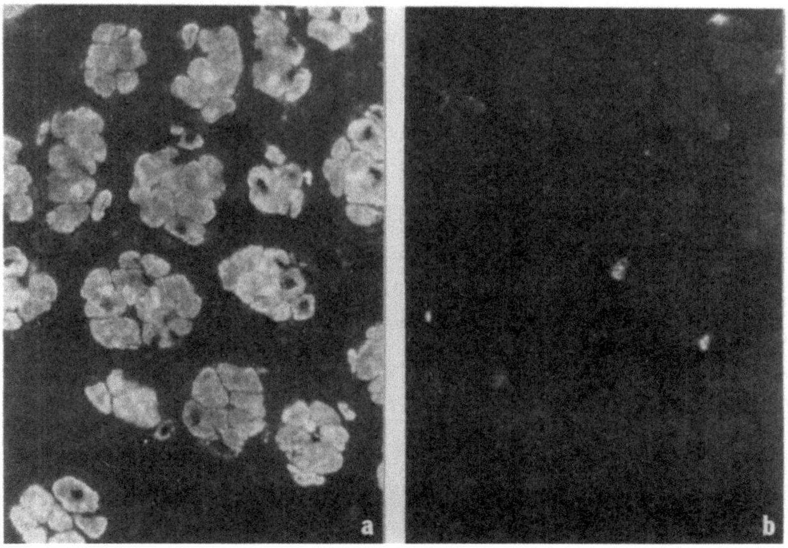

Figure 4 Regenerating muscle from denervated limb, 12 days after injury. All fibers stain with anti fast myosin (4a) and do not stain with anti slow myosin (4b).

to respond when the sciatic nerve was stimulated. When material from such regenerating muscle was stained with specific anti-myosins, all regenerating myofibers in both the innervated and denervated limbs stained intensely with anti-fast myosin and did not stain with anti-slow myosin (figures 3 and 4).

This pattern of staining persisted in the 30 days regenerating soleus from the denervated limb (Figure 6). By contrast, in the innervated limb, differentiation into fiber types could be recognized (Figure 5), the majority of fibers stained with anti-slow myosin and only a minority stained with anti-fast myosin. Those that did stain with anti-fast myosin occurred together in groups of at least 2 or 3 cells which is a pattern suggestive of type grouping characteristic of re-innervated muscle. The evolution of this pattern of differentiation corresponds very well with the studies on regenerating minced muscle by Carlson and Gutmann (1976) who found that primitive end plates did not appear until the end of the 3rd week post surgically. The development of slow muscle myosin can therefore be correlated with the expected time of re-innervation. This is supported by our observation that the 30 day regenerating muscles of innervated limbs, twitched upon stimulation of the sciatic nerve.

To further investigate the forms of myosin within these regenerating new muscle cells, their complement of light chains was studied using 2 dimensional, O'Farrell gels. At 12 days after surgery, the light chain complement from both the innervated and denervated limbs are virtually identical and dominated by the presence of LC1F and LC2F (Figure 7). Consistent with the findings in developing muscle, the relative proportions of LC3F are reduced compared to fully differentiated muscle. In most gels, there was a small satellite on the acidic side of LC1F. This corresponds to the satellite Whalen et al (1978) designate as an embryonic light chain. In addition, there was a small amount of slow myosin in the forms of LC1S in preparations from both the innervated and denervated limbs.

At 30 days (Figure 8), the complement of light chains from the regenerate in the denervated limb is identical to that at 12 days

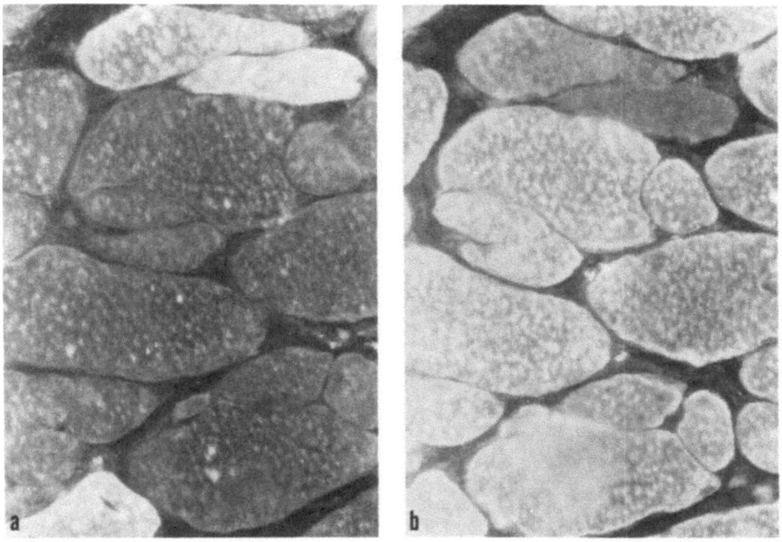

Figure 5 Regenerating muscle from innervated limb, 30 days after injury. The abnormal architecture is a result of the previous muscle damage. Fiber differentiation is evident. Very few stain with anti fast myosin (5a) and the majority stain with anti slow myosin (5b). x 500.

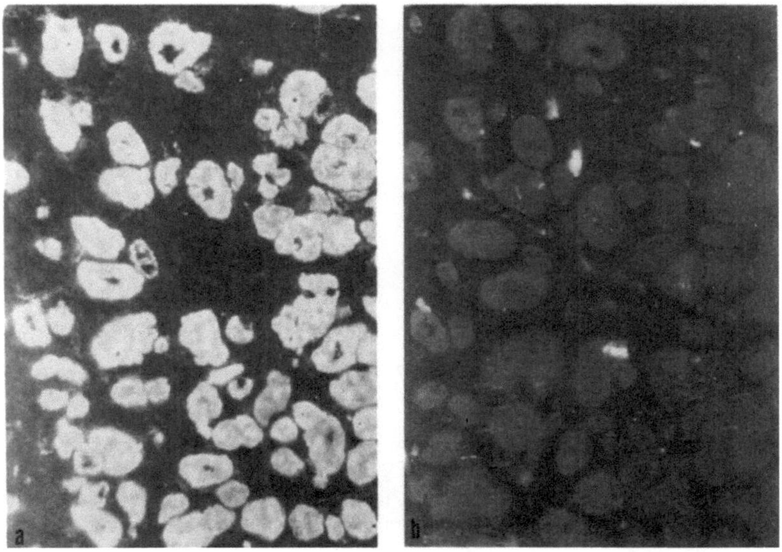

Figure 6 Regenerating muscle from denervated limb, 30 days after injury. Development has not advanced beyond that at 12 days. All fibers stain with anti fast myosin (6a) and they do not stain with anti slow myosin (6b). There are several brightly staining artefactual deposits in the field. x 300.

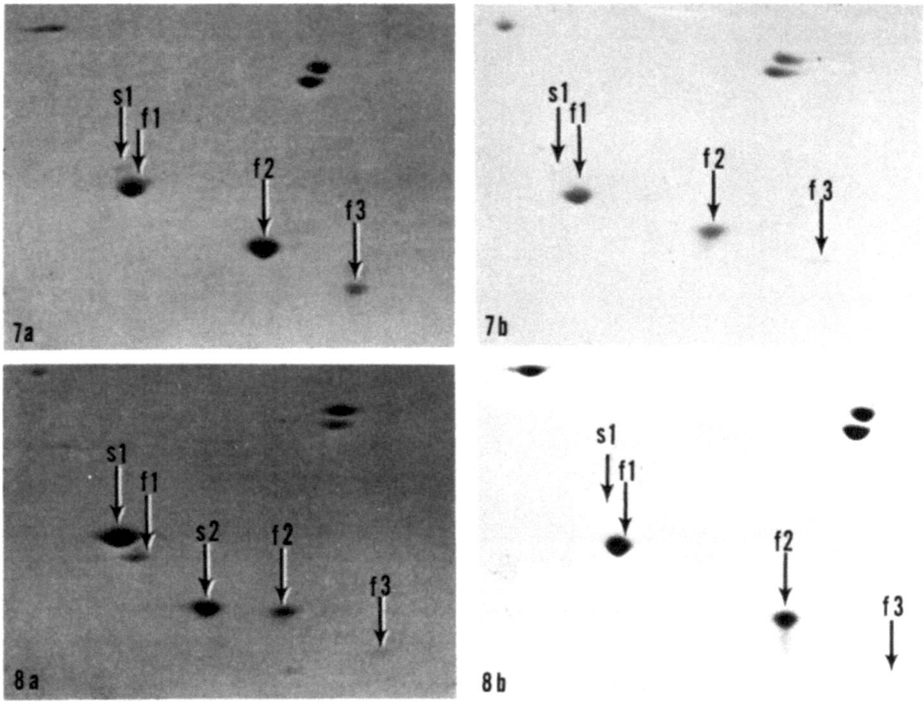

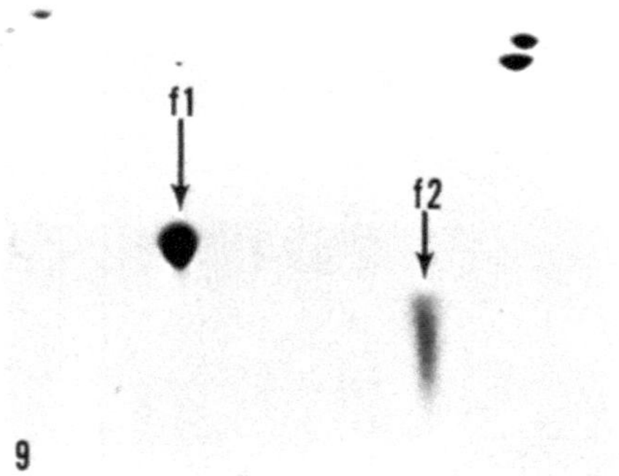

Figures 7, 8 & 9. Two dimensional electrophoresis of myosin light chains.

Figures 7a & b are from 12 day regenerating muscle of the innervated and denervated limbs, respectively. They are essentially identical and are dominated by LC1F and LC2F. There are very small proportions of LC3F and of LC1S.

Figures 8a & b are from 30 day regenerating muscle. There is a predominance of slow light chains LC1S and LC2S and only small proportions of the fast light chains in 8a, from the innervated limb. In 8b, from the denervated limb, there remains a predominance of fast light chains and very small proportions of the slow light chain LC1S.

Figure 9 is from 12 day regenerating EDL of an innervated limb. The light chains are exclusively of the fast type.

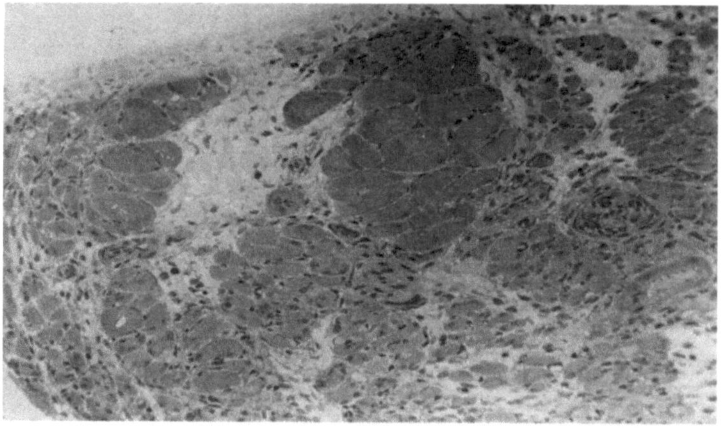

Figure 10 Denervated, regenerating muscle 6 days after injury. The majority of fibers are small and interpreted as newly forming muscle cells. Amongst them are 2 groups of larger fibers with comparatively normal morphology. These are interpreted as cells which escaped destruction at the time of surgery. x 250.

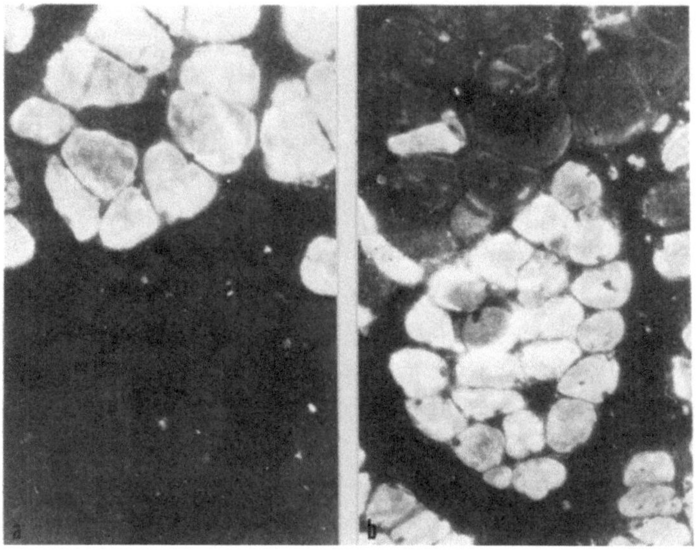

Figure 11 Denervated, regenerating muscle 12 days after injury. The smaller regenerating fibers in 11a do not stain with anti slow myosin where as the larger fibers are positive. Conversely in serial section 11b stained with anti fast myosin, the small fibers are positive and the large fibers are negative. These larger, slow fibers probably escaped destruction at the time of surgery.

(Figure 8b). By contrast, the pattern of light chains in the innervated limb (Figure 8a) is almost reversed and is now dominated by the two slow light chains LC1S and LC2S. Concomitantly, the proportions of the fast light chains are reduced. Figure 9 is from the 12 day regenerating E.D.L. of an innervated limb. The light chain complement is comparable to that of the soleus at this stage of regeneration, consisting of LC1F, LC2F and very small proportions of LC3F. Distinct from the regenerating soleus, however, there is no evidence of slow myosin light chains in the gel.

An important question remains concerning the small amounts of slow myosin light chains present in the regenerating soleus of the denervated limb. Their source is apparent when sections are taken from the extremities of the regenerating muscle within 5 to 6 days of experimental injury. In these regions, we consistently found between 30 and 100 large fibers (1 to 3% of the original population) which clearly had escaped destruction at the time of the surgery (Figure 10). These cells persisted in small groups and at 12 days were positive when reacted with anti-slow myosin (Figure 11). A comparable problem of incomplete myofiber destruction has confronted others who have used this type of approach to study muscle regeneration (Clark and Bloomfield, 1945).

To be certain these surviving fibers were the only source of slow myosin in the early regenerating soleus we combined samples from a denervated, regenerating soleus and from a normal adult rat E.D.L. muscle in a single block which was sectioned and stained with anti-slow myosin. Myofibers of this regenerate demonstrated no difference in their lack of affinity for the antibody compared to fast fibers of the normal E.D.L. Hence, we find no evidence to suggests that in the absence of innervation, new, regenerating myofibers of the soleus produce myosin other than a type of fast muscle myosin.

The changes in light chain complement of the regenerating soleus correlate very well with the companion immunofluorescence studies. They support the previous findings of Rubinstein and Holtzer (1979) on cultured chicken muscle in which fast myosin is the exclusive

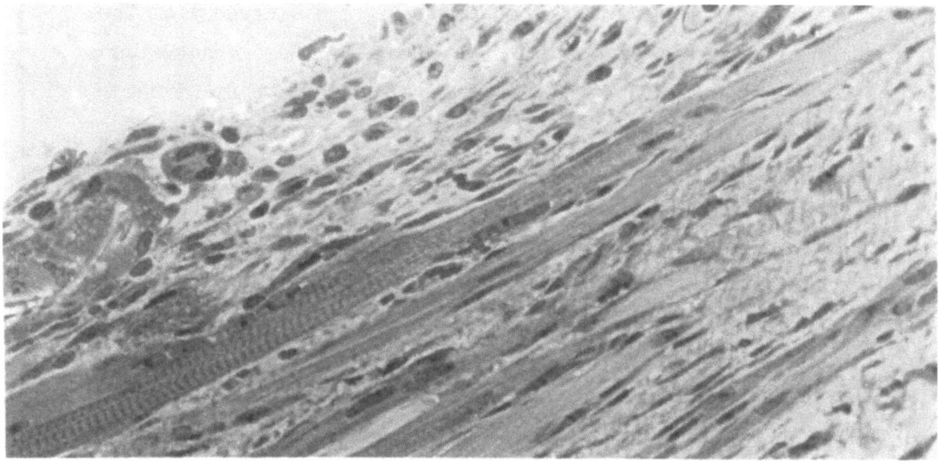

Figure 12 Longitudinal section through continously regenerating myofibers after local crush injury. The injured area is to the right. Cross striations are not evident towards the regenerating tip of the fibers. H & E. x 400.

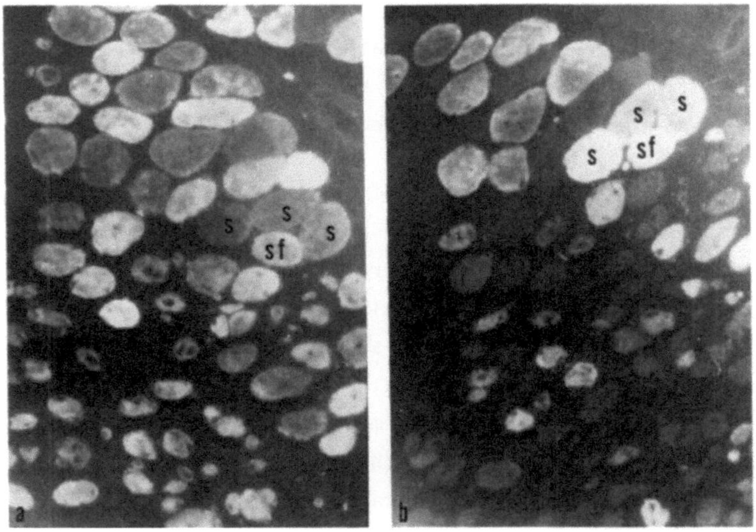

Figure 13 Transverse section through locally regenerating muscle just proximal to the crush injury. There is a predominance of staining with anti fast myosin (13a). All of the small fibers and a number of the larger fibers react positively with this anti serum. Staining with anti slow myosin occurs amongst the larger fibers (e.g. those in 13b marked s) and a limited number of small fibers. Some fibers stain with both fast and slow myosin (e.g. fiber marked sf).

gene product, regardless of whether the cells are derived from the primordium of a fast or a slow muscle. Based on these observations we conclude that during normal development the primordium of the soleus initially produces a form of fast myosin and then, under the influence of the nervous system, this is gradually replaced by slow myosin. That the 18 day fetal soleus contains handsome amounts of slow as well as fast myosin (Rubinstein & Kelly 1979) therefore indicates that specific neural control of myosin synthesis commences very early in ontogeny.

(b) Regeneration of the soleus after local crush injury.

When a myofiber is locally injured the area of altered sarcoplasm may be small and indeed may involve only two or three sarcomeres. Commonly a central core of necrosis develops, but it is rare for the lesion to extend along the entire length of a myofiber (Adams 1974). Regeneration proceeds rapidly after local injury and Schmallbruch (1976) has proposed that is both by formation of new small muscle cells and by repair of the damaged myofiber which grows from the healthy region, back into the area of necrosis along the empty, basement membrane, sheath.

We have locally injured the soleus of adult rats by pulling a loop of fine suture material tightly around the muscle distal to the point of innervation. Regeneration has then been investigated using our specific antibodies. Figure 12 is an H & E stained plastic section which shows myofibers continously regenerating into a traumatized area, 5 days after surgery. Cross striations are present in the myofibers but they are not evident in the sarcoplasm near the regenerating tip of the fiber. Figures 13a and b are transverse, serial sections through an area of regenerating muscle just proximal to the crush. Figure 13b is stained with anti-slow myosin and 13a with anti-fast myosin. Some fibers, mainly large ones, stain exclusively with anti-slow myosin. All of the small regenerating fibers stain with anti-fast myosin and a number of them stain with anti-slow myosin as well. The predominance

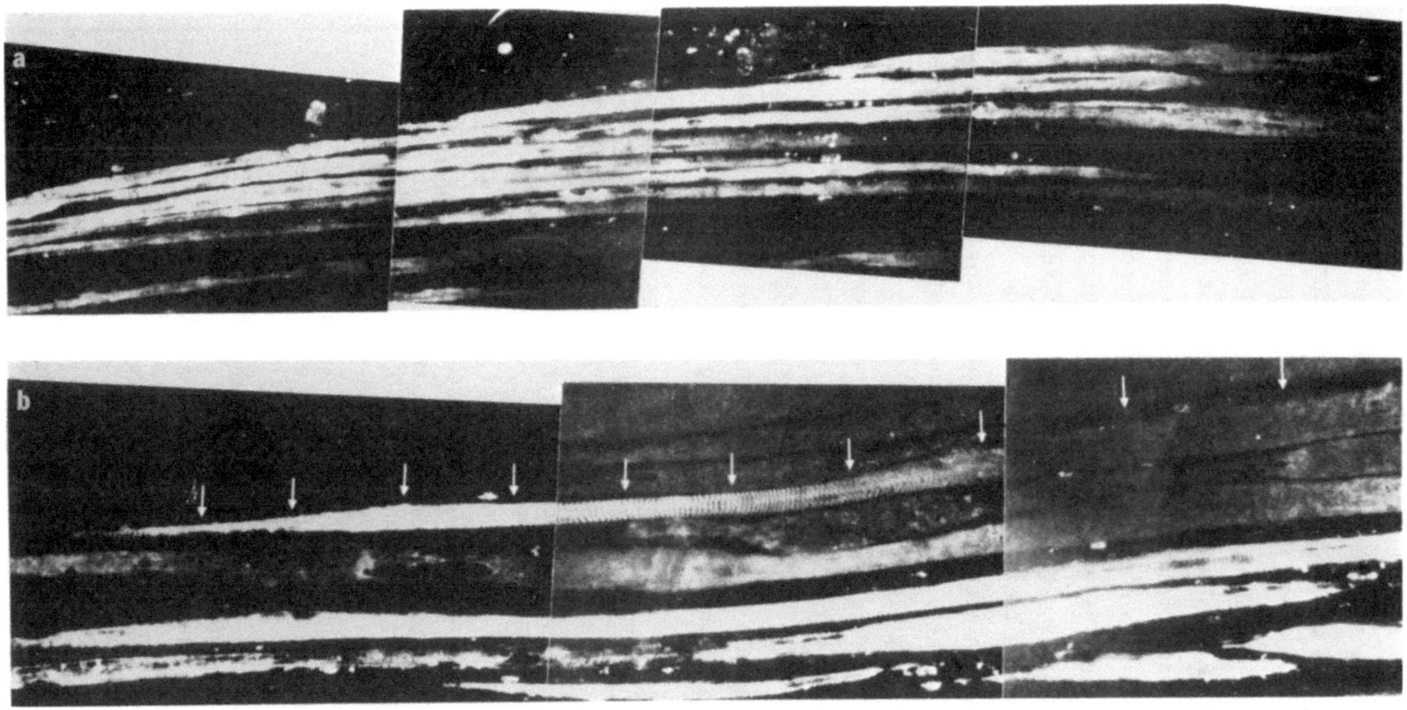

Figure 14 Longitudinal sections through regenerating muscle proximal to a local crush injury. The crush was inflicted to the muscle beyond the left side of the illustrations. Both sections are stained with anti fast myosin. There is a predominance of fast myosin amongst fibers neighboring the crush in 14a. The intensity of fast myosin staining appears to decline within individual fibers as distance from the injury is increased. This is particularly evident in 14b where the course of one fiber is depicted with arrows.

of fast myosin in these small fibers is undoubtedly due to the regenerative response. But how are the small fibers related to the slow twitch fibers which constitute over 70% of this muscle. Some may be new fibers comparable to those described in the previous section. Not all of them are however. Figures 14a and 14b are montages of longitudinal sections through the injured area. They are stained with anti-fast myosin. Both reveal that as the traumatized area is approached there is a local shift towards fast myosin within a single fiber. This phenomenon requires considerable further investigation as it may be accounted for in several ways. It does however, suggest that when precursor cells are committed to myosin synthesis even in the presence of innervation, they must initially synthesize a myosin with antigenic determinants in common with fast myosin before gaining the capacity to synthesize slow myosin. In normally developing muscle this may explain why so many fibers of the differentiating soleus contain a form of fast myosin as well as slow myosin. In this muscle, satellite cell fusion is extremely active, much more so than in the developing E.D.L. (Kelly 1978). Satellite cells which fuse and are newly incorporated into a developing slow fiber may initially produce a primitive form of fast myosin before maturing to slow myosin synthesis.

We thank Ms. J. Collenberg and Mrs. Z. Paltzman for their technical assistance. These studies were supported by N.I.H. grants HL 15835, NS, 14332 and the Muscular Dystrophy Association of America.

References

1. Adams, R.D.,: Pathological reactions of the skeletal muscle fiber in man. In Disorders of Voluntary Muscle (ed. J. H. Walton, 1974 189-190 Churchill Livingstone, Edinburgh and London (1974).

2. Carlson, B.M., Gutmann, E.: Contractile and histochemical properties of sliced muscle grafts regenerating in normal and denervated rat limbs. Exp. Neural 50, 319-329 (1976).

3. Clark, W.E. Le Gros, Bloomfield, L.B.: The efficiency of intra-muscular anastomosis with observations on regeneration of devascularized muscle. J. Anat. 79, 15-33 (1945).

4. Kelly, A.M.: Satellite cells and myofiber growth in the rat soleus and extensor digitorum longus muscles. Dev. Biol. 65, 1-10 (1978).

5. Mong, F.S.F.: Nervous influences on minced muscle regeneration in rats. Anat. Rec. 181,429 (1975).

6. O'Farrell, P.H.: High resolution, two-dimensional electrophoresis of proteins. J. Biol. Chem 10, 4007-4021 (1975).

7. Pearson, J., Sabarra, A.: A method for obtaining longitudinal cryostat sections of living muscle without contraction artefacts. Stain Technol. 49, 143-146 (1976).

8. Rubinstein, N.A., Holtzer, H.: Fast and slow muscles in tissue culture synthesize only fast myosin. Nature 280, 323-325 (1979).

9. Rubinstein, N.A., Kelly, A.M.: The sequential appearance of fast and slow myosin during myogenesis. In Plasticity of Muscle, (D. Pette, ed.), pp. 147-159, De Gruyter, Berlin New York, (1980).

10. Schmalbruch, H.: The morphology of regeneration of skeletal muscle in the rat. Tissue and Cell 8, 673-692 (1976).

CONTRACTILE PROTEIN ISOZYMES IN MUSCLE DEVELOPMENT: THE EMBRYONIC PHENOTYPE

Robert G. Whalen
Département de Biologie Moléculaire, Institut Pasteur,
25, Rue du Dr. Roux, 75724 Paris, France

Introduction

The plasticity of muscle is reflected in the succession of phenotypes that occur as muscles develop and differentiate. A number of different phenotypes are represented by the several fiber types that can be detected histologically; two major types can be defined with respect to the contractile proteins. It is becoming clearer that some of the proteins synthesized by the first muscle fibers (myotubes) are qualitatively different from the adult forms (1-4). Underlying transitions between these phenotypes are variations in gene activity. It would seem that "turning on" the definitive (i.e. muscle specific) genes is only one part of the problem. A type of "on-off" regulation also appears to be important, for example in the case of embryonic muscle proteins. This situation suggests the existence of multiple genetic control mechanisms acting at different times within muscle fibers.

Some of the phenotypic transitions can be observed in cell cultures of myoblasts. In general, the myoblasts obtained from fetal muscle by mechanical or enzymatic dissociation ("primary cultures") seem to recapitulate several of the processes which accompany muscle differentiation in the animal (5). Several permanently established myoblast cell lines have been derived (6). One of them, myogenic rat cell line L6, exhibits interesting differences when compared to primary rat cultures (1,2).

Such variants may be useful in understanding normal differentiation, although no extensive "genetic" investigations have as yet been possible.

In this paper I examine some of the evidence concerning the phenotypic transitions apparently occurring in adult and developing muscle. Particular emphasis is placed on the documentation of an embryonic phenotype.

A. The Myoblast To Myotube Transition

The observation that myoblasts will undergo fusion in cell culture conditions has permitted the study of the biochemical phenomena attending this morphological act. Nucleic acid hybridization data have shown that fused myotube cultures express more cytoplasmic information (as reflected by the messenger RNAs present) than unfused myoblast cultures (7). The increase in gene products expressed has been confirmed by direct analysis of protein synthesis in differentiating cultures (8-11). By analyzing purified myotube preparations, Garrels (11) has also shown that the synthesis of several proteins dramatically decreases in passing from unfused myoblast to fused myotube. It is therefore clear that myotubes acquire a different phenotype from that expressed by the myoblast.

The contractile proteins have been the subject of detailed qualitative and quantitative study. Myosin, actin, tropomyosin and α-actinin have all been found in non-muscle cells; they have in general been shown to represent isozymes different from those found in muscle tissue (12). A precise characterization of these proteins in myoblasts has yet to be undertaken, although it seems clear that the muscle forms are not present in large amounts (8-11,13). Devlin and Emerson (10) have shown that differentiating quail myoblast cultures begin to synthesize many of the major contractile proteins in a coordinate fashion, i.e. their rates of synthesis increase with the same time course. In contrast, Daubas (13) has found that whereas the muscle specific contractile proteins are indeed substantially induced after fusion of bovine myoblasts, there is a significant delay (approx. 24 hrs) between, for example, the increase in synthesis of the several myosin light chains and the increase in synthesis of α- and β-tropomyosin. These latter results would seem to indicate that at least for some cell systems, more than one "coordinated event" might take place. The existence of embryonic contractile protein isozymes (discussed below) would also imply the necessity for multiple points of control.

B. Aspects of the Adult Phenotypes

Immunological studies and experiments on isolated fibers have shown that different contractile proteins are present in the two major adult fiber types (14-16). Certain muscles are composed predominantly or uniformly of one of the types of fibers, and such muscle can be shown to have a relatively "fast" or "slow" time course of contraction and relaxation(17). The contractile protein isozymes found in these muscles are referred to as fast-type or slow-type proteins.

An important functional relationship has been documented between the physiology of the muscle and the proteins present. Barany (18) showed that in adult muscles the maximum velocity of contraction is directly proportional to the ATPase activity of the actomyosin extracted from the muscle. Whereas the absolute values for fast and slow contraction speeds (and ATPase values) vary among species, the relationship nevertheless holds in describing the correlation between the two parameters.

Early on, Buller, Eccles and Eccles (19) had shown that cross-reinnervation of fast and slow muscle nerves led to an interchange of their respective physiological properties. Subsequently, Salmons and Vrbova (20) demonstrated that continuous electrical stimulation of the nerve contacting a fast muscle could transform some of its physiological properties into those of a slow muscle. This latter system has been exploited to show convincingly that continuous electrical activity imposed on a fast muscle nerve will convert that muscle into one possessing all the physiological and biochemical properties normally attributed to a slow muscle (21).

It is important to know whether, during nerve-stimulation induced transformation, the biochemical changes take place within the pre-existing muscle fibers or whether the slow muscle characteristics arise from newly formed muscle fibers. To investigate these possibilities, Rubinstein et al. (22) prepared antibodies to adult fast and slow myosins and reacted them with sections of muscle undergoing stimulation. Fibers were found which bound both antibodies. This result was interpreted to mean that the initial fast fibers were being transformed into slow ones and at some point both myosin types were present. However, an alternative explanation could be that regeneration was in fact occurring; in this case fibers containing both fast and slow proteins could conceivably also be found (see below). Experiments of Pette and Schnez (23), who find both fast and slow myosin light chains in single fibers during transformation, would be subject to the same difficulty in interpretation. Both types of experiments do show however that fast and slow proteins can co-exist in a single fiber. Whether other, cross-reacting or biochemically similar, forms are also present remains to be established.

It would be useful to repeat the experiments mentioned above during a slow to fast transition, e.g. by stopping electrical stimulation after initial conversion of fast to slow. This model would presumably not be subject to the same ambiguity. The possibility of regeneration may be significant in the transformation occurring after nerve cross-reinnervation.

In an analogous experimental system, Lømo et al. (24) have found that direct electrical stimulation of a slow muscle, after having sectioned its nerve, leads to the acquisition of some fast muscle properties. This model has been less extensively studied than the one mentioned above. Therefore, a fundamental question remains to be answered definitively: is the pattern of electrical activity (i.e. synaptic or post-synaptic membrane) the important parameter, as Salmons and Sreter suggest (21), or in fact do nerve "trophic" substances have a role, as a number of other observations suggest (25). The elegant and useful model developed by Salmons and colleagues does not settle this point. Since the nerve receives direct stimulation in these experiments, it is equally possible to envision that stimulation alters the metabolism of the nerve (and, for example, the substances its secretes) as to consider that activity directly alters the muscles biochemistry.

C. Development of the Slow Muscle

A number of lines of evidence suggest that during the differentiation of slow muscles, fibers pass through a stage in which initially only fast proteins are present. The various observations can be summarized as follows.

1. Developing slow muscles contain both fast and slow type myosin light chains (1,26-30).
2. Cell cultures of myoblasts from future fast and slow muscles both yield myotubes which accumulate fast light chains (31).
3. Denervation of slow muscles in neonatal rats prevents the accumulation of slow myosin light chains (29).
4. Rat solei muscles contain fast light chains in early stages of regeneration but eventually regain the slow myosin proteins (32).

None of these observations alone proves the point. Finding a mixture of fast and slow light chains does not demonstrate that only fast forms were present initially. In fact, the future slow ALD muscle of chickens contains predominantly slow light chains at 11 days *in ovo* (28), presumably before innervation has occurred. In contrast, cultures of myotubes from embryonic ALD muscles synthesize predominantly fast light chains (31); whether they synthesize only fast proteins at any stage *in vivo* is open to question. The potentiality for slow light chain synthesis might be suppressed in cell culture.

The fact that denervation blocks the production of slow light chains has been taken as evidence that nerve activity is needed to induce the synthesis of these proteins. Denervation probably provokes a number of changes in the muscle; furthermore, it is important to verify that reinnervation by a fast nerve has not occurred. Finally, during regeneration the processes which occur are not necessarily those taking place in development.

The various observations discussed above make a case for the idea that all fibers have fast light chains initially and that nerve influence accompanies the differentiation of the slow fibers. Modifications in this scheme may be expected once more details are known.

D. The Embryonic Phenotype

It is known that embryonic muscles contract and relax slowly(17). A study of actomyosin ATPase activity of developing rat muscle showed that it was low, in correlation with the slow speed of contraction (33). However, fast type myosin light chains have been found in embryonic or newborn muscles (1,26-30). This observation does not necessarily mean that the Barany relation cannot be extended to developing muscles (28) since Barany had in fact demonstrated only the correlation between contraction speed and ATPase activity. Nevertheless, the problem of the nature of myosin from embryonic muscle was posed.

In 1972, Huszar (34) reported the amino acid sequence of a myosin peptide isolated from three-week old rabbit muscle tissue. He demonstrated the presence of a myosin which was a distinct polypeptide compared to adult fast or cardiac myosin. This result was attributed to the presence of fetal myosin. His data thus established the polymorphic nature of myosin at the gene level; however the corresponding peptide sequence of slow myosin was never obtained. This left open the question of whether developing muscles contained slow type myosin (or a mixture of fast and slow types), as was suggested by the ATPase data and, later, immunological studies.

Antibodies to adult myosins have also been used to investigate developing muscle and its myosin. Masaki et al. (35) and Gauthier et al. (36) found that both anti-slow and anti-fast antibodies reacted with embryonic muscle whereas Rubinstein *et al.* (28,31) claimed that only anti-fast myosin antibody reacted with myosin in embryonic tissue (future fast) or cultured cells. These latter authors claimed that since adult fast myosin and radioactive tissue culture myosin are both found in the same Ouchterlony precipitation line, they must be identical. However, this is precisely the result expected if the two myosins simply share some antigenic sites. Thus it has not been possible to determine if in fact one or several adult types are present, or whether a cross-reacting form is present.

We have studied the myosin synthesized by rat muscle cells in culture (1,2,4). The culture system permits access to newly formed myotubes which should express the earliest manifestations of the muscle phenotype. The analysis of such cultured cell myosin by two dimensional gel electrophoresis reveals the presence of a polypeptide which is not found in adult fast or slow muscle (1,2). This protein is similar in molecular weight but differs slightly in charge compared to the $LC1_F$ light chain. Characterization of this protein and its metabolism has led to the following observations.

1. The myosin of the rat L6 cell line contains this protein as its only LC1-type light chain; it is present in the stoichiometric ratio expected for an LC1 light chain (1,2).
2. The myosin purified from rat primary cultures contains both the adult $LC1_F$ and this additional protein (1).
3. This protein is part of the myosin subfragment-1 complex in both primary culture and L6 myosin, as expected for an LC1-type light chain (1).
4. This protein can be detected in the two dimensional gel analysis of total cellular extracts prepared from primary cultures or L6 cultures, showing that it is not the result of proteolysis during the myosin preparation (1).
5. In primary rat cultures, the additional protein is synthesized more abundantly than $LC1_F$ in the first several days after fusion; by six days after fusion the synthesis of $LC1_F$ is predominant (2).
6. This protein is present in embryonic or newborn rat muscle tissue; it disappears in the first two weeks after birth (1).

We have interpreted these results in the following manner. There exists in embryonic myosin a novel form of light chain, called embryonic LC1 or $LC1_{emb}$, which has several of the characteristics of an LC1 light chain and which is the predominant form present in the early stages after cell fusion. In primary myotube cultures the relative rates of synthesis of $LC1_{emb}$ and $LC1_F$ change with time. This leads to a transition from the embryonic to adult light chain phenotype (2). Unexpectedly, L6 myosin has $LC1_{emb}$ as its only LC1-type light chain. The availability of this myosin in reasonable quantities has allowed us to show that $LC1_{emb}$ is indeed a light chain (1). The observation that L6 myotubes do not undergo the embryonic to adult light chain transition suggests that L6 might be a variant cell line which does not progress beyond the earliest stages of myotube differentiation.

The $LC1_{emb}$ protein has also been found in fetal and newborn muscle tissue, demonstrating that it is not present uniquely in cultured myotubes (1). The fact that $LC1_{emb}$ is less prominent (relative to $LC1_F$) in the tissue preparations than, for example in primary cultures, can be attributed to the rapid evolution of the synthesis of $LC1_{emb}$ and $LC1_F$ in the period just following myotube formation. It has not yet been possible to carry out a detailed study of the rates of synthesis of the light chains in developing muscle tissue; nonetheless, the presence of $LC1_{emb}$ in both tissue and culture situations suggests that the primary cultures are a reasonable model for the events occurring in tissue.

Rubinstein and Holtzer (31) have been unable to find an analogous protein in chicken myotube cultures. We have observed that an LC1 protein found in chicken cultures can be separated from the adult $LC1_F$ by two dimensional electrophoresis (37). However, the separation in the molecular weight dimension is so small (an R_f difference of about 1%) that it is not surprising that Rubinstein and Holtzer failed to detect it. A protein equivalent in many respects to the rat $LC1_{emb}$ has also been found in other mammals that we have studied (mouse, calf and rabbit). The position of the other $LC1_{emb}$ proteins relative to the adult $LC1_F$ is not always the same as in rat, but such species differences are perhaps not surprising when using a high resolution technique such as two dimensional gels.

In order to investigate the nature of the myosin heavy chain in cultured myotubes or embryonic tissue, we purified rat myosins from a number of sources and denatured them with sodium dodecyl sulfate (SDS). These preparations were then subjected either to partial proteolytic cleavage in the presence of SDS (38), or to immunological analysis employing complement fixation, after removal of excess SDS (39).

The partial proteolytic cleavage of denatured myosin can be performed using one of several enzymes. We have used chymotrypsin and the protease from S. aureus (4). One dimensional gel electrophoretic analysis of the partial proteolytic reactions shows, as expected, that fast and slow myosin clearly differ in their cleavage patterns. Cardiac myosin is very similar to slow skeletal myosin, while diaphragm myosin can be seen to be a mixture of the fast and slow types. The myosins isolated from primary myotube cultures, L6 myotubes or embryonic (future fast) muscle tissue all show a similar cleavage pattern which is different from any of the adult types. These results (4) and others (see below) have led us to conclude that three major types of muscle myosin heavy chain can be identified: MHC_{fast}, $MHC_{slow/cardiac}$ and $MHC_{embryonic}$ (4).

It is possible to analyze these partial proteolytic products by the standard two dimensional technique (see also 40). This type of analysis reveals additional detail compared to the one dimensional gels. We have noted heterogeneity in the bulk tissue (i.e. "fast") myosin preparations, and have shown that the differences between slow and cardiac myosin are very small (4). In addition, the two dimensional analysis allows rather clear visualization of mixtures of the different myosin isozymes. For example, the embryonic tissue preparation is composed of largely the embryonic isozyme; however, several peptides are seen in the cleavage pattern which seem to correspond to those found in the adult "fast" myosin. The presence of such a mixture suggests that a transition is taking place between (or among) isozymes.

The existence of a distinct, embryonic myosin heavy chain isozyme has been confirmed by immunochemical analysis (4). Antibodies produced against SDS-denatured rat cardiac heavy meromyosin (HMM)(39) were reacted against a number of SDS-denatured rat myosins. The antigen-antibody reactions were detected by complement fixation, thus a quantitative comparison was possible. Soleus and cardiac myosins were found to cross-react at an antibody dilution of 1:4500 whereas neither bulk skeletal nor L6 myosin react to any significant extent. After concentrating the antiserum some 15-19 fold, slight reaction can be obtained with bulk or L6 myosin. It is possible to distinguish between these two myosin species however, employing an antibody against denatured rabbit fast skeletal HMM (39). This antibody, at a dilution of 1:120, reacts strongly with rat bulk myosin, but only slightly with L6 myosin.

A simple interpretation of previous immunological studies is to consider that Masaki et al. (35) and Gauthier et al. (36) employed antibodies which cross-reacted with the embryonic myosin heavy chain isozyme. The antibodies used by Rubinstein *et al*. (28,31) would differ, according to this interpretation, in that only their anti-fast myosin antibody cross-reacted with MHC_{emb} whereas their anti-slow antibody did not. In this context, we have demonstrated (4) that antibody to *native* cardiac HMM will cross-react extensively, as determined by complement fixation, with extracts containing MHC_{fast}, MHC_{slow} and MHC_{emb}. Such cross reactions could explain, at least partially, the results of other immunological investigations (28,31,35,36).

Conclusion

We have shown by biochemical and immunological analysis that cultured muscle cells and embryonic muscle tissue possess contractile proteins which are muscle specific and yet which differ from the isozymes found in adult muscle tissue. These proteins have been called "embryonic" isozymes, in part for lack of a better term. Specifically, our results support the existence of distinct forms of an LC1-type light chain and a myosin heavy chain in developing muscle. A unique heavy chain isozyme occurring in newborn rabbit muscle has recently been reported (3). The presence of these proteins defines an embryonic phase of muscle development.

The embryonic phenotype appears to be only one of several that occur in the passage from mononucleate myoblast to adult muscle fiber. As far as is known, the cell cycle producing a fusion-competent myoblast is the last mitotic (or DNA synthetic) event to occur in the nucleus which is incorporated into a myotube (41). Any further phenotypic transitions must therefore take place in the absence of mitosis (or DNA synthesis) in contradistinction to the idea that phenotypes are separated by mitotic events in some obligatory fashion (42). Nonetheless, it seems apparent that some phenotypic transitions do occur.

A simplified description of some of the myosin phenotypes that can be observed is shown in figure 1. This scheme illustrates several points. First, it appears that $LC1_F$ and $LC2_F$, two "adult" light chain proteins, can both associate with a heavy chain polypeptide (MHC_{emb}) which differs from that with which they are associated in the adult. Second, it seems likely that the different embryonic isozymes may turn over at different rates. Twenty-day embryonic rat muscle tissue contains mostly the adult $LC1_F$, however its myosin heavy chain remains predominantly MHC_{emb}. Such differences in turnover suggest that separate mechanisms may control the metabolism of the different myosin subunits.

The embryonic phenotype shown in figure 1 is represented by L6 myotubes. These cultured muscle cells have only the embryonic forms of both myosin heavy and light chains. The embryonic phenotype can also be observed in primary cultures. In the first several days following cell fusion, the $LC1_{emb}$ light chain is predominant (2). Therefore it is likely that a myosin molecule composed of $(MHC_{emb})_2(LC1_{emb})_2(LC2_F)_2$ will be found in these myotubes. The L6 culture system permits access to this embryonic myosin molecule, and we have exploited this situation to measure the ATPase activity of the "pure" embryonic myosin

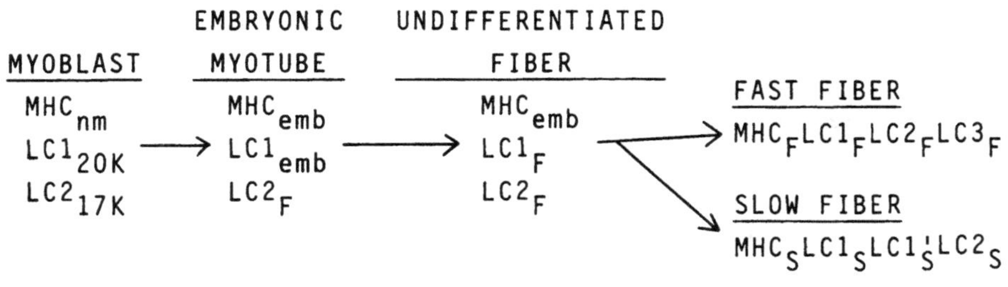

Figure 1. Schematic illustration of various myosin isozymes occurring in development. Myoblasts are assumed to have a non-muscle (nm) type myosin heavy chain. See refs. 1,2 and 4.

---

isozyme. We find that the actin-activated ATPase activity of L6 myosin is on the order of that given by slow (soleus) myosin (A.G.Weeds and R.G.Whalen, unpublished results). Thus the low ATPase value of embryonic myosin is correlated with the slow contraction speed of embryonic muscle, although it remains to be seen if this relationship has the same meaning as in adult muscle.

A number of questions persist. What isozymic forms of other contractile proteins are present in developing muscles? What is the functional utility of embryonic isozymes? Especially, what controls the passage from one phenotype to another; that is, what mechanisms influence muscle gene expression? Indeed, the plasticity of muscle, once documented, leads, as usual, to even more questions.

Acknowledgements

I am extremely grateful to Dr. Gillian Butler-Browne, Dr. Ketty Schwartz and Ms. Susan Sell, whose contributions to this work have enriched its contents beyond measure. I thank Prof. François Gros for support and encouragement at numerous points. I have been supported by a fellowship from the Muscular Dystrophy Association of America.

References

1. Whalen, R.G., Butler-Browne, G.S., Gros, F.: J. Mol. Biol. 126, 415-431 (1978).
2. Whalen, R.G., Butler-Browne, G.S., Sell, S., Gros, F.: Biochimie 61, 625-632 (1979).
3. Hoh, J.F.Y., Yeoh, G.P.S.: Nature 280, 321-322 (1979).
4. Whalen, R.G., Schwartz, K., Bouveret, P., Sell, S.M., Gros, F.: Proc. Natl. Acad. Sci. USA 76, in press (1979).
5. Hauschka, S.D.: Growth, Nutrition and Metabolism of Cells in Culture, Vol. 2, pp. 67-131, Academic Press, New York (1972).
6. Pearson, M.L.: The Molecular Genetics of Development, Academic Press, New York (1979).
7. Affara, N.A., Jacquet, M., Buckingham, M.E., Robert, B., Gros, F.: 12th FEBS Meeting: Gene Functions, Vol. 51, pp.233-243, Pergamon Press, New York (1979).
8. Whalen, R.G., Butler-Browne, G.S., Gros, F.: Proc. Natl. Acad. Sci. USA 73, 2018-2022 (1976).
9. Storti, R.V., Horovitch, S.J., Scott, M.P., Rich, A., Pardue, M.L.: Cell 13, 589-598 (1978).
10. Devlin, R.B., Emerson, C.P.: Cell 13, 599-611 (1978).
11. Garrels, J.I.: Dev. Biol., in press (1979).
12. Clarke, M., Spudich, J.A.: Ann. Rev. Biochem. 46, 797-822 (1977).
13. Daubas, P.: Thèse de Troisième Cycle, Univ. of Paris (1979).
14. Pette, D., Schnez, U.: Histochemistry 54, 97-107. (1977).
15. Gauthier, G.F., Lowey, S.: J. Cell Biol. 81, 10-25 (1979).
16. Dhoot, G.K., Perry, S.V.: Nature 278, 714-718 (1979).
17. Close, R.I.: Physiol. Rev. 52, 129-197 (1972).
18. Barany, M.: J. Gen. Physiol. 50, 197-218 (1967).
19. Buller, A.J., Eccles, J.C., Eccles, R.M.: J. Physiol. 150, 417-439 (1960).
20. Salmons, S., Vrbova, G.: J. Physiol. 210, 535-549 (1969).
21. Salmons, S., Sreter, S.: Nature 263, 30-34 (1976).
22. Rubinstein, N.A., Mabuchi, K., Pepe, F., Salmons, S., Gergely, J., Sreter, F.: J. Cell Biol. 79, 252-261 (1978).
23. Pette, D., Schnez, U.: FEBS Letters 83, 128-130 (1977).
24. Lømo, T., Westgaard, R.H., Dahl, H.A.: Proc. R. Soc. Lond. B. 187, 99-103 (1974).

25. Gutmann, E.: Ann. Rev. Physiol. 38, 177-216 (1976).

26. Syrovy, I., Gutmann, E.: Pflügers Archiv 369, 85-89 (1977).

27. Pelloni-Müller, G., Ermini, M., Jenny, E.: FEBS Letters 67, 68-74 (1976).

28. Rubinstein, N.A., Pepe, F.A., Holtzer, H.: Proc. Natl. Acad. Sci. USA 74, 4524-4527 (1977).

29. Rubinstein, N.A., Kelly, A.M.: Dev. Biol. 62, 473-485 (1978).

30. Pette, D., Vrbova, G., Whalen, R.G.: Pflügers Archiv 378, 251-257 (1979).

31. Rubinstein, N.A., Holtzer, H.: Nature 280, 323-325 (1979).

32. Whalen, R.G., Plaghki, L., Maréchal, G. : Abstracts, European Muscle Club, Heidelberg (1979).

33. Drachman, D.B., Johnston, D.M.: J. Physiol. 234, 29-42 (1973).

34. Huszar, G.: Nature New Biology 240, 260-264 (1972).

35. Masaki, T., Yoshizaki, C.: J. Biochem. (Tokyo) 76, 123-131 (1974).

36. Gauthier, G.F., Lowey, S., Hobbs, A.W.: Nature 274, 25-29 (1978).

37. Whalen, R.G.: submitted for publication (1979).

38. Cleveland, D.W., Fischer, S.G., Kirschner, M.W., Laemmli, U.K.: J. Biol. Chem. 252, 1102-1106 (1977).

39. Lompre, A.M., Bouveret, P., Leger, J., Schwartz, K.: J. Immunol. Methods 28, 143-148 (1979).

40. Brevet, A., Whalen, R.G.: Biochimie 60, 459-466 (1978).

41. Stockdale, F.E., Holtzer, H.: Exptl. Cell Res. 24, 508-520 (1961).

42. Holtzer, H., Rubinstein, N., Fellini, S., Yeoh, G., Chi, J., Birnbaum, J., Okayama, M. : Qt. Rev. Biophys. 8, 523-556 (1975).

TRANSITION IN MEMBRANE MACROMOLECULAR COMPOSITION AND IN MYOSIN ISOZYMES DURING DEVELOPMENT OF FAST-TWITCH AND SLOW-TWITCH MUSCLES

A. Margreth, G. Salviati, L. Dalla Libera, R. Betto, D. Biral and S. Salvatori.

Centro di Studio per la Biologia e la Fisiopatologia Muscolare del C.N.R., Istituto di Patologia generale, Università di Padova. 35100 Padova, Italy.

1. Introduction

It is now widely accepted that the pattern of activity of the muscles, linked to the firing pattern of the motorneurones, determines the twitch characteristic of fast and slow muscles, through specific effects on the synthesis of different polymorphic forms of myosin (see Ref. 41) and of the regulatory proteins (25), as well as by influencing the macromolecular composition and Ca^{2+}-pump activity of the sarcoplasmic reticulum (SR) (see Refs. 18, 6). As reviewed recently by Pette and Dölken (26), some though not all, metabolic features of skeletal muscles appear to be tightly matched to their contractile properties, such as, for instance, concerning the LDH isozyme pattern (14), and the presence in fast muscle, whether fast-red or fast-white muscle, of the glycerol-1-phosphate and FDP-ase-PFK substrate cycles (21, 17, 35).

Experimental studies and, in particular, the study of the fast-to-slow transformation of muscles following chronic stimulation at low frequencies, have shown a time-sequence of biochemical changes in the mitochondria, SR and myofibrils, which are providing interesting clues as to the trigger events and mediators involved in the overall change (26, 27, 6). As a contribution to this problem, we are reporting here data on the ontogeny of the biochemical changes leading to separation of fast-twitch and slow--twitch muscles.

Materials and Methods

Leg slow and fast muscles of the rat and rabbit were as described in previous work and other muscles, as reported in the text. Subcellular fractionation of the muscle homogenates was according to Refs.16,20,31. Myosin was isolated from rat muscles as in Ref.16, and, in the case of rabbit muscles, by either the method of Trayer and Perry (38) with subsequent purification by DEAE column chromatography (29), or according to Ref.2. Tryptic heavy meromyosin (HMM) was prepared according to Ref.1 under the conditions reported previously (4), and chymotryptic subfragment-1 (S-1), according to Ref.42. Myosin light chains were dissociated from the myosins of skeletal and cardiac (heart ventricle) of the rabbit and were fractionated by DEAE column chromatography by methods (3) derived from Ref.12. Soluble and mitochondrial enzyme activities were determined as described in Ref.16, and mGPDH, when measured from the reduction of potassium ferricyanide, according to Ref.10. The steady-state level of phosphoenzyme, c-AMP receptor protein and endogenous membrane phosphorylation of the isolated SR were determined according to Ref.20, and cytochrome b_5 and stearyl-CoA desaturase activity as in Ref.31. The K^+-EDTA and Ca^{2+}-ATPase activities of myosin, tryptic HMM and of chymotryptic S-1, were measured at 25° C (4) or at 37° C, (16). Actin-activated ATPase activity of S-1 was measured in the presence of 3 mM KCl. F-actin was purified according to Ref.36. Disc-gel electrophoresis of myosin and of tryptic HMM were carried out according to Weber and Osborn (39), or with borate buffer (24). Two-dimensional electrophoresis was carried out by the method of O'Farrel (22). Slab-gel electrophoresis was according to Laemmli (11). Antisera against fast-muscle myosin light chains were raised in chickens, using purified LC1F as the immunogen. Anti LC1F was purified from antisera obtained after a 9-month immunization period, by affinity chromatography on columns prepared by coupling purified LC1F to Sepharose 4B-CNBr. The immunological cross-reactivities among the several types of light chains were tested by a micro--ELISA assay (3).

A. Elements of Biochemical Uniformity between Future Slow and Fast Muscles at Early Stages of Development.

It is well known that at early stages of development, such as at or around the time of birth, in mammalian species which are born relatively immature, e.g. the rabbit and the rat, future fast and slow muscles are all slowly contracting muscles and that, by several biochemical properties, they are also similar to adult slow muscles.

Among the elements of biochemical uniformity which would support the view according to which postnatal differentiation involves mainly the fast muscles, the principal ones concern specific aspects in the macromolecular composition of muscle intracellular membranes. This appears to be the case, for instance, of mitochondrial glycerol-1-P dehydrogenase (mGPDH) which can be regarded as a specific marker of the inner mitochondrial membrane related to the contractile characteristics of the muscles.

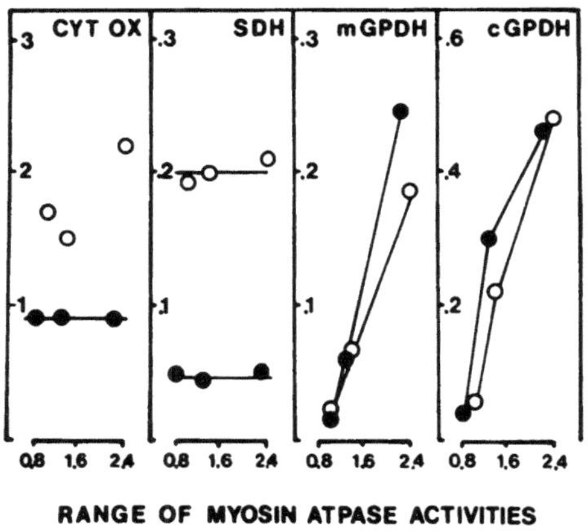

Fig. 1 - Relationship between the specific activities of Cyt ox, SDH, m-GPDH and c-GPDH, and the K^+-EDTA activated ATPase activity of the isolated myosin from leg slow, diaphragm and leg fast muscle of the rat (open circles) and the rabbit (filled circles), reported in the stated order on the abscissa. Enzyme activities are expressed as µAO consumed (Cyt ox), µmol cyt.c red. (SDH), µmol potassium ferricyanide red. (m-GPDH), or µmol NADH ox (c-GPDH), per min per mg of mitochondrial or of soluble protein, all measured at 30° c. Myosin ATP--ase activity was measured at 37°C. Mean values of several determinations.

Thus, the average mitochondrial content of m-GPDH shows a positive correlation with myosin ATPase activity and is relatively indipendent of the animal species, at variance with other membrane-bound enzyme activities, such as succinic dehydrogenase and cytochrome oxidase (Fig. 1). Furthermore, no appreciable differences in the enzymic composition of the mitochondria of future fast and slow muscles can be detected in newly born rabbits, when an initial separation is already apparent with respect to the concentration of cytoplasmic GPDH, as well as other glycolitic enzymes, e.g. LDH, in the soluble sarcoplasm (Table I).

TABLE I

COMPARISON BETWEEN FAST AND SLOW MUSCLES OF NEWLY BORN (1-7 days) AND ADULT RABBITS WITH RESPECT TO THE BIOCHEMICAL COMPOSITION OF ISOLATED SUBCELLULAR FRACTIONS

		IMMATURE MUSCLE	ADULT MUSCLE
A. MITOCHONDRIA			
SDH	nmol.min^{-1}mgpr^{-1}	59 (F)	52 (F)
	(cyt.c red.)	53 (S)	50 (S)
GPDH	nmol.min^{-1}mgpr^{-1}	28 (F)	244 (F)
	($K_3Fe(CN)_6$red.)	17 (S)	15 (S)
B. SOLUBLE SARCOPLASM			
LDH_t	µmol.min^{-1}mgpr^{-1}	2.44 (F)	10.60 (F)
	(NADH ox.)	0.95 (S)	4.90 (S)
LDH		1 (F)	1 (F)
	i.r.(low/high pyr.)	1 (S)	2.2 (S)
GPDH	µmol.min^{-1}mgpr^{-1}	0.11 (F)	0.47 (F)
	(NADH ox.)	0.04 (S)	0.04 (S)
C. SR MEMBRANES			
Phosphoenzyme	nmol^{32}P.mgpr^{-1}	0.58 (F)	4.13 (F)
		-	0.53 (S)
cAMP-receptor	pmol^{3}H-cAMP.mgpr^{-1}	1.77 (F)	0.26 (F)
Protein		-	1.98 (S)
Protein kinase	nmol^{32}P.mgpr^{-1}	1.24 (F)	0.18 (F)
Phosphorylation		-	1.41 (S)
Stearyl-CoA	nmol.mgpr^{-1}min^{-1}	0.01 (F)	0 (F)
Desaturase	(oleyl-CoA formed)	-	0.10 (S)
Cyt. b_5	nmol.mgpr^{-1}	0.16 (F)	0.01 (F)
		-	0.27 (S)

Mean values of several determinations. [1]Data from previously published work (20), except for cytochrome b5 and stearyl-CoA desaturase. For experimental details see Materials and Methods.

Likewise, the SR membranes derived from immature muscles, not only have a uniformly low content of Ca^{2+}-ATPase, as determined from the steady-state level of phosphorylated intermediate, but are endowed with a full complement of enzymatic systems characteristic of slow muscle SR, such as an endogenous cAMP-activated protein kinase and endogenous protein substrates for phosphorylation, as well as containing appreciable levels of the components of the NADH-cytochrome b_5 reductase-stearyl-CoA desaturase system (Table I).

B. Postnatal Differentiation of Fast Muscles.

It is well known from the early studies of Kendrick-Jones and Perry in the rabbit (9), as well as from our own studies in the rat (14,15), that phosphorylase and the enzymes of the glycolitic pathway increase steadily in predominantly fast muscles over a critical period between one and three weeks after birth, coincident with the beginning of motile activity and reach adult levels only after one month of age. As far as LDH is concerned, which has been mostly investigated (14), there is not observed in fast muscle the characteristic transition from the synthesis of M to H subunits found in developing slow muscle (Table I). For this reason, as well as because of their time-course, these changes can be visualized as, essentially, maturation changes associated with the sarcoplasmic growth of the muscle fibers (32).

On the other hand, from the time-course of accumulation of mGPDH in the mitochondria (Fig. 2), it appears that the rate of synthesis and assembly of the enzyme in the inner mitochondrial membrane occurs at a much faster rate than is the increase in concentration of cGPDH in the soluble sarcoplasm, and that, at two weeks after birth, the original mitochondrial population is completely substituted by newly formed mitochondria with an enzymic composition similar to that of the mitochondria of adult fast muscle. It is not unusual that different components of a metabolic or functional system evolve at different rates during ontogenesis, as discussed recently by Pette et al. (28). The interesting feature is, however, that mitochondrial differentiation is somehow time-correlated with the increase in Ca^{2+}-uptake activity of the SR, which is in fact the only earlier change found during postnatal development of fast muscle (Fig. 3).

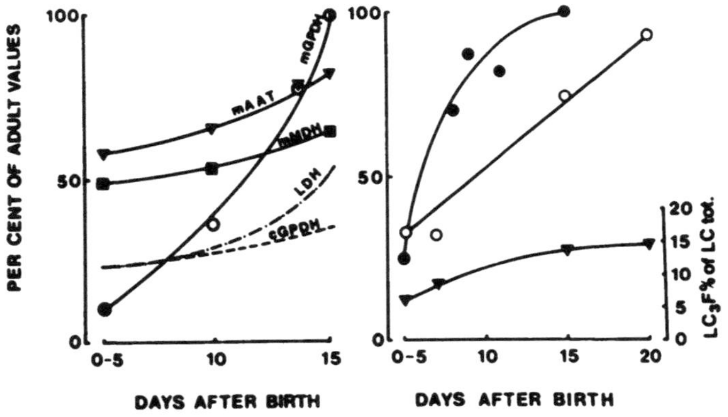

Fig. 2 - Developmental changes in the enzymic composition of the mitochondria and soluble fraction of rabbit fast muscles. The specific activities of each enzyme are the means of 3-9 determinations and are expressed as the percentage values of the activities in the corresponding fractions of adult fast muscle reported in Table I. The mean specific activities (µmol NADH ox. $min^{-1} mgpr^{-1}$) of AAT and MDH in fast-muscle mitochondria were 0.61 and 3.50, respectively. m-GPDH was measured from the reduction of either cyt. c (open circles) or of potassium ferricyanide (filled circles).

Fig. 3 - Developmental changes of the maximal Ca^{2+}-uptake activity (filled circles) of the isolated SR from rabbit hind leg muscles and concomitant changes of the Ca^{2+}-activated ATPase activity (open circles) and LC3F content (triangles) of myosin. Ca^{2+}-uptake activity values were recalculated from Ref.18. Myosin ATPase activity was assayed at 37° C, under the conditions reported in Materials and Methods. The amount of LC3F, expressed as the percentage of total light chain material, was determined by densitometry of electrophoretic gels, after staining with Coomassie-blue.

Furthermore, the developmental transition in the macromolecular composition both of the inner mitochondrial membrane and of the SR membranes, appear to preceed considerably the changes in the molecular properties of the myosin, as shown by the increase in Ca^{2+}- activated ATPase activity and in the myosin content of LC3F (Fig. 3). This time-sequence of changes is precisely that that has been observed in the transformation of muscles induced by either cross-reinnervation or chronic stimulation (See e.g. Refs.26,27,6).

Concerning the molecular basis for the changes in the catalytic properties of fast muscle myosin during postnatal development, it is not yet clear whether that should be explained by the concomitant accumulation of LC3F and transition from a pure LC1F myosin homodimer to a mixture of LC1F and LC3F-homodimers and a LC1F-LC3F myosin hybrid form (see Ref. 5). As shown in Table II, the observed differences in ATPase activity between the myosin

TABLE II

Comparison of ATPase activities of myosin and of myosin fragments between fast and slow muscles of 1 wk-old and adult rabbits

	Activator	1 wk Activity	1 wk Alkaline stability (%)	Adult Activity	Adult Alkaline stability (%)
MYOSIN	K^+, EDTA	1.25(F)	106	2.46(F)	100
		0.90(S)	88	0.85(S)	27
	Ca^{2+}	0.45(F)	89	0.60(F)	98
		0.24(S)	92	0.18(S)	17
HMM	K^+, EDTA	0.93(F)	105	2.00(F)	135
		0.59(S)	105	0.60(S)	53
S-1	K^+, EDTA	0.52(F)	135	1.08(F)	172
		0.66(S)	-	0.59(S)	64
	Mg^{2+}, actin	0.21(F)	-	0.49(F)	-
		-	-	0.09(S)	-

Mean values of several determinations are reported for the ATPase activities of myosin and of tryptic HMM,and representative values in the case of S-1. ATPase activities with K^+-EDTA and Ca^{2+} as activators were determined under conditions identical to those reported by Dalla Libera et al. (4) and with Mg^{2+}-actin in the presence of 3mMKCl and at an actin-to-S-1 ratio (wt/wt) of 0.5. All activities were measured at 25°C.

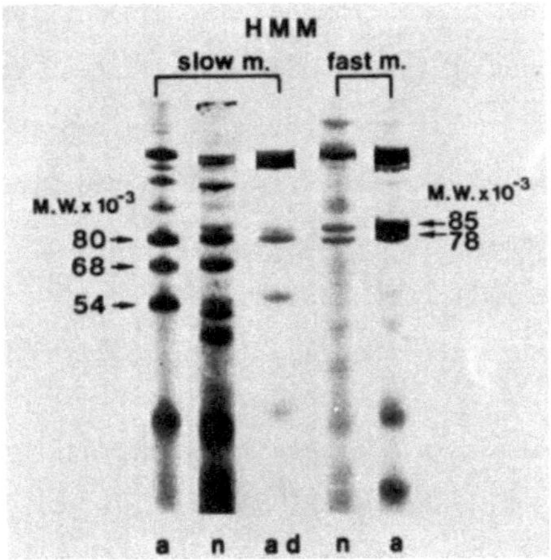

Fig. 4 - SDS-gel electrophoretic pattern of tryptic digests of myosins isolated from fast and slow muscles of adult (a) and newly born (n) rabbits and from two-years denervated rabbit soleus (ad). Electrophoresis was carried out according to Weber and Osborn (39), with 6% acrylamide.

of immature and adult fast muscle, apply equally to the heavy meromyosin fragment and subfragment-1, and whether Ca^{2+}, K^{+}-EDTA, or actin is the activator. Electrophoretic analysis of the main peptide constituents of tryptic HMM does not show any obvious change from one week to the adult stage (Fig. 4), but evidence has been reported recently (7) showing structural differences between the heavy chains of neonatal and adult muscle.

C. Postnatal Differentiation of Slow Muscles

Beside the above reported transition of LDH isozymes (see Table I), other changes related to protein polymorphism which more directely relate to the differentiation of slow muscles, concern the composition of the thin and the thick filaments of the myofibrils. As far as myosin in concerned, in particular, many studies have appeared (34,23,30,28), since our original observations (13,19) on myosin light chain composition, showing that both fast and slow myosin genes are expressed in the fibers of future slow muscles, at early stages of development. The fast-to-slow transition of myosin light chain isozymes, as slow muscle grows to maturity, is unattended by

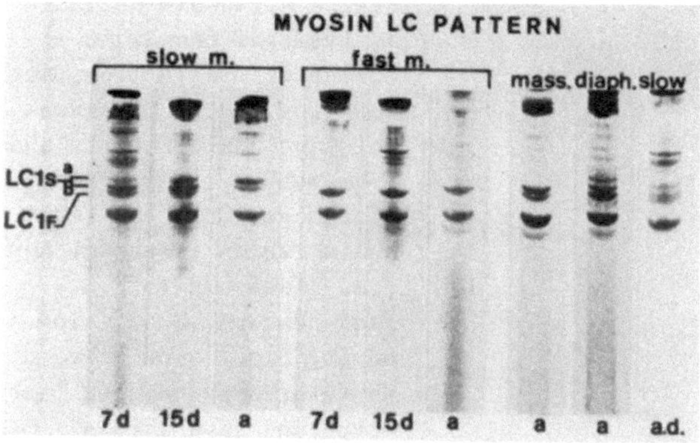

Fig. 5 - SDS-gel electrophoretic pattern of the light chains of myosins isolated from fast and slow muscles, masseter, diaphragm and from 2--years denervated soleus of rabbits. Electrophoresis was carried out with borate buffer, pH 7.0 and with 12.5% acrylamide. Key: a, adult; a.d., adult denervated; 7 d, 7 day-old; 15 d, 15 day-old.

changes in the ATPase activities of either the intact myosin, tryptic HMM or chymotryptic S-1, though it is marked by striking modifications in their susceptibility to alkali inactivation (Table II).

The alkali-stability of myosin ATPase in immature slow muscle, which is a property shared by rabbit embryonic and fast myosin (37), is unlikely to be due simply to the presence of fast-fetal isozyme in this myosin. Electrophoretic analysis of the HMM, obtained by mild tryptic digestion of the myosin of slow muscle from 1-week-old rabbits, reveals the presence of peptides characteristic of each type of adult muscle (Fig. 4). Therefore, hybrid heavy chain isozyme forms of myosin, might hypothetically occur, beside light chain hybrids (28).

The explanation for the increasing alkali-sensitivity of myosin ATPase associated with postnatal differentiation of slow muscle, may be more complicated than that, however. Thus, we have found that, parallel to the progressive loss from this myosin of the alkali-light chain characteristic of fast-fetal myosin, i.e. LC1F, the two functionally homologous light chains characteristic of adult slow myosin (40) accumulate at different rates. Whereas in slow myosin of adult rabbits, the amount of LC1Sa appears to be

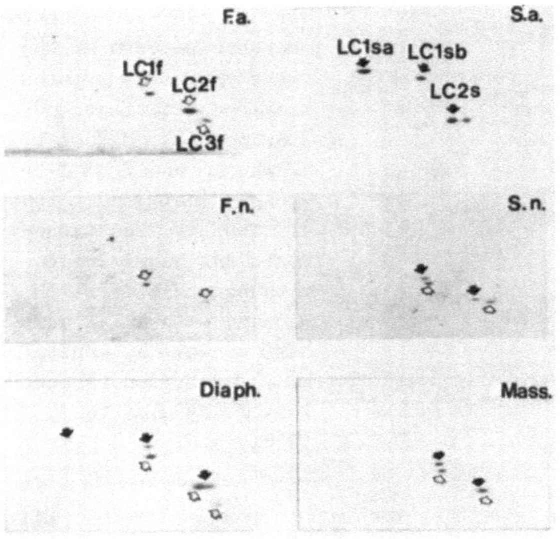

Fig. 6 - Two-dimensional analysis of the light chains of myosin from fast and slow muscle of 1-week--old and adult rabbits and from rabbit masseter and diaphragm. Only the light chain region is shown. Key: F.a., fast adult; F.n., fast newborn; S.a., slow adult; S.n., slow newborn; white arrows indicate fast--type and dark arrows slow--type light chains.

somewhat in excess to that of LC1Sb, the opposite appears to be true until 2-weeks of age (Fig. 5). As further shown by two-dimensional electrophoresis (Fig. 6), the myosin from slow muscle of 1-week-old rabbits, by its light chains composition consisting essentially of LC1F, LC2F and LC1Sb, LC2S, is easily distinguished from the myosin of either pure slow or fast muscle, as well as from the myosin of adult mixed muscle (diaphragm) which contains all the light chains characteristic of each type of muscle. Consequently, from what is known about myosin light chain isozymes, and by analogy with the developmental transition seen in fast muscle (see Refs.5,41), it is tempting to speculate that the earliest appearing isozyme form of slow muscle myosin in the course of differentiation corresponds to a LC1Sb homodimer, and that homo- and hetero-dimer forms with respect to LC1Sa are formed only at subsequent stages of postnatal development, i.e. at a time somehow coincident with the accumulation of LC3F in fast muscle myosin (see Fig. 3). In agreement with this interpretation, we have obtained circumstantial evidence from ion-exchange chromatography of S-1 (42) from rabbit slow muscle myosin, for the existence of a LC1Sa and LC1Sb species of S-1 with distinct alkali-stability properties.

Concerning the neural dependence of the overall developmental changes of myosin, i.e. both the switching-off of fast myosin genes and turning-on of the slow genes, it is relevant that chronic denervation (2-years) of rabbit soleus leads to a reversal of the myosin composition to that found in the newborn animal, with regard to both the heavy chains (Fig. 5) and the light chains (Fig. 6).

D. LC1 Light Chains from Slow-Twitch Muscle Myosin

It was shown by Weeds (40) that LC1Sa and LC1Sb from slow twitch muscle of the cat and the rabbit are structurally related, though phenotypically distinct light chains by their thiol peptides sequences and that, in both species, LC1Sb is more closely related to cardiac LC1. Rabbit LC1Sa and LC1Sb are more markedly heterogenous by isoelectric point than by molecular weight, and, accordingly, their electrophoretic separation, in the presence of SDS, is obtained only under certain conditions, such as with borate buffer or with tris-glycine, but not with phosphate buffer as in the conventional method of Weber and Osborn (39). Two-dimensional electrophoresis of the LC1Sa and LC1Sb light chains (Fig. 7) reveals that LC1Sb, by its overall properties, is more similar to rabbit LC1F and even more closely related to the LC1 light chain from myosin of heart ventricle muscle of the rabbit.

Since, by all these chemical differences, LC1Sa and LC1Sb light chains are, by definition, products of different genes, a pertinent question is whether the late expression of LC1Sa during differentiation of slow-twitch muscles may be triggered by some critical event associated with the maturation of tonic motorneurones, which, according to physiological data in the cat (8), is still incomplete at birth. If so, the appearance of LC1Sa would represent a basic feature of the differentiation process of these muscles, link-

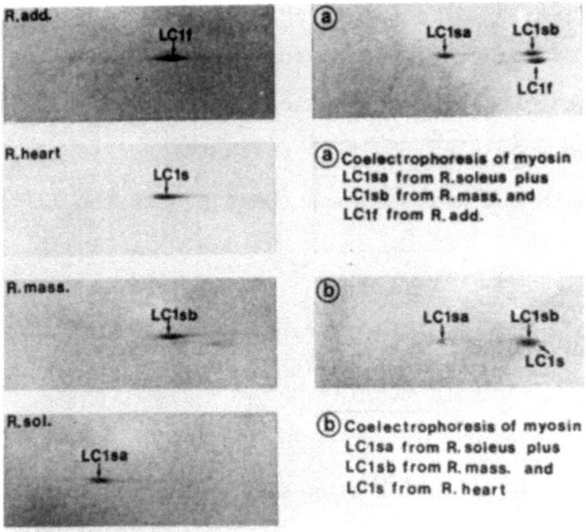

Fig. 7 - Two-dimensional analysis of purified light chains from rabbit skeletal and cardiac (heart ventricle) muscles.

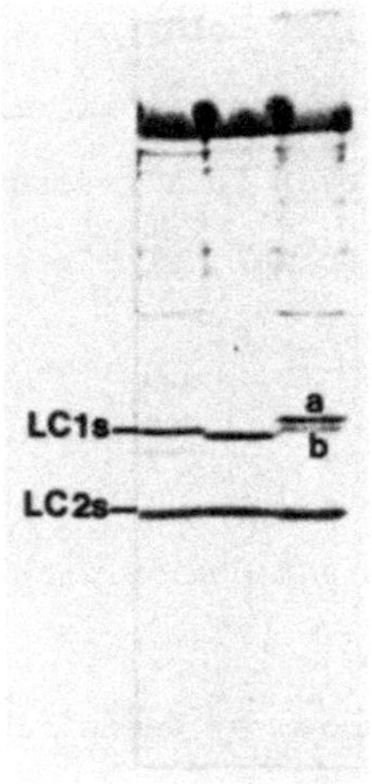

Fig. 8 - Slab-gel electrophoresis, according to Laemmli (11) of isolated myosins (2) from macacus monkey,rat and rabbit (left to right).

ed to the definitive attainment of their contractile characteristics. The available information from the literature, as well as our own experimental data, appear to indicate, however, that the existence of LC1S as a doublet is not an invariable feature of the myosin of slow-twitch muscle. As shown in Fig. 8, the myosin of rat soleus contains a single or, one greatly preponderant type of LC1, with an electrophoretic mobility similar to that of rabbit LC1Sb, and this appears to be the case also for the myosin of the homologous muscle of macacus monkey. Thus the myosin light chain pattern of the these slow-twitch muscles does not apparently differ from that reported for slow-tonic ALD of the chicken (28). Furthermore, the extent of myosin polymorphism, with respect to the LC1S light chain, can vary between the slow-twitch fibers of leg muscles and of muscles at other anatomical sites. For instance, we have found that the myosin of rabbit masseter, a muscle composed of a mixed fast-red, slow-red fiber population (17,18), has a light chain pattern which is consistent with its mixed composition, but

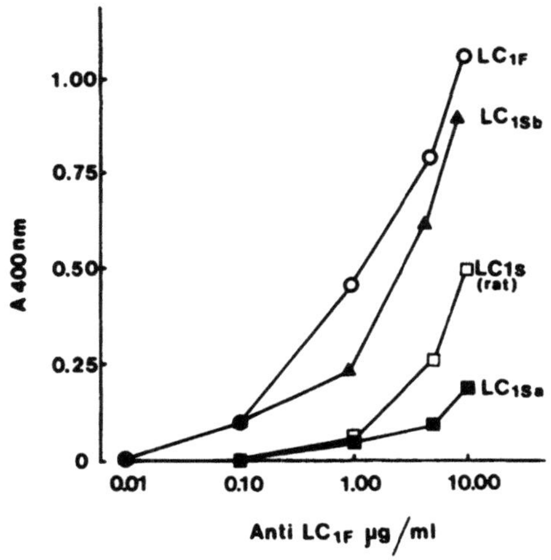

Fig. 9 - Immunological cross-reactivities between LC1F, LC1Sa and LC1Sb isolated from rabbit muscle myosins (see Materials and Methods). Each type of light chain (5µg/ml) was tested with anti-LC1F.

at variance with the diaphragm, lacks completely LC1Sa (Figs. 5, 6).

Concerning the evolution of myosin LC1S light chains, it is known (40) that the presence of an unusual proline-rich N-terminal peptide is a common feature of both these light chains, of cardiac LC1, and of fast myosin LC1F, whereas LC3F characteristically lacks this segment. Using anti-LC1F, we have found that LC1Sb is the most cross-reactive with the immunogen by a highly sensitive immunosorbent assay, (Fig. 9). On the other hand, the immunological cross-reactivity between LC1F and LC1Sa appears to be not only much less extensive, but even lower than that found for rat LC1S and for rabbit cardiac LC1 (not shown in the Figure). The antigenic differences between LC1Sa and LC1Sb, are even more remarkable since the long immunization period (9 months) should have helped in raising a population of cross-reactive antibodies. Consequently, the sharing of a larger number of antigenic determinats between LC1F and LC1Sb might imply that these two polypeptide light chains are more structurally interrelated. However, in view of earlier immunological studies (33, 3), the common antigenic determinants may be configurational rather than sequence determinants. Consequently, the relationship between LC1F and LC1Sb, on the one hand, and inferentially, between LC3F and LC1Sa, on the other hand, rests mainly based on pure biological thinking, at the present time.

This work was supported by funds from the Consiglio Nazionale delle Ricerche and, in part, by a grant from Muscular Dystrophy Association.

References

1. Balint, N., Srèter, F.A., Wolf, I., Nagy, B., Gergely, J.: J. Biol. Chem. 250, 6168-6177 (1975).

2. Barany, M., Close, R.I.: J. Physiol. 213, 455-474 (1971).

3. Biral, D., Dalla Libera, L., Franceschi, C., Margreth, A.: J. Immun. Methods, in the press.

4. Dalla Libera, L., Margreth, A., Mussini, I., Cerri, C., Scarlato, G.: Muscle and Nerve 1, 280-291 (1978).

5. Gauthier, G.F., Lowey, S., Hobbs, A.W.: Nature 274, 25-29 (1978).

6. Heilmann, C., Pette, D.: Eur. J. Biochem. 93, 437-446 (1979).

7. Hoh, J.F.Y., Yeon, G.P.S.: Nature 280, 321-323 (1979).

8. Huizar, P., Kuno, M., Myata, Y.: J. Physiol. 252, 465-479 (1975).

9. Kendrick-Jones, J., Perry, S.V.: Biochem. J. 103, 207-214 (1967).

10. Klingenberg, M.: Eur. J. Biochem. 13, 247-252 (1970).

11. Laemmli, U.K.: Nature 227, 683-685 (1970).

12. Lowey, S., Holt, J.C.: Cold Spring Harbour Symp. Quant. Biol. 37, 19-28 (1972).

13. Margreth, A.: Recent Advances in Myology, Amsterdam-Oxford, Excerpta Medica, 307-310 (1975).

14. Margreth, A., Angelini, C., Valfrè, C., Salviati, G.: Arch. Biochem. Biophys. 141, 371-377 (1970).

15. Margreth, A., Di Mauro, S., Tartarini, A., Salviati, G.: Biochem. J. 122, 597-599 (1971).

16. Margreth, A., Salviati, G., Di Mauro, S., Turati, G.: Biochem. J. 126, 1099-1110 (1972).

17. Margreth, A., Salviati, G., Carraro, U., Mussini, I.: Calcium binding proteins, Elsevier Scientific Pub. Co., Amsterdam, PWN-Polish Scientific Pub., Warsawa, 519-545 (1974).

18. Margreth, A., Salviati, G., Mussini, I., Carraro, U.: Exploratory concepts in Muscular Dystrophy II, Excerpta Medica, Amsterdam, 406-415 (1974).

19. Margreth, A., Carraro, U., Salviati, G.: Pathogenesis of human muscular dystrophies, Excerpta Medica, Amsterdam-Oxford, 161-170 (1977).

20. Margreth, A., Salviati, G., Salvatori, S.: Methodological surveys in biochemistry 6, 25-37, Ellis Horwood Ltd. Chichester, U.K. (1977).

21. Newsholme, E.A., Start, C.: Regulation in Metabolism, Wiley and Sons, London, New York, (1973).

22. O'Farrell, P.H.: J. Biol. Chem. 250, 4007-4021 (1975).

23. Pelloni-Müller, G., Ermini, M., Jenny, E.: FEBS Lett. 67, 68-74 (1976).

24. Perrie, W.T., Smillie, L.D., Perry, S.V.: Biochem. J. 135, 151-164 (1973).

25. Perry, S.V.: Exploratory concepts in Muscular Dystrophy, Excerpta Medica, Amsterdam, 319-328 (1974).

26. Pette, D., Dölken, G.: Advances Enzyme Regulation 13, 355-377 (1975).

27. Pette, D., Müller, W., Leisner, E., Vrbova, G.: Pflügers Arch. 364, 103-112 (1976).

28. Pette, D., Vrbova, G., Whalen, R.C.: Pflügers Arch. 378, 251-257 (1979).

29. Richards, E.G., Chung, C.S., Menzel, D.B., Olcott, H.S.: Biochemistry 6, 528-540 (1967).

30. Rubinstein, N.A., Kelly, A.M.: Develop. Biol. 62, 473-485 (1978).

31. Salviati, G., Betto, R., Salvatori, S., Margreth, A.: Biochim. Biophys. Acta 574, 280-289 (1979).

32. Schiaffino, S., Margreth, A.: J. Cell Biol. 41, 855-877 (1969).

33. Silberstein, L., Lowey, S.: Biochemistry 16, 4403-4408 (1977).

34. Syrovy, I., Gutmann, E.: Pflügers Arch. 369, 85-89 (1976).

35. Spamer, C., Pette, D.: Histochemistry 60, 9-19 (1979).

36. Spudich, J.A., Watt, S.: J. Biol. Chem. 246, 4866-4871 (1971).

37. Sreter, F.A., Balint, M., Gergely, J.: Dev. Biol. 46, 317-325 (1975).

38. Trayer, I.P., Perry, S.V.: Biochem. Z. 345, 87-100 (1966).

39. Weber, K., Osborn, M.: J. Biol. Chem. 244, 4406-4410 (1969).

40. Weeds, A.G.: Eur. J. Biochem. 66, 157-173 (1976).

41. Weeds, A.G.: Nature 274, 417-418 (1978).

42. Weeds, A.G., Taylor, R.S.: Nature 257, 54-56 (1975).

SARCOPLASMIC RETICULUM AND SARCOLEMMA DURING DEVELOPMENT

E. Zubrzycka-Gaarn and M.G. Sarzala
Department of Biochemistry of Nervous System and Muscle
Nencki Institute of Experimental Biology, 3 Pasteur St.,
Warsaw, Poland

Introduction

Sarcoplasmic reticulum (SR), the membrane system which regulates contraction - relaxation cycle of skeletal muscle by changing the concentration of calcium ions in the sarcoplasm (1), has been used in the studies on the molecular organization and assembly of biological membranes (2,3). The advantage of using the SR membrane as a model in these kinds of studies is due to its relatively simple protein and lipid composition (2). The major SR intrinsic protein, Ca^{2+}-transporting ATPase, of molecular weight (M_r) 105 000, accounts for about 70% of the total protein. The amphipathic molecules of the ATPase are inserted asymmetrically in the membrane (2,4). The hydrophobic part of each molecule spans the full width of the lipid bilayer; the hydrophylic part exposed to the sarcoplasm forms on the surface a 4 nm projection. Another intrinsic protein, a proteolipid of M_r about 12 000, makes up not more than a few percent of the membrane protein (5). Two highly acidic extrinsic proteins, calsequestrin (CS) and the high-affinity Ca^{2+}-binding protein (HACBP) have the M_r 45 000 and 55 000, respectively (5). CS constitutes for about 15 % of the total protein, whereas the HACBP for less than 10%. The function of these proteins is not clear. Most probably they play a role in sequestration of Ca^{2+} transported across the membrane (2). The experiments performed during the last few years in our laboratory have dealt with the characterization of the SR vesicles obtained from rabbit skeletal muscle during development, as well as from different types of mature muscle. For comparison the investigations were also performed on skeletal muscle sarcolemma isolated at different stages of development.

Results and Discussion

Preparation of the microsomal fractions. Muscle homogenates obtained from rabbit mixed developing (leg and back) muscle as well as from mature fast-twitch muscle (longissimus dorsi) and slow-twitch muscle (soleus, semitendinosus and crureus) were fractionated by differential centrifugation. In the earlier studies a fraction sedimenting between 8 000 and 50 000 x g was purified by 1 h centrifugation at 2 500 x g on a continuous sucrose-density gradient (6,7). The fraction taken from the top of the gradient (called through-out this paper "microsomal fraction"), and used in the earlier part of the studies, was free from the contaminations by myofibrils and mitochondria. In the recent studies (7-10) the fraction sedimenting between 8 000 and 50 000 x g was further fractionated by 1 h centrifugation at 131 000 x g on a continuous sucrose-density gradient. Two subfractions (called through-out this paper "heavy" and "light" subfraction) were obtained by this procedure (Fig. 1). Prolongation of the centrifugation time up to 4 h did not change the results of separation. Fractionation pattern shows (Fig. 1) that among the subfractions analyzed, those obtained from mature fast-twitch skeletal muscle are the heaviest, whereas those from undeveloped muscle are the lightest. Under identical conditions of centrifugation, a density of the fraction containing sarcolemmal vesicles essentially corresponds to that of the light subfraction from muscle of 3-4-day-old animals.

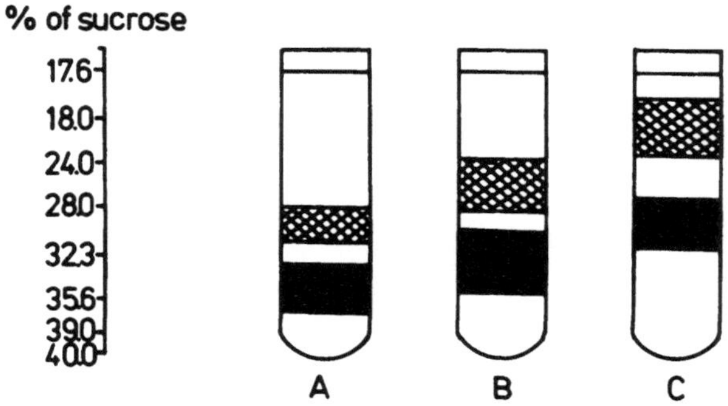

Fig. 1 Sucrose-density gradient centrifugation of the microsomal fractions from mature fast-twitch (A), slow-twitch (B) and 4-day-old (C) muscle. The microsomal fraction sedimenting between 8 000 und 50 000 x g was applied on a continuous sucrose-density gradient (17.6 - 40.0% sucrose) and centrifuged for 1 h or 4 h at 131 000 x g.

Purity of the subfractions was checked by determination of marker-enzyme activities (7-10). All heavy subfractions were essentially free from the contaminations by endoplasmic reticulum, sarcolemma, nerve endings and glycogen-protein particles. Virtually whole activity of marker-enzymes was found in the light subfractions.

Changes in the protein pattern and enzyme activities during development. The protein composition of the microsomal fraction was characterized at several steps of development (Fig. 2). During embryonic and the beginning of postnatal life the protein pattern of the vesicles differs significantly from that of the SR vesicles obtained from mature fast-twitch muscle. The 105 000 dalton protein makes up only few percent of the total protein and its content increases during development. The vesicles contain also several proteins of the M_r lower than 105 000, among them the 45 000 and 55 000 dalton proteins as well as a proteolipid (6).

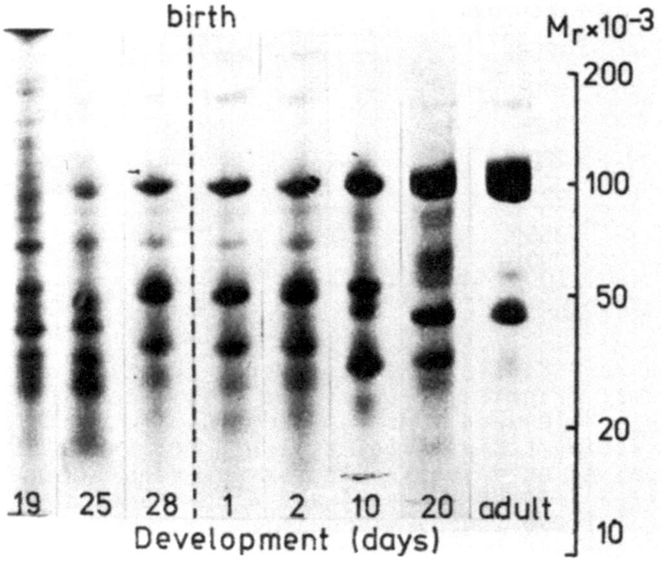

Fig. 2 Changes in the protein pattern of the microsomal fraction. From (6).

The activity of Ca^{2+}-dependent ATPase of the microsomal fraction is very low at the early stages of development (Fig. 3). Low activity of the ATPase is correlated with the low capacity to take up Ca^{2+} (6). A continuous increase of both activities is observed during development. In contrast to that, the activity of Mg^{2+}-dependent ATPase is very high at the beginning of postnatal life and then declines to a very low value in the vesicles from mature fast-twitch muscle (Fig. 3).

Changes in the ATPase activities of the microsomal fraction were compared with those of sarcolemma (Fig. 3) and (11). The activity of Ca^{2+}-dependent ATPase in sarcolemma also increases continuously during development. This activity is, however, still very low in the adult animals. The Mg^{2+}-dependent ATPase activity increases markedly up to twentieth day of postnatal life and then decreases to the value of 0.8 μmoles $P_i \times mg^{-1}$ protein $\times min^{-1}$ in adult animals. The activity of Mg^{2+}-dependent ATPase in sarcolemma reaches a maximum about two weeks later than that in the microsomal fraction (Fig.3). These results rather

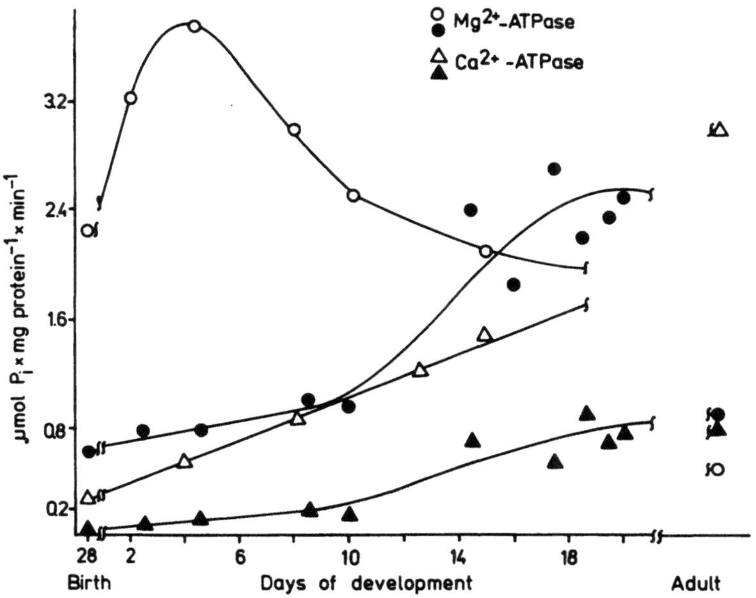

Fig. 3 Changes in the ATPase activities of the microsomal fraction and sarcolemma during development. The ATPase activity of the microsomal fraction △ ○ and sarcolemma ▲ ● were determined at 37°C according to the method described previously (6,11). Ca^{2+}-dependent ATPase (Ca^{2+}-ATPase) activity was calculated as a difference between total ATPase activity (determined in the presence of Ca^{2+} and Mg^{2+}) and Mg^{2+}-dependent ATPase activity (determined in the presence of Mg^{2+} and EGTA).

exclude the possibility that a very high Mg^{2+}-dependent ATPase activity found in the microsomal fraction just after birth, originates from the contaminations by sarcolemma.

Protein composition, ATPase activities and Ca^{2+} uptake in the microsomal subfractions. The microsomal fractions from mature fast and slow-twitch, as well as from undeveloped muscle of 4-day-old animals were further fractionated on a sucrose-density gradient. The subfractions obtained by this procedure were used in the recent studies.

Surprisingly, SDS-gel electrophoresis did not reveal any essential differences between the two subfractions obtained from 3-4-day-old animals (Fig. 4). In order to characterize the microsomal proteins, the vesicles of both subfractions were treated with EDTA (9). Following proteins were released by EDTA from the light subfraction: 160 000, 98 000, 60 000,

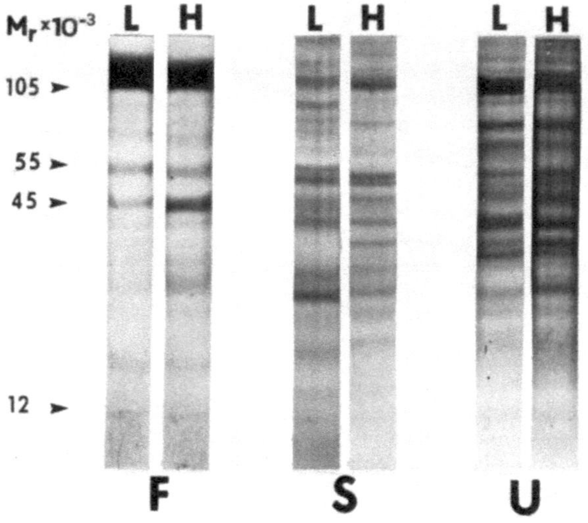

Fig. 4 Protein composition of the microsomal subfractions. SDS-gel electrophoresis of the microsomal light (L) or heavy (H) subfractions from mature fast-twitch (F), slow-twitch (S) or 4-day-old (U) muscle was carried out on 7.5% polyacrylamide slab-gels according to Weber and Osborn (12).

55 000 and 45 000 dalton. These proteins most probably correspond to the enzymes of glycogen-protein particles (13). Relatively high activity of phosphorylase b (M_r 98 000) and a high content of glycogen found in the light subfraction confirm this assumption. In the case of the heavy subfraction, EDTA caused the removal of the 80 000, 55 000 and 45 000 dalton proteins. The 55 000 and 45 000 dalton proteins are probably identical with HACBP and CS, respectively. Several proteins of M_r ranging from 105 000 to 12 000 were not released by EDTA

from the microsomal vesicles. We assume that these proteins belong to a group of intrinsic membrane proteins (9,14). Similar treatment with EDTA of the subfractions from mature fast or slow-twitch muscle also shows that all contaminations by glycogen-protein particles (if there were any) were always present in the light subfractions (7-10).

Although the protein pattern of the light and heavy subfractions from fast-twitch muscle is essentially similar (Fig. 4), the two subfractions differ, however, from each other in respect to the ratio of CS and HACBP. The heavy subfraction contains twice as much CS as HACBP, the light one contrary, more HACBP than CS (Fig. 4). In the heavy subfraction CS is located on the interior of the vesicles, whereas the HACBP is bound to the external surface (15). In the vesicles of light subfraction the localization of both proteins is opposite. The results seem to indicate that the light subfraction contains inside-out vesicles (8).

It has been recently reported (16,17) that the protein pattern of the SR vesicles obtained from slow-twitch muscle is more complicated than that of the vesicles from fast-twitch muscle. The results of the present work confirm these findings (10). The protein composition of the heavy subfraction isolated from slow-twitch skeletal muscle strongly resembles that of the heavy subfraction obtained from few-day-old animals (Fig. 4). Similarly, this subfraction is composed of several proteins of M_r between 105 000 and 17 000. It is worth mentioning that among them the 105 000 dalton protein represents only about 20% of the total protein.

After solubilization of the microsomal vesicles with detergents the proteins present in the extract were identified with the specific antibodies raised against Ca^{2+}-dependent ATPase and CS from fast-twitch muscle. The results indicate that the 105 000 and 45 000 dalton proteins present in the subfractions isolated from undeveloped and mature slow-twitch muscle are immunologically identical with the two major proteins of SR from fast-twitch muscle (9,10).

Table 1 shows that the activity of Ca^{2+}-dependent ATPase in the subfractions obtained from few-day-old and mature slow-twitch muscle, when calculated on a basis of the total protein, is 3-4 times lower, than that in the heavy subfraction from fast-twitch muscle. The same activity is, however, almost identical in all heavy subfractions when expressed on the basis of the content of 105 000 dalton protein (Table 1). It remains to be solved, why in spite of the fractionation on a sucrose-density gradient, the heavy subfractions from few-day-old and mature slow-twitch muscle still exhibit a high activity of Mg^{2+}-dependent ATPase (Table 1).

Table 1. ATPase activities and ability to take up Ca^{2+} by the heavy microsomal subfractions.
The ATPase activity was determined at 25°C.
Values obtained from scannings of the protein bands.

Heavy subfraction	Relative content of ATPase	Mg^{2+}-dependent ATPase	Ca^{2+}-dependent ATPase		Ca^{2+}-accumulation
	%	μmoles Pi/mg of total protein/ min	μmoles Pi/mg of total protein/ min	μmoles Pi/mg of ATPase protein/ min	μmoles Ca^{2+} /mg of total protein/ 5 min
mature fast - twitch muscle	70	0.14	0.90	1.20	3.00
undeveloped 3-4-day-old muscle	20	0.43	0.25	1.25	0.80
mature slow - twitch muscle	20	0.96	0.25	1.25	0.50

Morphology of the microsomal vesicles. Increase of the activity of Ca^{2+}-dependent ATPase and Ca^{2+}-uptake ability observed in the microsomal vesicles during development, is well correlated with the changes in their ultrastructure.

Negative staining reveals that at the beginning of development the microsomal fraction represents a heterogeneous population of the vesicles. At that time, thick-contour vesicles without any surface projections appear predominantly in the microsomal fraction. In addition, this fraction contains some thin-contour vesicles. A number of the vesicles covered with 4 nm surface projections and a number of these projections in each vesicle increase during development (6,9). It has been reported (3,18) that the number of intramembranous particles representing molecules of Ca^{2+}-dependent ATPase, visible in the freeze-fractured vesicles, also increases parallelly with age. After fractionation of the microsomal fraction from 3-4-day-old rabbit on a gradient the smooth and thick-contour vesicles are found in the light subfraction (9). Similar vesicles are present in the preparations of sarcolemma obtained at different stages of development (11). In contrast to that, the heavy subfractions from few-day-old animals and mature slow-twitch muscle are composed of thin-contour vesicles covered with 4 nm projections (9,10).

Lipid composition. In view of the requirement of phospholipids for the main function of SR (19) we have also measured the developmental changes in the lipid composition of this membrane.

The content and composition of lipids present in the microsomal fraction change considerably during development (3,6). These changes are attributed to the continuous decrease of the content of both neutral lipids and phospholipids. Thus, the ratio of total phospholipids to protein decreases during development from about 1.2-1.4 in the early embryonic life to

0.65-0.7 in the SR from mature fast-twitch muscle (6). The relative proportions of particular classes of phospholipids, expressed as a percentage of the total phospholipids, also change during development. Phosphatidylcholine, the major phospholipid of the SR membrane, corresponds only to about 50% of the total phospholipids in the microsomal fraction at the early embryonic stage and increases progressively to about 75% in adult animals. On the other hand, a percentage of phosphatidylethanolamine declines during ontogenesis from about 27% in embryos to about 14% in adult animals. The changes in phospholipid pattern continue after birth, e.g. in 30-day-old animals a percentage of phosphatidylethanolamine is still higher than that in the mature SR (6). Besides, the marked changes in the fatty acid composition of the membrane lipids, namely the decrease in palmitate and increase in linolate content, are observed. Fig. 5 shows that also a decrease of the content of monoenoic and increase of dienoic species of phosphatidyl-

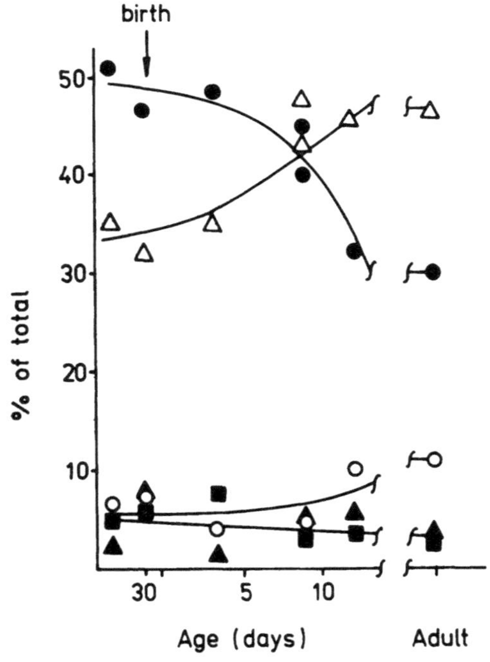

Fig. 5 Developmental changes in the content of molecular species of phosphatidylcholine isolated from microsomal fraction.

Molecular species were isolated by argentation thin-layer chromatography.

■ saturated, ● monoenoic, △ dienoic, ○ tetraenoic, ▲ hexaenoic.

From (20)

choline takes place during development.

In parallel to other determinations, the lipid composition of both subfractions from mature and few-day-old animals has been measured. The phospholipid (PL) to protein ratio is higher in the light than in the heavy subfractions from both, mature fast-twitch and few-day-old mixed muscle. The phospholipid pattern and cholesterol content are very similar in the two subfractions from mature fast-twitch muscle, whereas they differ from each other in the subfractions from undeveloped muscle (Table 2).

Table 2. Lipid content and composition of the microsomal subfractions and sarcolemma.

Abbreviations: PC, phosphatidylcholine; PE, phosphatidylethanolamine;others: PS, phosphatidylserine; PI, phosphatidylinositol; Sph, sphingomyelin.

Fraction	Age of animals	PL/protein mg/mg	% of total lipid			cholesterol
			PC	PE	others	μg/mg protein
microsomes:[1]	adult					
light		1.00	72.0	12.5	15.5	15.5
heavy		0.70	75.0	13.5	11.5	20.0
microsomes:[2]	3-4-					
light	day-	1.50	48.0	25.0	27.0	110.0
heavy	old	0.83	62.0	18.0	20.0	55.0
sarcolemma[3]	adult	0.68	50.5	27.0	22.5	200.0
sarcolemma[3]	3-4-day	0.83	44.5	27.5	28.0	205.0

From: 1/ (8), 2/ (9), 3/ (11).

The cholesterol content is significantly higher in the subfractions from undeveloped muscle than in those from mature fast-twitch muscle. Since the cholesterol content in any intracellular membrane is very low, the amount of cholesterol found in the microsomal subfractions might be due to contaminations by sarcolemma.

For comparison, the lipid content and composition of sarcolemmal vesicles is presented in Table 2. Similarly to all plasma membranes, sarcolemma contains considerable amounts of neutral lipids (40%) and cholesterol. In the vesicles isolated from few-day-old as well as from mature animals the molar ratio of cholesterol to phospholipids is about 0.6. Phosphatidylcholine accounts for about 50% and phosphatidylethanolamine for 27%. In contrast to the SR membrane, phospholipid content and composition of sarcolemma does not change significantly during development.

Phospholipid synthetizing enzymes. In view of the changes in the lipid content and composition occurring in the SR during development, we have also studied the phospholipid metabolism of this membrane.

The presence of all enzymes involved in the phospholipid synthesis described for other tissues has been found in the SR at all stages of muscle development studied (21). Fig. 6 shows developmental changes in the activity of investigated enzymes present in the microsomal fraction. The highest activity of glycerolphosphate acyltransferase(s) is observed already during early postnatal life. Thus, the most active synthesis of phosphatidic acid, a key intermediate in the synthesis of both neutral lipids and phospholipids, takes place at the time when the rapid increase of muscle mass occurs. The activity of lysolecithin acyltransferase is also the highest during the first days after birth.

The pattern of developmental changes in the lysolecithin acyltransferase activity coincides rather closely with the changes in the composition of fatty acids present in the phospholipids (3,20).

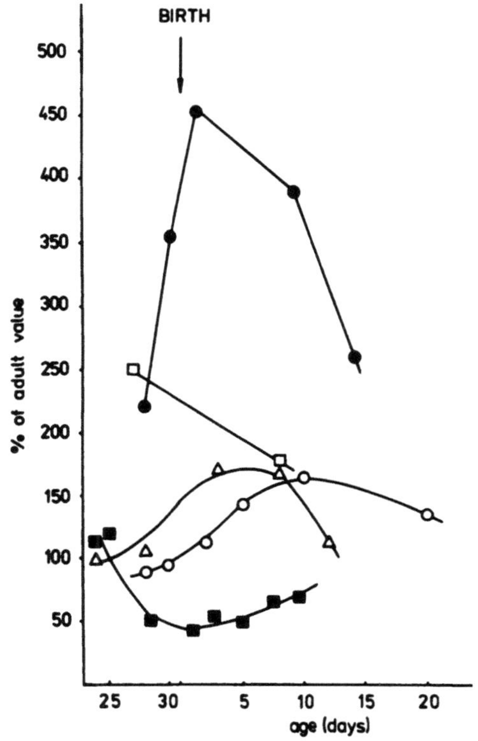

Fig.6. Developmental changes in the activities of enzymes involved in the phospholipid synthesis. The activity of each enzyme is expressed as a percentage of the activity found in the SR isolated from adult animals. ● lysolecithin acyltransferase, ○ glycerolphosphate acyltransferase(s) ■ cholinephosphotransferase, □ ethanolaminephosphotransferase, △ phosphatidylethanolamine methyltransferase.

From (21).

In contrast to acyltransferases, the activities of choline and ethanolamine phosphotransferases are the highest at the earliest period of development. Then they decrease continuously until about the 2nd day of postnatal life, and then again increase slightly. This decrease found at the time of intensive muscle development, when massive synthesis of the two major phospholipids takes place, is very difficult to explain. One of the possible explanations could be that decrease of the activity of both phosphotransferases may be compensated by the high activity of acyltransferase(s) observed when the synthesis of unsaturated species of phosphatidylcholine takes place.

All the results presented indicate that the appearance of 4 nm surface projections, as well as the marked increase of the activity of Ca^{2+}-dependent ATPase and capacity to take up

Ca^{2+} observed in the microsomal vesicles during development, are well correlated with the continuous increase of the content of 105 000 dalton protein. On the basis of the immunological properties we assume that this protein is identical with the Ca^{2+}-dependent ATPase protein of the SR membrane from mature fast-twitch muscle.

The lipid content, phospholipid composition, as well as the degree of phospholipid unsaturation, also change during development. The question arises, whether these changes have any influence on the function of the membrane enzymes, especially on the activity of Ca^{2+}-dependent ATPase. It has been shown recently that the Ca^{2+}-dependent ATPase activity of sarcoplasmic reticulum depends on rotational motion of the enzyme, or lipid fluidity, or both (22). However, the specific activity of Ca^{2+}-dependent ATPase in the microsomal fraction of undeveloped muscle is proportional to the content of the Ca^{2+}-dependent ATPase protein in the membrane. Furthermore, it has been shown that although the phospholipid content and composition of the SR differ at early stages of development, the fluidity of the lipid bilayer is reasonably constant (3).

The ability to phospholipid synthesis is an intrinsic property of the SR. The activity of the enzymes involved in phospholipid metabolism is probably sufficient for the synthesis in situ of all phospholipids essential for the membrane structure and function.

A comparison between the heavy subfractions from different kind of muscle leads to the conclusion that all of them represent SR-enriched material. There is a big similarity between the heavy subfractions from mature slow-twitch and 3-4-day-old muscle. This may suggest that in respect with the properties of SR, slow-twitch muscle is closer related to undeveloped than to mature fast-twitch muscle.

References

1. Ebashi, S., Endo, M., Ohtsuki, I.: Q.Rev.Biophys.2, 351-384 (1969).
2. MacLennan, D.H., Holland, P.C.: The Enzymes of Biological Membranes, A. Martonosi, New York,3, 221-259 (1976).
3. Martonosi, A.: Biochim.Biophys.Acta 415, 311-333 (1975).
4. MacLennan, D.H., Campbell, K.P.: TIBS 4, 148-151 (1979).
5. MacLennan, D.H., Zubrzycka, E., Jorgensen, A.O., Kalnins, V.I.: The Molecular Biology of Membranes, S. Fleischer, I. Hatefi, D.H. MacLennan, and A. Tzagaloff, New York, 309-320 (1978).
6. Sarzala, M.G., Pilarska, M., Zubrzycka, E., Michalak, M.: Eur.J.Biochem. 57, 25-34 (1975).
7. Michalak, M., Sarzala, M.G., Drabikowski, W.: Acta Biochim. Polon. 24, 105-116 (1977).
8. Sarzala, M.G., Michalak, M.: Biochim.Biophys.Acta 513, 221-235 (1978).
9. Zubrzycka, E., Michalak, M., Kosk-Kosicka, D., Sarzala, M.G.: Eur.J.Biochem. 93, 113-121 (1979).
10. Korczak, B., Zubrzycka, E.: manuscript in preparation.
11. Kosk-Kosicka, D.: Doctoral Thesis, Warszawa (1979).
12. Weber, K., Osborn, M.: J.Biol.Chem. 244, 4406-4412 (1969).
13. Wanson, J., Drochmans, P.: J.Cell Biol. 54, 206-224 (1972).
14. Sarzala, M.G., Zubrzycka, E., Michalak, M.: Calcium Transport in Contraction and Secretion, E. Carafoli, Amsterdam, 329-338 (1975).
15. Zubrzycka, E., Korczak, B., Sarzala, M.G., Drabikowski, W.: FEBS Lett. 90, 215-217 (1978).
16. Heilmann, C., Brdiczka, D., Nickel, E., Pette, D.: Eur.J. Biochem. 81, 211-222 (1977).
17. Margreth, A., Salviati, G., Salvatori, S., Libera, D.: Calcium Transport in Contraction and Secretion, E. Carafoli, F. Clementi, W. Drabikowski, A. Margreth, Amsterdam, New York, 383-399 (1975) .
18. Baskin, R.J.: J.Ultractruct.Res. 49, 348-371 (1974).
19. Martonosi, A., Donley, J., Halpin, R.A.: J.Biol.Chem. 243, 61-70 (1968).
20. Pilarska, M.: Doctoral Thesis, Warszawa (1975).
21. Sarzala, M.G., Pilarska, M.: Biochim.Biophys.Acta 441, 81-92 (1976).
22. Hoffman, W., Sarzala, M.G., Chapman, D.: PNAS USA 76, 3860-3864 (1979).

COMPARATIVE STUDIES OF VARIOUS KINDS OF TROPOMYOSIN

R. Dabrowska, J. Sosiński and W. Drabikowski
Department of Biochemistry of Nervous System and Muscle,
Nencki Institute of Experimental Biology, Warsaw, Poland

Introduction

Tropomyosin is a structural protein present in all types of muscle and also in non-muscle cells. It has been established that tropomyosins isolated from different sources have very similar amino acid composition and, in consequence, chymotryptic fingerprints. All tropomyosins are rod-shaped and highly α-helical and form dimers in which two subunits are twisted in coiled-coil manner. Usually the two subunits of tropomyosin isolated from various sources are very similar but not identical in respect to the molecular weight; the smaller subunit is called α, and the larger β. Subunits of tropomyosin from vertebrate smooth muscle are somewhat bigger and those from non-muscle cells much smaller than tropomyosin subunits of skeletal muscle (1-4). It has been found that in the skeletal muscle the proportion of the two subunits changes during development. At early stages of development the β subunit, characteristic for slow-twitch adult muscle, predominates. On the other hand in mature fast-twitch muscle subunit α considerably exceeds the amount of β subunit (5-7).

It is already well established that in all kinds of muscle the threads formed from the molecules of tropomyosin upon their aggregation in head-to-tail fashion lay in the two grooves of double-stranded F-actin filaments. In the skeletal muscle of vertebrates each tropomyosin molecule binds one molecule of troponin (8).

It has been generally accepted that tropomyosin plays an important role in the regulation of skeletal muscle contraction. In the relaxed vertebrate muscle the threads of this protein block sterically these sites on actin which interact with myosin. During contraction, as a result of binding of Ca^{2+} to troponin, tropomyosin threads alter their position in the grooves of the thin filament allowing myosin heads to interact with actin monomers (9).

Contrary to skeletal muscle the biological function of tropomyosin in smooth muscle remains still unclear, especially in view of the current concepts postulating that troponin is not involved in the regulatory mechanism of this system (10). Our knowledge on the localization and role of tropomyosin in non-muscle cells is even more scarce.

One of the main questions which still requires more study is the problem of specificity of each kind of tropomyosin in respect to its biological function. It has been suggested that in the skeletal muscle the proportions of the two subunits vary according to the speed of muscle contraction. Thus, the general question arises to which extent the difference in the activity of actomyosin ATPase and in its mechanism of regulation existing in various muscle and non-muscle systems is related to the particular kind of tropomyosin. In order to shed more light on these problems in the present work the comparative studies of various kinds of tropomyosin, isolated from skeletal and smooth muscle as well as from brain were undertaken. One of the aims was to establish whether the differences in the properties, possibly occuring between various tropomyosins, have any relation to their possible different functions.

Another part of our studies was focused on the composition of dimers present in various kinds of tropomyosin. Since in the previous studies (11-13) only $\alpha\alpha$ and $\alpha\beta$, but not $\beta\beta$, dimers of tropomyosin were found in adult fast-twitch muscle, the analysis of dimers in the tropomyosin isolated from fœtal muscle, where β subunit predominates, was performed. We have assumed that this approach can help to explain some problems of the assembly of tropomyosin threads during development and also the presence of homo- or heterodimers of tropomyosin in the single muscle fibres.

Results

Fig. 1 shows that all three kinds of tropomyosin: skeletal (rabbit), smooth (chicken gizzard) and non-muscle (bovine brain) used in this work are composed of two nonidentical subunits. The chain weight of gizzard tropomyosin is 37 000 and 39 000, that of skeletal muscle 34 000 and 36 000 and that of brain 29 000 and 30 000, in agreement with the previous studies (1, 15).

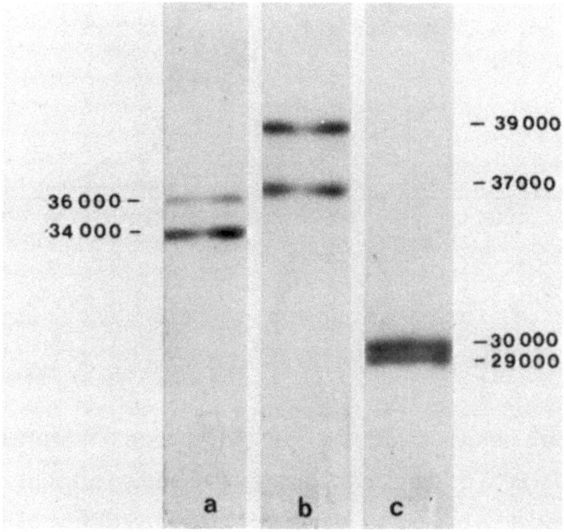

Fig. 1. Subunit pattern of various kinds of tropomyosin. SDS-gel electrophoresis on 10% acrylamide according to (14). a - rabbit skeletal muscle; b - chicken gizzard muscle; c - bovine brain.

At low ionic strength, the solutions of rabbit skeletal and gizzard muscle tropomyosin are viscous, as a result of their polymerization, with the viscosity of the latter kind of tropomyosin even somehow higher (16). The relative viscosity of both kinds of tropomyosin decreases sharply as the ionic strength is increased (Fig. 2). Brain tropomyosin, however, has low viscosity even in the absence of salt, which indicates that it is essentially unable to aggregate (Fig. 2) (17).

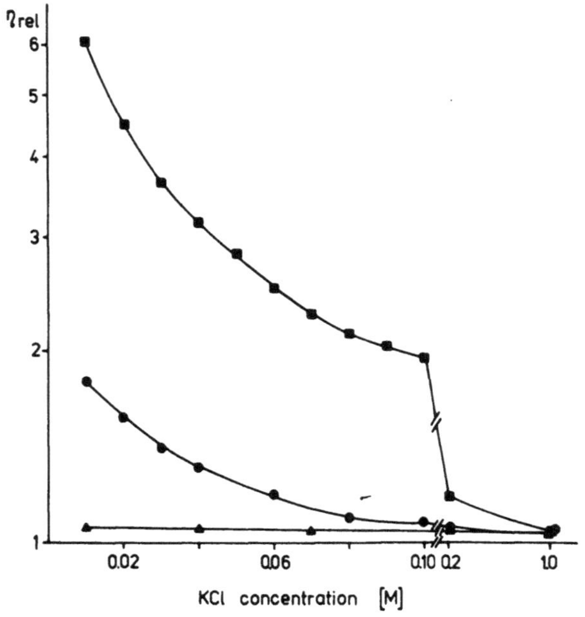

Fig. 2. Effect of ionic strength on the viscosity of various kinds of tropomyosin (16). Circles - rabbit skeletal muscle; squares - chicken gizzard muscle; triangles - bovine brain.

It has been shown that removal of four residues from C-terminal of rabbit skeletal tropomyosin leads to the loss of its polymerizability (18). Also in case of chicken gizzard tropomyosin, parallelly with the digestion with carboxypeptidase A a significant drop of the viscosity has been observed (16). All these observations suggest that the mechanism of polymerization of gizzard tropomyosin is similar to that of skeletal one. The lack of the ability of brain tropomyosin to polymerize indicates that either one or both N- and C-terminal sequences differ from those of the respective muscle protein, similarly as it has been recently found for platelet tropomyosin (19). These differences probably account for inability of non-muscle tropomyosin to form head-to-tail aggregates at low ionic strength.

The main function of skeletal muscle tropomyosin is its participation, together with troponin, in regulation of the contraction-relaxation cycle. Although troponin has been found only in striated muscle, it is important to establish whether smooth muscle, as well as non-muscle tropomyosin, retain the ability to interact with this protein. One of the approaches in our studies on the interaction of gizzard muscle and brain tropomyosin with the skeletal muscle troponin was to analyse the effect of the addition of the latter protein on the viscosity of tropomyosin. The large increase of viscosity that occurs when troponin is added to skeletal muscle tropomyosin is caused by the promotion of tropomyosin aggregation as a result of the interaction of the two proteins (20). The same promoting effect is observed when troponin is added to chicken gizzard tropomyosin (16). The relative enhancement of viscosity is even higher as in case of the skeletal muscle tropomyosin under the same conditions (Fig. 3). However, as shown in this figure, troponin is unable to cause aggregation of brain tropomyosin (17). Côté et al. (2) found also the lack of the effect of troponin on the viscosity of platelet tropomyosin and suggested that therefore proteins do not interact. On the other hand, the same authors found that platelet tropomyosin contains a region where the amino acid sequence is surprisingly almost identical to that of a region in rabbit skeletal tropomyosin (residues 197-217) postulated to be the binding site for troponin (2).

An important property of the skeletal muscle tropomyosin which has been extensively studied is its ability to bind

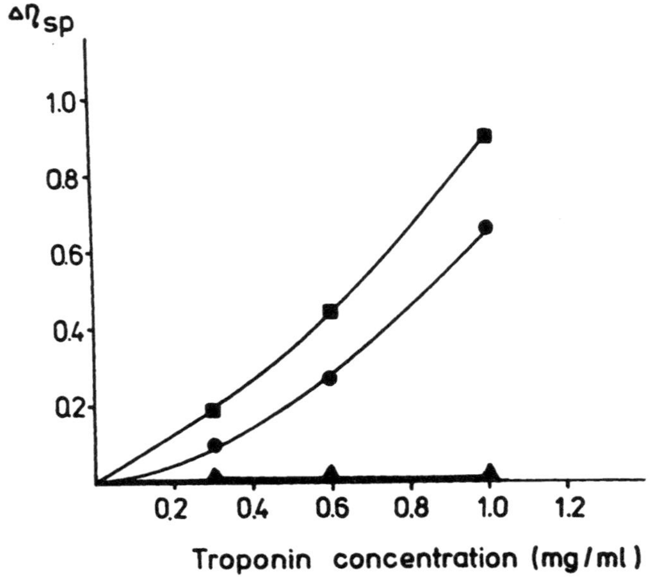

Fig. 3. Effect of troponin on the viscosity of various kinds of tropomyosin (16). Circles - rabbit skeletal muscle; squares - chicken gizzard muscle; triangles - bovine brain.

to F-actin. It is apparent that the primary structure of skeletal α tropomyosin shows repeating sequences which can be correlated to the seven actin monomers covered by one tropomyosin molecule (21). Sequence data for tropomyosin from smooth muscle are not available but in view of virtual identity of the stoichiometry of actin-tropomyosin interaction for both skeletal and smooth muscle tropomyosin, one could predict similar periodical distribution of amino acids in both tropomyosins. Indeed both proteins bind to F-actin at the ratios approaching the theoretical values (16). Preliminary experiments show that the binding of brain tropomyosin to skeletal muscle actin is somewhat weaker and the observed molar ratio found in the pellet is lower than in the case with both muscle tropomyosins. This difference can be explained in terms of the shorter length of the molecule of brain tropomyosin and the lack of its polymerizability. It is not known as yet whether molecules of brain tropomyosin polymerize on the thin filaments, i.e. form the threads, or bind to F-actin as the monomers.

Pellets formed after ultracentrifugation of the skeletal F-actin in 0.1 M KCl with gizzard muscle, as well as brain tropomyosin in the presence of troponin contain all three proteins. This indicates that at least in the presence of F-actin, troponin binds not only to smooth muscle but also to brain tropomyosin.

Substitution of gizzard muscle and brain tropomyosin for skeletal muscle tropomyosin in the regulatory system operating in the skeletal muscle also indicates that both proteins can interact with the skeletal troponin to confer calcium sensitivity on Mg^{2+}-stimulated skeletal muscle actomyosin ATPase (Fig. 4 A).

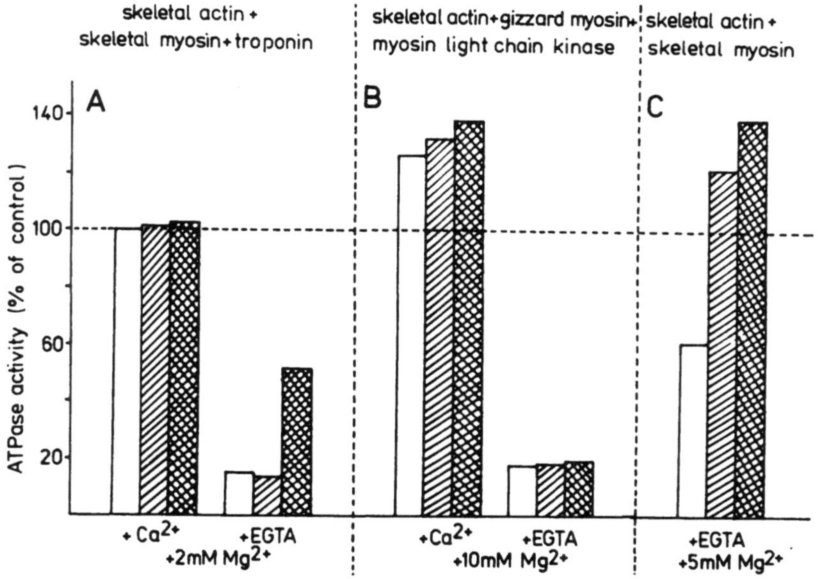

Fig. 4. Effect of various kinds of tropomyosin on the ATPase activity of actomyosin. White bars - rabbit skeletal tropomyosin; dashed bars - chicken gizzard tropomyosin; checkered bars - bovine brain tropomyosin. Incubation mixture contained 0.24 mg/ml of either skeletal or gizzard myosin. The ratio of actin to skeletal myosin was 1:4 (w/w), that of actin to gizzard myosin 1:2 (w/w). The ratio of tropomyosin to troponin to actin was 1:1:7 (molar ratio). KCl concentration: 30 mM in A and C, 50 mM in B.

At the same 1:1 molar ratio of troponin to tropomyosin the effect of skeletal and smooth muscle tropomyosin on the acto-myosin ATPase in the absence of Ca^{2+} is the same. The effect of brain tropomyosin is lower. However, more detailed studies show that one can also obtain the same inhibition of the ATPase activity as with skeletal and gizzard tropomyosin, but the amount of troponin and brain tropomyosin must be much higher (about three and two moles, respectively, per seven moles of actin monomer). These results suggest that shorter tropomyosin, perhaps in non-polymerized form, is unable to transmit the effect of one molecule of troponin on six actin monomers.

Brain tropomyosin, similarly to skeletal one, can also substitute gizzard tropomyosin in activation of actomyosin ATPase of gizzard muscle in the presence of myosin light chain kinase and calcium ions (Fig. 4 B).

Although under the above conditions smooth muscle tropomyosin does not differ from skeletal one, it behaves similarly to the non-muscle tropomyosin in respect to the effect on the skeletal muscle actomyosin ATPase measured in the presence of an excess of $[Mg^{2+}]$ over that of [ATP] (Fig. 4 C) under the conditions used by Eaton et al. (22). In contrast to the skeletal tropomyosin which in this case inhibits actomyosin ATPase gizzard and brain tropomyosins stimulate its activity.

The analysis of protein pattern in the precipitates of reconstituted actomyosin obtained from skeletal actin and both skeletal and smooth muscle myosin under all conditions of ATPase assay described above shows the presence of all three kinds of tropomyosin as well as troponin. The degree of binding is in all cases comparable to that found in F-actin pellets, i.e. the same for skeletal and smooth muscle tropomyosin and somehow lower for brain one. Fig. 5 shows an example of these experiments.

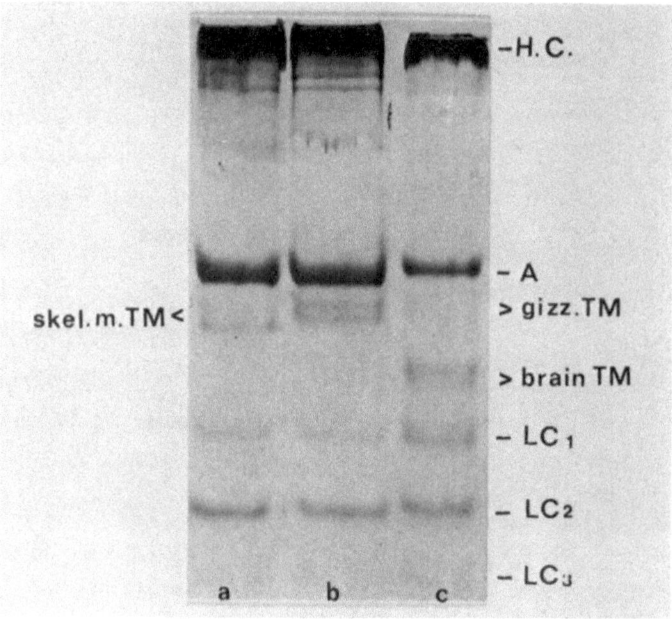

Fig. 5. Binding of various kinds of tropomyosin to rabbit skeletal actomyosin measured in the presence of 5 mM Mg^{2+} and 2.5 mM ATP under conditions of ATPase assay (see Fig. 4 C). a - rabbit skeletal tropomyosin; b - chicken gizzard tropomyosin; c - bovine brain tropomyosin. SDS-gel electrophoresis of the precipitates on 10% acrylamide according to (23). HC-myosin heavy chain; LC_1, LC_2, LC_3-myosin light chains, A-actin, TM-tropomyosin.

Tropomyosin threads in a single thin filament can be composed of one or two species of subunits. Consequently, one could expect the existence of either two kinds of thin filaments, containing only αα or ββ homodimers of tropomyosin or even three kinds of thin filaments, when in addition, αβ heterodimers do exist. In this work we have analysed the composition of dimers of various kinds of tropomyosin fixed by cross-linking of the subunits by S-S bonds. In these experiments cow muscle have been also used as a source of skeletal muscle tropomyosin. During development of bovine skeletal muscle (m. latissimus dorsi) the ratio of α:β subunit of tropomyosin changes from 0.6:1 and 0.8:1 in three and six month old foetus, respectively, to 1.6:1 in the muscle of adult cow (Fig.6).

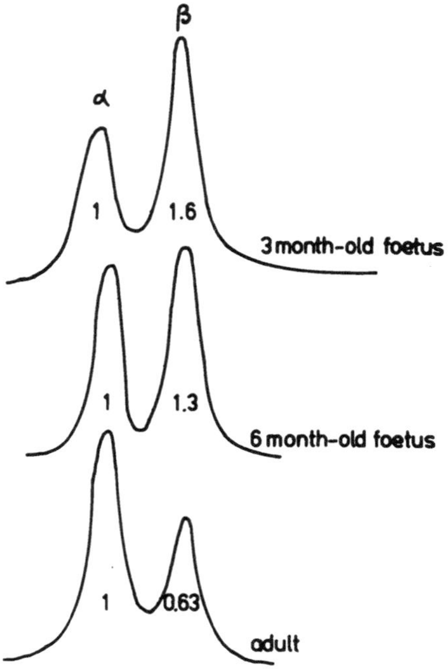

Fig. 6. Changes in the ratio of bovine skeletal muscle tropomyosin subunits during development. Scans of the gels are presented. Numbers refer to the ratio of α:β subunit.

In the tropomyosin preparations isolated from both adult and foetal cow muscle homo- and heterodimers are present in the proportions close to those calculated for the random model from the total ratio of α:β subunits present in the particular tropomyosin (Fig. 7 and Table 1). The only exception is the αα dimer. In all preparations of cow skeletal tropomyosin some deficiency of that dimer is noticed; in case of foetal muscle it is even virtually absent. The deficiency of ββ dimer in rabbit skeletal muscle seems to be much less pronounced than in cow skeletal muscle. When a mixture of foetal and adult cow tropomyosin with the resulting ratio of α:β subunits equalling about 1:1, after reduction with DTT and unfolding in 8M urea, is reoxidized after removal of urea, the

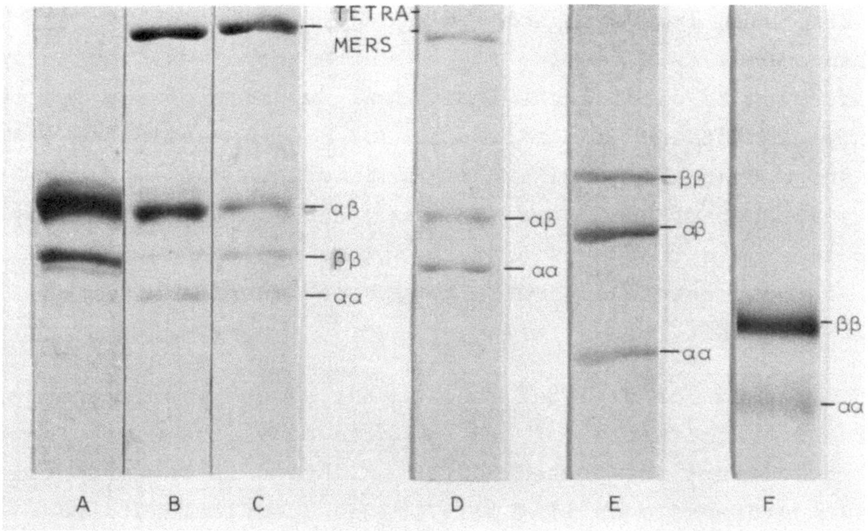

Fig.7. Dimers of various kinds of tropomyosin. a - 3 month-old cow foetus; b - adult cow skeletal muscle; c - mixture of a and b; d - rabbit skeletal muscle; e - chicken gizzard muscle; f - bovine brain. Conditions of SDS-gel electrophoresis as in Fig. 1.

Table 1. Relative proportions of dimers of various kinds of tropomyosin.

Source of tropomyosin	Ratio of α:β	Percentage of					
		αα		αβ		ββ	
		theoretical	found	theoretical	found	theoretical	found
Cow skeletal muscle	1.6:1	38.0	23.2	47.0	63.0	15.0	13.8
3-month old cow foetus	0.6:1	15.0	0	47.0	68.0	38.0	32.0
rabbit skeletal muscle	3.8:1	63.0	60.7	33.0	39.3	4.0	0
chicken gizzard muscle	0.9:1	23.0	25.7	50.0	52.0	27.0	22.3
bovine brain	0.4:1	10.0	28.0	42.3	0	48.5	72.0

proportions of dimers do somehow change, indicating partial rearrangement. As a result, the amount of αα dimers increases. The attempts to oxidize the individual subunits of cow muscle tropomyosin obtained after fractionation on CM-cellulose (24) have shown that the formation of αα dimers is much more difficult than that of ββ dimers. This may explain the deficiency of αα dimers in tropomyosin preparations and can be the result of the presence of two SH groups per one β subunit and one in α subunit.

In contrast to the skeletal muscle, the smooth muscle tropomyosin, with the ratio of α:β subunits equalling 0.9:1, homo- and heterodimers are present after oxidation in the proportion αα : ββ : αβ dimers equalling 1:1:2, i.e. identical with the values calculated for the random model. Unexpectedly, in case of brain tropomyosin only homodimers are present with the ratio of αα : ββ in favour of the latter one.

The present studies show that in spite of the different roles tropomyosin plays in skeletal and smooth muscle the properties of gizzard and skeletal muscle tropomyosin are very similar. Both proteins form polymers at physiological ionic strength and both are able to bind troponin. This binding reflects the ability of the complex of troponin with smooth muscle tropomyosin to function _in vitro_ as the regulatory system characteristic for skeletal muscle. Physicochemical and functional properties of brain (non-muscle) tropomyosin are more distinct as also shown by (25). This protein does not aggregate itself and troponin is unable to promote the aggregation. The mixture of troponin and brain tropomyosin, however, confers Ca^{2+} sensitivity on skeletal actomyosin ATPase which indicates the binding of brain tropomyosin and troponin. It is possible that for that binding the presence of actin is necessary. In this respect brain tropomyosin behaves similarly to the skeletal tropomyosin treated with carboxypeptidase A (26). It appears that the lack of polymerizability does not exclude the ability of tropomyosin to function together with troponin in the regulatory system of skeletal muscle. The only case when smooth muscle tropomyosin behaves similarly to the brain tropomyosin but differently to the skeletal muscle one is the effect on actomyosin ATPase determined in the presence of an excess of Mg^{2+}. The reason for this phenomenon is difficult to explain at the moment.

In respect to the role of tropomyosin in smooth muscle, i.e. its requirement for the full activation of actomyosin ATPase (27), this property seems to depend entirely on the interaction of tropomyosin with actin and therefore is common to all three kinds of tropomyosin studied. To sum up, all the present results indicate a high plasticity of actomyosin system in respect to tropomyosin.

The analysis of the composition of various tropomyosin preparations in respect to the presence of polymorphic forms of dimers reveals several very interesting features. Both muscle tropomyosins studied, but not the brain one, form heterodimers in addition to homodimers and the proportions of all dimers with the exception of αα homodimer of bovine skeletal muscle follow a statistical distribution. Thus the existence of ββ dimers, predicted at least in foetal skeletal muscle from the low ratio of α:β subunits, but not detected by previous workers, has been unequivocally demonstrated in the present studies. The ββ dimers are present also in the mature muscle. Our results also show that the recent suggestion (28) postulating the presence of α subunit in the fast-twitch fibers and the β subunit in the slow-twitch fibers may not be valid. One can expect that the observed tropomyosin subunit pattern reflects the situation <u>in vivo</u>. Preliminary experiments indicate that for the possible rearrangement of tropomyosin subunits in the dimers very drastic conditions are necessary. The present results indicate, however, that several points still have to be elucidated. One of the most intriguing questions which requires further study is the problem of localization of genes responsible for the synthesis of subunits of tropomyosin in the individual muscle cells, and the mechanism of the control of their expressions.

References

1. Cummins, P., Perry, S.V.: Biochem. J. 141, 43-49 (1974)

2. Côtê, G.P., Lewis, W.G., Pato, M.D., Smillie,L.B.: FEBS Lett. 94, 131-135 (1978).

3. Fine, R.E., Blitz, A.L.: J. Mol. Biol. 95, 447-454 (1975).

4. Smillie, L.B.: Trends in Biochem. Sci 4, 151-155 (1979)

5. Amphlett, G.W., Syska, H., Perry, S.V.: FEBS Lett. 63, 22-26 (1976).

6. Roy, R.K., Sreter, F., Sarkar, S.: Develop. Biol. 69, 15-30 (1979).

7. Dabrowska, R., Sosiński, J., Szpacenko, A.: Proc. V Meet. of the European Muscle Club, Bressanone, Italy. Abstr. of Commun. p. 68 (1976).

8. Ebashi, S.: Essays: in Biochemistry, 10, 1-36 (1974)

9. Huxley, H.E.: Cold Spring Harbor Symp. Quant. Biol. 37, 361-376 (1972).

10. Dabrowska, R., Sherry, J.M.F., Hartshorne, D.J.: Motility in Cell Function, John M. Marshall Symp. on Cell Biol., Philadelphia, 1977. Academic Press, New York (in press).

11. Strasburg, G.M., Greaser, M.: FEBS Lett. 72, 11-14 (1976).

12. Lehrer, S.S.: Proc. Natl. Acad. Sci, U.S.A. 17, 3377-3381 (1975).

13. Ookubo, N.: J. Biochem. Tokyo 81, 923-931 (1977).

14. Laemmli, U.K.: Nature 227, 680-685 (1970).

15. Bretscher, A., Weber, K.: FEBS Lett. 85, 145-148 (1978).

16. Dabrowska, R., Nowak, E., Drabikowski, W.: Comp. Biochem. Physiol. (in press).

17. Drabikowski, W., Nowak, E.: Proc. VI Int. Biophys. Congr. Kyoto, Japan, Abstr. of Commun. p. 154 (1978).

TROPOMYOSIN TRANSITIONS IN AVIAN AND MAMMALIAN SKELETAL MUSCLES

R.K. Roy, F.A. Sreter, M.G. Pluskal and S. Sarkar

Department of Muscle Research, Boston Biomedical Research Institute, and Harvard Medical School, Boston, Massachusetts 02114, U.S.A.

Introduction

It is currently believed that differences in functional activity, as judged by physiological and histochemical properties between adult avian and mammalian skeletal fast and slow-twitch muscles, are also reflected at the biochemical level in the structure and composition of the contractile proteins present in the fibers. The majority of the myofibrillar proteins present in adult skeletal muscles exist in multiple isozymic forms coded by different structural genes. The presence and relative proportion of these isozymes have been correlated with the muscle fiber type (for a review see reference 1). Changes in isozymic forms of myofibrillar proteins take place in skeletal muscle under a variety of conditions such as during development (2-13) and during transformation of typical fast or slow fibers to the opposite type by cross-reinnervation (14-19) or, in the case of a fast fiber, by chronic electrical stimulation of the intact nerve (20-25). Transformation of muscle fibers also produces at the biochemical level changes in the Ca^{2+}-transporting activity of the sarcoplasmic reticulum and enzymes of energy metabolism (22,27,28). These changes, as shown for myosin subunits (24,25) take place within pre-existing fibers suggesting that the type of innervation strongly influences the biochemical and physiological characteristics of adult muscle fibers.

With regard to the question whether or not embryonic muscles, which resemble adult slow-twitch muscles in many physiological properties (29,30), contain isozymic forms of myofibrillar proteins characteristic of the adult fast or slow-twitch muscles, the literature appears to be controversial. According to several reports (13,31) both fast and slow myosins are synthesized initially in embryonic fibers. Other investigators have reported that embryonic myosin is identical to adult fast myosin with regard to light chain patterns and histochemical and immunological properties (2-4,7,8,32,33). A third view is that embryonic myosins are distinct forms, which differ from adult myosins in primary structure of the heavy chains (2,11,34), and altered electrophoretic mobility of one of the light chains (35). How the synthesis and assembly of myofibrillar proteins are regulated during various stages of muscle development is an important question in myogenesis. We have chosen myosin light chains and tropomyosin (Tm) subunits in order to study in parallel the developmental changes in isozymic forms of these proteins and the relationship of these changes to the functional type of muscle. In this paper we shall review evidence showing that:(a) distinct isozymic changes involving these proteins take place in rabbit and chicken skeletal muscles during development leading to different distribution of the oligomeric proteins; (b) the embryonic fibers, in terms of these changes, can not be strictly considered as definitive fast or slow; (c) in chicken skeletal muscles the developmental changes of Tm subunits and myosin light chains do not occur in a strictly parallel manner; and (d) the isozymic changes of these myofibrillar proteins could be due to an intrinsic developmental program in all embryonic skeletal muscles regardless of their maturation into fast or slow fibers.

Methods

Both myosin and Tm were isolated from chicken and rabbit skeletal and cardiac muscles at various stages of development. The details of the preparation and purification of the proteins and assay of their biological activities have previously been published (6,7).

Denervation in young adult rabbits was performed by removing 10 mm sections of the sciatic nerve in one leg of the animals under aseptic conditions. After a period of 6 weeks the animals were sacrificed, the soleus, extensor digitorum longus (EDL), and tibialis anterior (TA) muscles were removed and immediately processed for the isolation of myosin and Tm.

The intramolecular disulphide crosslinking of Tm subunits by reaction of the reduced Tm with 5,5'-dithiobis(2-nitrobenzoate) ($(NbS)_2$) was carried out by the method of Lehrer (36).

In order to study the *in vitro* synthesis of the tropomyosin-troponin (Tm-Tn) complex, salt-washed free polysomes were prepared from 14-day-old chick embryonic muscles by our previously described procedure, except that 0.5M NH_4Cl was used instead of KCl as the monovalent cation in the preparative buffers (for details see also reference 37). The free polysomes were fractionated by centrifugation through a linear 15-40% sucrose gradient, as previously described (37) and the polysome fraction obtained from the middle portion of the gradients containing about 5-30 ribosomes (37,38) was used to program the *in vitro* synthesis of proteins using the pre-incubated wheat germ S-30 system (39). The standard incubation mixture in a total volume of 100 µl contained the following: 100 mM KCl; 20 mM Hepes, pH 7.4; 3 mM $MgCl_2$; 2.5 mM dithiothreitol; 1 mM ATP; 0.1 mM GTP; 6 mM creatine phosphate; 25 µg creatine kinase; 15 µCi [^{3}H]-leucine (New England Nuclear, specific activity 60 Ci/mmol); 3 µM each of 19 amino acids except leucine; 40 µl pre-incubated wheat germ S-30 and 1.5 A_{260} units of polysome. Incubations were carried out at 25° for 90 min. The *in vitro* products were mixed with nonradioactive Tm-Tn complex isolated from chick embryonic muscles (6). After removing the ribosomes by centrifugation at 160,000 xg for 120 min, the supernatant was chromatographed on a column of DEAE-cellulose for the isolation of Tm-Tn complex (6). The peak fractions eluted from the columns and containing the Tm-Tn complex were further purified by $(NH_4)_2SO_4$ fractionation as previously described (6).

Sodium dodecyl sulfate (SDS)-polyacrylamide gel electrophoresis in 10% gels, densitometric scans of the Coomassie-blue-stained gels and estimation of the molar stoichiometry of the myofibrillar proteins were carried out as previously described (6,7,40).

Results

1) Subunit patterns of Tm during development of rabbit and chicken skeletal and cardiac muscles. Fig. 1 shows the schematic representation of the

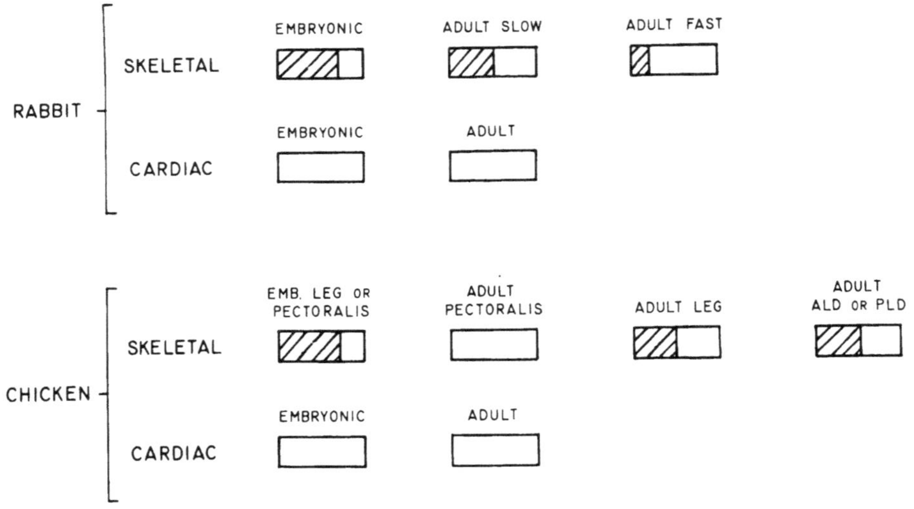

Fig. 1. Schematic representation of the relative proportions of the α and β subunits of Tm in various striated muscles of chicken and rabbit. ▭ , α subunit of Tm; ▨, β subunit of Tm.

relative ratio of the Tm subunits calculated from the densitometric scans of gel runs of the purified proteins isolated from adult and early embryonic skeletal and cardiac muscles of chicken and rabbit.

The β subunit is the major component present in both chicken and rabbit in early embryonic stages. It amounts to about 70% of the total subunits in 20-day-old rabbit embryonic skeletal muscles and 12-day-old chick embryonic leg and breast muscles. During development it is progressively replaced by the α subunit (for details see also reference 7 and Fig. 2). These changes depending on the species and the muscle type lead to various distinct patterns in adult skeletal fibers, such as: (i) the single α type in chicken m. pectoralis; or (ii) to a predominantly α type (about 80%) in rabbit fast muscles; or (iii) to a pattern consisting of about 50% α in rabbit slow and chicken leg muscles. The cardiac muscles of both chicken and rabbit contain only the α subunit which remains invariant during development, in contrast to the changes observed in skeletal muscles.

The α/β ratio of Tm subunits has been correlated with the speed of contraction of vertebrate skeletal muscles, the slow muscles having a

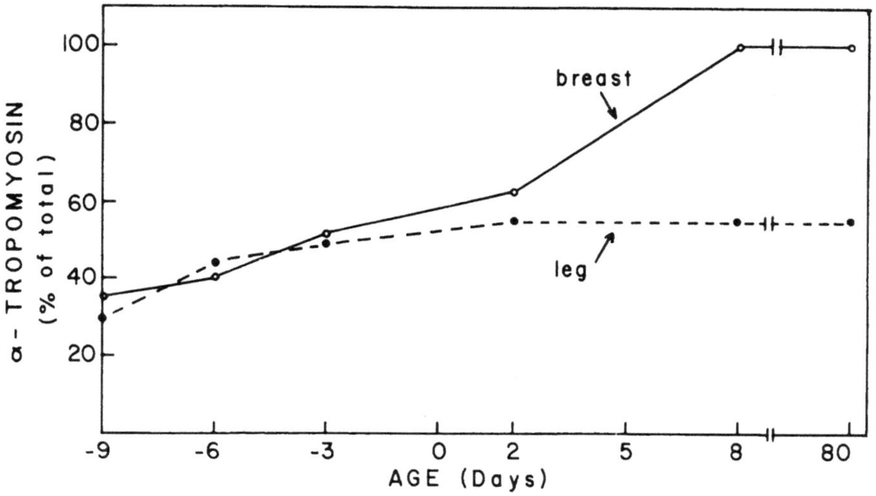

Fig. 2. Time course of developmental changes in chicken leg and breast muscle Tm. Reproduced from reference 7 with the permission of Academic Press, Inc.

higher content of β subunits than the fast muscles (1,41). Although, in general, this correlation appears to be true, Tm isolated from chicken anterior latissimus dorsi (ALD) and posterior latissimus dorsi (PLD), which are considered as typical avian slow and fast muscles, respectively, gave surprisingly identical subunit patterns (α:β = 55:45). It should be noted that myosins isolated from these two muscles, in terms of light chain patterns and immunological properties of the heavy chains, are either the slow or fast types respectively (3,8,33), and thus reflect the speed of contraction of the muscles.

The time course of developmental changes of Tm subunits in chicken leg and breast muscles is shown in Fig. 2. The rates of changes in both muscles during the early stages, from 12-day-old embryos to 2-day old chickens, were very similar. However, whereas in leg muscles no further change was observed, in breast muscles the β subunit decreased progressively till day 8 post hatching, when the single α subunit, characteristic of the adult stage, was reached. Thus, the transition of Tm subunit patterns from embryonic to adult types takes place at different rates in different skeletal muscles of the same species.

2) Myosin light chain patterns of chicken and rabbit skeletal muscles during development. In parallel studies we have examined the light chains of myosins of chicken and rabbit skeletal fast muscles during development. As judged by mobilities of the three light chains, LC_1, LC_2 and LC_3, myosins of both embryonic chicken and rabbit skeletal muscles resemble the myosins isolated from adult fast muscles of the corresponding species (for details see reference 7), except that LC_3 is either absent or present in reduced amounts in early stages of development (2,3,7,8,11,35). Fig. 3 shows that in chicken breast muscles, while LC_2 remains unchanged at about 2 moles/mole of myosin during development, the molar ratio of LC_1/LC_3 changed significantly between the period of hatching and day 8 post hatching.

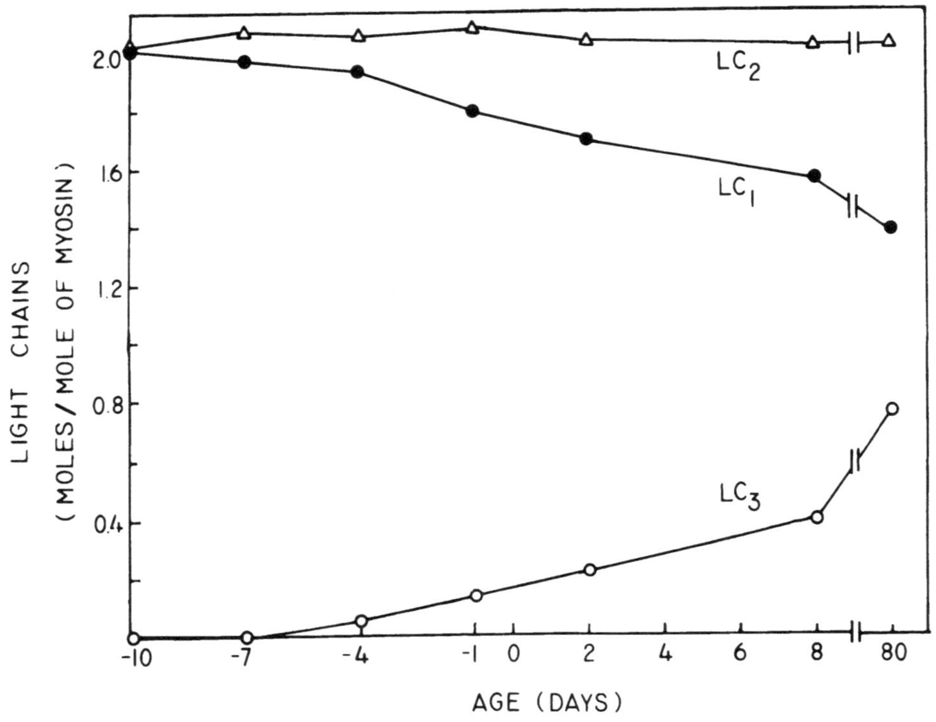

Fig. 3. Time course of developmental changes in the myosin light chain patterns of chicken breast muscles. Reproduced from reference 7 with the permission of Academic Press, Inc.

In terms of the concept of homo- and heterodimers of myosin isozymes containing LC_1 and LC_3 (40,42), these developmental changes indicate that LC_1-containing homodimers are the major species present in embryonic fast muscles (7,43). During development changes in the myosin isozyme populations occur until the pattern characteristic of the adult fast fiber is reached. These results are in agreement with recent studies by others who have shown the presence of multiple myosin isozymes, both homo- and heterodimers with respect to LC_1 and LC_3, in embryonic chicken pectoralis and newborn rabbit longissimus dorsi muscles (11,44). With regard to the myosin type of developing skeletal slow muscles, other investigators have shown that in both rabbit and chicken at early stages of development they contain light chains of both adult fast and slow myosins (4,8,11,44).

Comparison of the results shown in Figs. 2 and 3 indicates that developmental changes in myosin light chain and Tm subunit patterns of chicken pectoralis muscle are not coordinately regulated, the Tm subunit composition characteristic of the adult muscle being reached earlier. Interestingly, we have recently observed (26) that during transformation of rabbit fast muscles by chronic electrical stimulation, the Tm subunit ratio ($\alpha:\beta = 80:20$) is quantitatively altered to that characteristic of slow muscles ($\alpha:\beta = 55:45$) at 3 weeks - a period earlier than the 6-8 weeks required for the completion of corresponding changes involving myosin light chains. These results suggest that isozymic changes of different myofibrillar proteins in skeletal muscles take place in a temporal manner during both embryonic development and transformation of the adult fibers.

3) The synthesis and assembly of the regulatory proteins in stoichiometric amounts during embryonic muscle development. In order to test whether the components of the Ca^{2+}-regulatory complex are synthesized and assembled in embryonic muscles in the same stoichiometric ratios as in adult muscles, we have used two experimental approaches. The first one is to isolate the whole Tm-Tn complex from chick embryonic muscles and estimate the relative proportions of the components from densitometric scans of SDS-gel runs of the purified complex (6). The second approach is to analyze the *in vitro* products obtained in a cell-free translation system programmed with polysome fraction containing the mRNAs coding for the

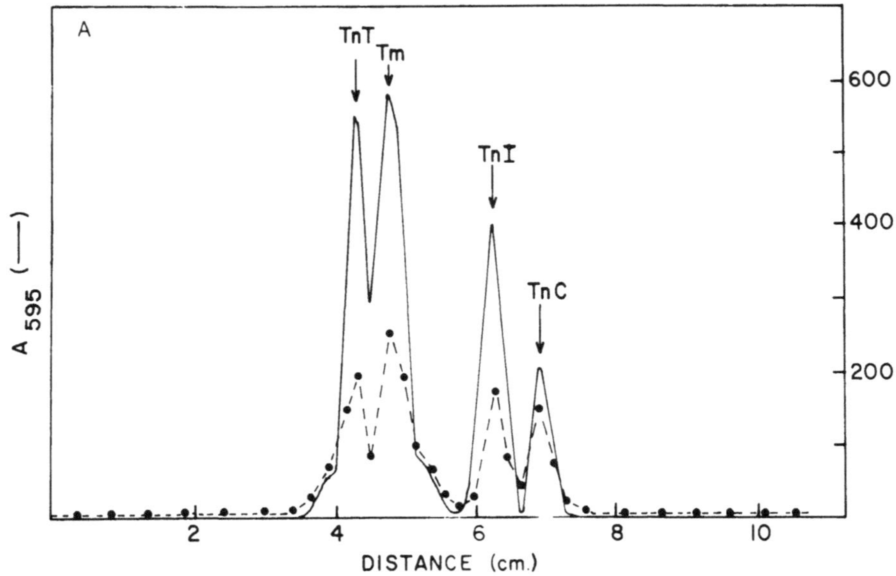

Fig. 4. SDS-polyacrylamide gel electrophoresis of in vitro products copurified with carrier Tm-Tn by DEAE-cellulose and Sephadex G-200 chromatography. For details see Methods and reference 6. The actin-free purified Tm-Tn complex (6) containing about 2300 cpm was analyzed in 10% gels as previously described (37).

polypeptide components of Tm-Tn complex. Fig. 4 shows the SDS-gel electrophoresis of the in vitro products after copurification with non-radioactive embryonic Tm-Tn complex. The correspondence between the densitometric scans and the radioactive peaks indicates that not only the individual components of the regulatory complex are synthesized in vitro but they are also assembled to form a complex which resembles Tm-Tn isolated from embryonic muscles. The relative ratio of the components of Tm-Tn isolated from embryonic muscles and the in vitro synthesized products (Fig. 4) is shown in Table I. The results indicate that the polypeptide components of Tm-Tn are synthesized both in vivo and in vitro in molar stoichiometric amounts.

4) Native subunit patterns of chicken embryonic and adult skeletal Tm. The high β subunit content of embryonic skeletal Tm suggests that the ββ dimer, which is absent in adult skeletal fast muscles (36,45,46), should

TABLE I. Stoichiometry of Components of Chick Embryonic Muscle Tm-Tn Synthesized Both *in vivo* and *in vitro*

Preparation	Relative Ratio of Components			
	TnT	TnI	TnC	Tm
(i)[a] *In vivo* synthesized; isolated from 14-day-old chick embryonic muscles	1.55	1.15	1.0	1.05
(ii)[b] *In vitro* synthesized by chick embryonic muscle polysomes in wheat germ S-30 system	0.83	1.0	1.0	0.80

[a]Calculated as average of six gel scans of the purified Tm-Tn complex. The values for each component are based on TnC as 1.0. The higher proportion of TnT may be due to the fact that the peaks of TnT and Tm are not completely resolved in gel runs (6).

[b]Calculated from the radioactivities in peaks corresponding to each component of Tm-Tn (Fig. 4) and their known leucine contents. The results are expressed with respect to TnC as 1.0.

be present in embryonic stages. We have examined the subunit pairing of Tm by SDS-gel electrophoresis of disulphide crosslinked (36) embryonic and adult chicken skeletal Tm. The 12-day-old chick embryonic Tm showed three bands in the dimer region (Fig. 5, gel a). The two faster moving bands which are not always clearly resolved, are tentatively assigned as αβ and αα in order of increasing mobility, based on the migration of the

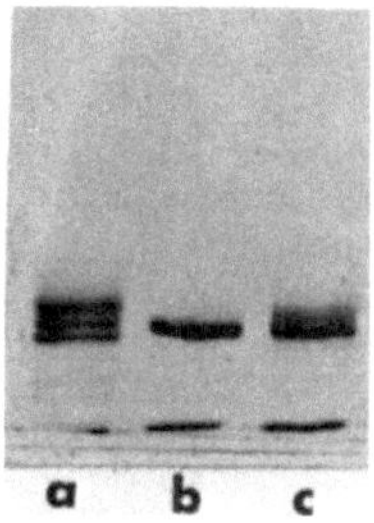

Fig. 5. SDS-polyacrylamide gel electrophoresis of chicken embryonic and adult skeletal Tm crosslinked with $(NbS)_2$. For details see reference 36. 6 to 8 μg proteins were applied on 5% gels. Gel a: 12-day-old embryonic chicken breast TM; gel b: adult chicken breast Tm; gel c: adult chicken leg Tm.

single αα dimer band obtained with a sample of crosslinked adult breast muscle Tm (gel b). The slowest of the three bands obtained with embryonic crosslinked Tm (gel a) should therefore be due to ββ. Adult chicken leg muscle Tm (α:β = 55:45) gave two bands, the major and faster moving band having mobility similar to the αα dimer (gel c vs. gel b). Since the αα and αβ bands are not always resolved, particularly if they are present as the major species, it is quite likely that the major band of adult chicken leg Tm (gel c) is due to αα and αβ dimers. By comparison (gel a vs. gel c) the slower moving minor band obtained with adult chicken leg Tm is due to ββ. Further work to characterize the subunit populations of adult and embryonic skeletal Tm are now in progress.

5) <u>Tm subunit patterns in denervated adult skeletal muscles</u>. We also tested the possibility that the characteristic Tm subunit patterns of adult fast and slow muscles may revert to the embryonic type upon removal of the neural influence by denervation. After six weeks of denervation in adult

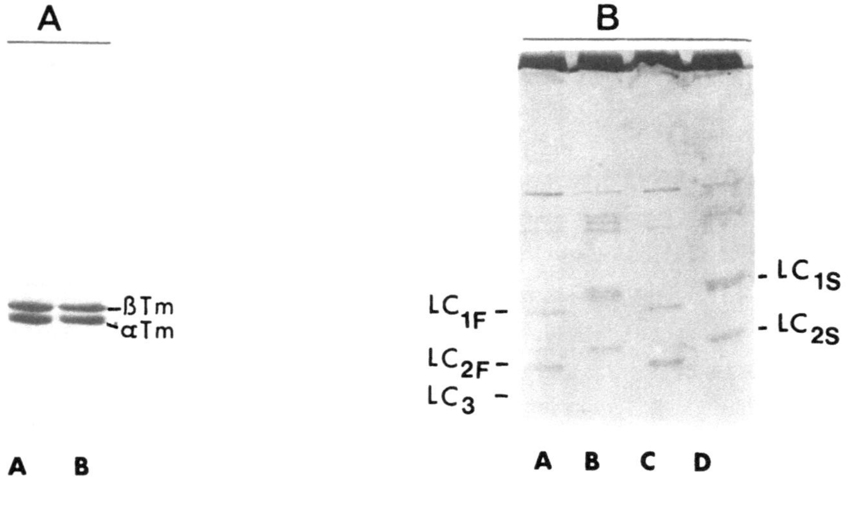

Fig. 6. SDS-polyacrylamide gel electrophoresis of Tm and myosin samples from denervated and contralateral skeletal muscles of adult rabbits. 10% gels were used. Panel A: A, Tm from denervated soleus muscle; B, Tm from contralateral soleus muscle. Proteins loaded, 8 μg. Panel B: A and B, myosins from contralateral fast and slow muscles respectively; C and D, myosins from denervated fast and slow muscles respectively. Proteins loaded, 30 μg.

rabbits the subunit patterns of Tm isolated from denervated and contralateral soleus were found to be similar, with α:β ratio of 55:45 (Fig. 6, panel A). Similarly, no significant change in the subunit patterns (α:β = 80:20) was observed between the contralateral and denervated fast muscles (results not shown here). With respect to myosin light chains both soleus and EDL muscles of the denervated animals essentially gave the same patterns as compared to the corresponding contralateral muscles (Fig. 6, panel B). Thus it appears that the Tm subunits and myosin light chain patterns of adult skeletal muscles are not significantly altered by denervation.

Discussion

The results presented here and the conclusions derived from studies on isozymic changes of myofibrillar proteins during fiber transformation and on regulation of myogenesis in cultured muscle cells may best be summarized by a model of muscle differentiation and growth shown in Fig. 7. The presumptive myoblast cells after two to three cycles of cell proliferation ultimately give rise to postmitotic mononucleated myoblast cells, which by a process of fusion form the embryonic multinucleated myotube (for a review see reference 47). The muscle-specific contractile proteins such as myosin, actin, troponin, tropomyosin, etc. are hardly detectable in replicative myoblasts. The latter, like most eukaryotes, contain the nonmuscle type of contractile proteins which is believed to be coded by a genetic program distinct from the one that codes for the sarcomeric contractile proteins in postmitotic myoblasts or myotubes. The isozymic changes of myofibrillar proteins during muscle development, as reported here and in the literature, strongly suggest that there exists an intrinsic developmental program in all embryonic skeletal fibers regardless of its future maturation into fast or slow type of muscle (7,8,32,33,48). This intrinsic program leads to the expression of specific kinds of myofibrillar proteins, e.g. myosin light chains of the adult fast muscle type; high levels of the β subunit of Tm; the fast form of TnI, and possibly the "embryonic" form of myosin. Subsequently, under the influence of exogenous factors or stimuli, such as neural influence, activity pattern, etc., other genes coding for different isozymic forms of the myofibrillar proteins are turned on, and this leads to the differentiation of individual muscle fibers into specific types. The appearance of the LC_3 light chain of the fast myosin, the switch from fast to slow type of myosin and TnI in slow muscles and the change in $\alpha:\beta$ ratio of tropomyosin subunits during muscle development may be due to the response of the fiber to extrinsic factors. These responses may be all-or-none-type switches, or alternatively they may involve quantitative changes in protein levels. Since the isozymic changes of myofibrillar proteins during development take place in terminally differentiated fibers, these changes are most likely to be due to a switch in gene expression in the same pre-existing fiber. This view is strongly supported by the work of Kugelberg (49) which showed that during development of rat soleus muscle type II fibers are gradually altered into type I fibers due to progressive differentiation of one and the same motoneuron.

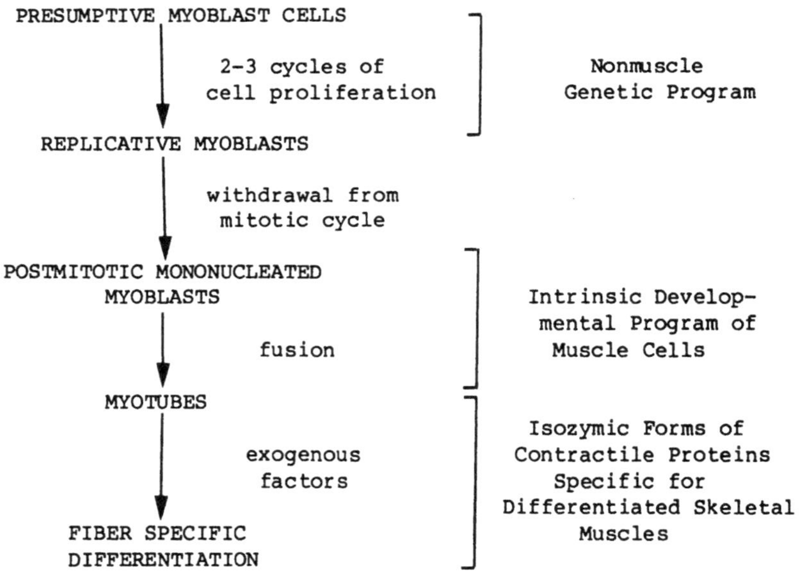

Fig. 7. A scheme for gene expression during differentiation and growth of skeletal muscle fibers.

Acknowledgment: This work was supported by grants from the National Institutes of Health and the Muscular Dystrophy Associations of America, Inc.

References

1. Perry, S.V.: Exploratory Concepts in Muscular Dystrophy, vol. II, A.T. Milhorat, ed., pp. 319-328, Excerpta Medica, Amsterdam, 1974.

2. Sreter, F.A., Balint, M., Gergely, J.: Develop. Biol. 46, 317-325 (1975).

3. Sreter, F., Holtzer, S., Gergely, J., Holtzer, H.: J. Cell Biol. 55, 586-594 (1972).

4. Pelloni-Muller, G., Ermini, M., Jenny, E.: FEBS Lett. 67, 68-74 (1976).

5. Amphlett, G.W., Syska, H., Perry, S.V.: FEBS Lett. 63, 22-25 (1976).

6. Roy, R.K., Potter, J.D., Sarkar, S.: Biochem. Biophys. Res. Commun. 70, 28-36 (1976).

7. Roy, R.K., Sreter, F.A., Sarkar, S.: Develop. Biol. 69, 15-30 (1979).

8. Rubinstein, N.A., Pepe, F.A., Holtzer, H.: Proc. Nat. Acad. Sci. USA 74, 4524-4527 (1977).

9. Syrovy, I., Gutmann, E.: Pfluegers Arch. 369, 85-89 (1977).

10. Dabrowska, R., Sosinsky, J., Drabikowski, W.: FEBS Lett. 79, 295-300 (1977).

11. Hoh, J.F.Y., Yeoh, G.P.S.: Nature 280, 321-323 (1979).

12. Pette, D., Vrbova, G., Whalen, R., Pfluegers Arch. 378, 251-257 (1979).

13. Gauthier, G., Lowey, S., Hobbs, A.: Nature 274, 25-29 (1978).

14. Amphlett, G.W., Perry, S.V., Syska, H., Brown, M.D., Vrbova, G.: Nature 257, 602-604 (1975).

15. Barany, M., Close, R.I.: J. Physiol. 213, 455-474 (1971).

16. Sreter, F.A., Gergely, J., Luff, A.R.: Biochem. Biophys. Res. Commun. 56, 84-89 (1974).

17. Weeds, A.G., Trentham, D.R., Kean, C.J.G., Buller, A.J.: Nature 247, 135-139 (1974).

18. Weeds, A.G., Burridge, K.: FEBS Lett. 57, 203-208 (1975).

19. Mommaerts, W.F.H.M., Seraydarian, K., Suh, M., Kean, C.J.G., Buller, A.J.: Exp. Neurol. 55, 637-653 (1977).

20. Salmons, S., Vrbova, G.: J. Physiol. 201, 535-549 (1969).

21. Sreter, F.A., Gergely, J., Salmons, S., Romanul, F.C.A.: Nature 241, 17-18 (1973).

22. Pette, D., Müller, W., Leisner, E., Vrbova, G.: Pfluegers Arch. 364, 103-112 (1976).

23. Salmons, S., Sreter, F.A.: Nature 263, 30-34 (1976).

24. Pette, D., Schnez, U.: FEBS Lett. 83, 128-130 (1977).

25. Rubinstein, N., Mabuchi, K., Pepe, F., Salmons, S., Gergely, J., Sreter, F.: J. Cell Biol. 79, 252-261 (1978).

26. Roy, R.K., Mabuchi, K., Sarkar, S., Mis, C., Sreter, F.A.: Biochem. Biophys. Res. Commun. 89, 181-187 (1979).

27. Sreter, F.A., Luff, A.R., Gergely, J.: J. Gen. Physiol. 66, 811-821 (1975).

28. Heilmann, C., Pette, D.: Eur. J. Biochem. 93, 437-446 (1979).

29. Buller, A.J., Eccles, J., Eccles, R.: J. Physiol. 150, 399-416 (1960).

30. Close, R.: J. Physiol. 173, 74-95 (1964).

31. Masaki, T., Yoshizaki, C.: J. Biochem. 76, 123-131 (1974).

32. Rubinstein, N.A., Kelly, A.M.: Develop. Biol. 62, 473-485 (1978).

33. Rubinstein, N.A., Holtzer, H.: Nature 280, 323-325 (1979).

34. Huszar, G.: Nature New Biol. 240, 260-264 (1972).

35. Whalen, R., Butler-Browne, G., Gros, F.: J. Mol. Biol. 126, 415-431 (1978).

36. Lehrer, S.S.: Proc. Nat. Acad. Sci. USA 72, 3377-3381 (1975).

37. Sarkar, S., Cooke, P.H.: Biochem. Biophys. Res. Commun. 41, 918-925 (1970).

38. Heywood, S.M., Rich, A.: Proc. Nat. Acad. Sci. USA 59, 590-597 (1968).

39. Roberts, B.F., Paterson, B.M.: Proc. Nat. Acad. Sci. USA 70, 2330-2334 (1973).

40. Sarkar, S., Cold Spring Harbor Symp. Quant. Biol. 37, 14-17 (1972).

41. Cummins, P., Perry, S.V.: Biochem. J. 133, 765-777 (1973).

42. Holt, J.C., Lowey, S.: Biochemistry 16, 4398-4402 (1977).

43. Takahashi, M., Tonomura, Y.: J. Biochem. 78, 1123-1133 (1975).

44. Hoh, J.F.Y.: FEBS Lett. 98, 267-270 (1979).

45. Ookubo, N.: J. Biochem. 81, 923-931 (1977).

46. Strasburg, G.M., Greaser, M.L.: FEBS Lett. 72, 11-14 (1976).

47. Holtzer, H., Sanger, J.W., Ishikawa, H., Strahs, K.: Cold Spring Harbor Symp. Quant. Biol. 37, 594-566 (1972).

48. Roy, R.K., Sreter, F.A., Sarkar, S.: Aging in Muscle, G. Kaldor and W.J. DiBattista, eds., pp. 23-48, Raven Press, New York, 1978.

49. Kugelberg, E.: J. Neurol. Sci. 27, 269-289 (1976).

FACTORS DETERMINING THE EXPRESSION OF THE GENES CONTROLLING THE SYNTHESIS OF THE REGULATORY PROTEINS IN STRIATED MUSCLE

G.K. Dhoot and S.V. Perry

Departments of Immunology and Biochemistry, University of Birmingham, P.O.B. 363, Birmingham B15 2TT.

Most skeletal muscles can be shown to be composed of several different fibre types by classical histochemical procedures, particularly by staining for the myosin ATPase reaction (see 1 for review). The nomenclature used to indicate these different cell types has varied from author to author. In general three main types can be distinguished on the basis of their morphological and histochemical properties but only two types based on the time course of the isometric twitch and the kinetic properties of the myosin ATPase at neutral pH (2). Certain skeletal muscle cells, the number of which varies in different muscles are intermediate in staining properties (1). It has been suggested that the three types should be called fast-twitch white, fast-twitch red and slow-twitch intermediate fibres. The ATPase activities after incubation at pH 9.4 of red and white fast-twitch muscle fibres are high and they are usually referred to as type II fibres, whereas the fibres staining weakly for ATPase activity under these conditions are called type I fibres (3). The latter nomenclature will be used throughout this discussion.

Only two main types of motor units can be distinguished in most muscles and the present view is that the type I fibres are innervated by the slow motor nerves and the type II fibres by the fast motor nerves.

The time course of the contractile response that distinguishes the different muscle fibre types will depend on a number of factors. One of these is the nature of the myosin and this is reflected in the way in which different cell types stain histochemically for this enzyme. A particularly important factor will be the nature of the regulatory proteins of the I filament for clearly their properties will determine the way in which the ATPase activity of the myofibril, and hence the contractile response, changes as a result of the Ca^{2+} liberated into the sarcoplasm when muscle is stimulated.

The I filament consists of five proteins, actin, tropomyosin, troponin C, troponin I and troponin T, all of which are known to play a part in affecting the Mg^{2+}-stimulated ATPase activity of the myosin, although only the tropomyosin and the troponin complex are usually referred to as the regulatory proteins. Each of the regulatory proteins have been shown, by sequence or immunochemical studies, to exist in two forms in skeletal muscle of mammals (see 4 for review). The different forms of each of the troponin components and tropomyosin have unique primary sequences and are, therefore, clearly different gene products (Table 1).

	Type I Fibre	Type II Fibre	Ref.
Myosin	Slow A_1 homodimer	Fast A_1 homodimer (*FM_3)	
	Slow A_1A_2 heterodimer	Fast A_1A_2 heterodimer (*FM_2)	5-8
	Slow A_2 homodimer	Fast A_2 homodimer (*FM_1)	
Actin	α ?	α	9
Tropomyosin	β	α (very similar, if not identical with cardiac form)	4
Troponin C	Slow (= cardiac)	Fast	10,11
Troponin I	Slow	Fast	12
Troponin T	Slow	Fast	10

*Nomenclature of Hoh (8)

Table 1. Distribution of the different forms of the major myofibrillar proteins in adult mammalian skeletal muscle.

The relative amounts of the two forms of each of the regulatory proteins are fairly constant in each muscle in a given species but they do vary between different types of muscles in the same species in a manner that suggests that there is a correlation with the muscle speed. Thus it would appear that the genes controlling the synthesis of the regulatory proteins are in some way affected by the activity pattern of the muscle. Study of the distribution of the polymorphic forms of the regulatory proteins has therefore been undertaken in attempt to understand the factors that determine the way that muscle cells change in composition and physiological properties in response to their environment.

Preparation of antigens and antibodies

In general antibodies to the different forms of the tropomyosin and the components of the troponin complex are monospecific for the particular form of protein used as antigen but they are not species specific. For this study α and β tropomyosin and the various forms of each of the components of the troponin complex were isolated from rabbit or chicken muscle in as pure a form as possible and injected into guinea pigs. Serum was collected, the specificities of the antisera carefully monitored and sheep anti-guinea pig immunoglobulin conjugated to type VI horseradish peroxidase (see 10 and 12 for details). The different forms of the regulatory proteins were located by staining frozen serial sections (6 μ thick) using the immunoperoxidase technique as described by Dhoot _et al._ (10).

Distribution of different forms of the troponin components in normal adult skeletal muscle

Study of cross sections of normal adult muscles by the immunoperoxidase technique indicated that the two forms of each of the troponin components were located in different cell types. The fast skeletal muscle forms of troponin I, troponin C and troponin T were restricted to type II fibres and the corresponding slow skeletal forms of the components of the troponin complex to type I fibres (Fig.1a, 1b, 2a, 2b and 2c). Usually this was the normal appearance in all muscles examined from the human, rabbit, rat, mouse, hamster, rhesus monkey and baboon (12). The only exceptions to this distribution were noted in the soleus at certain stages of its development. Although about 2 weeks after birth the two

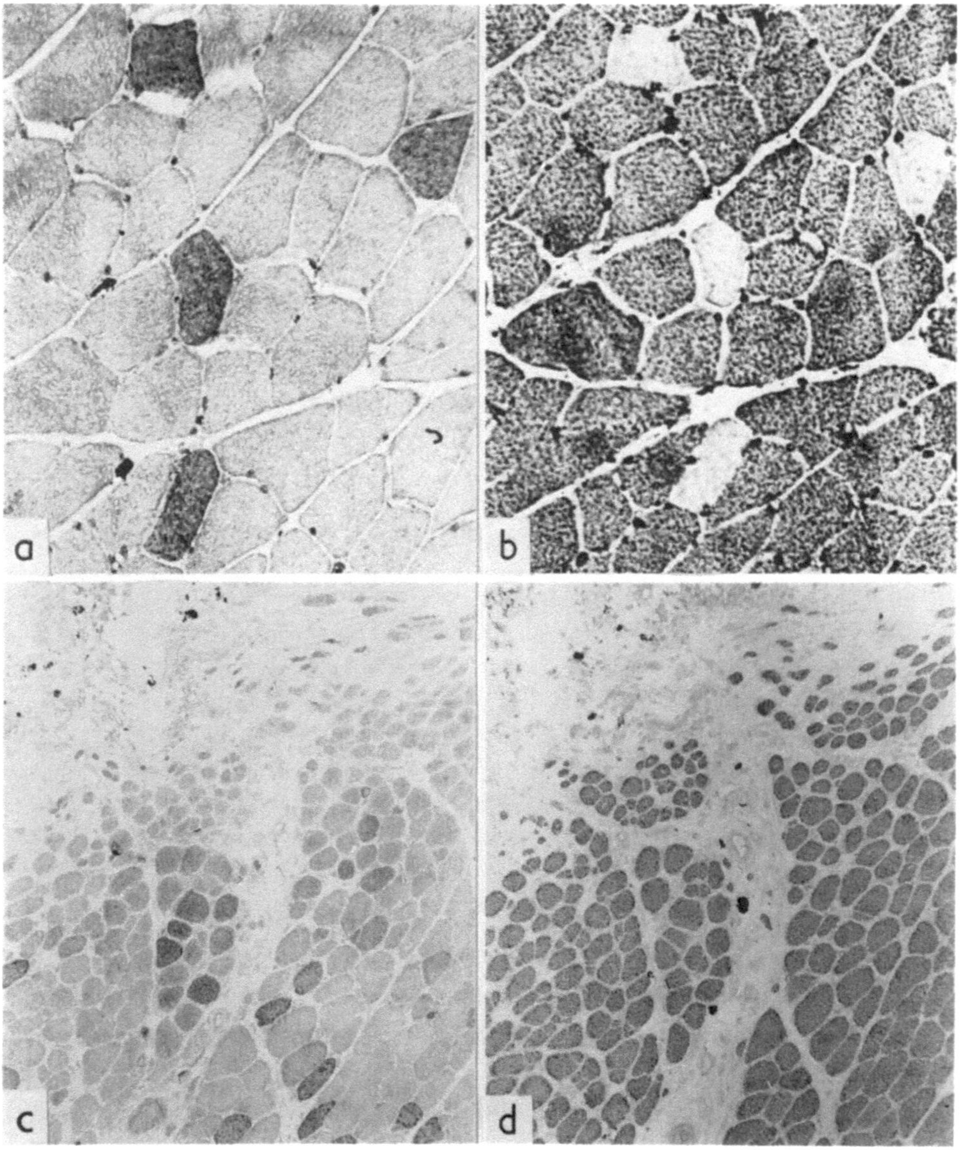

Fig.1. Sections of adult rat gastrocnemius muscles stained with antibodies to troponin I. Figures (a), (b) and (c), (d) are pairs of serial sections. Injury produced by injection of 0.25 ml of an emulsion of oil-alcohol into the muscle (18) and examined after 19 days. Stained by the immunoperoxidase sandwich technique (12). (a) Normal muscle, slow troponin I; (b) Normal muscle, fast troponin I; (c) Injured muscle, slow troponin I; (d) Injured muscle, fast troponin I.

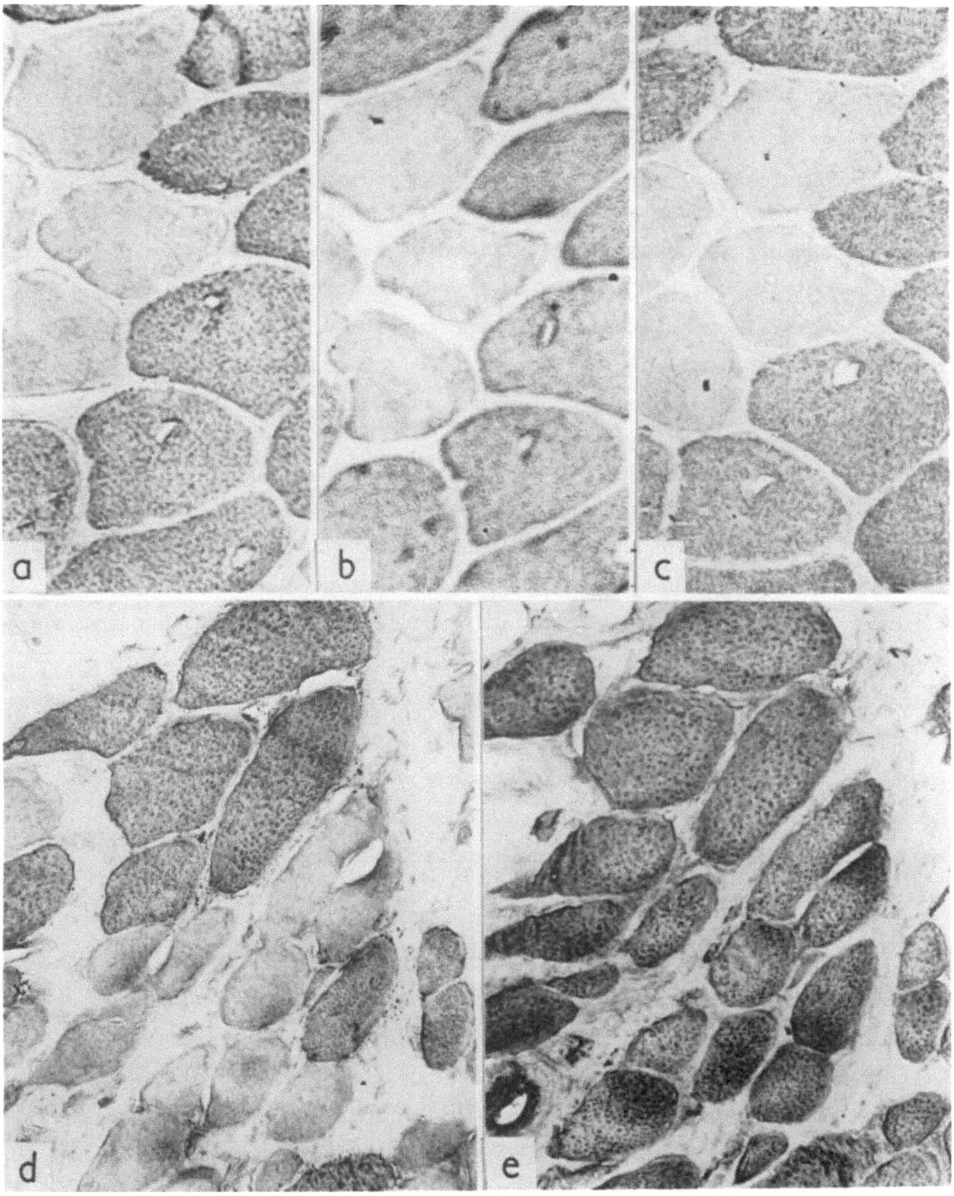

Fig.2. Sections of normal and diseased human muscle stained with antibodies to the troponin components. Staining procedures as described in legend to Fig.1. Normal semispinalis capitis: (a) slow troponin I; (b) slow troponin T; (c) slow (cardiac) troponin C; serial sections. Duchenne muscular dystrophy, quadriceps muscle, six year old boy: (d) slow troponin I; (e) fast troponin I; serial sections.

forms of the regulatory proteins were confined to different cell types in the rat soleus, some weeks later a significant number of cells were observed that stained for both fast and slow forms of the troponin components. It was noted, as has been reported by others (13), that the proportion of type I fibres steadily increases with ageing in the rat. The interconversion was apparent with both myosin ATPase staining and the immunoperoxidase procedure using antibodies to the components of the troponin complex but the changes observed with both techniques were not exactly parallel at all stages. It was concluded that the appearance of both forms of the troponin components is a consequence of a slow process of interconversion of fibre types that was occurring in the rat soleus over the period under investigation.

The distribution of the α and β forms of tropomyosin has not been so intensively investigated but preliminary studies indicated that α form is restricted to type II cells and the β form to type I cells, at least so far as muscles such as quadriceps and gastrocnemius from man and rat are concerned (G.K. Dhoot, N. Frearson & S.V. Perry, unpublished observations). If this is the case the ratio of α to β tropomyosin in these muscles should be similar to the ratio of the type I to type II cells, assuming both cell types contain the same amount of tropomyosin. This does not appear to be the case in all muscles studied. Table 1 summarizes the distribution of the myofibrillar proteins in the two skeletal muscle fibre types.

With the exception of the soleus which appeared to be unusual in that the proportion of fibre types changed with growth and ageing, cells that contained both forms of each component of the troponin system were rarely seen in normal adult skeletal muscle. Indeed if a significant number of such cells were present this was usually an indication that the muscle had been perturbed in some way. Cells containing both forms, however, were consistently observed during development, after cross innervation, during regeneration and at certain stages of muscle disease. These situations were therefore studied in more detail in attempt to throw light on the factors concerned with the expression of the genes that controlled the synthesis of the troponin components.

Developing muscle

The components of the troponin complex are present in lower amounts in the leg muscle from the rat foetus than in the adult tissue and therefore the cells stained less strongly with the immunoperoxidase technique. At 18 days' gestation all cells stained for the fast skeletal muscle form of troponin I but some cells also stained for the slow form. The latter cells were assumed to be presumptive type I fibres and the cells that stained only for fast troponin I were considered to be presumptive type II fibres. As the muscle developed the amount of fast and slow troponin I in the presumptive type II and type I cells respectively increased as judged by the intensity of the staining obtained under standard conditions of the immunoperoxidase procedure. At the same time the amount of fast troponin I in the presumptive type I cells decreased until about 10-11 days after birth it could no longer be detected by the immunoperoxidase technique (14). At that stage the muscle assumed the appearance of the adult tissue in that the fast and slow forms of troponin I were restricted to type II and type I fibres respectively (Fig.3).

Although studied less intensively, the distribution of the fast and slow skeletal muscle forms of troponin C and troponin T during development followed a pattern similar to that observed with troponin I. This indicates that the expression of the genes controlling the synthesis of the fast and slow forms of the three components of the troponin complex are under some kind of coordinated control. Thus, in the rat 10-11 days after birth, the pattern of expression of the genes controlling the synthesis of troponin components was similar to that of adult muscle even though the speed of contraction at this stage is about 50-60% of the adult value (15,16). The observations also indicate that the slow forms of the troponin components are being synthesized in the presumptive type I cells at 18 days' gestation. With more sensitive methods of detection it is possible that the slow skeletal muscle troponin components would be detected even earlier. Even though the motor nerve in the rat has established contact with the muscle at about 16 days' gestation (17), it is unlikely at this stage that there would be significant differences

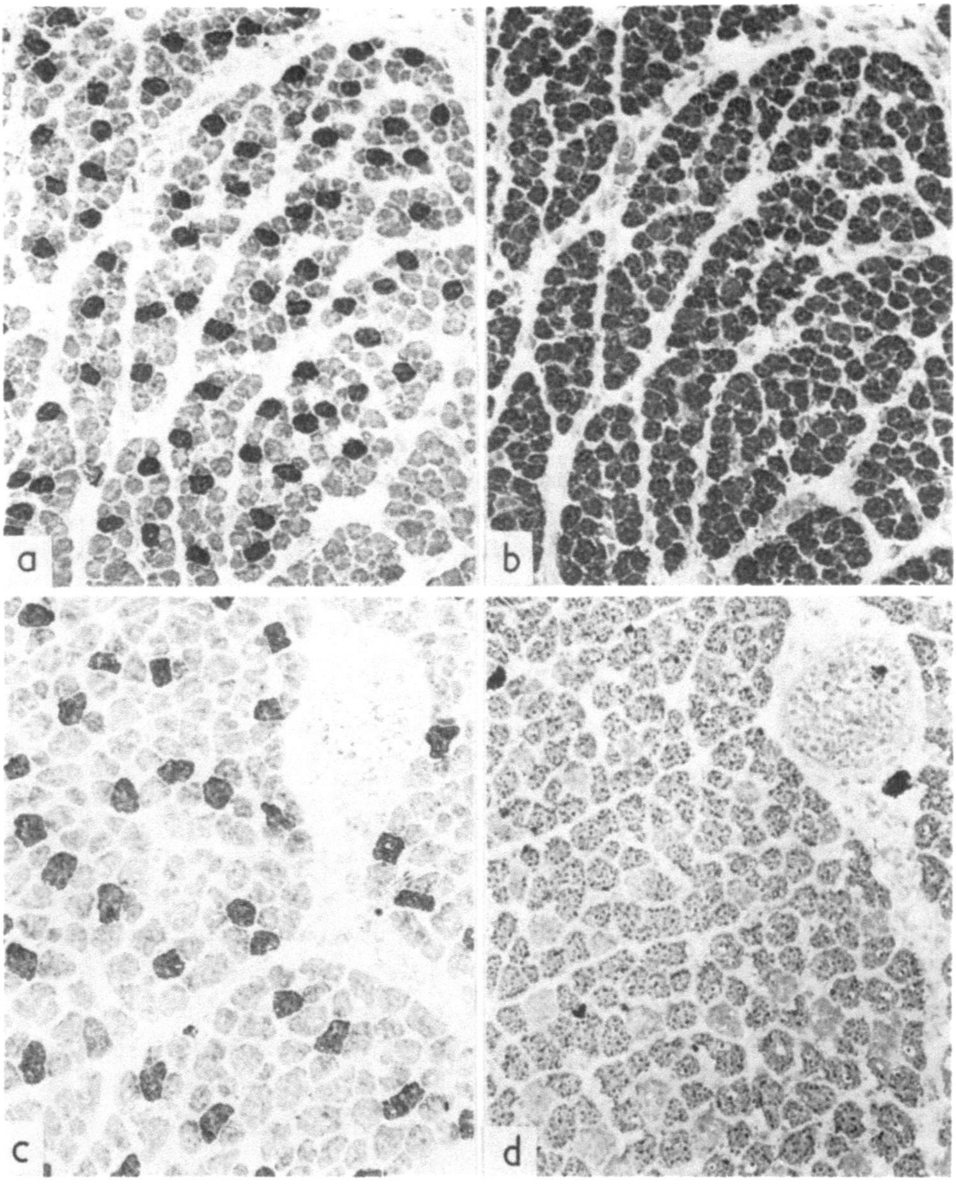

Fig.3. Serial sections of rat quadriceps muscle during post-natal development stained with antibodies to troponin I. Staining procedures as described in legend to Fig.1. Figures (a), (b) and (c), (d) are pairs of serial sections. (a) 5 day old, slow troponin I; (b) 5 day old, fast troponin I; (c) 10 day old, slow troponin I; (d) 10 day old, fast troponin I.

in the impulse frequencies of different nerves.

It would appear therefore that in early foetal muscle all cells synthesize the fast forms of the troponin complex but only those designated to become type I fibres are also able to produce simultaneously the slow form. The question whether the stimulus for the synthesis of this slow form is the differentiation of the innervation is still open. It is possible that a certain type and level of impulse activity is required for the suppression of the synthesis of the fast form of troponin I so that in the fully differentiated type I fibre only the slow troponin complex exists.

Regeneration

Similar changes in distribution of the polymorphic forms of the regulatory proteins to those that occurred during development were observed during muscle regeneration. If rat skeletal muscle was subjected to injury by the injection of oil and alcohol emulsion (18) new cells could be observed in regions adjacent to the site of injury (Fig.1c,d). These new fibres were of smaller diameter than the normal adult fibres and had a central nucleus (presumably myotubes). They also differed from the adjacent uninjured tissue in that all stained for the fast troponin components and slow troponin components were present only in certain of the cells which were very similar in this respect to the presumptive type I cells observed in developing muscle. As differentiation proceeded, the latter cells finally lost the fast troponin components and became normal type I cells containing only slow skeletal muscle troponin components. In preliminary experiments alcohol injury was also induced in rat leg muscles that had been denervated at the same time by removing a section of the sciatic nerve. In these muscles all regenerating myotubes stained for fast troponins and presumptive type I figres synthesized slow troponins as well.

Denervation and cross innervation

The studies on development and regeneration suggest that innervation per se may not be the only factor that determines the expression of the genes controlling the synthesis of the regulatory proteins. Nevertheless by extrapolation from the studies on myosin the frequency of stimulation, whether changed by direct means or by cross innervation, would be expected to be an important regulatory factor.

In adult muscle in which the fast and slow forms of the components of the troponin complex are compartmentalised in the different cell types, preliminary experiments suggest that in the short term (studied up to 6 weeks in the rat) simple denervation does not lead to the synthesis of both forms in a given cell type. Thus the action of the nerve in these circumstances is probably not a simple suppression of the synthesis of one of the forms of the troponin complex as is suggested from the study of the development of type I cells.

Cross innervation experiments carried out in collaboration with Dr. G. Vrbova clearly indicated that type I cells in rabbit soleus muscle that originally synthesized only the slow forms of troponin C, troponin I and troponin T can be transformed into cells that synthesize only the fast forms of the troponin components when the normal nerve supply is replaced with that which innervates a fast muscle. In the rabbit the process is complete in about 15 weeks and cross-innervated muscle cells pass through a stage of synthesizing both polymorphic forms of the troponin components. In this case the new innervation is clearly associated with the production of the fast forms of the components of the troponin complex and at the same time the suppression of the synthesis of the slow forms until they can no longer be detected in the original type I fibre by the immunoperoxidase technique. A similar, but reverse, process occurred when the normal nerve of the fast muscle was replaced with a nerve that normally supplied a slow skeletal muscle. In cross innervation experiments, the synthesis of the polymorphic forms of the troponin components which were not previously present in the cell were detectable about 7-8 weeks after the cross innervation. Unlike the observations during development and regeneration, innervation in this situation appears to be the signal for the expression of the gene that was previously suppressed or active at a very low level. In common with the situation postulated during development the activity pattern imposed by the nerve could also lead to suppression of the synthesis of the form of troponin inappropriate for the new type of innervation.

Discussion

In mammalian skeletal muscle the regulatory protein system appears to exist in two forms namely the homocomplex of fast troponin components with α-tropomyosin and the slow troponin homocomplex with β-tropomyosin. The immunoperoxidase technique indicates that the normal adult skeletal muscle fibre contains the complement of those proteins appropriate to its function. It is possible that traces of the other forms may be present but are not detectable by this technique.

The results of all the studies fit in with the hypothesis that if both forms of the components of the troponin complex are present the cell is usually undergoing some change in function. This can be due to development, disease, regeneration or cross innervation. It can also be observed in certain normal adult muscles, such as soleus, associated with ageing. The studies suggest that all three components of the troponin complex change simultaneously implying that their synthesis is under some kind of coordinated control.

The level at which the production of regulatory proteins appropriate for the function of the fibre type is controlled, is a matter for speculation. Each skeletal muscle cell clearly has the capacity to synthesize both forms of each component, but it is unlikely that levels are regulated by the rates of degradation. The two polymorphic forms are too similar in primary sequence to expect that the cell contains enzymes of the necessary specificity to degrade selectively the two forms of each protein. It is more likely that control occurs at the gene itself or at least at the level of translation.

It is clear that in the adult muscle the expression of the genes responsible for the fast or slow polymorphic forms of each component are associated with fast and slow innervation respectively. Further as has been shown for myosin, it is likely that the important factor is the frequency of the impulse reaching the muscle rather than innervation per se (19-21). The importance of the type of innervation in determining the expression of the genes controlling the synthesis of the regulatory proteins is reinforced by the results of the cross innervation experiments. All muscle cells during the foetal development stages have the capacity to synthesize the fast polymorphic forms of the troponin complex. From the development and regeneration studies it can be concluded that certain fibres are able to make both forms of each component of the troponin complex even before differentiation of nerve into different types is presumed to have occurred or indeed in the absence of nerve. In view of these findings the possibility cannot be excluded that the developing muscle cells of the foetus may convey some sort of signal to the developing nerve and thus stimulate differentiation of the nerve fibre. In adult muscle a particularly important role of the activity pattern imposed by the innervation may be the suppression of synthesis of the polymorphic form of the regulatory protein which is inappropriate to the activity pattern of the fibre.

Our studies suggest that although the type of innervation is important, other factors may be involved in controlling the genes responsible for the synthesis of the proteins of the myofibril.

Acknowledgements

This work was supported by a grant from the Muscular Dystrophy Group of Great Britain. We are grateful to Professor P.G.H. Gell for discussion and advice. We also wish to thank Dr. G.W. Pearce for biopsy samples and Dr. N. Frearson for providing some of the muscle proteins.

References

1. Close, R.: Physiol. Rev. 52, 129-197 (1972).
2. Barnard, R.J., Edgerton, V.R., Furukawa, T., Peter, J.B.: Am. J. Physiol. 220, 410-414 (1971).
3. Padykula, H.A., Herman, E.: J. Histochem. Cytochem. 3, 170-195 (1955).
4. Dhoot, G.K., Perry, S.V.: Nature, 278, 714-718 (1979).
5. Pette, D., Schnez, U.: Histochemistry, 54, 97-107 (1977).
6. Rubinstein, N.A., Mabuchi, K., Pepe, F.A., Salmons, S., Gergely, J., Sreter, F.: J. Cell Biol. 79, 252-261 (1978).
7. Gauthier, G.F., Lowey, S.: J. Cell Biol. 81, 10-25 (1979).
8. Hoh, J.F.Y., Yeoh, P.S.: Nature, 280, 321-323 (1979).
9. Whalen, R.G., Butler-Browne, G.S., Gros, F.: Proc. Natl. Acad. Sci. (USA), 73, 2018-2022 (1976).
10. Dhoot, G.K., Frearson, N., Perry, S.V.: Exp. Cell Res. In the press.
11. Wilkinson, J.M. Eur. J. Biochem. Submitted for publication.
12. Dhoot, G.K., Gell, P.G.H., Perry, S.V.: Exp. Cell Res. 117, 357-370, (1978).
13. Kugelberg. E.: J. Neurol. Sci. 27, 269-289 (1976).
14. Dhoot, G.K., Perry, S.V.: Exp. Cell Res. Submitted for publication.
15. Close, R.: J. Physiol. 173, 74-95 (1964).
16. Brown, M.D.: Nature, 244, 178-179 (1973).
17. Diamond, J., Miledi, R.: J. Physiol. 162, 393-408 (1962).
18. Walton, J.N., Adams, R.D.: J. Path. Bact. Vol.LXXII, 273-298 (1956).
19. Pette, D., Müller, W., Leisner, E., Vrbova, G.: Pflügers Arch. 364, 103-112 (1976).
20. Salmons, S., Sreter, F.A.: Nature, 263, 30-34 (1976).
21. Pette, D., Schnez, U.: FEBS Lett. 83, 128-130 (1977).

Section III. Nerve Muscle Interaction

NERVE-MUSCLE INTERACTIONS DURING EARLY DEVELOPMENT.

R.A.D. O'Brien and G. Vrbová.
Dept. of Anatomy and Embryology, University College London, Gower Street, London WC1E 6BT, England.

The Development of the Motor Unit and Its Size

In mammals the excitability of a motoneurone is correlated to its size and in a given voluntary or reflex movement motoneurones will be recruited in an orderly fashion; small motoneurones will be activated first and larger ones will be activated with increasing force of contraction (Henneman and Olson, 1965; Henneman, Somjen and Carpenter, 1965 a and b; Mendell and Henneman, 1967). Neurones are also specified according to their input into different functional groups (Sherrington, 1910 and 1939) and it appears that within each functional group the smallest cells are recruited first (Clamann, Gillies, Skinner and Henneman, 1974). Small motoneurones are believed to supply fewer muscle fibres and therefore constitute small motor units, whereas large cells supply a large number of muscle fibres and form large motor units (Burke, Levine, Tsairis and Zajac, 1973). The development of muscle tension is thus graded: low levels of excitation will activate only small motor units. With increasing excitatory input larger units are recruited, each increment in tension being greater as larger units, which develop more tension, become active (Wuerker, McPhedran and Henneman, 1965). This being so, large motor units, which are the most wasteful in terms of energy expenditure, will be used only in maximal muscular effort.

In newborn animals the sizes of motoneurones are similar and only later can cells of different sizes be distinguished (Conradi, 1976). The mechanisms that lead to the development of the different sizes of motoneurones and motor units are poorly understood. In this article an attempt will be made to suggest that the activity of individual motoneurones during early development may be of importance in determining the size of their motor units.

During early stages of postnatal development all motor units are large, i.e. individual motoneurones supply a greater number of muscle fibres in newborn animals than during later life (Bagust, Lewis and Westerman, 1973; Brown, Jansen and van Essen, 1976). This greater size of motor units is due to the fact that, at this stage, each muscle fibre is supplied by branches from more than one axon. Thus individual muscle fibres contribute to more than one motor unit. The sizes of individual motor units are reduced during postnatal development by a process of elimination of superfluous contacts, so that finally each muscle fibre is supplied by only one axon and one motoneurone (Redfern, 1970; Bagust, Lewis and Westerman, 1973; Benoit and Changeux, 1975; Brown, Jansen and van Essen, 1976; O'Brien, Östberg and Vrbová, 1978). The size of the motor unit will then finally depend on the number of contacts that the motorneurone will be able to maintain with individual muscle fibres, and will be inversely proportional to the number of contacts that become eliminated during development, i.e. the larger the number of connections that will become eliminated during development the smaller the final size of the motor units will be. Thus, to understand the factors that determine the sizes of motor units during development it may be helpful to consider the mechanisms that lead to the elimination of contacts between nerve endings and muscle fibres during development.

Effect of activity. The final stage in the elimination of polyneuronal innervation - the appearance of muscle fibres innervated by one axon - takes place between 10 and 16 days of age in the rat, and coincides with an increase in the general activity of the animal (Redfern, 1970). Artificial reduction of muscle activity, either with tenotomy (Benoit and Changeux, 1975) or nerve-block (Thompson, Kuffler and Jansen, 1979), can delay this process for several days. These observations indicated that the removal of the superfluous nerve-muscle contacts depends on neuromuscular activity, but the final proof should come if the elimination process could be accelerated by increasing this activity. This was tested by implanting stimulating electrodes around the sciatic nerve in one leg of rats aged 6 to 8 days, and stimulating them at 8 Hz for 4 to 6 hours a day (O'Brien,

Östberg and Vrbova, 1978). After 2 to 4 days of stimulation the soleus muscles were removed and tested for polyneuronal innervation using the electrophysiological method of Redfern (1970), in which different axons innervating an end-plate can be distinguished from their different stimulus thresholds. As expected, the vast majority of muscle fibres from control muscles (taken from normal animals or ones which had been implanted with electrodes but not stimulated) were innervated by two or more axons up to 10 days of age (Fig. 1). However, the extent of polyneuronal

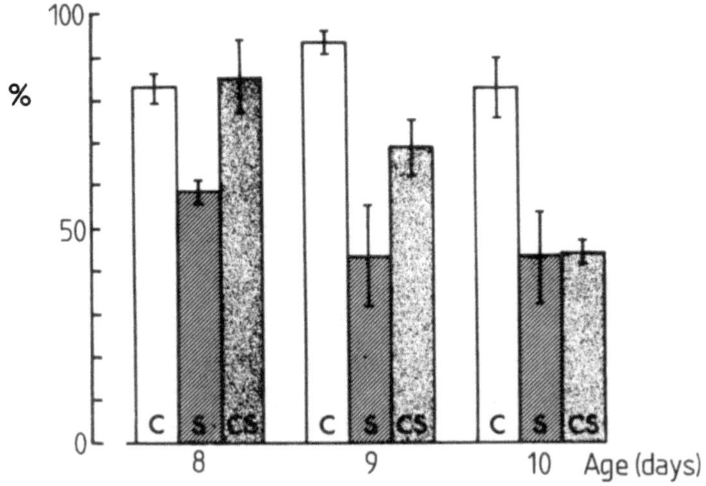

Fig. 1. Polyneuronal innervation measured electrophysiologically in control animals (C), in stimulated muscles (S), and in muscles contralateral to the stimulated side (CS). The number of polyneuronally innervated muscle fibres is expressed as a percentage of the total number of fibres examined in each muscle. Values are mean ± S.E.M.

innervation in the stimulated muscles was markedly reduced even at 8 days of age - after only a few hours of stimulation. A further reduction in polyneuronal innervation was not observed, probably because of the limited duration of the stimulation periods. It was noticed that the contralateral leg of the stimulated rats showed a crossed-extensor response, so the contralateral soleus muscles were also tested for polyneuronal innervation. Elimination was also accelerated in these muscles, suggesting that activity conducted by the crossed-extensor reflex has a similar effect to direct stimulation of the nerve. It may therefore be concluded that neuromuscular activity is vital to the removal of superfluous synapses during postnatal development, and thus to the reduction in motor unit size.

In the above experiments the same pattern of activity was artificially imposed on all the axons to soleus muscle, and preliminary observations indicate that this leads to a relatively homogeneous distribution of motor unit sizes in the adult muscle (Lowry and Vrbová, unpublished). In the normal animal, however, the axons innervating a muscle will have a range of activity patterns, and one might expect the more active neurones to end up with smaller motor units.

Proteolytic enzymes. So far we have only partly answered the question of what mechanism is involved in the elimination of polyneuronal innervation; if activity is important, how does it work? Muscle fibres can be induced to accept more than one permanent synapse if the end-plates are at least 1 mm apart (Brown, Jansen and van Essen, 1976; Frank, Jansen, Lømo and Westgaard, 1975), and this suggested to us that there could be some diffusible inhibitory substance released from an active end-plate. This substance might bring about the digestion of the nerve terminals, and the terminal that survives at an end-plate after the elimination process could be the one which is most resistant to the digestion. Poberai, Sávay and Csillik (1972) demonstrated the release of proteolytic enzymes at the end-plates of stimulated or eserinized muscles, and acetylcholine (ACh) has been shown to evoke the release of these enzymes from a wide variety of tissues (e.g. Ignarro, 1975). We have shown that treatment with ACh will cause muscles to release substances that digest the gel layers of a photographic film (O'Brien, Östberg and Vrbová, 1978). More recently, we have found that incubating soleus muscles from young rats with ACh *in vitro* leads to a reduction in the average number of nerve terminal profiles seen by electron microscopy at the end-plate. If calcium is included with the ACh in the incubation, the reduction in the number of nerve terminals is so severe that the muscle becomes partially denervated. However, if the proteolytic enzyme inhibitors Leupeptin and Pepstatin are also included, this effect of ACh is abolished (O'Brien, Östberg and Vrbová, 1979).

To summarize these results, it is clear that activity influences the removal of superfluous nerve-muscle contacts during post-natal development, possibly via an ACh-mediated release of proteolytic enzymes at the end-plate. It would therefore be expected that the activity of a motor nerve axon would influence the size of its motor unit.

Mechanisms That Determine Contractile Characteristics of Muscle.

It is well known that in adult animals the motoneurone determines the contractile and metabolic properties of the muscle fibres it supplies (Buller, Eccles and Eccles, 1960 a and b). The mechanism by which it accomplishes this has recently been elucidated. It was found that the activity pattern imposed on the muscle by its motor nerve determines the contractile properties of the muscle fibres (Salmons and Vrbová, 1969). In this way a perfect matching between the requirements of the motoneurone and its muscle fibres is achieved. Motoneurones that fire at low frequencies supply muscle fibres that contract and relax slowly and develop their maximal tension at low rates of stimulation. Motoneurones that fire at high frequencies supply muscle fibres that develop maximal tetanic tension at high rates of firing. Moreover, muscle fibres of motor units that are active for long periods of time fatigue less than muscle fibres of motor units that are active for only short periods of time (see Burke and Edgerton, 1975; Vrbová, Gordon and Jones, 1978).

In newborn animals the different properties of muscle fibres from individual motor units are not yet clearly defined (Close, 1964; Bagust, Lewis and Westerman, 1973). Muscle fibres develop their various different properties only several weeks after birth (Gutmann, Melichna and Syrovy, 1974). This may be due either to the fact that they are not activated appropriately by their neurones, or that the muscle fibres have to reach a certain degree of maturity before they are able to specialize in response to particular activity patterns.

As stated above, muscle fibres of newborn animals are supplied by more than one axon. This polyneuronal innervation disappears with age so that in rats at 16 days all muscle fibres have lost their superfluous nerve endings and are supplied by only a single axon. If motoneurones innervating the same muscle fibres have different activity patterns then individual muscle fibres during early development could be receiving more than one type of activity pattern. If this was the case then it could be argued that differentiation can only begin once each muscle fibre is supplied by only a single axon and that the absence of differentiation during early development is due to this 'promiscuity' of muscle fibres. Indeed, it was argued by Gauthier, Lowey and Hobbs (1978) that this may be the case and could explain their findings that muscles of new-born animals synthesise both slow and fast type of myosin. Muscle fibres from newborn animals combine a set of characteristics which is different from that in any muscle fibre type in later life. Although they contract and relax slowly (Buller, Eccles and Eccles, 1960b) they contain predominantly myosin light chains characteristic of fast muscles (Rubinstein, Pepe and Holtzer, 1977; Pette, Vrbová and Whalen, 1978) as well as high levels of ATPase (Gutmann, Melichna and Syrovy, 1974). In addition, all fibres contain a mixture of "slow" and "fast" type of myosin (Gauthier, Lowey and Hobbs, 1978). This peculiar combination of slow contractile speeds with predominantly fast type of contractile proteins shows that in immature muscles the slow rate of rise of tension is not associated with the type of contractile proteins, but with other components of the muscle fibre.

Buller, Eccles and Eccles (1960b) reported that, in kittens after birth, both slow and fast muscles first become faster contracting and then the slow muscles gradually achieve their slow rate of tension development. A similar situation was found in rats, where after birth both soleus and EDL muscles increase their speed on contraction with age (Close, 1964; Brown, 1973). The time course of this increase in speed of contraction is gradual from birth until about 20-30 days of age. Although polyneuronal innervation is completely eliminated by 16 days of age, the changes of twitch speed continue at the same rate until 20-30 days after birth, so that there is no correlation between the time course of the elimination of polyneuronal innervation and the change in contractile properties.

The increase in speed of contraction after birth is dependent on innervation for, when EDL and soleus muscles were denervated at birth, the increase in the speed of contraction of soleus and EDL muscles seen after birth was much reduced in both muscles (Lewis, 1973; Brown, 1973). It is possible that the nerve influences the muscle by imposing a particular "phasic" activity pattern on the muscle fibres it supplies, and that this activity pattern induces the increase in speed. In adults, tonic or slow frequency activity reduces the rate of tension development of fast and slow muscles.

The possibility that the increase in speed seen after birth could be prevented by slow frequency activity was tested. Muscles of 8 to 25 day old rats were stimulated via implanted electrodes at 10 Hz for 6 hours a day. Isometric contractions of control and stimulated muscles were recorded in anaesthetized rats. The time to peak twitch tension of T.A. and E.D.L. muscles from 10 to 15 day old control rats was 35.8 $\pm$ 1.42 msec (8 muscles) and that from control rats aged 20-33 days was 26.6 $\pm$ 1.9 msec (9 muscles). Thus between these times the muscles become faster contracting. Time to peak tension of muscles from 20-33 day old rats stimulated at 10 Hz was 27.3 $\pm$ 0.93 ms (9 muscles) which is not different from the unstimulated control muscles, but is significantly faster than that of muscles from animals aged 10-15 days (see Fig. 2).

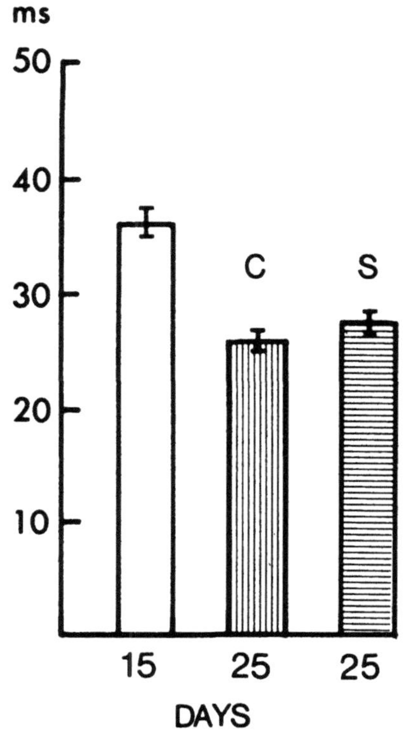

Fig. 2. Time to peak twitch tension (msec) of tibialis anterior (TA) and extensor digitorum longus (EDL) muscles of control rats aged 12-15 days (empty column), 18-25 day old control rats (vertical shading), and muscles stimulated for 1 to 2 weeks from 18-25 day old rats (horizontal shading). Values are mean $\pm$ S.E.M.

Thus, an activity pattern which, when applied to adult fast muscles has a slowing effect on the time course of contraction, did not have the same effect on immature muscles.

In these experiments the innervation of the muscles was left intact so that it is still possible that the natural activity can override the effects of electrically imposed stimulation. Experiments therefore were performed on denervated muscles by M. D. Brown (1973). In these studies rabbits were used because of their more convenient size. In rabbits, as in rats, the T.A. and E.D.L. muscles contract and relax slowly at birth and become faster contracting and relaxing with age. If these muscles are denervated at birth the increase in speed of contractions is prevented, and their contractile properties are similar to those from newborn rabbits. Electrodes were implanted so as to stimulate these denervated muscles directly and the pattern of activity used was similar to that of a slow muscle, i.e. the muscles were stimulated by trains of pulses at 10 Hz. This frequency was chosen because (a) it produced an almost fused tetanus of the denervated muscles, and (b) faster rates of stimulation increased the rate of atrophy of the denervated muscles. Fig. 3 shows that after 9 to 14 days of such stimulation the contractions of these muscles became significantly faster than those of unstimulated denervated muscles. Thus although this type of activity has a slowing effect on the development of tension in normal adult fast muscles in young animals it brought about an increase of the speed of contraction.

From these results it appears that an increase in the speed of contraction of developing muscles is part of their maturation process. It is probably brought about by activity and, at this stage, the pattern of activity is not important, for the muscle can only respond to activity by increasing its speed of contraction. Only when this initial maturation is completed can further differentiation into muscle fibre types take place. This differentiation will be brought about by the different activity patterns that the motoneurones impose upon the now mature muscle fibres.

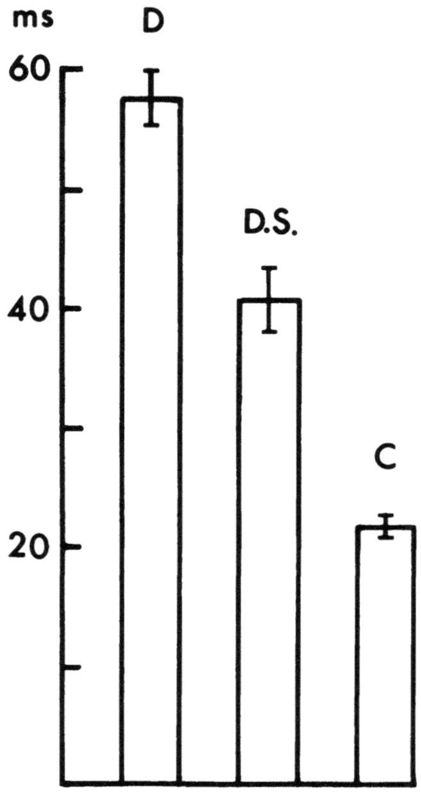

Fig. 3. Time to peak twitch tension (msec) of rabbit TA and EDL muscles denervated at birth (D), denervated at birth and stimulated from 7 to 10 days onwards (DS), and controls (C). The muscles were tested at 3 weeks of age. Values are mean ± S.E.M. The differences between each group are statistically significant. Results are from Brown (1975).

In summary, the postnatal development of the neuromuscular system proceeds in this order: (1) first the size of the motor unit is determined, and the activity of the motoneurone seems to play an important part in this process; (2) the maturation of the muscle fibre is brought about by the motor nerve and its activity; (3) differentiation into types of muscle fibres that match the requirements of motoneurones that supply them is accomplished by particular activity patterns imposed on the mature muscle fibres by the motoneurone that supplies them.

References

1. Bagust, J., Lewis, D.M., Westerman, R.A.: Polyneuronal innervation of kitten skeletal muscle. J. Physiol. 229, 241-255 (1973).

2. Benoit, P., Changeux, J.P.: Consequences of tenotomy on the evolution of multi-innervation in developing rat soleus muscle. Brevin Res. 99, 354-358 (1975).

3. Brown, M.C., Jansen, J.K.S., van Essen, D.: Polyneuronal innervation of skeletal muscle in new-born rats and its elimination during maturation. J. Physiol. 261, 387-422 (1976).

4. Brown, M.D.: Role of activity in the differentiation of slow and fast muscles. Nature 244, 178-179 (1973)

5. Brown, M.D.: The role of the motor nerve and muscle activity in the development of some properties of skeletal muscle. PhD thesis, University of Birmingham (1975).

6. Buller, A.J., Eccles, J.C., Eccles, R.M.: Differentiation of fast and slow muscles in the cat hind limb. J. Physiol. 150, 399-416 (1960a).

7. Buller, E.J., Eccles, J.C., Eccles, R.M.: Interactions between motoneurones and muscles in respect of the characteristic speeds of their responses. J. Physiol. 150, 417-439 (1960b).

8. Burke, R.E., Edgerton, V.R.: Motor unit properties and selective involvement in movement. Exercise and Sport Sci. Rev. 3, 31-81 (1975).

9. Burke, R.E., Levine, D.M., Tsairis, P., Zajac, C.E.: Physiological types of histochemical profiles in motor units of the cat gastrocnemius. J. Physiol. 234, 723-748 (1973).

10. Clamann, H.P., Gillies, J.D., Skinner, R.D., Henneman, E.: Quantitative measures of output of a motoneuron pool during monosynaptic reflexes. J. Neurophysiol. 37, 1328-1337 (1974).

11. Close, R.: Dynamic properties of fast and slow skeletal muscles of the rat during development. J. Physiol. 173, 74-95 (1964).

12. Conradi, S.: Functional anatomy of the anterior horn motor neurone. In The Peripheral Nerve, ed. Landon, D.M. Chapman & Hall, London. pp 279-329 (1976).

13. Frank, E., Jansen, J.K.S., Lømo, T., Westgaard, R.H.: The interaction between foreign and original motor nerves innervating the soleus muscle of rats. J. Physiol. 247, 725-743 (1975).

14. Gauthier, G.F., Lowey, S., Hobbs, A.W.: Fast and slow myosin in developing muscle fibres. Nature, 274, 25-29 (1978).

15. Gutmann, E., Melichna, J., Syrovy, I.: Developmental changes in contraction time, myosin properties and fibre pattern of fast and slow skeletal muscles. Physiol. Bohemoslov. 23, 19-27 (1974).

16. Henneman, E., Olson, C.: Relations between structure and function in the design of skeletal muscles. J. Neurophysiol. 28, 581-598 (1965).

17. Henneman, E., Somjen, G., Carpenter, D.O.: Functional significance of cell size in spinal motor neurones. J. Neurophysiol. 28, 560-580 (1965a).

18. Henneman, E., Somjen, G., Carpenter, D.O.: Excitability and inhibitability of motoneurons of different sizes. J. Neurophysiol. 28, 599-620 (1965b).

19. Ignarro, L.J.: Regulation of lysosomal enzyme release by prostaglandins, autonomic neurohormones and cyclic nucleotides. In Lysosomes in Biology and Pathology, Vol. 4, ed. Dingle & Dean. North Holland, Amsterdam (1975).

20. Lewis, D.M.: The effect of denervation on the differentiation of twitch muscles in the kitten hind limb. Nature New Biology, 241, 285-286 (1973).

21. Mendell, L.M., Henneman, E.: Terminals of single Ia fibers: distribution within a pool of 300 homonymous motor neurons. Science, 160, 96-98 (1968).

22. O'Brien, R.A.D., Östberg, A.J., Vrbová, G.: Observations on the elimination of polyneuronal innervation in developing mammalian skeletal muscle. J. Physiol. 282, 571-582.

23. O'Brien, R.A.D., Östberg, A.J.C., Vrbová, G.: Structural and functional changes in developing rat skeletal muscle induced in vitro. J. Physiol. In Press (1979).

24. Pette, D., Vrbová, G., Whalen, R.C.: Independent development of contractile properties and myosin light chains in embryonic chick fast and slow muscle. Pflugers. Arch. 378, 251-257.(1979).

25. Poberai, M., Sávay, G., Csillik, B.: Function-dependent proteinase activity in the neuromuscular synapse. Neurobiology, 2, 1-7 (1972).

26. Rubinstein, N.A., Pepe, F.A., Holtzer, H.: Myosin types during the development of embryonic chicken fast and slow muscles. Prof. Nat. Acad. Sci. U.S.A. 74, 4524-4527.

27. Salmons, S., Vrbová, G.: The influence of activity on some contractile characteristics of mammalian fast and slow muscles. J. Physiol. 201, 535-549 (1969).

28. Sherrington, C.: Flexion reflex of the limb, crossed extension reflex and reflex stepping and standing. J. Physiol. 40, 28-21 (1910).

29. Sherrington, C.: Selected writings of Sir Charles Sherrington (compiled and edited by Denny-Brown). Hamish Hamilton, London (1939).

30. Thompson, W., Kuffler, D.P., Jansen, J.K.S.: The effect of prolonged reversible block of nerve impulses on the elimination of polyneuronal innervation of new-born rat skeletal muscle fibres. Neuroscience 4, 271-281 (1979).

31. Vrbová, G., Gordon, T., Jones, R.: Nerve-muscle interaction. Chapman and Hall, London.

32. Wuerker, R.B., McPhedran, A.M., Henneman, E.: Properties of motor units in a heterogeneous pale muscle (m.gastrocnemius) of the cat. J. Neurophysiol. 286, 85-99 (1965).

REMATCHING OF NERVE AND MUSCLE PROPERTIES IN CAT MOTOR UNITS AFTER REINNERVATION

Tessa Gordon and R.B. Stein
Department of Physiology, University of Alberta, Edmonton, Canada

Although the early development of nerve and muscle occurs independently, the survival of nerves and muscles thereafter depends on their mutual interaction. Even in the adult animal, mutual interaction between peripheral nerves and their end-organs is essential for their structural integrity since the nerves and their end-organs atrophy following nerve section. The process of muscle atrophy has been extensively studied (8). Although the shrinkage of the central stump of cut peripheral nerves has been described (9), the dependence of peripheral nerves on their end-organs is far less well understood.

A chronic recording method developed in this laboratory (5, 25, 26) has provided a valuable and unique tool for studying the long-term mutual interactions of nerves and their end-organs. With these techniques to follow the time course of atrophy of the central stump of severed nerves, we found that the atrophy is not a continuous process, but that the compound action potentials (CAP's) in the central stump of ligated nerves reached a stable non-zero level (5). If the nerves were permitted to regrow back to their end-organs, the decline in the amplitude of the CAP's was halted and began to recover soon after reinnervation. Although the trend of decay of the CAP's was reversed by reinnervation, recovery was far from complete within 200 days after resuture of the nerve either to the distal stump of the cut nerve or directly into the muscle. In contrast, the twitch tension of the reinnervated muscle recovered to approximately control levels within that time, indicating that perhaps reinnervation reversed the atrophic changes in the muscle more completely than those in the peripheral nerves. Some of the discrepancy between the recovery of the muscle twitch tension and the neural potential could be accounted for 1) by the greater degree of

Plasticity of Muscle

atrophy of the sensory fibres following section (13), and 2) to the contamination of the recovery of the nerve fibres that had successfully remade connections, by the continuing decay of nerve fibres that had not managed to grow back to the periphery and innervate their end-organs (5). In order to follow the recovery of nerves and muscles following reinnervation more closely, we have now recorded unit potentials with their associated electromyographic activity (EMG) and twitch tension and followed their recovery during reinnervation.

Chronic Recording of Single Motor Units

Chronic recording electrodes contained in silastic cuffs (26) were implanted aseptically in cats around the sciatic nerve and one or more of its branches. A bipolar electrode was also implanted on the surface of relevant muscles. This permitted both stimulation and recording at two points on each nerve and on the muscle. The leads were externalized via a twelve-pin socket within a percutaneous carbon connector (24), which was secured in the skin on the hip. A flexible cable coupled the socket to an external stimulator or preamplifier for stimulation and recording respectively. Measurements were made of the evoked CAP's and of unit potentials under fluothane anaesthesia. Either the medial gastrocnemius (MG) or the lateral gastrocnemius-soleus (LGS) nerve was maximally stimulated and the compound triphasic neural potential was recorded on the sciatic nerve. Compound potentials were also recorded on the MG or LGS nerves by stimulating the sciatic nerve. Compound and unit potentials on the nerves were recorded using a tripolar configuration as described in detail previously (25, 26) in order to reduce the EMG contamination of neural signals.

The evoked EMG was recorded simultaneously with the evoked neural signals and with the muscle tension. To measure tension the foot was placed into a specially designed boot which was free to rotate about a line co-axial with the ankle joint. A strain gauge built into the boot measured the ankle torque while the tibia was stabilized non-invasively at two points, and the angle of the ankle was adjusted to give maximum torque. The ankle torque values given in the figures are in units of Newtons, measured through a lever arm of 4.5 cm. Evoked unit potentials of single motor fibres were evoked in an all-or-none fashion by intramuscular microstimulation of nerve branches using a bipolar concentric needle inserted percutaneously into the muscle, as described by Taylor and Stevens (29) in humans. Unit neural potentials were recorded on the relevant nerve and higher up on the sciatic. The triphasic electrode configuration was coupled by a step-up transformer with an approximate gain of 20 (Hammond 585D) to a low noise preamplifier (Grass P-15) with a broad band pass of 10 Hz to 10 KHz. The calibrated voltage levels of neural signals shown in the Figs. are after amplification by the transformer. The transformer provided valuable improvement of the signal-to-noise ratio which was essential for recording single unit neural potentials. Since unit responses were small compared with the evoked responses of the whole nerve and muscle (Fig. 1) and to improve the signal-to-noise ratio further, 25 to 40 responses were averaged using a laboratory computer (Digital Equipment Corp., PDP11-34). The size of the nerve fibres

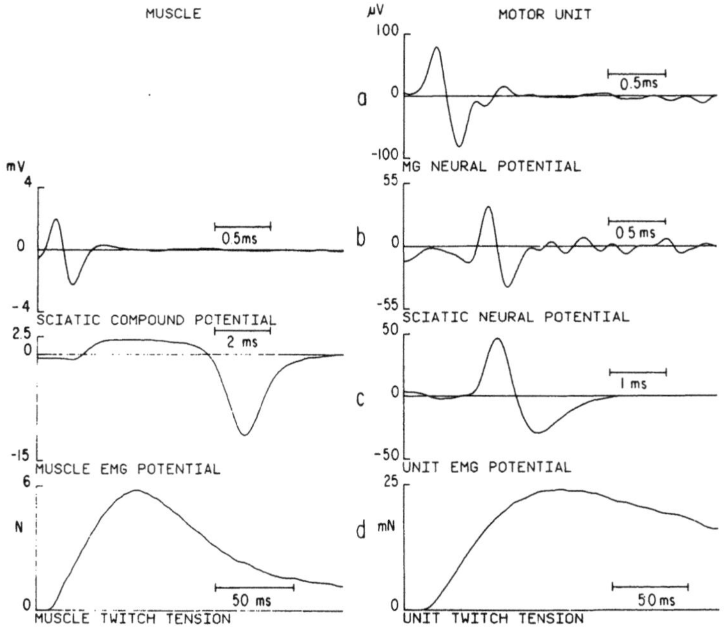

Fig. 1. Computer averaged records of the triphasic compound action potential recorded on the sciatic nerve in response to maximal stimulation of the medial gastrocnemius (MG) nerve, together with the evoked EMG and twitch. Averaged unit neural potentials recorded on the MG and the sciatic nerves, evoked by intramuscular microstimulation, together with the all-or-none unit EMG potential and twitch contraction. The neural and EMG traces have been delayed in time to eliminate the stimulus artifact.

was assessed by the peak-to-peak amplitude of the unit potentials recorded at the two sites and the size of the motor unit by the EMG potential amplitude and the twitch tension. The latency of the unit neural potentials varies considerably, particularly in reinnervated muscles since it depends on the site of microstimulation and the conduction velocity in the distal stump.

Recovery of Nerve and Muscle Following Reinnervation

Nerves were sectioned and the proximal stump was either resutured to the distal stump of the nerve (*nerve-nerve suture*) or directly into the muscle (*nerve-muscle suture*). Muscles were reinnervated within 30 to 50 days after resuture. The amplitude of the CAP recorded in the nerve proximal to the site of section and resuture declines with a time constant of about 45 days (5). Soon after reinnervation the decline is halted and the potential begins to recover. However, the amplitude of the CAP shows less recovery than its latency. In 8 resutured nerves the average CAP amplitude reached 45.2% of control 200 days after suture, as compared with 90% of control for its latency. The discrepancy between the recovery of amplitude and latency is seen in the example in Fig. 2. Time to the first peak of the CAP began to shorten after reinnervation although the amplitude of the triphasic potential continued to decline (Fig. 2A, B). This indicates that some fibres do begin to recover but that CAP amplitude continues to decline because of the wide dispersion of latencies of fibres that have and have not been suceessful in remaking peripheral connections. A better indication of the recovery of successful nerves is obtained when the population of nerve fibres that has remade connections is sampled at intervals during reinnervation by stimulating single nerve fibres. Soon after reinnervation the nerves are small relative to pre-operative values but they recover completely by 200 days after suture, and sometimes the mean value exceeded pre-operative values. This shows that the reformation of nerve-muscle connections reversed the atrophic changes in the central stump of the nerves completely.

The time course of recovery in the muscle was similar to that of the nerve. The recovery in average twitch tension of muscle units is compared with the recovery of tension in the whole muscle in Fig. 3. When the muscle was first

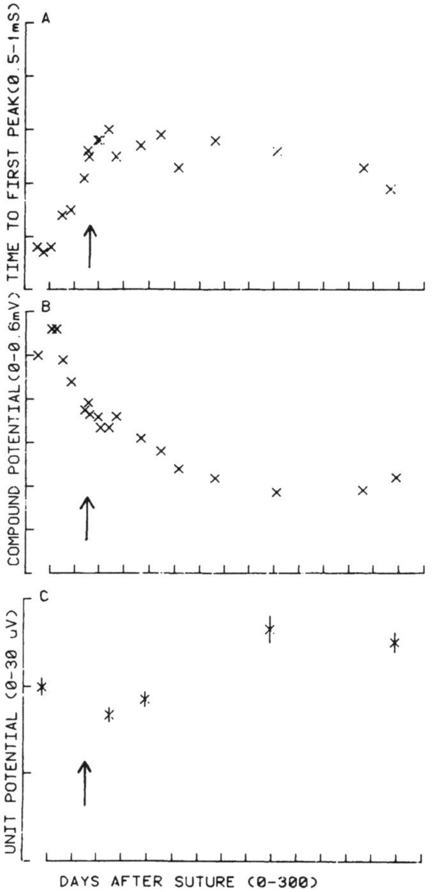

Fig. 2. Changes in A) the time to the first peak of the evoked compound action potential, B) the amplitude of the evoked compound action potential, and C) the mean and standard error of the unit neural potentials with time after nerve-nerve suture of the MG nerve distal to the recording electrodes. Reinnervation was first observed one month after suture.

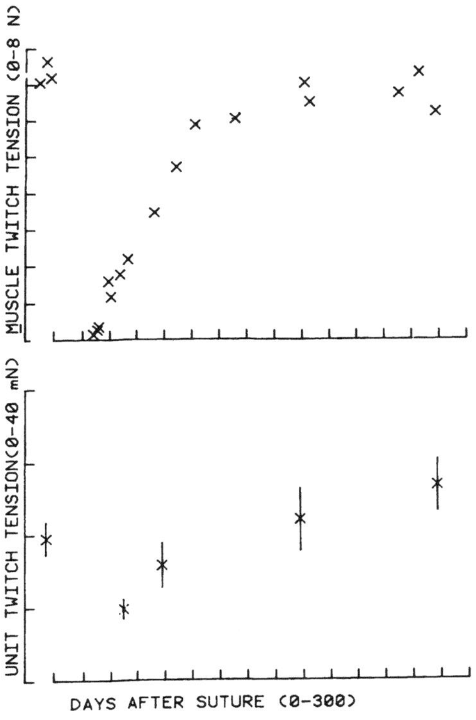

Fig. 3. Changes in muscle tension with time after nerve-nerve suture are compared for single maximal shocks to the MG nerve and microstimulation of single units. The mean ± S.E. of unit twitches are shown.

reinnervated, the size of the motor units was smaller than normal, reflecting the denervation atrophy of the fibres prior to reinnervation. However, the tension increased with time and recovered to control values within 200 days after suture. The recovery of the unit tension followed the recovery of whole muscle tension which implies that recovery of the reinnervated muscle tension represents the recovery of muscle fibre diameter rather than increasing numbers of muscle fibres reinnervated with time. Nerve fibres that will be successful in growing past connective tissue barriers at the site of suture should grow at a similar rate into the muscle. The period in which they will make functional connections with muscle would then be in the order of a few days rather than months.

In the example of reinnervation shown in Figs. 2 and 3, and in 4 other cases of nerve-nerve sutures, the muscles nearly regained their pre-operative tension levels. However, particularly where the nerve was sutured directly into muscle, the pre-operative tension levels were not always regained in the reinnervated muscles. Yet even in these muscles the motor unit size returned to pre-operative levels (Fig. 4). Although the muscle regained only 45% of its pre-operative tension, the average unit

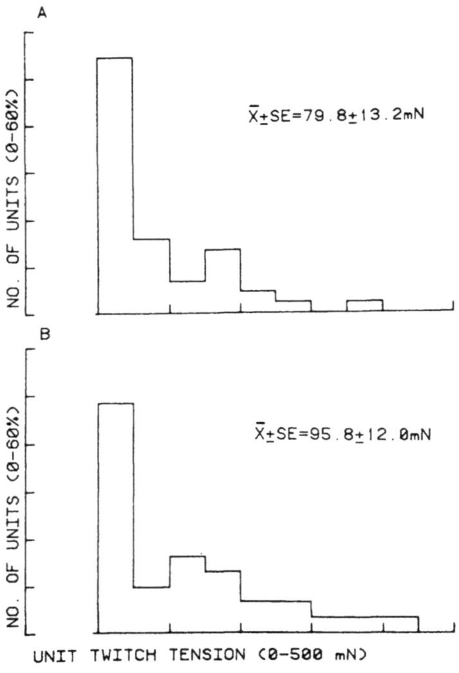

Fig. 4. Twitch tension of MG units A) before, and B) one year after suturing the nerve directly back into the muscle. The mean twitch tension $\bar{X}$ and its standard error (SE) are also given.

tensions before and one year after reinnervation, were not significantly different. This implies that many muscle fibres were not reinnervated at all and remained denervated. This situation differs considerably from that in which a muscle is partially denervated clinically, as in motoneurone disease, or experimentally, in which case the intact motor nerves sprout and can reinnervate many of the available denervated muscle fibres (3, 14, 15). In the case of collateral sprouting there is far less muscle atrophy and growth of connective tissue because the intact nerve fibres sprout and re-innervate adjacent denervated muscle fibres in a short time of their denervation. Although nerves retain the capacity for regeneration for long periods (1), the success of reinnervation in muscle fibres depends particularly on the ability of the nerves to grow back to the muscle unimpeded by physical barriers of connective tissue and is inversely related to delay of time before reinnervation (10). Despite the ability to sprout and supply many muscle fibres, axons after resuture appear to be unable to overcome connective tissue barriers to enlarge their units and compensate for the inability of some nerve fibres to grow back. Consequently, some muscle fibres may remain denervated.

In conclusion, the mutual interaction of nerve and muscle after reinnervation reverses the atrophic changes in both which followed their isolation after nerve section. When muscles do not regain their former tensions, the lack of success of reinnervation is accounted for not by the inability for nerve and muscle to grow and recover, but by mechanical barriers that prevent the growth of nerve fibres back to the muscle.

Rematching of the Size and Properties of Nerve and Muscle Following Reinnervation

The histochemical properties of all muscle fibres belonging to a single motor unit are homogeneous and this homogeneity is preserved after reinnervation. However, muscle fibres innervated by a single axon are grouped together in clumps after reinnervation rather than intermingled with muscle fibres innervated by a number of different axons (15). Thus, newly formed motor units must be assembled from muscle fibres of different histochemical and contractile properties which were formerly innervated by different motor nerves. After reinnervation there must be a rematching of the histochemical properties of muscle fibres to the motor nerve that innervates them.

It was shown above that the nerves and muscle fibres innervated by them appeared to recover synchronously after reinnervation. Normally there is a direct relationship between the size of the motor unit in terms of its tension output, the size of the axon supplying that unit and its contractile speed (20, 29). On the basis of this size relationship, an orderly recruitment of motor units has been demonstrated in a number of muscles (6, 23). Since the smaller units are recruited first, the increment in tension remains fairly constant over the entire range of tension (12). Following reinnervation in man, Milner-Brown, Stein and Lee (22) noted that the order of recruitment of motor units in the first dorsal interosseus muscle was irretrievably lost when the ulnar nerve was accidentally severed and surgically resutured. One explanation may be that the relationship between the size of the nerve and the motor unit was not restored after reinnervation. In young cats this relationship was found 6 months after reinnervation, but there appeared to be no relationship between size of nerve and muscle and contractile speed (2). The loss of recruitment order may be alternatively related to the high chance of mismatching of nerves to different muscles in resuturing a large nerve trunk, such as the ulnar nerve, to its distal stump. Then nerve fibres would grow down inappropriate channels, and muscle fibres within one muscle would become innervated by nerves from several motor pools instead of one.

We used chronic recording methods to follow the nerve and muscle properties of single units in individual animals before and during the course of reinnervation to distinguish between these two possibilities. We were also interested in the extent to which the contractile properties of reinnervated muscle fibres in adult animals also become rematched to their motor nerves, even though they now occupy different positions in the muscle. The size of the nerve fibres could be assessed from the potential recorded at either site since these were directly correlated and remained well correlated throughout reinnervation. Either EMG or tension indicated the size of the motor unit, but tension proved the more reliable measure.

Before nerve section, the size of a motor unit was directly related to nerve size. When unit twitch tension is plotted as a function of the neural potential amplitude on logarithmic scales, the points can be fitted with a linear regression line with a slope of approximately 2, which is statistically different from zero. Furthermore, larger units are faster contracting than smaller units as contraction time of the units is inversely correlated with twitch tension (Fig. 5). Thus, larger muscle units are innervated by larger nerves and are faster contracting, while smaller units are supplied by smaller nerves and have longer contraction times. These results are consistent with the size principle first described by Henneman and his colleagues for this muscle (30). The relationships could not be demonstrated when the muscle first becomes reinnervated after section and resuture (Fig. 5). At this time, when both nerve and muscle are still

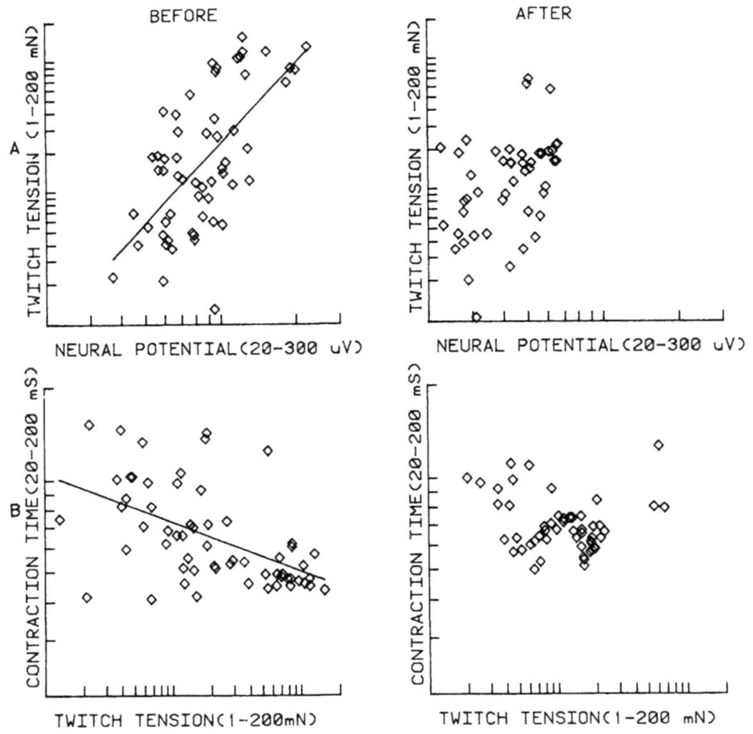

Fig. 5. Unit twitch tension plotted as a function of the unit neural potential and contraction time as a function of the unit twitch tension on logarithmic scales. Computer linear regression lines were fitted to the data A) before, and B) 2 months after nerve-nerve suture. The slope of the lines was significantly different from zero before but not 2 months after resuture.

atrophic, there was no statistically significant relationship between the size of nerve and muscle or its contractile speed. The size of nerves and muscles recovered and the properties of nerve and muscle became rematched with time (Fig. 6). The slope of the regression lines relating twitch tension and neural potential and relating contraction time and twitch tension were computed on logarithmic scales at different times during the

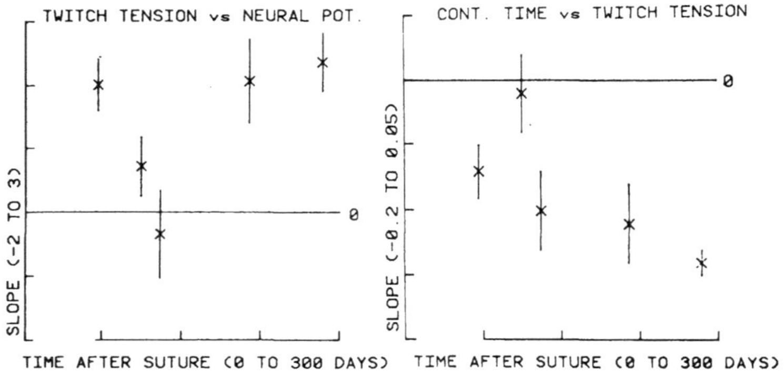

Fig. 6. The slope ± S.E. of the computed regression lines of A) twitch tension vs neural potential of single MG motor units, and B) unit contraction time vs unit twitch tension on logarithmic scales, are plotted as a function of time after MG nerve-nerve suture.

course of reinnervation. The regression lines have been plotted in Fig. 6 against time to follow the rematching of nerve and muscle properties. The positive correlation between unit twitch tension and neural potential amplitude of intact units was not restored immediately following reinnervation, the slope of the regression line computed not being significantly different from zero (Fig. 6A). By 3 months after resuture, unit tension and neural potential were still unrelated, although they had both begun to recover in amplitude. By 5 months after reinnervation the units had regained their former size, and motor unit twitch tensions and neural potential amplitudes were again correlated. Rematching of nerve and muscle size was excellent since the slope of the regression line of unit tension against neural potential amplitude returned close to control levels.

Motor unit size was unrelated to contractile speed when muscle first became innervated, but this relation returned within 2 months after reinnervation. This is seen as the return of a regression line with negative slope which was significantly different from zero. If anything, the slope of this reestablished relationship was greater.

Thus, although regrowth of nerves back to muscles is relatively unspecific and nerve fibres reinnervate adjacent fibres with no apparent preference for their original muscle fibres (15, 16), muscle properties again became rematched with time after reinnervation.

Discussion

The cross-innervation studies of Buller, Eccles and Eccles in 1960 (4) clearly established that muscle properties were determined by their motor nerve, for when a fast muscle was supplied by a slow nerve it became slowly contracting. Similarly, cross-innervated slow muscles became faster when cross-innervated by fast nerves. Properties of nerves, on the other hand, are not altered by their target organ; for example, the shortened after-hyperpolarization of the axotomized slow motoneurones again becomes long in duration relative to fast motoneurones after reinnervation, whether the target muscle is fast or slow (17). Even differentiation of motoneurones appears to be independent of muscle influence (7).

The present results show that atrophic changes of nerve and muscle subsequent to their mutual isolation is reversed once functional connections are remade in the periphery. When nerves first make contact with muscle, the denervated muscle fibres are in a dedifferentiated form (20) and nerves supply muscle fibres that formerly belonged to several different motor units with different histochemical and contractile properties (15). At this time there is no relationship between nerve and muscle properties. Within the first months after reinnervation, muscle properties become determined by the motor nerve for contraction time became inversely related to tension output. Muscle fibres within a newly reinnervated motor unit are therefore probably becoming more uniform, but there is as yet no relationship between muscle and nerve sizes. Soon after reinnervation there was also no correlation between EMG amplitude and muscle tension, which possibly reflected the different grouping of muscle fibres which would alter EMG recorded on the muscle surface. Reorganization of motor units was further indicated by return of the correlation of EMG potential amplitude and muscle tension. With time, the size of nerve and muscle again became interrelated, and nerve and muscle properties became rematched despite the different location of motor units within the muscle.

Since the properties of nerve fibres remain that of the parent axons whatever the target muscle (17), it appears that it is the nerve that imposes its influence on the muscle fibres supplied and the muscle properties become rematched to the nerve supplying them. The nerve determines at the motor unit level in a mixed muscle such as the medial or lateral gastrocnemius muscle the contractile speed of muscle fibres it supplies in the same way as a slow or fast nerve determines the contractile speed of the muscle it supplies (4). Our results show that size of the motor unit in terms of its tension output as well as its contractile speed is related to the nerve that reinnervates it. Whether this rematching of sizes of nerve and unit tension is due to a mutual interaction between nerve and muscle, or whether the nerve determines motor unit size in a similar way to which it determines the contractile speed, is not yet known.

The results indicate that the loss of orderly recruitment of motor units in reinnervated human muscles (22) is probably not due to an inability of the nerve to re-specify properties within a single muscle. It may rather be due to the high probability of mismatching of individual nerve fibres after suturing the proximal stump of a large severed nerve to its distal stump such that the nerve fibres from different motor pools would be misdirected to alien muscles. This would be particularly serious in muscles used for finely controlled movements, as is the case for human hands. The loss of orderly recruitment and mismatching of muscles could explain why patients find it extremely difficult to relearn fine motor skills. Our results are consistent with findings that regrowth of nerve fibres in the adult are not specific but that any cholinergic nerve will make connections with skeletal muscle (19). Although there is some evidence to indicate that the growth of motor nerves to muscle in the embryo may be specified (18), even this is uncertain, and as in the adult animal, the mechanical influences and the time factors of outgrowth may be important factors for the establishment of nerve-muscle connections (31). While in the developing animal the central connections of motoneurones develop some time after those in the periphery (11), the central connections in the adult may be such that the animal cannot compensate for new connections in the periphery (28), making it difficult to relearn controlled and fine movements with the same mechanisms which they used prior to reinnervation. These remain interesting problems for future investigations.

Acknowledgements

This research was supported by the Medical Research Council and the Muscular Dystrophy Association of Canada.

References

1. Aitken, J.T., Sharman, M., Young, J.Z.: J. Anat. *81*, 1-22 (1947).
2. Bagust, J., Lewis, D.M.: J. Physiol. *237*, 91-102 (1974).
3. Brown, M.C., Ironton, R.: J. Physiol. *278*, 325 (1978).
4. Buller, A.J., Eccles, J.C., Eccles, R.M.: J. Physiol. *150*, 417-439 (1960).
5. Davis, L.A., Gordon, T., Hoffer, J.A., Jhamandas, J., Stein, R.B.: J. Physiol. *285*, 543-559 (1978).
6. Desmedt, J.E., Godaux, E.: J. Physiol. *264*, 673-693 (1977).
7. Gallego, P., Huizar, P., Kudo, N., Kuno, M.: J. Physiol. *281*, 253-265 (1978).
8. Gutmann, E., Hník, P.: in *The Denervated Muscle*, ed. Gutmann, E. Publishing House of the Czechoslovak Academy of Sciences, Prague, pp. 13-56 (1962).
9. Gutmann, E., Sanders, F.K.: J. Physiol. *101*, 489-518 (1943).
10. Gutmann, E., Young, J.Z.: J. Anat. *78*, 15-43 (1944).
11. Hamburger, V.: *The Developmental History of the Motoneurone*. F.O. Schmidt Lecture in Neurosciences, Neuroscience Research Program Bulletin, M.I.T., Cambridge, Mass.
12. Henneman, E., Somjen, G., Carpenter, D.O.: J. Neurophysiol. *28*, 560-580 (1965).
13. Hoffer, J.A., Stein, R.B., Gordon, T.: Brain Res., in press.
14. Hoffman, H.: Aust. J. Exp. Biol. Med. Sci. *28*, 383-397 (1950).
15. Karpati, G., Engel, W.K.: Neurol. *18*, 447-455 (1968).
16. Kugelberg, E., Edström, L., Abbruzzese, M.: J. Neurol. Neurosurg. Psychiat. *33*, 319-329 (1970).

17. Kuno, M., Miyato, Y., Muñoz-Martinez, E.J.: J. Physiol. *242*, 273-288 (1974).

18. Landmesser, L., Morris, D.G.: J. Physiol. *249*, 301-326 (1975).

19. Langley, J.N., Anderson, H.K.: J. Physiol. *31*, 365-391 (1904).

20. Lewis, D.M.: J. Physiol. *222*, 51-75 (1972).

21. McPhedran, A.M., Wuerker, R.B., Henneman, E.: J. Neurophysiol. *28*, 71-84 (1965).

22. Milner-Brown, H.S., Stein, R.B., Lee, R.G.: J. Neurol. Neurosurg. Psychiat. *37*, 665-669 (1974).

23. Milner-Brown, H.S., Stein, R.B., Yemm, R.: J. Physiol. *230*, 359-370 (1973).

24. Stanitski, C.L., Mooney, V.: J. Biomed. Mater. Res. *4*, 97-108 (1973).

25. Stein, R.B., Charles, D., Davis, L., Jhamandas, J., Mannard, A., Nichols, T.R.: Can. J. Neurol. Sci. *2*, 235-244 (1975).

26. Stein, R.B., Nichols, T.R., Jhamandas, J., Davis, L., Charles, D.: Brain Res. *128*, 21-38 (1975).

27. Stein, R.B., Charles, D., Gordon, T., Hoffer, J.A., Jhamandas, J.: IEEE Trans. Bio-Med. Engng. *25*, 532-537.

28. Sperry, R.W.: Quart. Rev. Biol. *20*, 311-369 (1945).

29. Taylor, A., Stephens, J.A.: Brain Res. *117*, 331-335 (1976).

30. Wuerker, R.B., McPhedran, A.M., Henneman, E.: J. Neurophysiol. *28*, 85-99 (1965).

31. Vrbová, G., Gordon, T., Jones, R.: *Nerve-Muscle Interaction*. Chapman and Hall, London (1978).

DIFFERENT STIMULATION PATTERNS AFFECT CONTRACTILE PROPERTIES OF DENERVATED RAT SOLEUS MUSCLES

T. Lømo, R.H. Westgaard and L. Engebretsen
Institute of Neurophysiology, University of Oslo, Norway, and Institute of Work Physiology, Oslo, Norway

Introduction

Cross-reinnervation experiments (1,2) have established that neural influences largely determine the differentiation of mammalian skeletal muscle into fast or slow types. More recent experiments using chronic nerve stimulation (3-5) show further that the pattern of nerve impulse activity is a major factor in these interactions. However, in these experiments innervated muscles are used and it is difficult to distinguish between transformations in the muscle caused by changes in muscle activity *per se* or by changes in flow of trophic substances resulting from altered nerve activity. To overcome this inherent difficulty we have earlier stimulated denervated muscles with different patterns of stimuli and shown that transformations occur in the absence of neural influences (6). In the present experiments we have confirmed this result and extended it by showing that the denervated muscle will become fast or slow, strong or weak, fatigue resistant or easily fatiguable depending on the particular pattern of activity which is imposed on it.

Material and Methods

The hind legs of one or both sides of male adult Wistar rats weighing about 250 g were denervated under barbiturate

Plasticity of Muscle

anaesthesia by cutting and reflecting the sciatic nerve in the thigh. Stimulating electrodes from external stimulators were implanted around the denervated soleus muscle on one side, first as previously described (6) and later by using teflon coated multistranded steel wires whose naked ends were placed across the ventral and dorsal surface of soleus. Stimulus duration and strength was first 1 msec and 5-10 mA and later 0.2 msec and 10-15 mA.

The acute experiments were done after 3 to 15 weeks of stimulation. The animals were then in good condition and had gained up to 150 g of weight. In initial experiments the soleus was freed extensively from the surrounding tissue (6). This sometimes interferred with the blood supply and made measurement of endurance unreliable. In later experiments measurements were therefore made after minimal dissection of the distal end of the soleus and then repeated after more extensive dissection. With the exception of the tests for endurance the two types of measurements gave corresponding results. The soleus was stimulated by supramaximal 0.5 msec long pulses through platinum electrodes encircling its distal end, at optimal length and at a surface temperature of 35-36°C maintained by running warm oxygenated Ringer's solution over it. In each experiment the soleus nerve trunk was stimulated and signs of reinnervation were never observed.

Results

A. Stimulation with "Fast" Impulse Pattern

If intermittent high frequency bursts of activity determine the contractile properties of fast muscle, a denervated slow muscle should become fast when similar activity is imposed on it by direct stimulation. Normal fast activity probably varies greatly and is impossible to mimic exactly. To cover at least a wide range of such activities we stimulated the denervated soleus with intermittent, brief (2/3 sec long) 100 Hz trains repeated either frequently (every 67 sec, mean frequency 1 Hz) or infrequently (every 1 hour, 57 min, 50 sec; mean frequency 0.01 Hz).

1) Frequent 100 Hz stimulation. After 7-9 weeks of such stimulation contraction time and half relaxation time of the isometric twitch was markedly reduced (Fig. 1C, Table 1). The twitch-tetanus ratio had decreased (Table 1), post tetanic potentiation of the twitch (Fig. 2C) and "sag" (7) during tetanic stimulation had appeared and the steep part of the tension vs. frequency curve had moved from about 15-20 Hz to about 10 Hz (Fig. 4). Similar changes occur in slow muscles after cross-reinnervation with a fast nerve (2) and the result confirms our previous report (6).

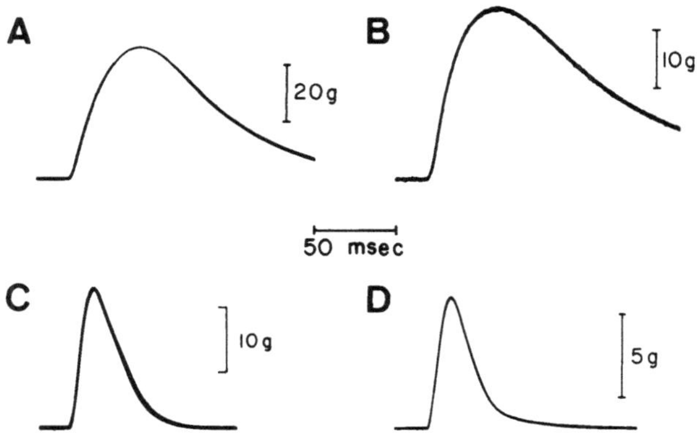

Fig. 1. Isometric twitch responses of normal (A) and chronically denervated and stimulated (B-D) soleus muscles. Stimulation had been nearly continuously at 10 Hz for 70 days in B, briefly (2/3 sec long trains) at 100 Hz every 67 sec for 48 days in C and briefly (2/3 sec long trains) at 100 Hz every 1 hour, 57 min and 50 sec for 61 days in D.

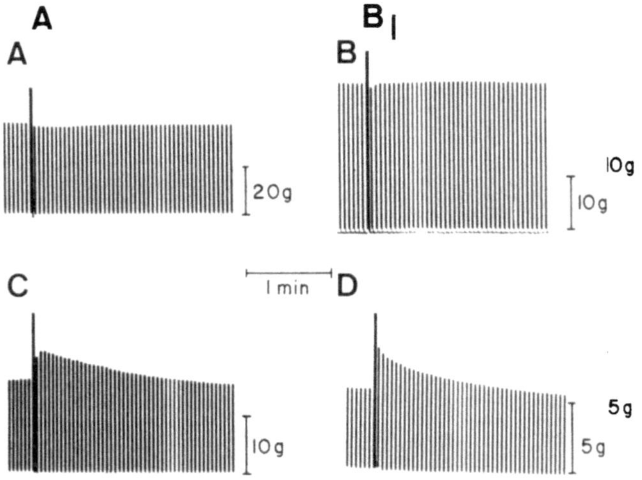

Fig. 2. Isometric twitches (one every 3.3 sec) before and after tetanic stimulation at 200 Hz for 1 sec of normal (A) and chronically denervated and stimulated (B-D) soleus muscles. Same muscles as in Fig. 1.

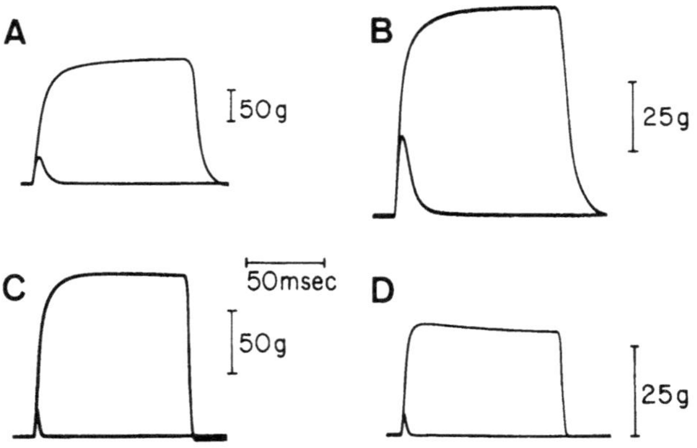

Fig. 3. Isometric responses to single and repetitive (125 Hz for 1 sec) stimuli of normal (A) and chronically denervated and stimulated (B-D) soleus muscles. Same muscles as in Figs. 1 and 2.

TABLE 1

Effects of chronic stimulation on contractile properties of denervated soleus muscles

	Normal	De-nervated	Denervated and Stimulated			
			100 Hz trains mean 1 Hz	100 Hz trains mean 0.01 Hz	10 Hz trains mean 9 Hz	10 Hz trains mean 1 Hz
Days		45-79	48-62	61-105	49-70	41-51
Twitch contraction time (msec)	40.4 (±5.1)	39.9 (±2.8)	14.4 (±0.6)	13.3 (±0.6)	44.7 (±7.0)	30.1 (±5.2)
½ relaxation time (msec)	50.5 (±6.4)	44.2 (±4.7)	18.7 (±2.4)	13.2 (±0.3)	75.8 (±17.6)	35.2 (±10.2)
Twitch-tetanus ratio	0.19 (±0.03)	0.49 (±0.05)	0.15 (±0.03)	0.19 (2) (±0.01)	0.39 (±0.09)	0.52 (±0.04)
Post tetanic potentiation	0.91 (±0.06)	0.97 (±0.04)	1.30 (±0.03)	1.39 (±0.13)	0.88 (±0.10)	1.07 (±0.04)
Fatigue index	0.67 (3) (±0.13)	0.71 (3) (±0.07)	0.54 (±0.15)	0.10 (±0.09)	0.67 (±0.25)	0.63 (±0.11)
Muscle weight (mg)	160 (3) (±20.2)	29 (±4.5)	100 (2) (±46)	83 (±7.6)	107 (±16.5)	51 (±27)
Maximal tetanic tension (g)	206 (±32.3)	3.3 (±0.8)	102 (±55)	38 (±7.4)	66 (±15.9)	31 (±12.3)
Number of muscles	8	6	3	3	3	3

±values within paranthesis are ±S.D. Number within paranthesis give number of muscles when this is different from the main group.

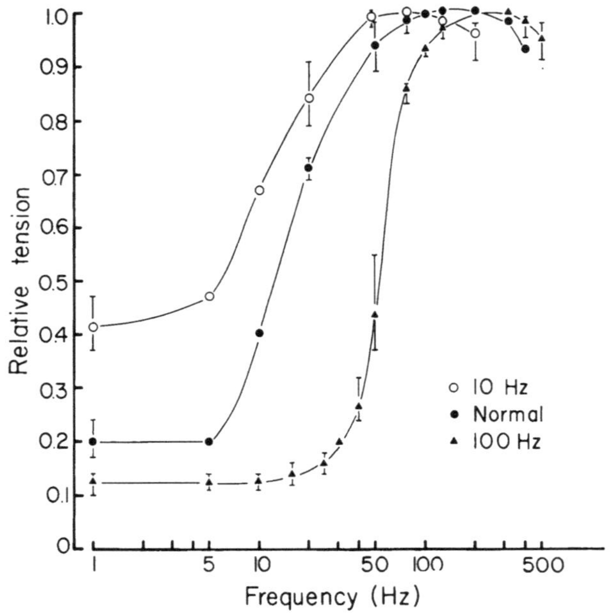

Fig. 4. Isometric tension as fraction of maximal tetanic tension plotted against stimulus frequency. Vertical bars represent entire range of 3-5 muscles. Some of the symbols without bars are from one muscle only. ● Normal soleus muscles. ○ Soleus muscles denervated and stimulated nearly continuously at 10 Hz for 49-70 days. ▲ Soleus muscles denervated and stimulated with brief (2/3 sec long) 100 Hz trains every 67 sec for 48-62 days.

The changes illustrated in Figs. 1-4 are likely to have been caused in part by changes in sarcoplasmic reticulum (SR) affecting Ca transport and active state and in part by changes in contractile proteins (6) but isometric measurements alone do not allow a clear distinction between these mechanisms. Recently, chronic 10 Hz stimulation of motor nerves has been shown to transform the SR in a rapid way (8) and myosin in a more slowly way (5,9) into types characteristic of slow muscle.

2) <u>Infrequent 100 Hz stimulation</u>. This pattern also induced properties characteristic of fast muscle (Figs. 1D-3D, Table

1). The magnitude of the changes were similar to those induced by the more frequent 100 Hz stimulation (Table 1), but the muscles had been stimulated a little longer (9-15 weeks vs. 7-9 weeks). The finding that widely different numbers of high frequency bursts effectively transform slow into fast muscle while muscles stimulated with low frequency trains remain slow (see below), suggest that the more important factor is the presence of high frequency bursts rather than any particular combination of such bursts. Normal fast motor units are probably used to very different degrees and yet may have similar contraction times. For example, in rat tibialis ant. the fast motor units have nearly all the same contraction time of 10-12 msec, yet they are very different with respect to fatique resistance, SDH and fibre size (10).

The two 100 Hz patterns had quite different effects on endurance. With frequent 100 Hz stimulation the fatigue index (7) was high (Fig. 5C, Table 1), whereas with infrequent 100 Hz stimulation it was very low (Fig. 5D, Table 1). Furthermore, in the infrequently stimulated muscle the profile of the response during repeated tetanic stimulation (Fig. 5D) showed an initial increase in tension and then a rapid decline similar to that observed in fast, easily fatiguable motor units or muscles. A long lasting depression of tension was also present after brief tetanic stimulations which made it impossible to obtain meaningful tension-frequency curves (Fig. 4) for these muscles. The result suggests that speed and endurance are independently regulated; endurance depending more on amount of activity than on the characteristic frequency within bursts.

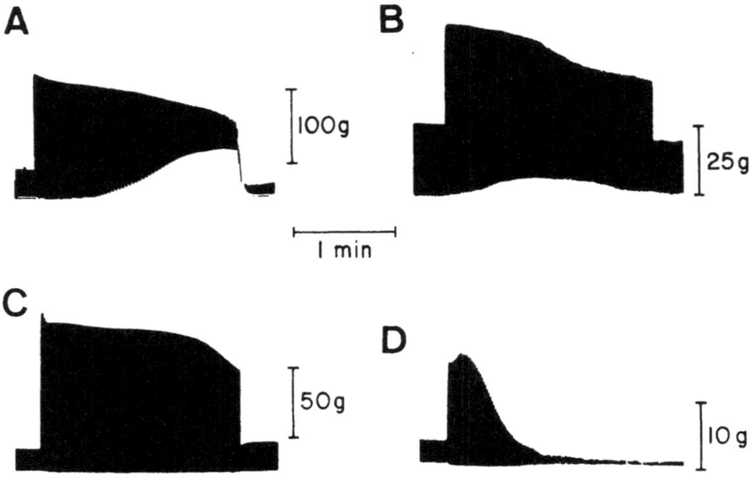

Fig. 5. Isometric responses to 1/3 sec long stimulus trains at 77 Hz repeated every sec for 2 min of normal (A) and chronically denervated and stimulated (B-D) soleus muscles. B,C and D same stimulation regime as for B,C and D in Figs. 1-3, but B and D are from different muscles. B had been stimulated for 56 days and D for 105 days.

It has recently been suggested that trophic factors rather than fast activity induce the contractile properties of a fast muscle (11). These factors, it is suggested, are released from both fast and slow motor nerves but their effect is overcome in slow muscle by tonic, slow motor neurone discharges, whereas in the fast muscle the activity is considered to occur too infrequently to have any effect *per se*. This would explain that spinal cord transection causes the slow soleus muscle to become faster if, as is assumed, the transection virtually eliminates motor neurone discharges. The dramatic effect of very infrequent high frequency stimulation shown here does not support this scheme, and the possibility should be considered that occasional high frequency bursts of activity may be generated in slow and fast motor neurones after spinalization.

B. Stimulation with "Slow" Impulse Pattern

If tonic low frequency impulse activity determines the contractile properties of slow muscle, similar activity imposed by direct stimulation would be expected to maintain the slow properties of denervated slow muscles. To test this we denervated the slow soleus and stimulated it directly at 10 Hz either nearly continuously (1 sec pause every 10 sec, mean frequency 9 Hz) or intermittently (7 sec trains every 70 sec, mean frequency 1 Hz).

1) Nearly continuous 10 Hz stimulation. This pattern resulted in very different contractile properties from those obtained with 100 Hz stimulation. After 7-10 weeks muscles stimulated at 10 Hz had contraction times and half relaxation times which were approximately 3 and 5 times longer than muscles stimulated at 100 Hz (Fig. 1B, Table 1). There was post tetanic depression rather than potentiation (Fig. 2B, Table 1), a slow increase rather than "sag" in tension during tetanic stimulation (Fig. 3B, Table 1) and a relatively high resistance to fatigue (Fig. 5B, Table 1). Evidently low frequency stimulation maintains the slow contractile properties of the denervated soleus muscle and in some respects makes the muscle even slower than normal. Thus the half relaxation time is considerably longer than normal (Table 1) and the tension-frequency curve is moved to the left so that the steep part of the curve occurs around 10 Hz (the imposed impulse frequency) rather than around the 15-20 Hz of normal soleus muscles (Fig. 4).

These results resemble those obtained by continuous 10 Hz stimulation of fast muscle through the nerve (4, 5). In both cases the stimulated muscles become as slow or slower than the normal soleus muscle. Moreover the muscles lose weight and develop larger twitch tetanus ratios (see also below).

2) Intermittent 10 Hz stimulation. This pattern produced muscles which were somewhat faster than normal soleus muscles but not nearly as fast as muscles stimulated at 100 Hz (Table 1). The different effects of continuous and intermittent 10 Hz stimulation suggest that contraction speed is influenced by number and pattern of impulse bursts in addition to impulse frequency within bursts. However, other factors must also be considered. For example, these muscles became very atrophic probably because of low loads and low levels of active tension during the low frequency stimulation (see below). It is also important to consider the time course of the induced changes and the extent to which different impulse patterns affect only the rate of change or also the final steady state level. Infrequent, intermittent stimulation, particularly at low frequency, is known to induce membrane changes which develop much more slowly than they do during more frequent stimulation although they may reach the same steady state level (12). Contractile properties also change faster with continuous than with intermittent 10 Hz stimulation (5).

C. Maintenance of Muscle Weight and Tetanic Force Output

Stimulation prevented the large reduction in muscle weight and tetanic force output which occurred in non-stimulated muscles (Table 1). For example, after 7-9 weeks of denervation muscles stimulated intermittently at 100 Hz weighed approximately 3 times more and produced approximately 30 times more tetanic force than non-stimulated muscles. Nevertheless, compared with normal muscles some reduction occurred (Table 1). This may be attributed in part to the reduced load on the soleus muscle caused by denervation of antagonists and by cocontractions of faster synergists (gastrocnemius) due to spread of stimulus current. Furthermore, in muscles stimulated intermittently at 10 Hz little or no summation of tension occurred at 10 Hz and chronic stimulation must have generated little active tension. These muscles where both active and passive tension generation was low, were also weakest and most atrophic (Table 1).

It is interesting that muscles stimulated at 100 Hz for less than a second only every 2 hours weighed more and produced more tetanic tension than muscles stimulated much more frequently at 10 Hz with 100 times as many stimuli (Table 1). This indicates that high frequency discharges are a potent stimulus to muscle growth and force generation. The effect may help to explain why the largest and fastest motor units have the largest and strongest muscle fibres although they are probably used least often. Usage must nevertheless play a role since muscles stimulated often at either 10 or 100 Hz weighed more and produced more tetanic tention than muscles stimulated less often at the same frequencies (Table 1).

D. Conclusion

The present work shows that the contractile properties of denervated muscle change dramatically when different patterns of activity are imposed on the muscle by direct stimulation. Intermittent high frequency (100 Hz) stimulation makes the normally slow muscle fast, while continuous low frequency (10 Hz) stimulation makes it slower than normal. High frequency stimulus trains induced fast contractile properties whether they were delivered often or very seldom. Fatigue resistance, on the other hand, required frequent stimulation and both 10 Hz and 100 Hz stimulation trains were effective. High frequency stimulation was most effective in preventing muscle atrophy and reduction in force output.

During chronic 10 Hz stimulation twitch duration and twitch-tetanus ratio (Fig. 5) increased markedly at the same time as the steep part of the tension-frequency curve (Fig. 5) moved to around 10 Hz (the imposed frequency). During chronic 100 Hz stimulation marked changes occurred in the opposite direction (Fig. 5). It seems likely that these changes are part of an adaptive response in which the time course of the active state is adjusted in order to bring the steep part of the tension-frequency curve to coincide with the prevailing impulse frequency that is imposed on the muscle. This ensures that the production and control of force becomes most efficient at the impulse rates which the muscle receives. A similar correspondence between motor neurone discharge rates and the steep part of the tension-frequency curve is observed in normal motor units (13).

All the changes observed in these experiments occurred in the absence of any neural influences, indicating that the pattern of muscle activity per se plays an important role in determining the contractile properties of skeletal muscle.

References

1. Buller, A.J., Eccles, J.C. and Eccles, R.M.: J.Physiol. 150, 417-439 (1960).
2. Close, R.: J.Physiol. 204, 331-346 (1969).
3. Salmons, S. and Vrbová, G.: J.Physiol. 201, 535-549 (1969).
4. Salmons, S. and Sréter, F.A.: Nature 263, 30-34 (1976).
5. Pette, D., Müller, W., Leisner, E. and Vrbová, G.: Pflügers Arch. 364, 103-112 (1976).
6. Lømo, T., Westgaard, R.H. and Dahl, H.A.: Proc.R.Soc.B. 187, 99-103 (1974).
7. Burke, R.E., Levine, D.N., Tsairis, P. and Zajac, F.E.: J.Physiol. 234, 723-748 (1973).
8. Heilmann, C. and Pette, D.: Eur.J.Biochem. 93, 437-446 (1979).

9. Sreter, F.A., Gergely, J., Salmons, S. and Romanul, F.: Nature New Biol. 241, 17-19 (1973).

10. Kugelberg, E. and Lindegren, B.: J.Physiol. 288, 285-300 (1979).

11. Gallego, R., Huizar, P., Kudo, N. and Kuno, M.: J. Physiol. 281, 253-265 (1978).

12. Lømo, T. and Westgaard, R.H.: J.Physiol. 252, 603-626 (1975).

13. Kernell, D.: Brain Res. 160, 159-162 (1979).

Support from the Muscular Dystrophy Association of America is gratefully acknowledged.

NEURAL CONTROL OF CONTRACTILE AND MEMBRANE PROPERTIES IN SLOW MUSCLE FIBRES OF THE FROG?

H. Schmidt

I. Physiologisches Institut der Universität des Saarlandes
D - 6650 Homburg/Saar, Germany

Introduction

Some skeletal muscles of the frog are composed of two types of muscle fibres, twitch and slow (1). The main functional differences between these two populations of muscle fibres are as follows. Upon direct or indirect stimulation twitch fibres respond with an action potential, while slow fibres are devoid of an action potential mechanism. If depolarizing agents (acetylcholine or K-rich solutions) are applied twitch fibres contract transiently, and complete relaxation occurs within less than a minute (2); slow fibres, however, develop a contracture which is maintained for many minutes (3,4,5). Both the contracture and the absence of an action potential in normal slow fibres are thought to be under the control of small(slow) motor axons which normally innervate the slow muscle fibres (6,7,8,9).

In mammalian muscles contractile as well as membrane properties clearly depend on the presence and the pattern of muscle activity (10,11,12,13). It is therefore of interest to investigate whether in frog slow muscle fibres factors other than innervation have an important influence on membrane and contractile properties.

Plasticity of Muscle

Methods

Experiments were performed on pyriformis, iliofibularis and rectus abdominis muscles of Rana temporaria. All operations were done under general ether anaesthesia.

Denervation: The motor nerves to the distal segments of the rectus abdominis muscle were crushed at a distance of 5-7 mm from the lateral edge of the muscle. Following the operation animals were kept in running tap water at 18-21° C. Experimental and contralateral muscles were removed 13 to 159 days following the operation; the distal pelvic segment of each muscle was dissected and mounted in a chamber for nerve stimulation and application of depolarizing agents.

Implantation of nerve cuffs: In another group of frogs procaine cuffs were placed around the right sciatic nerve in order to block impulse conduction for at leat 10 days. The cuffs were made by mixing Dow medical adhesive (0.5 g) with procaine-hydrochloride (0.16 g); their length was 2-3 mm, and their internal diameter about 1.5 mm. The outer surface of the cuffs was sprayed with polyethylene.

Transplantation: The left sartorius muscle was replaced by the right iliofibularis muscle (14). The sartorius nerve was joined to the nerve stump in connection with the transplanted muscle, and the skin was closed with several stitches. The animals were kept in running tap water at 18-23(mean 20)^{o} C for the following 17 months and fed weekly with meal worms.

Mechanical tension was measured with Devices mechano-electrical transducers connected to a double channel recorder. Usually both the experimental and control muscles were set side by side in the same bath. Contractures were elicited by rapidly exchanging the normal bath solution for Ringer's solution containing acetylcholine (10^{-5} g/ml) or a high concentration of K-ions (90 mM). Contractures were measured 10-15 min following application of the depolarizing solution.

Membrane potentials were recorded using conventional microelectrode techniques. Pyriformis and iliofibularis muscles were mounted as described previously (15). Slow muscle fibres were identified by their membrane time constants (16). In these experiments the Ca-concentration of Ringer's solution was raised to 7.2 mM.

Stimulation of nerve fibres: Usually the nerve was placed over Pt-electrodes in a small chamber sealed with vaseline; rectangular current pulses of 50-100 µsec duration were applied. Selective stimulation of small motor axons was performed with the method described earlier (17).

Solutions: Normal Ringer's solution had the following composition (mM): NaCl 110.4; KCl 2.5; $CaCl_2$ 1.8; Tris-HCl 5.0; pH = 7.2.

Results

A. Maintenance of Tension (Contracture)

1) Contracture of denervated slow fibres. Chronic denervation over several months does not lead to loss of the contracture in slow muscle fibres (6); this observation was confirmed in the present experiments using rectus abdominis muscles. There is, however, a reduction of the ability to maintain tension (18).

2) Effect of self-reinnervation on contracture. During reinnervation after crushing the motor nerve iliofibularis muscles of the frog lose their ability to maintain tension (6). This observation was explained by the assumption that the slow fibres become reinnervated by foreign(fast) axons. Later on it could be demonstrated that this assumption was correct (19). It remained, however, open whether the reappearance of the contracture was closely associated with restoration of neuromuscular transmission from small motor axons to slow muscle fibres. An attempt was therefore made

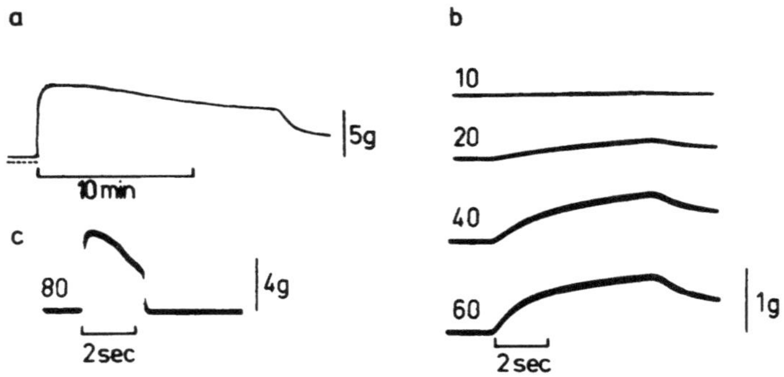

Fig. 1. Contractile responses of a rectus abdominis muscle to application of acetylcholine 10^{-5} g/ml (a), indirect stimulation with rectangular pulses (c) or selective stimulation of small axons (b). Numbers adjacent to records in b and c indicate stimulation frequencies (pulses per sec). Horizontal dashed line in a indicates zero tension.

to answer this question. Rectus abdominis muscles were denervated by crushing the nerves, and their ability to develop a contracture was correlated with their contractile response to selective stimulation of small nerve fibres. Fig. 1 shows the principle of this method and the results obtained with a normal muscle. Application of acetylcholine results in an isometric tension which is maintained at a level of more than 50 % of the maximum for more than 10 min (Fig. 1a). Stimulation of the motor nerve with rectangular pulses at high frequency (80/sec) results in the usual tetanus (Fig. 1c). The same result is obtained with triangular current pulses. If such pulses are, however, applied with the anode being placed near the muscle, a pulse strength can be found at which single pulses do not elicit mechanical tension because the action potentials of fast axons are blocked under the anode. Upon repetitive stimulation at more than 5/sec the muscle develops a slowly rising tension which is due to activation of slow muscle fibres only (Fig. 1b).

During a period of 2-4 weeks after crushing the nerves the ability of the muscles to maintain tension decreased to very

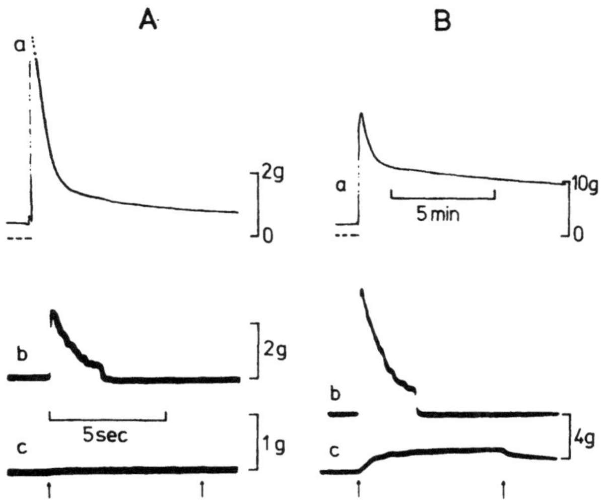

Fig. 2. Contractile response of rectus abdominis muscles to application of acetylcholine 10^{-5} g/ml (a), stimulation of fast axons with rectangular pulses at 80/sec (b), and selective stimulation of small axons at 40/sec (c) between arrows. Note presence of contracture, while selective stimulation of small axons (A,a and c) does not elicit appreciable tension. Muscles investigated 30 (A) and 50 (B) days following nerve crush. Zero tension indicated by dashed lines in a.

low values; this was already observed earlier in iliofibularis muscles (6). After approximately 4 weeks neuromuscular transmission was restored first in the twitch fibre system; in some of these muscles the contracture was, however, partially restored (Fig. 2A). The contracture then increased towards and above control values, and selective stimulation of slow axons was always associated with the slowly rising tension characteristic of the activation of slow fibres (Fig. 2B). Fig. 3 summarizes the results obtained during 5 months following nerve crush. It is important to note that the contracture recovers quickly as soon as the muscles are being reinnervated; most interestingly in a few muscles the contracture has recovered substantially even before selective stimulation of small motor axons results in mechanical tension.

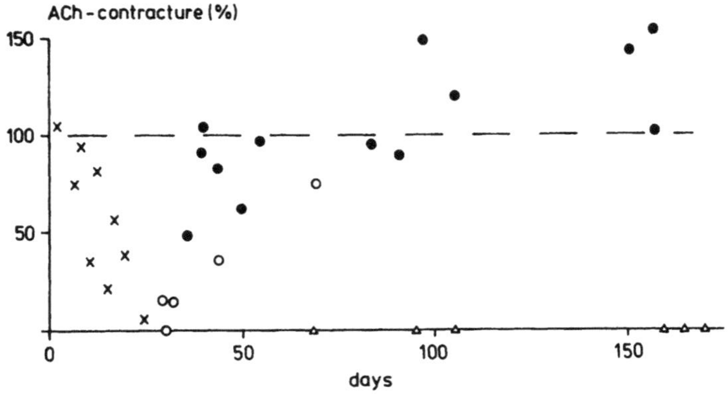

Fig. 3. Acetylcholine contractures of rectus abdominis muscles recorded 3-159 days following nerve crush. Contractures are given as % of those developed by contralateral control segments. x, no response to indirect stimulation; o muscles responding to stimulation of fast axons only; • muscles responding to stimulation of fast axons as well as to selective stimulation of small motor axons; Δ values obtained from transplanted iliofibularis muscles upon application of 90 mM K-Ringer's.

3) Contracture of transplanted muscles. No contracture was found in transplanted iliofibularis muscles reinnervated by the sartorius nerve (14). This was confirmed in the present experiments for the first 7 months following transplantation. 6 additional muscles were examined after 15-17 months, and 4 of them showed small contractures (Fig. 4) which amounted to 12-44 (mean 27) % of the contracture developed by the control muscles.

4) Contracture after reduction of contractile activity. Complete paralysis of iliofibularis muscles for 2-4 weeks produced by chronic curarization or by blocking the nervous impulses with procaine cuffs had no effect on the amplitude of contractures elicited by application of K-rich Ringer's solution.

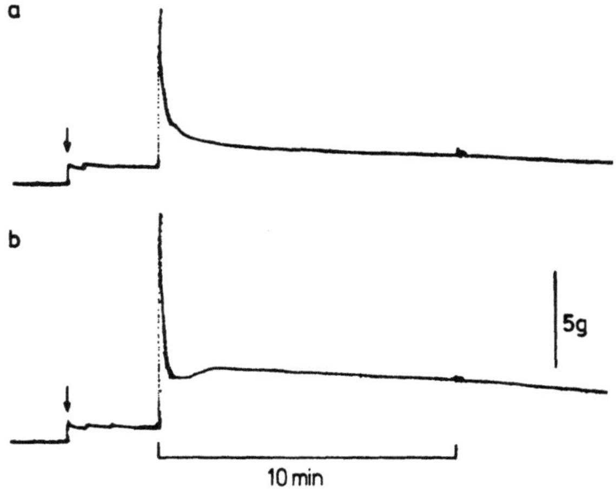

Fig. 4. Contracture of iliofibularis muscles elicited by application of 90 mM K-Ringer's (horizontal bar); at arrows muscles were stretched. Note that contracture 441 days after transplantation (a) amounts to 33 % of that developed by the control muscle (b).

B. Action Potential in Slow Muscle Fibres

Small motor axons seem to exert a continuous 'trophic' effect on the membrane of slow muscle fibres (7,8,9), and the interruption of this influence induces the action potential. The possible contribution to this transformation process of other factors were investigated in a variety of experiments.

1) Effect of nerve blockade. Impulse conduction in the sciatic nerve of 8 frogs was blocked by application of procaine cuffs for periods of time which, after denervation, are sufficiently long for the action potential to develop to full size (9). 10-15 days following implantation of the cuffs pyriformis and iliofibularis muscles were removed. A total of 48 slow fibres were identified; 44 slow fibres were clearly devoid of an action potential although only 23 of them responded with an endplate potential to nerve stimulation. Four slow fibres (3 of them innervated) had small action potentials; they were

found in the iliofibularis and pyriformis muscle of the same animal, together with a slow fibre which did not show an action potential. It is clear from these results that 'functional denervation' by interruption of impulse conduction in the motor fibres cannot mimic the effect of cutting the nerve. The appearance of action potentials in 3 innervated slow fibres is easily explained by the assumption that these fibres had been partially denervated (20) by some mechanical or toxic effect of the procaine cuffs. On the other hand, 21 slow fibres were found to be denervated, but did not produce an action potential upon direct stimulation. Obviously, the deleterious effect of the procaine cuffs developed slowly, and the denervation period was too short. Furthermore, in these muscles up to 68 % of the superficial twitch fibres were found to be denervated. Thus, the presence of degenerated nerve terminals did not induce the action potential in neighbouring slow fibres.

2) Action potential in slow fibres of transplanted muscles. 13 iliofibularis muscles were removed 5-17 months following transplantation; all muscles were reinnervated by the sartorius nerve. 22 slow fibres were identified electrophysiologically in 7 muscles examined between the 5th and 7th month. Only 14 slow fibres showed action potentials upon direct stimulation while the remaining 8 slow fibres were not excitable. The amplitude of the action potentials did never exceed 42(mean 18) mV, and was thus much smaller than observed in muscles denervated in situ (7,8,9). Another 17 slow fibres were found in 6 muscles 7-17 months following transplantation, but action potentials were no longer elicited by direct stimulation. Thus, the slow fibres lost their action potential not later than 7 months after transplantation, and 4 months after reinnervation of the muscle by the sartorius nerve. Resting potentials, effective membrane resistances and membrane time constants of twitch and slow fibres seemed essentially normal during the whole period examined.

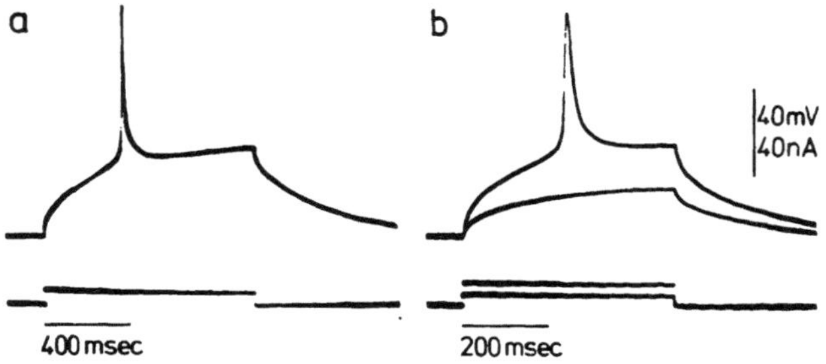

Fig. 5. Action potentials (upper records) recorded from slow muscle fibres upon intracellular application of depolarizing current pulses (lower records). a, slow fibre of an iliofibularis muscle denervated 308 days following transplantation, and examined 18 days later; b, slow fibre of control iliofibularis muscle, denervated at same time.

The loss of the action potential was, however, not irreversible. Two muscles were denervated together with their normal controls by cutting the sciatic nerve 297 and 308 days following transplantation. 18 and 19 days later 5 slow fibres were identified in the transplanted muscles, and all had action potentials of 40-80 (mean 58) mV. The action potentials recorded from 8 slow fibres in the control muscles were of similar amplitude (Fig. 5).

The experiments thus clearly show that upon chronic reinnervation of iliofibularis muscles by the sartorius nerve, the action potential of slow muscle fibres is not maintained; it disappears 4 months after reinnervation, but it can be made to reappear by cutting the nerve once more.

3) <u>Failure to induce the action potential by curarization or tenotomy.</u> Several frogs were chronically paralysed for 2-4 weeks by repeated subcutaneous injection of d-tubocurarine. 14 slow fibres were identified in pyriformis and iliofibularis muscles but none of them responded with an action potential to direct or indirect stimulation; all slow fibres were innervated and showed normal frequencies of miniature endplate potentials (~ 15/min).

In 4 frogs the iliofibularis muscle was tenotomized. 14-160 days later 13 slow fibres were identified but all were unable to produce action potentials.

Discussion

A. Contracture

Earlier (6,14) and present results show that slow fibres lose their ability to maintain a contracture following nerve crush or transplantation and reinnervation by a fast nerve. In both cases the slow fibres are disconnected from their own motor axons and subject to the effect of regenerating fast motor axons normally innervating only twitch fibres. Full recovery of the contracture is only observed during self-reinnervation. The present experiments show that the recovery process is closely related to the restoration of neuromuscular transmission from small motor axons to slow muscle fibres. We did, however, find a few muscles in which the contracture had recovered substantially while selective stimulation of small axons was still ineffective. This observation suggests that the recovery process does not require resumed synaptic function; the slow motor axons may therefore exert their influence prior to functional reinnervation of the slow muscle fibres (8).

Up to 7 months following transplantation (and reinnervation by the sartorius nerve) iliofibularis muscles did not show a contracture. This was certainly not due to degeneration of slow fibres because they could be easily identified electrophysiologically, and their membrane properties were unchanged. The absence of contracture must therefore be attributed to an influence of the nerve (14). A few muscles investigated 10 months later did, however, show a partial recovery of the contracture. This was not observed by Miledi

& Orkand (14) after a similar period of time. The reason for the discrepancy may be that our frogs were kept at a higher temperature; other factors (animal species, feeding etc.) may also be important.

The mechanism behind the partial recovery of the contracture in transplanted muscles reinnervated by the fast sartorius nerve is obscure. Several possibilities have to be considered: a slowly developing 'trophic' effect of the fast motor axons comparable to that presumably exerted by slow motor axons; attraction of slow motor axons by the slow muscle fibres; rejection of fast axon synapses by the slow muscle fibres. The latter effect may be sufficient because slow fibres maintain their contracture in the denervated state (6,18). More likely, the second process mentioned above could also take place (see below).

Attempts to modify the contracture of slow fibres by other means (curarization, blockade of nervous impulses) were unsuccessful. This also shows clearly that synaptic transmission and muscle contractions are no essential factors as regards the ability of slow fibres to maintain a contracture.

B. Action Potential

One of the most important results in this section is the observation that the action potential disappears from slow fibres of iliofibularis muscles reinnervated by the fast sartorius nerve. This result was quite unexpected; moreover, denervation of the iliofibularis muscle more than 10 months after transplantation made the action potential reappear. Obviously, it had been suppressed by some influence exerted by the sartorius nerve. Since slow fibres kept denervated for 6 months maintain their action potential (7), it is unlikely that rejection of fast axon synapses could be the only reason for the suppression of the action potential. A possible explanation would be that the slow fibres in transplanted muscles attract small motor axons which then would suppress the action potential. Small motor axons are presumably absent in the motor branch innervating the sartorius muscle (because the sartorius muscle is devoid of slow muscle fibres); they may, however, be present in another branch of the nerve and could be attracted towards the transplanted muscle by some neurotropic effect (22). Alternatively it could be assumed that fast motor neurones are transformed into slow ones by a retrograde effect (23) of the transplanted slow muscle fibres. In any case, we do not know at present whether the slow fibres in transplanted iliofibularis muscles are innervated by fast or by slowly conducting motor axons; this question has to be studied in further experiments.

It was also shown that the action potential in slow fibres of transplanted muscles never reached amplitudes similar to those of slow fibres denervated in situ. Since we did not examine the muscles earlier than 5 months following the operation it may be that a suppression process had already begun earlier. A bad state of the muscle fibres is unlikely, because their resting potentials and other membrane properties were essentially normal; moreover, these muscles developed maximal tensions only 24 % smaller than those of the control muscles, and the twitch fibres showed normal action potentials.

The experiments have also demonstrated that abolition of contractile activity cannot play an important role in the transformation process of the denervated slow fibre membrane. Neither interruption of impulse traffic nor block of neuromuscular transmission are able to induce the action potential (21). Similarly, loss of evoked transmitter release and loss of binding of acetylcholine to its postsynaptic receptors are unimportant. Furthermore, the presence of degenerating nerve terminals cannot be regarded as a factor producing signs of denervation as has been reported for mammalian muscles (24).

Concluding remarks. The present results further support the hypothesis (6,7,8,9,14) that both the contracture and the inexcitability of slow muscle fibres are due to a 'trophic' influence of small nerve fibres. Factors which interfer with contractile activity seem to play no role in the regulation of the slow fibre properties examined.

References

1. Kuffler, S.W., Vaughan Williams, E.M.: J. Physiol. (Lond.) 121, 289-317 (1953).
2. Hodgkin, A.L., Horowicz, P.: J. Physiol. (Lond.) 153, 386-403 (1960).
3. Sommerkamp, H.: Naunyn-Schmiedeberg's Arch. exp. Path. Pharmak. 128, 99-115 (1928).
4. Nasledov, G.A., Thesleff, S.: Acta physiol. scand. 90, 362-372 (1974).
5. Lännergren, J.: Acta physiol. scand. 69, 362-372 (1967).
6. Elul, R., Miledi, R., Stefani, E.: Acta physiol. latinoam. 20, 194-226 (1970).
7. Miledi, R., Stefani, E., Steinbach, A.B.: J. Physiol. (Lond.) 217, 737-754 (1971).
8. Schmidt, H., Stefani, E.: J. Physiol. (Lond.) 270, 507-517 (1977).
9. Schalow, G., Schmidt, H.: Pflügers Arch. 372, 17-22 (1977).
10. Lømo, T., Rosenthal, J.: J. Physiol. (Lond.) 221, 493-513 (1972).
11. Al-Amood, W.S., Buller, A.J., Pope, R.: Nature (Lond.) 244, 225-227 (1973).
12. Jones, R., Vrbovà, G.: J. Physiol. (Lond.) 236, 517-538 (1974).
13. Lømo, T., Westgaard, R.H.: J. Physiol. (Lond.) 252, 603-626 (1975).
14. Miledi, R., Orkand, P.: Nature (Lond.) 209, 717-718 (1966).

15. Stefani, E., Schmidt, H.: Pflügers Arch. 334, 276-278 (1972).

16. Stefani, E., Steinbach, A.B.: J. Physiol. (Lond.) 203, 383-409 (1969).

17. Forrester, T., Schmidt, H.: J. Physiol. (Lond.) 207, 477-491 (1970).

18. Lehmann, N., Schmidt, H.: Pflügers Arch. 1979, in press.

19. Schmidt, H., Stefani, E.: J. Physiol. (Lond.) 258, 99-123 (1976).

20. Schalow, G., Schmidt, H.: Nature (Lond.) 253, 122-123 (1975).

21. Miledi, R., Spitzer, N.: J. Physiol. (Lond.) 241, 183-199 (1974).

22. Edds, M.V., jr.: Q. Rev. Biol. 28, 260-276 (1953).

23. Sperry, R.W.: Growth Symp. 10, 63-87 (1951).

24. Cangiano, A., Lutzemberger, L.: Science 198, 542-545 (1977).

ABILITY OF ELECTRICALLY SILENT NERVES TO SPECIFY FAST AND SLOW MUSCLE CHARACTERISTICS

Lynn Eldridge and Wilfried Mommaerts
Department of Physiology, School of Medicine, University of California at Los Angeles, California, U.S.A.

Introduction

A basic question arising from the nerve-crossing experiments which demonstrated that nerves determine the type-specific biochemical and contractile properties of fast and slow muscles (2, 3, 5, 6, 15) regards the nature of the differentiating neural signal. Is the nerve controlling the muscle characteristics through the pattern of electrical stimulation which it delivers, or through non-electrical, trophic influences? Many experiments in which various patterns of muscle stimulation have been artifically imposed through the nerve have demonstrated altered contractile and biochemical characteristics of fast and slow muscles (1, 11, 16, 17, 18, 19, 20), but it is impossible to tell whether the changes were due to the electrical stimulation of the muscle or to a trophic substance secreted by the nerve in response to its own firing pattern. The few experiments in which the stimulation patterns were administered directly to a denervated muscle have shown that the activity of the muscle itself can change some contractile properties (14). The actual role of stimulation patterns in the normal physiological regulation of the muscle remains obscure, as relatively little is known about the ability of the nerves to exert control by non-electrical means. We have been studying the neurotrophic control by using cat preparations in which the motoneurones have been rendered permanently silent by surgical isolation from excitatory inputs in the spinal cord. We have examined the effects of two types of spinal isolation on the ability of the nerves to maintain fast and slow muscle properties and to effect reversal of these properties after a nerve cross.

Plasticity of Muscle

Ventral Horn Isolation (VHI)

In the first study, the ventral horn in the region from lumbar 6 to sacral 3, carrying the motoneurones to the calf muscles, was isolated not only from the ascending and descending tracts and dorsal roots, but also from the rest of the cord. A thin scalpel was inserted immediately lateral to the exit of the dorsal roots from one side of the cord and directed toward the center of the base of the cord. At this angle, it was moved longitudinally through the segments L6 to S3, thus separating this ventral horn from both dorsal horns and the contralateral ventral horn. At the rostral and caudal ends of this cut, the ventral horn on the affected side was pinched to eliminate descending and ascending input to it. The dorsal roots were severed unilaterally in the region of isolation.

From the description of this procedure, it should be obvious that the operation was imprecise. It was impossible to know at the time of operation that the longitudinal cut had been made at the proper angle and depth. As might be expected, most of the cats prepared in this way did not acquire the permanent and complete flaccid paralysis of the calf muscles required for the test of the ability of the silent nerves to maintain or reverse fast and slow muscle properties. However, in five cats, the operation did seem to be successful, in that over the entire nine months of the study, no tone, either reflex or voluntary, was ever felt in the calf muscles. In two of these five cats, the nerves to the flexor hallucis longus (FHL) and soleus were crossed 1 and 2 months after the isolation. Data from these cats were compared to those from one normal cat which had undergone a nerve cross at the same time as the two VHI cats and also to data from normal cats, both crossed and intact, which had been studied over several years using the same biochemical assays. The five VHI cats and the normal cat studied concurrently were sacrificed on schedules such that the VHI muscles had been paralyzed for about nine months and the nerve-crossed muscles had been crossed for seven months.

The raw data from these six cats are presented in Table 1 and Figure 1. The uncrossed leg of the normal cat showed values similar to those of other control cats assayed in this laboratory. In the uncrossed VHI legs,

Table 1. Summary of data on atrophy, twitch time, Ca^{++}-activated myosin ATPase activity, and lactate dehydrogenase (LDH) activity of muscles from cats with no operations (--) or ventral horn isolations (VHI), with and without nerve crosses (NC).

OPER.	CAT	% ATROPHY		MSEC TO PEAK TENSION		MYOSIN ATPase ACTIVITY $\mu M\ P_i/mg/min$		LDH ACTIVITY U/mg/min	
		FHL	SOL	FHL	SOL	FHL	SOL	FHL	SOL
--	1	-	-	40	120	.65	.15	11,000	1,960
VHI	2	48	65	40	65	.53	.31	10,500	1,940
VHI	3	71	74	35	100	.64	.15	9,920	1,970
VHI	4	59	50	33	90	.60	.15	4,950	2,000
NC	1	40	-6	70	60	.29	.20	5,000	2,900
VHI + NC	5	46	30	70	54	.24	.51	2,610	6,800
VHI + NC	6	75	59	58	48	.40	.28	2,600	2,120

the soleus muscles maintained much of the normal slow pattern with respect to lactate dehydrogenase (LDH) activity and isozymes, myosin ATPase activity, and time to peak twitch tension. One of the three VHI soleus muscles showed partial speeding in all of these parameters except LDH activity, and all of the three showed somewhat shortened times to peak twitch tension. The uncrossed VHI FHL's maintained a normal fast pattern, except for a decrease in LDH activity.

The two nerve-crossed VHI legs showed marked crossing effects, and these changes were at least as great as those in the crossed normal leg. In the crossed soleus and FHL muscles, all parameters studied changed significantly except the soleus LDH activity, which failed to rise in the normal cat and one VHI cat. As expected from the relatively short period of crossing, reversals of characteristics were incomplete, although the LDH activities in the FHL's of the two crossed VHI legs dropped to values very close to those of control soleus muscles.

This VHI study was carried out by Drs. Lynn Eldridge and George Szekeley. Technical assistance was provided by Mr. Nicholas Ricchiuti and Ms. Miwon Suh.

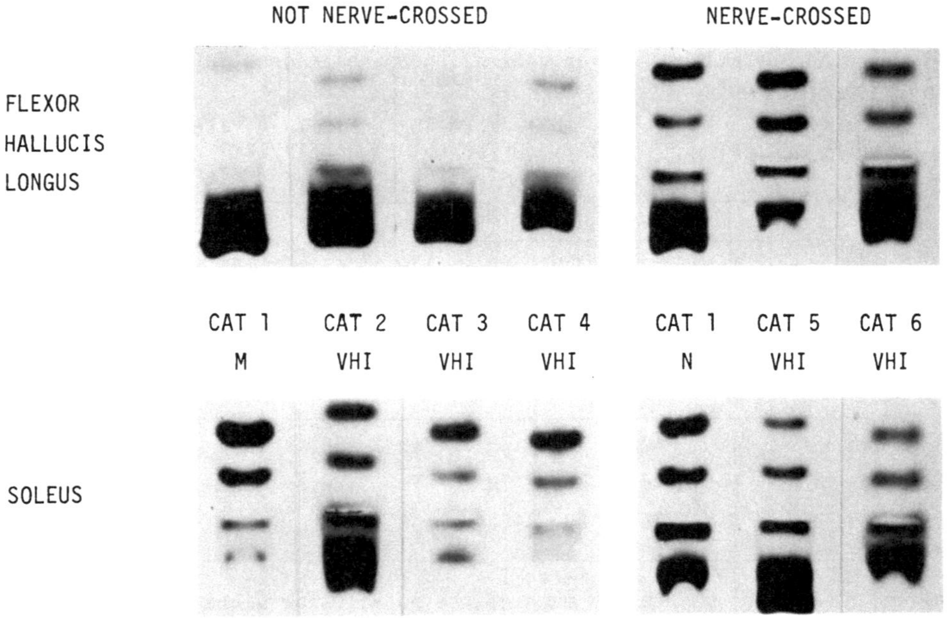

Figure 1. Agar gels of lactate dehydrogenase isozymes from muscles of cats undergoing ventral horn isolation (VHI) or no spinal operation (N), with or without nerve crossing.

Spinal Isolation

Methods

In the second study, the operation used to inactivate the motoneurones was a modification of the Tower (21) procedure, in which the spinal cord was transected and the dorsal roots below the transection were cut between the dorsal ganglia and the cord, leaving the motoneurones below the transection isolated from descending and reflex activation. In the original Tower preparation, activity of the supposedly isolated muscles was observed, but was determined to be the result of mechanical stimulation of the spinal cord, left relatively unprotected after the extensive laminectomy performed to allow the dorsal rhizectomy. The length of cord Tower isolated was large, increasing the possibility of complex activity resulting from resulting from stimulation of the spinal generator by accidental irritation of the cord. In the Eldridge modification of the basic Tower preparation, opportunities for such stimulation were minimized. The laminectomy in the region of isolation was kept very narrow (2mm), leaving the cord well protected by bone. To diminish the probability of spontaneous activity of the spinal generator in the cord, the length of cord isolated was restricted to the minimum needed to ensure inactivation of the specific muscles to be studied, the soleus and FHL. Extensive anatomical and physiological mapping of the spinal columns of 30 cats prior to the first isolation operation made it possible to carry out the surgery under the condition of low visibility imposed by the restrictions of the narrow laminectomy and localized isolation. To maximize the health of the isolated section of spinal cord, a third issue not mentioned by Tower was emphasized: maintaining the blood supply to the cord. In the cat, in the lumbosacral region involved in our isolation (lumbar 6 through sacral 4), the blood supply to the cord is carried primarily in the ventral artery. The transection of the cord in the Eldridge preparation is performed by progressive dissection, finally leaving the artery visible beneath the space between the isolated region and the rest of the cord.

The cats prepared with this isolation procedure have proved to be a chronically healthy and effective preparation; we have maintained some of them for three years. The calf muscles are in flaccid paralysis and are unresponsive to attempts to elicit reflexes. Electromyographic recording from various calf muscles has shown no more than an occasional single unit active. Neither pressure nor thumps on the back evoke muscle activity. The silent nerves regenerate after being severed in the process of a nerve cross, and evoke contraction in the muscles when electrically stimulated.

Results from eight cats which were maintained for 17 to 36 months after the spinal isolation operation are summarized here. In one leg of each of these, the nerves to the soleus and FHL were crossed at intervals of 14 to 19 months after the isolation. The nerves in the contralateral leg were either left intact or cut and resutured for self-reinnervation. At the time of the final testing, the cats had been nerve-crossed for 3 to 20 months. For purposes of comparison with these SI cats, a number of cats were prepared with other neurological lesions, including sciatic and ventral root denervation.

Muscles from the lesioned cats were examined in a variety of ways in an attempt to describe the effects of the spinal isolation itself, as well as those of the nerve cross. For these analyses, we used not only the intact or self-reinnervated FHL and soleus, but also several other muscles from the paralyzed legs. Drs. Eldridge, Reggie Edgerton, Joe Henry Steinbach, and Ms. Mary Liebhold studied electromyographic activity of the tibilias anterior, lateral gastrocnemius, soleus, and FHL in the awake, unrestrained cats at various times from immediately prior to the nerve cross to several weeks before sacrifice. Drs. Eldridge, Steinbach, and Ms. Liebhold tested isometric contraction properties of the FHL and soleus through electrical stimulation of the nerves on the day of sacrifice. Dr. Steinbach examined the innervation, neuromuscular junctions, and acetylcholine receptor distributions of the peroneal muscles, extensor digitorum longus, and vastus intermedius. Dr. Eldridge, Donald Simpson, and Ms. Liebhold analyzed FHL, soleus,flexor digitorum longus, tibialis anterior, and gastrocnemius for histochemical staining patterns and structural changes. Drs. Steinbach and David Schubert of the Salk Institute examined contractile proteins of the soleus, FHL, and tibialis anterior using 2-dimensional gels. Ms. Miwon Suh and Mira Marusich in our laboratories assessed ATPase and LDH activities and isozymes and calcium uptake by the sarcoplasmic reticulum vesicles in soleus and FHL.

Results

Effects of spinal isolation. The disused muscles showed profound changes in appearance and morphology. By the time of the nerve cross, which was the first opportunity of directly observing the muscles, the soleus muscles had become as pale as the fast muscles, and all muscles had substantially atrophied. At sacrifice, the soleus and FHL muscles weighed only 12 to 50% as much as muscles from a normal cat of the same body weight. Microscopic study of the muscles showed that the pattern of atrophy differed from muscle to muscle, with no apparent relationship to fiber type. In some muscles, all fibers atrophied equally. In others, there was great variation in fiber diameter, from area to area or within the same region, or both. There was an increased proportion of both connective tissue and fat, which appeared in histochemical cross section of the muscle, biochemical extraction procedures, and direct studies of connective tissue.

Alterations in innervation, neuromuscular junctions, and acetylcholine receptor distribution occurred in the SI cats. Disuse evoked sprouting of the motor nerves. Wandering axon sprouts were seen in silver stained preparations; these seemed to form separate junctions at an average of 200 μm from the original one. The average junctional length was greater in the disused muscles, with some junctions up to 4 times the mean of the normal lengths (measured by cholinesterase stains or fluorescent alpha-bungarotoxin binding). There was a small increase in extrajunctional ACh receptors, just over the normal undetectable level, as assessed by curare-protectable binding of radioactive alpha bungarotoxin. However, the extrajunctional receptor density was much lower than that seen on muscle fibers denervated for 1-8 months. The extrajunctional receptor density was not correlated with fiber diameter; e.g. the very atrophied fibers had no more extrajunctional receptors than more normal ones in a paralyzed muscle.

The SI muscles were very different from those which had been denervated for 1 to 8 months. In denervated muscles, the junction remained the same length but broke down. Patches of extrajunctional receptors appeared on denervated fibers, but not on SI fibers.

Contraction characteristics of the SI muscles changed in a number of ways. As would be expected from the atrophy, the muscles became much weaker, especially the soleus, which developed only 1 to 10% of the isometric tetanic tension typical of soleus muscles from cats of equal size. Disused FHL's pulled only 3 to 25% of normal tension. The loss of isometric tetanic strength in both muscles was greater than could be accounted for in terms of decrease in gross muscle mass, probably due to the increased proportion of fat and connective tissue. The most atrophied muscles also seemed to have lost more stamina than the larger disused muscles.

In time to peak twitch tension (Table 2), the disused soleus muscles became very similar to normal and disused fast muscles, which did not differ

Table 2. Means of data on muscles from the spinal isolation study.

OPER.	MSEC TO PEAK TENSION		MSEC TO HALF RELAX		CA^{++} UPTAKE M/mg/10 min		MYOSIN ATPase ACTIVITY μM P_i/mg/min		LDH ACTIVITY Units/mg/min	
	FHL	SOL	FHL	SOL	FHL	SOL	FHL	SOL	FHL	SOL
----	26	89	19	96	1.52	.24	.67	.15	12,500	2,000
NC	52	45	49	60	.89	1.25	.32	.27	5,000	4,000
SI	25	28	24	29	1.56	.25	.68	.56	8,040	2,730
SI+NC	23	31	22	29	1.53	.32	.67	.56	7,920	3,170
D	--	--	--	--	--	--	.67	.17	2,790	2,240

Key: ---- no operation; NC nerve-cross; SI spinal isolation; D denervation; LDH lactate dehydrogenase; MSEC to peak tension and half relaxation refer to twitch; MYOSIN ATPase is Ca^{++}-activated.

from each other. Mean time to peak tension was about 26 milliseconds for normal and disused FHL and 28 for disused soleus, whereas it was 89 for normal soleus. Mean half-relaxation times for the isometric twitches were 19 for normal FHL, 24 for disused FHL, and 29 for disused soleus, in contrast to 96 for normal soleus. Thus, by isometric twitch characteristics, the chronically disused muscles remained or became fast.

According to histochemical analysis of serial sections of the inactive muscles, there was a pronounced shift in fiber type composition toward the fast pattern (heavy staining for both ATPase and glycolytic enzyme activity). FHL and FDL, which normally contain about 90% fast fibers, showed 99% to 100% fast fibers after disuse. Tibialis anterior and gastrocnemius, normally mixed, also rose, sometimes to 100% fast. Soleus, normally 100% slow, became 75 to 97% fast. Denervated muscles, in contrast, retained their normal fiber type composition while degenerating.

Biochemical assays of Ca^{++}-stimulated ATPase activity were in agreement with the histochemical and isometric twitch results. There was still a difference between SI soleus and FHL, but the disused soleus was much closer to normal and disused FHL (which did not differ) than to a normal soleus. Mean values are presented in Table 2. Individual soleus values were at least twice normal, and in one cat, higher than that of a normal FHL. Disused soleus and FHL varied much more than their normal counterparts. Two dimensional gels separating tropomyosins and light chains of myosin showed a conversion (usually complete) to the fast pattern in the soleus; FHL retained its fast pattern. In 1 to 8-month denervated muscles, ATPase activities and two-dimension gel patterns remained normal in fast and slow muscles.

FHL and soleus maintained a considerable difference in LDH activity (see Table 2). Disused soleus showed values of about 300 to 5000, whereas normal values are from 1500 to 2500 in our system. Disused FHL's ranged from about 6000 to 13,500, in comparison to the normal 12,000 to 14,000. Denervated FHL's dropped to values typical of normal soleus; denervated soleus remained low. With respect to LDH isozymes (Figure 2), disused soleus converted to the fast pattern, although some muscles retained traces of normal slow isozymes. Disused FHL's retained the normal fast pattern. Again in contrast, denervated cats showed the normal patterns in both muscles.

Another biochemical parameter, the uptake of Ca^{++} by vesicles of the sarcoplasmic reticulum, was measured in three SI cats sacrificed 17 to 23 months after isolation. In all three cats, FHL and soleus retained the characteristic differences, with FHL uptake being almost 7 times higher (see Table 2).

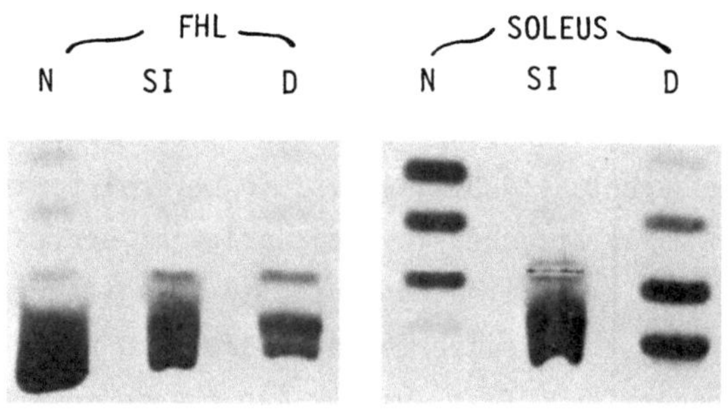

Figure 2. Agar gels of lactate dehydrogenase isozymes from muscles of cats undergoing no operation (N), spinal isolation (SI), or denervation (D).

Effects of silent nerve crossing. For the most part, the pattern of results described above did not vary as a function of the isolation-sacrifice interval, which ranged from 17 to 36 months. An exception is that there was a greater difference in ATPase activities between FHL and soleus in the cats sacrificed closer to the isolation. Since the soleus had become very similar to the FHL in many aspects by 17 months after the isolation, and the nerve crosses were performed at 14 to 19 months, there was little opportunity for observation of reversal of properties as a result of the crosses. In the two cats terminated the earliest, at 17 and 20 months, there did seem to be crossing effects, even though the nerves had been crossed only 3 and 4 months, respectively. In both cats ATPase activity in crossed (X) soleus rose much more than in self-reinnervated soleus. LDH activity of one X-soleus rose and that of one X-FHL dropped. Ca^{++} uptake by vesicles of the sarcoplasmic reticulum dropped in both X-FHL's and rose in both X-soleus muscles. The cats sacrificed 23 months or more after isolation showed no crossing effects even though LDH and ATPase activities still differed in the uncrossed FHL and soleus muscles and the nerves had been crossed for 8 to 20 months. Means for the groups are presented in Table 2.

Discussion

The results from the two studies involving long-term inactivation of the motoneurones have demonstrated that fast and slow muscles attached to electrically silent nerves can under some conditions reverse their properties. Slow muscles can eventually become fast in several contractile and biochemical parameters while attached to their original nerves. They can also become fast as a result of being connected to a fast nerve at time intervals when slow muscles not subjected to nerve crossing are retaining most of their normal properties. Fast muscles can change toward slow muscles when connected to a slow nerve, but otherwise retain their original properties, even after three years of inactivity.

In the VHI study, which demonstrated reversal of fast and slow properties after a silent nerve cross, it is clear that non-electrical aspects of neural function caused the reversal. In the SI study, which produced dramatic conversions of slow characteristics to fast in the absence of a nerve cross, it is not clear whether the inactivity of the muscle or some changed trophic property of the nerve was responsible for the muscle speeding. The fact that denervated slow muscle did not acquire fast characteristics indicates that the presence of the nerve was in some way involved in the speeding of the SI soleus, but the effect might have been permissive rather than specifying.

Experimental evidence from the literature on effects of inactivation and stimulation illustrates the complexity of the issue of mechanisms underlying alterations in muscle properties. Various methods of inactivating cat soleus muscles have speeded (4,10), slowed (7,9), or failed to change (9) them. Surgical denervation can slow both fast and slow twitch cat muscles (12,13). Obviously, activity of the muscle is not the only factor controlling its speed, although the fact that stimulation of denervated muscle in various patterns can speed or maintain the soleus twitch (14) certainly indicates that the muscle activity can play a part. That inactivity has been shown to affect the motoneurone itself (8,10) reinforces the idea that in any experiment involving changes in activity of a physically intact motor nerve-muscle system, trophic alterations in the nerve are confounded with activity pattern in the muscle, thereby making it difficult to determine the cause of any changes in muscle characteristics.

Partially due to these complexities, it is not possible to decide which of several possible factors is responsible for the discrepancy in effects of the nerve crosses in our VHI and SI studies. It seems most likely that the difference in time interval between the inactivation and the nerve cross was the critical variable. If a differentiating trophic capability is normally present in nerves but declines gradually when the neurones are deprived of their own electrical activity or of trophic influences from afferent or descending neural elements, then a nerve cross performed soon after the inactivating operation would be expected to be more effective in reversing muscle properties. It is also possible that the eventual similarity between the formerly slow and fast muscles in the long-term spinal isolation study made it impossible to detect trophic crossing abilities of the nerves. If the nerves which had been silent for a year and a half were emitting fast trophic patterning signals and had been reconnected to muscles which were still slow, perhaps they could have converted these muscles. This explanation does not seem quite adequate, as there were some differences in the soleus and FHL which remained even after three years, and should have provided an opportunity for a nerve cross to convert properties, at least in the slow → fast direction. A third possibility for the differences in nerve-crossing results in the two studies is the difference in the spinal isolation procedure. It may be that the motoneurones behave differently in the VHI preparation, in which the isolation is anatomically more stringent. Electrical inactivity and deprivation of trophic influences from the neural elements surgically removed are confounded in these isolation procedures, and if the trophic influences are important, then the motoneurones in the two preparations might be functionally quite dissimilar, even though they are electrically silent in both. Experiments are currently being conducted to distinguish among these possibilities. We are testing the ability of nerves rendered silent by the Eldridge spinal isolation procedure to effect reversal in muscle properties when crossed within four hours after the isolation.

References

1. Al-Amood, W.S., Buller, A.J. and Pope, R. Nature 244, 225-227 (1973).
2. Barany, M. and Close, R.J. J. Physiol. 213, 455-474 (1971).
3. Buller, A.J., Eccles, J.C. and Eccles, R.M. J. Physiol. 150, 417-439 (1960a).
4. Buller, A.J., Eccles, J.C., and Eccles, R.M. J. Physiol. 150, 399-416 (1960b).
5. Buller, A.J., and Mommaerts, W.F.H.M. J. Physiol. 205, 46-47P (1969).
6. Buller, A.J., Mommaerts, W.F.H.M., and Seraydarian, K. J. Physiol. 205, 581-597 (1969).
7. Cooper, R.R. J. Bone & Joint Surg. 54A, 919-953 (1972).
8. Czeh, G., Gallego, R., Kudo, N. and Kuno, M. J. Physiol. 281, 239-252 (1978).
9. Davis, C.J. and Montgomery, A. J. Physiol. 270, 581-594 (1977).

10. Gallego, R., Huizar, P., Kudo, N. and Kuno. M. J. Physiol. 281, 253-265 (1978).

11. Hudlicka, O., Brown, M., Cotter, M., Smith, M., and Vrbova, G. Pflugers Arch. 369, 141-149 (1977).

12. Kean, C.J.C., Lewis, D.M. and McGarrick, J.D. J. Physiol. 237, 103-113 (1974).

13. Lewis, D.M. J. Physiol. 222, 51-75 (1972).

14. Lomo, T., Westgaard, R.H and Dahl, H.A. Proc. Roy. Soc. Lond. B187, 99-103 (1974).

15. Mommaerts, W.F.H.M., Buller, A.J. and Seraydarian, K. Proc. Nat. Acad. Sci. 64, 128-133 (1969).

16. Pette, D., Muller, W., Leisner, E. and Vrbova, G. Pflugers Arch. 364, 103-112 (1976).

17. Pette, D., Smith, M.E., Staudte, H.W. and Vrbova, G. Pflugers Arch. 338, 257-272 (1973).

18. Salmons, S. and Vrbova, G. J. Physiol. 201, 535-549 (1969).

19. Salmons, S. and Sreter, F.A. Nature 263, 30-34 (1976).

20. Sreter, F.A., Gergely, J., Salmons, S., Romanul, F.C.A. Nature 241, 17-19 (1973).

21. Tower, S.S. J. Comp. Neurol. 67, 241-261 (1937).

Acknowledgements

This work has been supported by Muscular Dystrophy Association of America grants to Lynn Eldridge, Wilfried Mommaerts, David Schubert, and Joe Henry Steinbach; United States Public Health Service #AM 19996 to Wilfried Mommaerts; and National Science Foundation grant #BNS 77-01209 to Joe Henry Steinbach.

EFFECTS OF NERVE CROSS-UNION AND CORDOTOMY ON MYOSIN ISOENZYMES IN FAST-TWITCH AND SLOW-TWITCH MUSCLES OF THE RAT

J.F.Y. Hoh, B.T.S. Kwan, C. Dunlop and B.H. Kim
Department of Physiology, University of Sydney, Sydney, NSW 2006, Australia.

Introduction

Since the classical work on the neural transformation of contractile properties of fast-twitch and slow-twitch skeletal muscles by Buller *et al.* (1) and Close (2), many reports have appeared showing a correlation between contractile properties of cross-reinnervated muscles on the one hand and the ATPase activity or the light chain composition of their myosins on the other (3-9). In addition, the electrophoretic properties of intact myosins from fast-twitch and slow-twitch muscles have been shown to be altered by nerve cross-union (10).

The effects of nerve cross-union on the properties of skeletal myosins are of importance in molecular physiology from two main points of view. Firstly, the molecular basis for alterations in contractile properties of the fast-twitch and slow-twitch muscles under neural influence may reside principally in the changes in properties of their myosins. Secondly, the neural effects on myosin provides an interesting system for manipulating gene expression in eukaryotic cells. An understanding of both aspects of this problem requires the detailed characterization of the molecular structures of myosins in normal and cross-reinnervated muscles. Myosins of both fast-twitch and slow-twitch muscles are now known to be hexameric (11), each molecule consisting of 2 heavy chains and 4 light chains. They are also heterogeneous. Fast-twitch and slow-twitch myosins of the chick (12) or rabbit (13) can each be resolved by pyrophosphate gel electrophoresis (14) into isoenzymes which may differ in the composition of light chains. These recent findings pose the following questions regarding neural regulation of myosin: (1) To what extent, and in what manner, may the isomyosin profiles of individual fast-twitch and slow-twitch muscles be altered? (2) Can nerve cross-union bring about novel combinations of myosin subunits to form unusual isomyosins? This report attempts to answer these questions in the rat by examining isomyosin profiles in normal and cross-reinnervated muscles as well as by the analysis of light chains of individual isomyosins.

Cordotomy performed on developing mammals prevents the development of the slow-twitch muscle (15). When this operation is performed after muscle differentiation has taken place, slow-twitch muscle is transformed into fast-twitch muscle whereas fast-twitch muscle remains unchanged (16). This method of neural intervention results in a slow-to-fast muscle transformation not unlike that seen after nerve cross-union, but without involving denervation and cross-reinnervation. It is therefore of interest to compare the changes in myosin brought about by these two methods of neural intervention. Changes in isomyosin profiles and in light chain composition of isomyosins in fast-twitch and slow-twitch muscles of cordotomized rats are also reported in this paper.

Methods

Male Wistar rats were used throughout these experiments. All operations were performed 3-4 weeks postnatally, when developmental changes in muscle speed are virtually complete (17). Nerves to the fast-twitch extensor digitorum longus (EDL) and the slow-twitch soleus (SOL) were crossed as described previously (18), care being taken to avoid reinnervation of each muscle by its own nerve. Muscles were studied 16-18 weeks after operation. The innervation of each muscle was carefully checked during dissection. All muscles used were reinnervated only by the intended nerve.

Complete spinal cord transection at the lower thoracic level was performed with the help of a glass hook. The hook was maneuvered around the whole spinal cord which was completely transected so that the hook could be lifted through the region of transection. The operation induced flaccid paralysis. Muscles were studied 8-12 weeks after cordotomy.

Myosin extracted from individual muscles was analysed by pyrophosphate gel electrophoresis while light chains of purified isomyosins were analysed by SDS gel electrophoresis using methods previously described (12,14).

Results

Isomyosins in normal EDL and SOL muscles. The normal adult rat hindlimb muscles contained a total of 5 isomyosins which could be resolved on pyrophosphate gels. These are numbered 1-5 in the order of decreasing electrophoretic mobility. The profiles of these isomyosins in EDL and SOL muscles of a young adult are shown in Fig.1A. SOL contained only

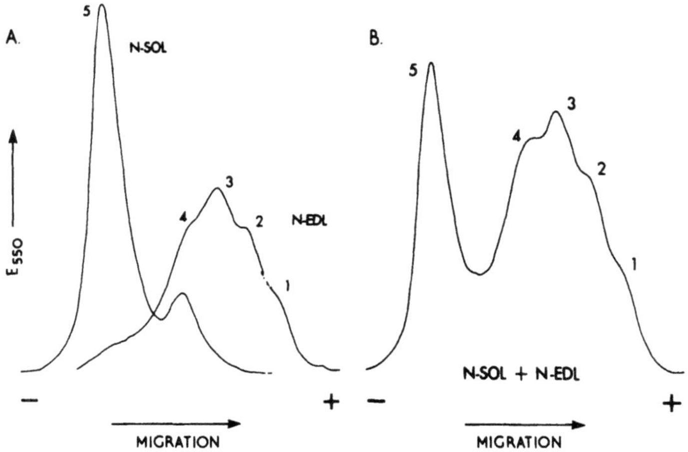

Fig.1. Isomyosin profiles from normal slow-twitch soleus (N-SOL) and normal fast-twitch extensor digitorum longus (N-EDL) muscles of the rat. Electrophoresis was performed at 2°C in 20 mM sodium pyrophosphate buffer, pH 8.8 with 10% glycerol, in 4% polyacrylamide gels and ran at 80 V for 9 hours. Gels were stained in Coomassie Brilliant Blue and scanned at 550 nm. (A) The profile from each muscle superimposed. (B) Result of co-electrophoresis.

isomyosins 4 and 5, the former constituting about 15% of the total myosin. EDL contained principally isomyosins 1-4, with a trace of isomyosin 5. When myosins from these muscles were co-electrophoresed, isomyosins 4 and 5 from SOL co-migrated with those from EDL (Fig. 1B).

The light chain compositions of some of the isomyosins have been analysed by SDS gel electrophoresis after electrophoretic purification in pyrophosphate gels. In this technique isomyosin bands were cut out under direct vision after rapid staining of pyrophosphate gels (12). Isomyosin 5 from SOL presented no difficulty because of its predominance and separation from isomyosin 4. The result (Fig.2A) showed the presence

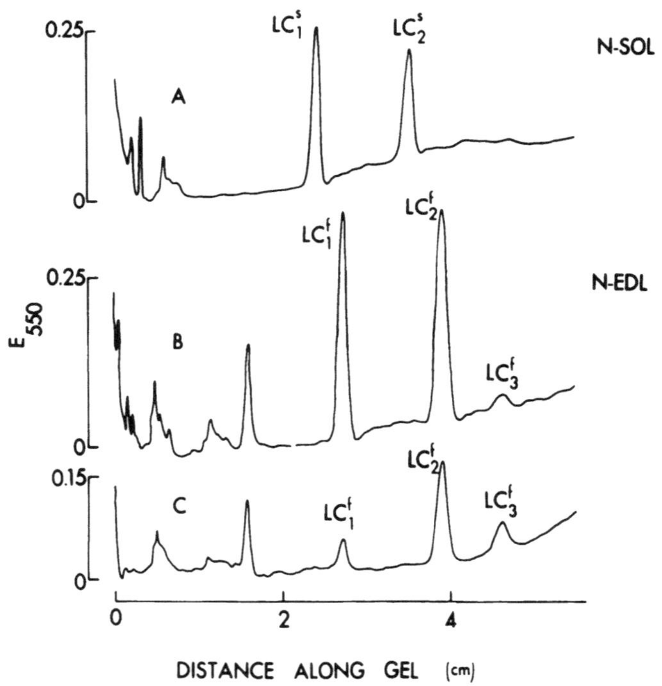

Fig.2. SDS gel electrophoresis of myosin light chains in isomyosins from normal SOL and EDL muscles after purification by pyrophosphate gel electrophoresis. Diagrams show densitometer scans of SDS gels for (A) isomyosin 5 from SOL, (B) isomyosins 3 & 4 from EDL (C) isomyosins 1 & 2 from EDL. (B) and (C) were obtained from adjacent slices of the same pyrophosphate gel containing EDL isomyosins. SDS gel electrophoresis was performed on 12.5% polyacrylamide gels at 80 V for 3 hours at 20°C.

of the two slow-twitch myosin light chains LC_1^s and LC_2^s in approximately equal proportions.

In contrast to chicken (12) and rabbit (13) fast-twitch isomyosins, the resolution on pyrophosphate gels obtainable between the rat isomyosins 1-4 was less favourable for unequivocal determination of the light chain distribution for individual isomyosins using the methods described. However, a reproducible pattern of light chain distribution has been obtained for isomyosins taken two at a time. Isomyosins 3 and 4 (Fig.2B) showed LC_1^f and LC_2^f in approximately equal proportions. A trace of LC_3^f was also present, but this was presumably due to overlapping isomyosin 2. This result is consistent with the hypothesis that the light chain compositions of isomyosins 3 and 4 in EDL are the same as that for fast-twitch isomyosin FM_3 in the chick (12) or rabbit (13). Rat isomyosins 1 and 2 (Fig.2C) showed all three fast-twitch light chains, the sum of LC_1^f and LC_3^f being approximately equal to LC_2^f. These observations are consistent with the view that rat isomyosins 1 and 2 have identical light chain compositions to FM_1 and FM_2, respectively, of chicken or rabbit fast-twitch muscles.

Isomyosins in cross-reinnervated muscles The isomyosins in cross-reinnervated muscles showed considerable variation in the extent of shift from the normal patterns. Fig.3 shows the profiles of five individual cross-reinnervated EDL muscles (A - E) and four cross re-innervated SOL muscles (F - I). The most transformed cross-reinnervated EDL (Fig.3A) differed from normal SOL only in having a small amount of isomyosin 3, while the most transformed cross-reinnervated SOL was indistinguishable from normal EDL. The profiles of the least transformed muscles were only moderately different from corresponding normal muscles. A noteworthy feature of the profiles in the partially transformed muscles is the prominence of isomyosins 4 and 3 relative to isomyosins 2 and 1.

The light chain compositions of isomyosins in two of these cross-reinnervated muscles were analysed. Fig.4 (A - C) shows the results from the cross-reinnervated EDL whose isomyosin profile is shown in Fig.3E. Isomyosin 5 (Fig.4A) consistently showed LC_1^f and LC_2^f

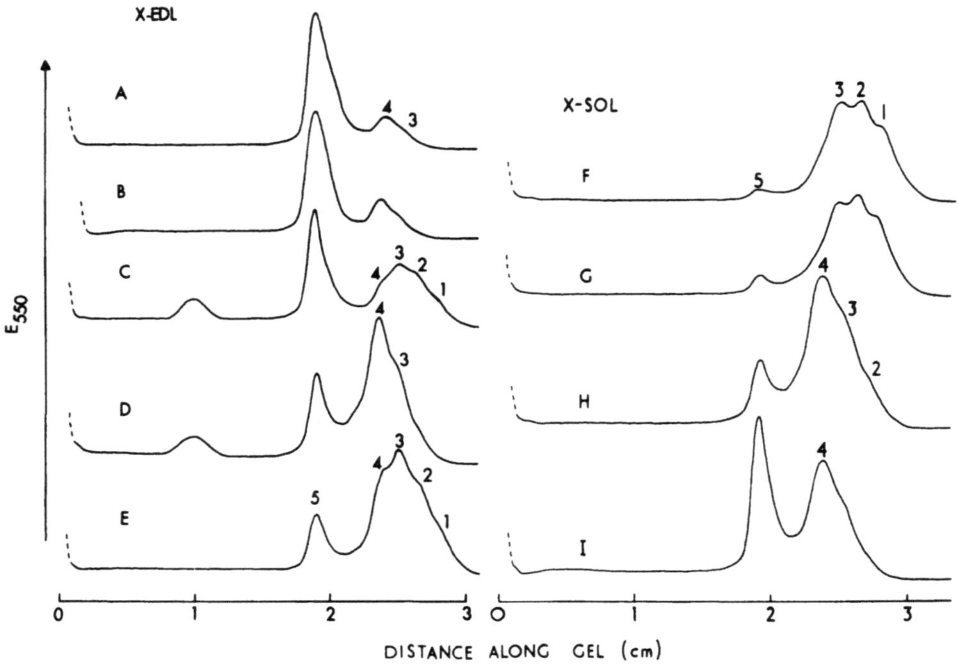

Fig.3. Isomyosin profiles of cross-reinnervated rat muscles. Diagrams show densitometer scans of pyrophosphate gels for 5 different cross-reinnervated EDL (X-EDL) muscles (A - E) and 4 different cross-reinnervated SOL (X-SOL) muscles (F - I). The profiles in each category are arranged in the order of decreasing extent of transformation from normal.

in addition to LC_1^s and LC_2^s. However, isomyosins 4 and 3 (Fig.4B) and 2 and 1 (Fig.4C) have light chain compositions identical to corresponding isomyosins in normal EDL.

The light chain compositions of isomyosins in the cross-reinnervated SOL whose isomyosin profile is shown in Fig.3H offered a striking contrast to those in cross-reinnervated EDL described above. Isomyosin 5 (Fig.4D) showed only LC_1^s and LC_2^s as in normal SOL. Because of the prominence of isomyosin 4 in this muscle, the light chains from this isomyosin could be analysed reliably on its own. This isomyosin reproducibly showed LC_1^s, LC_1^f and LC_2^f in the approximate ratio of 1:1:2.

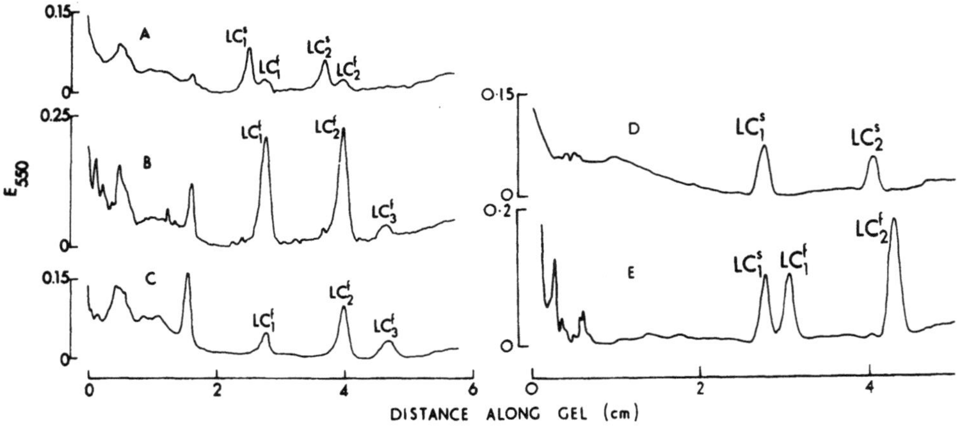

Fig.4. Light chain analysis of isomyosins in cross-reinnervated muscles. Isomyosin 5 (A), isomyosins 3 & 4 (B), and isomyosins 1 & 2 (C) from consecutive slices of pyrophosphate gels with myosin from the cross-reinnervated EDL illustrated in Fig.3E. Isomyosin 5 (D) and isomyosin 4 (E) from a pyrophosphate gel with myosin from the cross-reinnervated SOL illustrated in Fig.3H.

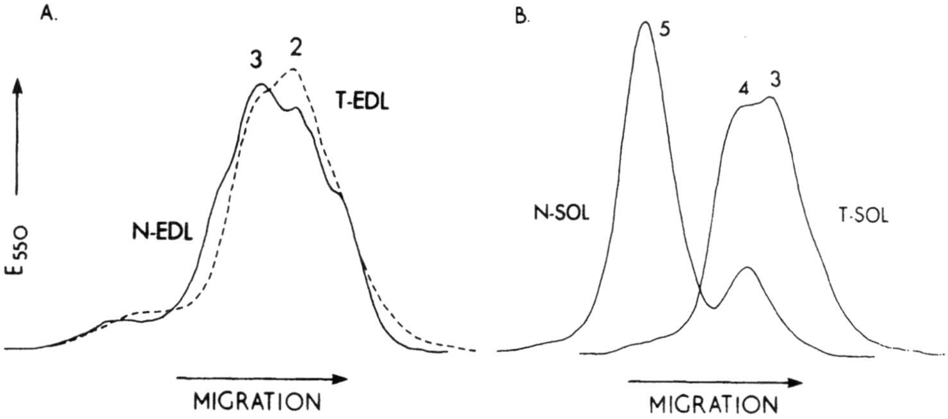

Fig.5. Isomyosin profiles of EDL and SOL muscles of cordotomized rat. Diagrams show (A) superimposed isomyosin profiles for EDL from normal (N-EDL) and cordotomized (T-EDL, broken line) rats, (B) superimposed isomyosin profiles for SOL from normal (N-SOL) and cordotomized (T-SOL) rats. The muscles were taken from a cordotomized rat 8 weeks post-operatively and from a normal rat of the same age.

<u>Isomyosins in muscles of cordotomized rats</u> Fig.5 shows the isomyosin profiles for EDL and SOL muscles of a cordotomized rat 8 weeks after operation. For comparison, the profiles for muscles from an unoperated

rat of the same age are also shown. While isomyosins of the EDL in the operated animal showed little deviation from normal, the profile for the SOL was drastically altered. There was virtually a complete suppression of isomyosin 5, the profile being dominated by isomyosins 4 and 3. Similar results were obtained from several rats examined 8 weeks post-operatively. Fig.6 shows results from another batch of cordotomized rats

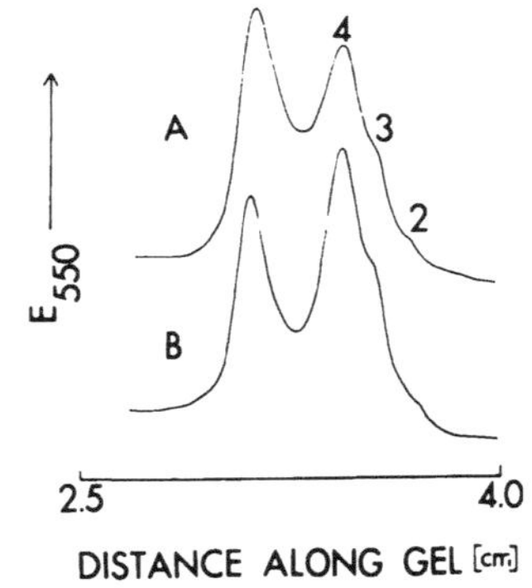

Fig.6. Isomyosin profiles of SOL muscles from two different cordotomized rats analysed 12 weeks post-operatively.

studied 12 weeks after operation. The shift in isomyosins in SOL of these animals was more moderate, the profiles resembling those from partially transformed cross-reinnervated SOL. Those of EDL (not shown) were not significantly different from normal.

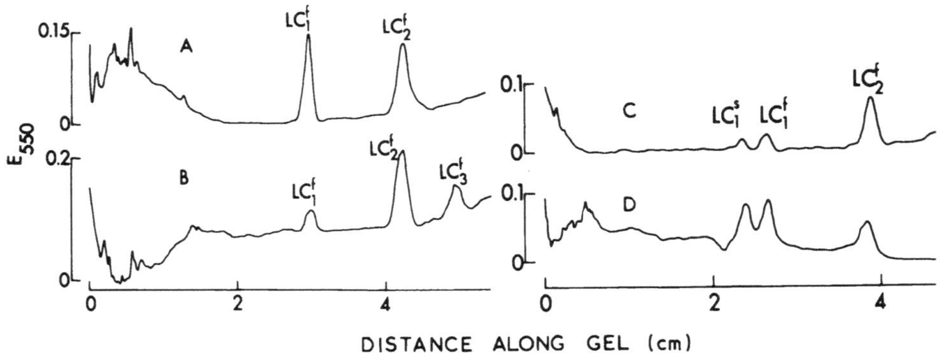

Fig.7. Light chain analysis of isomyosins in muscles of cordotomized rats. Diagrams show isomyosins 3 & 4 (A) and isomyosins 1 & 2 (B) from an EDL 12 weeks after cordotomy. (C) and (D) show results of 2 separate experiments on the same isomyosin 4 from the SOL of cordotomized rat illustrated in Fig.6B.

Light chain analysis of EDL isomyosins in cordotomized rats (Fig.7A & B) did not reveal any difference from those of normal EDL. However, isomyosin 4 of the SOL muscle of the cordotomized rat illustrated in Fig. 6B showed LC_1^s, LC_1^f and LC_2^f (Fig. 7C & D), i.e. the same combination of light chains as observed for isomyosin 4 in cross-reinnervated SOL. The stoichiometry of these light chains in the cordotomized rat was less clearly defined because isomyosin 4 formed a smaller proportion of the total myosin compared with the cross-reinnervated SOL analysed. Nevertheless, the available data suggest that LC_1^s and LC_1^f are present in equal amounts as observed for the cross-reinnervated SOL.

Discussion

An unusual feature of the rat EDL myosin compared with the fast-twitch myosin of the rabbit or chick is the presence of 4 isomyosins instead of 3. However, muscle bundles containing only isomyosins 1-3 can be dissected from the white portion of another fast-twitch muscle, the sternomastoid, while the red portion of this muscle contains isomyosins 1-4 (Dulhunty & Hoh, unpublished observations). Evidence cited below shows that isomyosin 4 is normally segregated in motor units of intermediate contraction speed. These observations suggest that only isomyosins 1-3 are normally present in rat fast-twitch fibres. The data from light chain analysis confirm that rat isomyosins 1-3 are respectively homologous to the three fast-twitch isomyosins FM_1 - FM_3 of the rabbit and chick. The fact that light chain composition of isomyosins 3 and 4 taken together is the same as for FM_3 suggests that the difference in electrophoretic mobility of isomyosins 3 and 4 is due to a difference in their heavy chains.

Light chain analysis shows that isomyosin 5 of the rat SOL is homologous to the slow-twitch isomyosin of the rabbit. Isomyosin 4, the minor component in rat SOL is absent from rabbit SOL. Its presence in the SOL of the rat is correlated with the existence of a population of motor units of intermediate contraction speed (19), with histochemically type II staining characteristics (20)and immunoreactivity against two specific antisera, those against fast-twitch myosin rod and slow-twitch myosin respectively (21). The proportion of isomyosin 4 in normal rat SOL matches the fraction of total tension contributed by intermediate motor units as well as the fraction of total cross-sectional area of fibres having these histochemical and immunocytochemical properties. These observations suggest that isomyosin 4 is segregated in these intermediate motor units. The absence of detectable amounts of fast-twitch isomyosins in the rat SOL shows that the intermediate nature of the contractile properties of these motor units is not due to a mixture of fast-twitch and slow-twitch isomyosins, but rather to the presence of a single species of myosin, viz. isomyosin 4.

The nature of the heavy chains of isomyosin 4 is germain to the problem of neural regulation of myosin. A possible model of isomyosin 4 is that its heavy chains are distinct from the heavy chains of both fast-twitch and slow-twitch myosins. This would mean that there is a new myosin heavy chain gene which is expressed in rat hindlimb muscles but is absent or not expressed in similar muscles of the rabbit. A simpler hypothesis not invoking a new gene is that the heavy chains of isomyosin 4 constitutes a hybrid between a fast-twitch myosin heavy chain and a slow-twitch myosin heavy chain. The presence of fast-twitch light chains unaccompanied by slow-twitch ones in isomyosin 4 of normal EDL is not inconsistent with this hypothesis since light chains of subfragment 1 preparations of fast-twitch and slow-twitch myosins can be experimentally interchanged (22). The hybrid nature of isomyosin 4 explains its intermediate electrophoretic mobility, the intermediate speed of contraction of fibres containing it and its immunochemical cross-reactivity. It could also account for the hybrid nature of the light chains in isomyosin 4 of SOL of cross-reinnervated and cordotomized rats (see below).

The analysis of isomyosins from individual cross-reinnervated muscles has revealed a wide range in the extent of transformation. This is correlated with the wide range in the transformation of physiological properties of cross-reinnervated muscles (2,18,23). This variability is not related to post-operative duration, nor to incomplete cross-reinnervation. Transient denervation and subsequent reinnervation encountered in nerve cross-union were also unlikely to be the cause of this variability since a similar variation in the extent of transformation was seen in the SOL after cordotomy. The variability probably reflects an intrinsic property of the neural influence that transforms the isomyosin profiles.

The most interesting result to emerge from the light chain analyses was the finding that isomyosin 4 of partially transformed SOL had LC_1^s, LC_1^f and LC_2^f in the ratio of 1:1:2. The absence of LC_2^s eliminates the possibility that LC_1^s arose from overlapping isomyosin 5. It is possible that 2 populations of isomyosin 4 molecules existed, with light chain compositions $(LC_1^f, LC_2^f)_2$ and $(LC_1^s, LC_2^f)_2$, but the electrophoretic method failed to separate them. However, the equal stoichiometric ratios for LC_1^s and LC_1^f in both cross-reinnervated SOL and cordotomized rat SOL suggest that a single molecular species exists, with light chain structure (LC_1^s, LC_2^f) (LC_1^f, LC_2^f), i.e. a light chain heterodimer. This heterodimer hypothesis would be more compatible with the postulated hybrid nature of the heavy chain core for isomyosin 4, assuming that each type of LC_1 would have a greater affinity for its homologous heavy chain.

The unusual combination of light chainsin isomyosin 4 described above implies that the two types of light chains are synthesized in the same cell. Thus, the nerves, while able to induce the synthesis of LC_1^f and LC_2^f, had failed to suppress the synthesis of LC_1^s in spite of the long post-operative period. The prominence of isomyosin 4 in partially transformed SOL muscles suggests that the incomplete transformation did not arise from a mechanism whereby some fibres remained unaltered while others were completely transformed. It points rather to some mechanism whereby most fibres are partially transformed. In other words, the neural influence on myosin, as studied by nerve cross-union and cordotomy, is not an all-or-none phenomenon, but may be graded.

The significance of the apparent presence of LC_1^f and LC_2^f in isomyosin 5 of partially transformed cross-reinnervated EDL is more difficult to assess. It could mean that there exists a population of isomyosins with slow-twitch heavy chains, but fast-twitch light chains, thus indicating incomplete suppression of LC_1^f and LC_2^f by the nerve to the SOL. Unfortunately it has not been possible to resolve isomyosin 5 into two components to support this hypothesis. Alternatively, fast-twitch light chains may have arisen from overlapping isomyosin 4. Further work on another species in which isomyosins are better resolved than those of the rat may help to clarify this problem.

Cordotomy in the rat brought about isomyosin shifts in the SOL which were very similar to those seen in the cross-reinnervated SOL. The fact that the SOL nerve after cordotomy and the EDL nerve after cross-union exerted essentially the same influence on myosin suggests that the neural influence is not due to the specific differentiated state of the fast-twitch and slow-twitch motoneurones, but rather to some signal which both types of motoneurones are equally capable of carrying. Much evidence is now available to show that this signal is the pattern of impulse activity (9, 24-28). Such an impulse activity hypothesis would satisfactorily explain the differential effects of cordotomy on the isomyosin profiles of SOL and EDL muscles by postulating that the operation removed or reduced the difference in impulse pattern normally received by these muscles. The apparent graded nature of the neural influence may be due to impulse frequencies or patterns of activity which are intermediate between those characteristic of normal fast-twitch and slow-twitch motoneurones. Much work remains to be done to clarify the nature of the patterns of impulse activity which are necessary (a) to maintain the normal complement of isomyosins in fast-twitch, slow-twitch and intermediate motor units and (b) to shift the isomyosin distribution in various directions.

Acknowledgement

This work was supported by the National Health and Medical Research Council of Australia.

References

1. Buller, A.J., Eccles, J.C., Eccles, R.M.: J. Physiol.(Lond.) 150, 417-439 (1960).

2. Close, R.: J. Physiol.(Lond.) 204, 331-346 (1969).

3. Buller, A.J., Mommaerts, W.F.H.M., Seraydarian, K.: J. Physiol.(Lond.) 205, 581-597 (1969).

4. Samaha, F.J., Guth, L., Albers, R.W.: Exp. Neurol. 27, 276-282 (1970).

5. Barany, M., Close, R.I.: J. Physiol.(Lond.) 213, 455-474 (1971).

6. Sreter, F.A., Gergely, J., Luff, A.R.: Biochem.Biophys.Res.Commun., 58, 84-89 (1974).

7. Weeds, A.G., Trentham, D.R., Kean, C.J.C., Buller, A.J.: Nature, 247, 135-139 (1974).

8. Sreter, F.A., Luff, A.R., Gergely, J.: J.Gen. Physiol. 66, 811-821 (1975).

9. Salmons, S., Sreter, F.A.: Nature, 263, 30-34 (1976).

10. Hoh, J.F.Y.: Biochemistry, 14, 742-747 (1975).

11. Lowey, S., Risby, D.: Nature, 234, 81-85 (1971).

12. Hoh, J.F.Y.: FEBS Lett. 90, 297-300 (1978).

13. Hoh, J.F.Y., Yeoh, G.P.S.: Nature, 280, 321-323 (1979).

14. Hoh, J.F.Y., McGrath, P.A., White, R.I.: Biochem.J. 157, 87-95 (1976)

15. Buller, A.J., Eccles, J.C., Eccles, R.M.: J. Physiol.(Lond.) 150, 399-416 (1960).

16. Hoh, J.F.Y., Dunlop, C.: III International Congress on Muscle Diseases, pp. 50-51, Congress Series No. 334, Excerpta Medica, Amsterdam 1974.

17. Close, R.: J. Physiol. (Lond.) 173, 74-95 (1964).

18. Hoh, J.F.Y., Salafsky, B.: J. Physiol. (Lond.) 216, 171-179 (1971).

19. Close, R.: J. Physiol. (Lond.) 193, 45-55 (1967).

20. Kugelberg, E.: J. Neurol. Sci. 27, 269-289 (1976).

21. Gauthier, G.F., Lowey, S.: J. Cell Biol. 81, 10-25 (1979).

22. Wagner, P.D., Weeds, A.G.: J. Mol. Biol. 109, 455-473 (1977).

23. Hoh, J.F.Y.: Exp. Neurol. 45, 241-256 (1974).

24. Sreter, F.A., Gergely, J., Salmons, S., Romanul, F.: Nature New Biol. 241, 17-19 (1973).

25. Al-Amood, W.S., Buller, A.J., Pope, R.: Nature, 244, 225-227 (1973).

26. Lomo, T., Westgaard, R.H., Dahl, H.A.: Proc. Roy. Soc. B. 187, 99-103 (1974).

27. Pette, D., Müller, W. Leisner, E., Vrbova, G.: Pflügers Arch. 364, 103-112 (1976).

28. Hoh, J.F.Y. & Fitzsimons, R.: Proc. Aust. Physiol. Pharmacol. Soc. 9, 155p (1978).

Section IV. Influence of Usage

MUSCLE AND MOTOR UNIT PROPERTIES OF EXERCISED AND NON-EXERCISED CHRONIC SPINAL CATS

V.R. Edgerton, L.A. Smith, E. Eldred, T.C. Cope* and L.M. Mendell
Dept. Kinesiology, Anatomy and Brain Res. Inst., UCLA, Los Angeles,
*Dept. Physiology, Duke University, Durham, USA

Introduction

Many experiments have been performed using a variety of models to identify the mechanisms involved in the neural regulation of the kind of protein and perhaps the amount of a given protein that is synthesized in a muscle fiber. Buller et al (1) cross united nerves to fast and slow muscles and showed that the control mechanisms for regulating the contractile properties were dictated in some manner by their motoneurons. It is now clear that the type of contractile protein can also be reversed by crossing these peripheral nerves (2).

There is growing support for the concept that the firing frequency of motoneurons is involved in this control of the muscle's physiological and biochemical properties (3, 4, 5, 6). Data from several models other than chronic electrical stimulation suggest that factors other than frequency play a critical role in this regulation of muscle protein.

This paper deals primarily with spinalized animals, some of which have been exercised following the lesion. This model was used because of the reports that the contraction time (CT) of a slow muscle like the soleus (SOL) shortens with the spinal transection (7) and that the histochemical staining profile converts toward one that resembles fast muscle. In addition this model is useful because of the possibility to control the neuromuscular activity level using a treadmill to facilitate spinal locomotion (8).

Methods

Observations were made on 19 cats, 8 of which were spinalized at the T_{12} level at two weeks of age and the others at 12 weeks. Following cord section four of the 2-week lesioned kittens were assigned to an exercise program, while the other four were treated similarly but without specific exercise. Of the 12-week lesioned cats five were subjected to exercise and six were kept without treadmill exercise. Terminal observations were made 12-14 weeks after spinalization. The cats from each litter were assigned equally to exercised and non-exercised groups.

The spinal lesion was induced under ketamine (100mg/kg) anesthetic. After the laminectomy at T_{12} and opening of the dura, xylocaine was placed topically on the spinal cord. The section was then made by lifting the cord with a curved hook and the cord, including the ventral artery, then was cut transversely. Complete transection was assured since the two ends of the cord suspended by the instrument fell into place. Gel-foam was then placed between the ends of the cord.

The animals were housed in 1x2x2 meter cages, two to five being housed together to provide social interactions. The animals were observed to frequently engage in playful wrestling with one another. In fact, this arrangement provided so much cutaneous stimulation that some of our "unexercised" kittens may have differed little in activity from the exercised ones. This was especially true for the 2-week lesioned cats. Absorbant corrugated paper lined the bottom of the cage along with scattered shredded newspaper, which seemed to provide considerable cutaneous stimulation which was manifested in alternating rhythmic movements of the hindlimbs.

Exercise consisted of walking on a motor driven treadmill for 5-30min. three times a week. The period of time for each exercise session was increased gradually from the beginning to the end of the 12-14 week period. Exercise was initiated as soon as possible following recovery from the spinal surgery, and lasted about three weeks for the 2-week old kittens and six weeks for the 12-week old cats. Treadmill speed was about 0.23m/s but for the last few minutes of each exercise session of the treadmill was increased to 0.46m/s. Knee and ankle joints in some cats were found to undergo ankylosis in the hyperextended position, while in others the joints though hyperextended maintained the full capability of passive motion. Non-exercise cats showed greater susceptability to such joint problems, and these usually developed during the 3-6 weeks following surgery.

Contractile properties of whole medial gastrocnemius (MG) and soleus (SOL) muscles were studied in terminal observations under choralose-urethane anesthesia. The muscle tendon was tied with nylon cord to a strain gauge whose output was displayed both on a polygraph and on an oscilloscope, photographed by a kymographic camera. Force was measured from the polygraph trace. For the whole muscle isometric responses were obtained by stimulation of the muscle nerve at about 2x the maximal strength and a 0.2ms pulse duration. Tension was recorded at stimulation frequencies of 1, 5, 10, 15, 20, 25, 30, 50, 100, and 200 Hz. Contraction times of twitches were measured from the point of the rise from the baseline to the time peak tension occurred.

For the purpose of observing responses of single motor units, complete hindlimb denervation was performed except for the muscle to be studied, and ventral root filaments teased until a single unit responded to stimulation. The muscle was kept at the length at which a maximal whole muscle twitch tension had been obtained, and was stimulated at selected frequencies ranging from 1 to 200 Hz. Potentiation by previous tetanic stimulations was avoided. Following the determination of the frequency-tension responses at the maximum rate of 200 Hz the motor unit was stimulated for two minutes at 40 Hz with a train of impulses each lasting 330ms and delivered once per second as described by Burke et al (9). The fatigue index was defined as the ratio of active tension produced at the end of two minutes to the mean of the peak tensions produced during the first five contractions of the fatigue series. Muscle temperature was maintained at $37^{\circ} \pm 1^{\circ}C$.

Some animals were shipped to Duke University where intracellular recordings from motoneurons were taken in addition to examination of whole muscle contractile properties as just described. Individual excitatory postsynaptic potentials (EPSP) produced by the stretch evoked activity of single MG Ia afferent fibers in homonymous MG motoneurons were studied using the spike triggered averaging technique (10, 11, 12). The conduction velocity of each motoneuron axon was determined from antidromic conduction delays from the site of stimulation in the popliteal fossa to the ventral horn. These neuronal studies were performed on the left side of the animal, and muscle properties were observed in the right hindlimb.

The histochemical procedures were standard (13) for staining myofibrillar adenosine triphosphatase with an acid and basic preincubation, reduced nicotinamide adenine dinucleotide diaphorase, succinic dehydrogenase, and α-glycerophosphate dehydrogenase activity. Muscle fibers were classified as described by Peter et al (14).

Results

Behavioral Observations. Six animals which were free of joint complications could stand unassisted in a quadrupedal position. Four of these demonstrated spontaneous walking movements consisting of 5-8 continuous steps. Such walking always terminated by the cat losing the balance of the hindquarters. On the treadmill this difficulty was avoided by having the torso supported by a vest attached to supports on both sides of the treadmill. At higher speeds of the treadmill trotting and galloping patterns in the hindlimbs occurred. Differences between regularly exercised and non-exercised preparations were much greater in animals transected at 12 weeks than those transected at 2 weeks. For example, 3 of the 5 12-week lesioned exercised cats could support their weight during treadmill locomotion, but none of the nonexercised, 12-week cats could support their weight.

Muscles. The musculature of the hindlimbs was maintained in visibly good condition for the period of three months following complete cord transection. The mean values given in Table I were less impressive because of the lack of success in cats with joint complications. These data suggest that the treadmill exercise was more critical for the 12-week than the 2-week lesioned cats. In all cats the SOL muscle was more difficult to maintain after transection than the predominantly fast muscles. Figure 1 shows that there was a significant correlation between the rated motor

TABLE I. Muscle weight/body weight ratios (Mean ± SD x 10^{-3})

Group	n	SOL	Plantaris	Gast	TA
2 wk Ex	4	0.45 ±0.24	0.99 ±0.14	3.28 ±0.44	1.71 ±0.24
2 wk NE	4	0.42 ±0.15	1.09 ±0.13	3.52 ±0.63	1.72 ±0.32
12 wk Ex	5	0.64 ±0.19	1.61 ±0.35	5.39 ±0.42	2.11 ±0.18
12 wk NE	3	0.41 ±0.16	1.07 ±0.48	3.78 ±1.74	1.45 ±0.75
Normal Adult	12	1.09*		5.98*	1.65**

SOL= Soleus; Gast= Gastrocnemius; TA= Tibialis Anterior

Ex= exercised for 12 weeks after the spinal lesion; NE= not exercised following the lesion. N= number of cats. From reference *15 and **16.

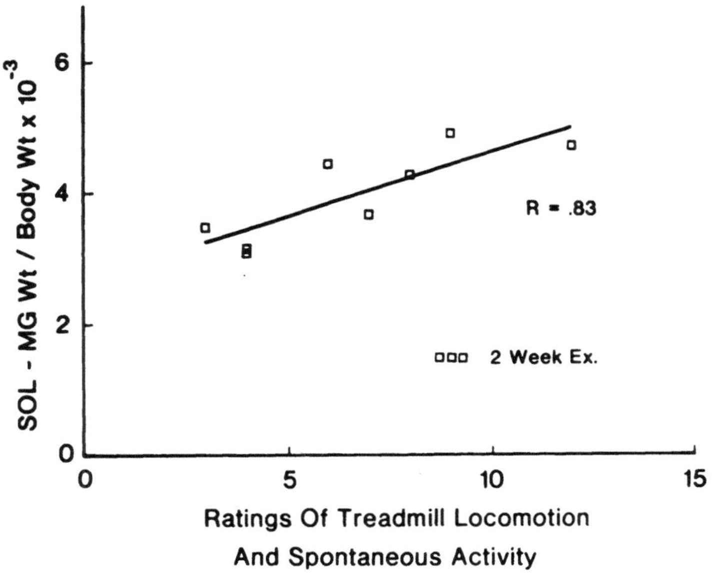

Fig. 1

Correlations of the wet weight of the MG and SOL muscles combined with subjective evaluation of cat's ability to walk on the treadmill and execute spontaneous movement.

capacity and the SOL + gastrocnemius muscle mass relative to body weight for the 2-week lesioned cats. This relationship was not apparent in the 12-week lesioned cats. Details of the kinematics and electromyography (EMG) during locomotion and the rating system used here have been reported (15).

Examples of the EMG activity of the SOL and LG muscles during treadmill locomotion are shown in Figure 2. As is true in normal cats (15), the SOL in the spinal cat was more active than the LG muscle during stepping. The only obviously different feature of the EMG was the presence of repetitive bursts of activity at 10-14 c/sec during the support phase of the step cycle.

The contractile measures for the whole MG and SOL muscles are given in Table II. These values show that the CT and 1/2 RT of the SOL shorten after transection whereas MG CT and 1/2 RT are normal. In accordance with this, the variation of these values was much greater in SOL than in the MG (TABLE II). The magnitude of the difference between the SOL in normal and transected preparations was greater in the 2-week than the 12-week lesioned cats. In the one case in which the CT and 1/2 RT of the MG differed from normal, these values (80 and 140 ms respectively) were similar to those of a normal SOL (79 and 91 ms). This was the least active cat. The P_t/P_o ratios were within the normal range for a slow and fast muscle (17) except for a slight increase in the SOL of the 12-week nonexercised and 2-week exercised cats.

Maximal tetanic tension of the MG was near normal in one exercised cat. The exercise program did not appear to be as essential in the 2-week as in the 12-week lesioned cats for retarding atrophy (TABLE II). Neither the group mean of twitch or tetanic tension of the MG expressed as absolute or relative to muscle weight or body weight, suggested a positive exercise effect.

The CT and 1/2 RT of single motor units corresponded closely to observations made on whole muscle. The mean CT of 41 motor units collected from MG muscles of two 2-week nonexercised cats was 33ms (fig. 3), while the mean whole muscle CT for this group was 30ms. Similarly, the mean CT

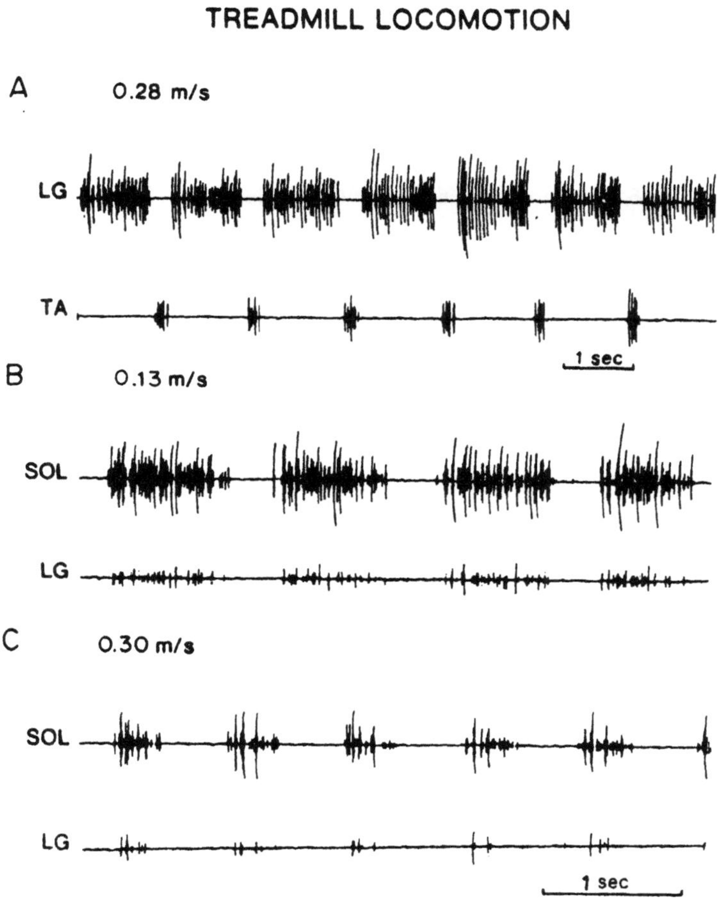

Fig. 2

EMG of selected muscles during treadmill locomotion of spinalized cats (T_{12})
LG= lateral gastrocnemius; TA= tibialis anterior; SOL= soleus. ( ) cat identification.

TABLE II. Whole Muscle Contractile Properties (Mean ± SD)

Group	CT (ms)		1/2 RT (ms)		P_t (kg)		P_o (kg)	
	SOL	MG	SOL	MG	SOL	MG	SOL	MG
2 wk Ex (4)	30 ±6	34 ±5	27 ±11	29 ±10	0.137 ±0.083	0.418 ±0.298	0.489 ±0.322	0.771 ±0.330
2 wk NE (4)	51 ±27	30 ±10	*62 ±21	20 ±6	*0.123 ±0.083	0.677 ±0.317	**0.556 ±0.215	2.842 ±1.849
12 wk Ex (5)	45 ±12	33 ±6	41 ±12	27 ±12	0.224 ±0.061	0.567 ±0.304	0.748 ±0.218	3.206 ±2.653
12 wk NE (3)	55 ±14	52 ±20	56 ±24	66 ±28	0.082 ±0.090	0.372 ±0.260	0.292 ±0.331	3.783 ±2.676
Normal Adult	86	30	104	23	0.750	2.680	3.310	12.500

( ) = number of cats

*N=3, **N=2

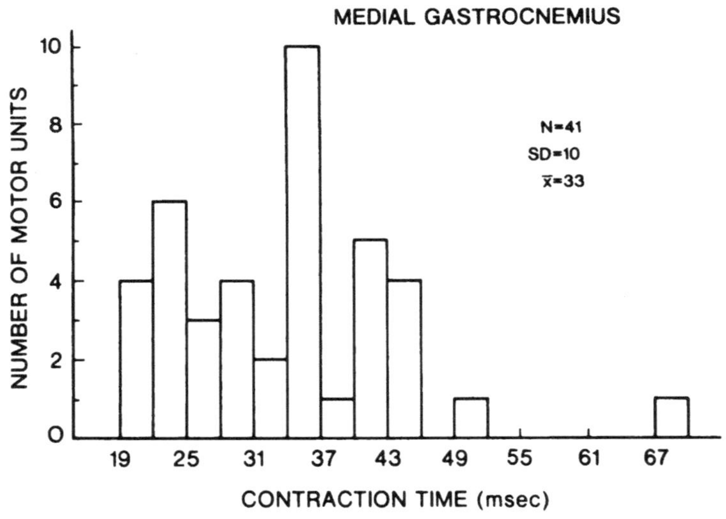

Fig. 3

Single motor unit contraction times from the MG of two nonexercised cats lesioned at 2 weeks.

of 60 SOL motor units from four 12-week exercised cats was 48ms (fig. 4). Contraction times for the MG motor unit ranged from 19 to 58ms, and in the SOL from 25 to 80ms. There did not appear to be any exercise effect on motor unit CT. Mean motor unit twitch and tetanic tensions for each group are shown in Table III. The relationship of the mean tension relative to P_0 at 20 Hz and the CT of single motor units and whole muscles is shown in figure 5. The frequency at which 1/2 P_0 was reached in motor units was elevated in the SOL muscle of spinalized cats approaching the value obtained for a normal fast muscle.

Conduction velocities of MG motoneurons, Ia-EPSPs, and the number of motoneurons to which a given afferent projects (projection frequency) were studied (Table IV). The conduction velocities in the exercised cats were significantly greater ($p < 0.01$) than those in nonexercised kittens. On the other hand the mean EPSP amplitude values for each kitten were within the normally wide range of values reported for intact or chronically T_{13}-transected adult cats (10, 11, 19). The projection frequencies (ie. percentage of homonymous motor pool in which EPSPs were observed) for both exercised and nonexercised groups are similar to those reported for chronically transected cats (19).

The percentage of fibers that stained darkly with alkaline myofibrillar ATPase and α-GPD in each of the experimental groups are shown in Table V. No marked difference in the percentages of high ATPase fibers was observed among the groups but all cats had a greater than normal percentage of high alkaline ATPase staining fibers. No histochemical changes were observed in the MG muscle. In the fibers of the SOL that stained darkly with ATPase (alkaline medium) α-GPD also stained more darkly than surrounding fibers with a histochemical profile which characterizes a normal SOL muscle. Occasionally a slight decrease in the staining for NADH-D was also evident in the high ATPase staining fibers but they were still classified as FOG as opposed to FG fibers. The fatigue properties of motor units in the SOL supported this observation in that motor units even of the least active cats showed a high percentage of motor units that were resistant to fatigue (Smith, L.A., Edgerton, V.R., and Eldred, E., unpublished observation).

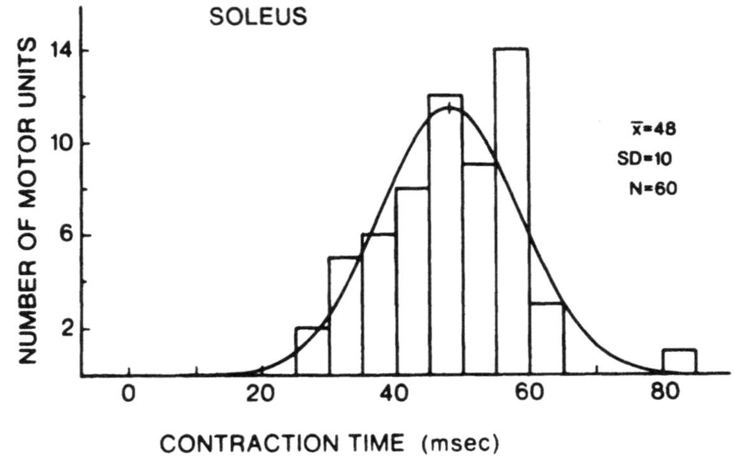

Fig. 4

Contraction times of single motor units of the SOL from four 12-week lesioned cats that were trained for 12 weeks post lesion.

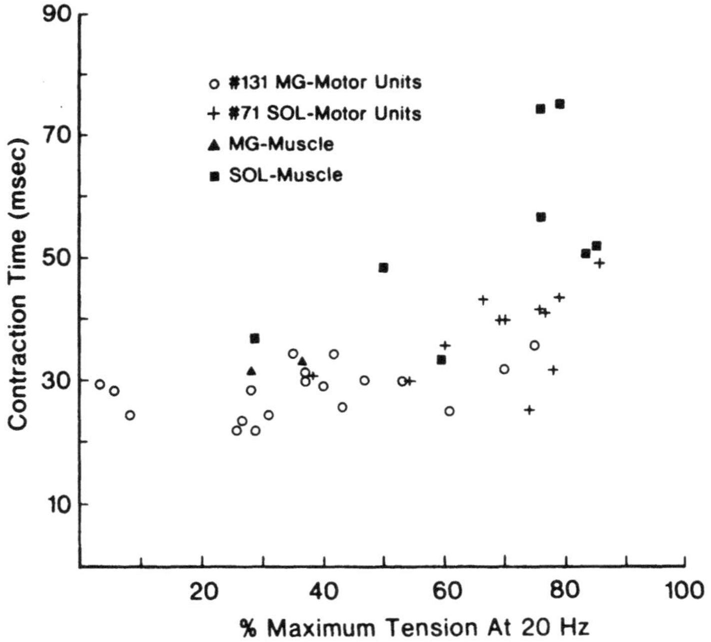

Fig. 5

Relationship of the peak tension produced when stimulated at 20 Hz, and the contraction time of motor units and whole muscle in spinalized cats.

TABLE III. Motor Unit Mean Twitch and Tetanic Tension

	N (n)	P_t (g)	P_o (g)
2 wk MG	3 (67)	3.0	16.4
12 wk MG	2 (17)	4.3	15.8
12 wk SOL	5 (66)	1.5	5.4

N = Number of cats (n) = Number of motor units
Exercised and nonexercised cats are combined.

TABLE IV. Motoneuronal and Ia Afferent Properties

Group	Conduction Velocity (m/s)	EPSP Amplitude (μV)	Projection Frequency
Not Exercised	67±2* (3)	126±46 (3)	100% (3)
Not Exercised	57±1 (11)	61±13 (10)	91% (11)
Exercised	75±1 (18)	157±27 (18)	100% (18)
Exercised	75±0 (16)	106±16 (14)	87% (16)

EPSPs produced by MG Ia afferents in homonymous motoneurons. Animals were spinalized at age 2 weeks and measurements taken 13-14 weeks later. Each line represents data from one cat.

*Std. error of mean

( ) number of motoneurons

TABLE V. Alkaline Myofibrillar ATPase and α-Glycerophosphate Dehydrogenase Histochemical Staining of Soleus Muscle

	% Dark Staining	
	Alkaline ATPase	α-GPD
2 wk Ex	27 (110)	60 (49)
2 wk NE	36 (367)	58 (278)
12 wk Ex	23 (172)	23 (81)
12 wk NE	33 (220)	-
Normal Adult*	0	0

( ) Number of fibers

*Reference 14 and 20

Discussion

These findings suggest that only a minimal amount of activity is required to maintain the muscle mass of the hindlimbs. It was also apparent that the CT and half RT of an initially slow twitch muscle became shorter. The major effects of exercise seemed to be maintenance of muscle mass and to retard the development of complications of knee and ankle joints.

Quantity of activity. It is certain that there was considerable EMG activity in the hindlegs of all animals, even those not exercised. It was also clear from the EMG in the spinal cats that rhythmic alternating activation of flexors and extensors of the hindlimb was common, and that the EMG activity patterns were normal, except perhaps for the presence of varying degrees of tremor. Thus spinally transected (T_{12}) cats cannot be assumed to be a model for neuromuscular inactivity.

The changes detected in CT and half RT of the SOL did not seem to be related to the quantity of neuromuscular activity. The SOL muscle in the cat with the least active hindlimbs and the most atrophy demonstrated a normal CT and half RT. In addition the MG of this same cat was the only one which had a long CT and half RT as a normal SOL. Also the mean CT and half RT for cats exercised regularly on the treadmill was shorter than for nonexercised animals. These observations are inconsistent with the prevailing concept that low level muscle use results in faster muscle and vice versa. It must be conceded that these decreases in SOL CT and half RT provide no absolute proof of an increase in the velocity of shortening of this muscle; however, since P_t/P_o was unchanged these reductions in CT could not simply be accounted for by reduced twitch tension (17) which could reduce CT without affecting intrinsic shortening.

Impulse frequency. Some observations have suggested that the pattern of impulses to a muscle may dictate the fiber type and its consequent properties. Most of these have shown that chronic stimulation of a fast muscle at 10 Hz causes a lengthening of CT and half RT of an initially fast muscle (3-5). The converse experiments, i.e. stimulation of a slow muscle with a high frequency impulse, were not as conclusive (5). Sreter and Salmons (6) found that indirectly stimulating the SOL muscle at 10 Hz after cross-reinnervation by a "fast" nerve did not allow the muscle to develop a short CT and RT. If the muscle was not stimulated, it developed a shortened CT and RT. Previous studies have shown the utility of chronic spinal transected cats as a model to study the stimuli that regulate muscle protein metabolism and muscle fiber type conversion. There are numerous experiments that suggest the frequency of impulses is a critical variable (3-6). It is known that the duration of after-hyperpolarization of a motoneuron is closely related to and may be the factor that determines the firing frequency characteristics of a motoneuron. But Gallego et al (10) found that the decrease in duration of after-hyperpolarization that occurs after spinal transection (which is commensurate with a higher frequency of impulses) was not correlated with changes that occurred in the CT of the SOL muscle. These data taken collectively suggest that changes in frequency of excitation could not explain the reduced CT that occurs in the SOL following spinal transection.

There are other reasons why the concept that the firing frequencies of a motoneuron may not be as important a factor in protein regulation as is commonly believed. The assumption that slow muscle fibers are exposed primarily to a slow frequency (10 Hz) and fast fibers to a higher frequency (40 Hz) may not be appropriate in normal cats. Kernell (21) has demonstrated, for example, that although the minimum discharge rate of a motoneuron and the CT of that motor unit correlates well, extensive overlap in frequencies exists between units with a CT of <45 and >45 ms. Estimating from the slope reported, a motor unit with a CT of 65 ms would have a minimum frequency of about 10 Hz, while one with a CT of 20ms would discharge at slightly less than 20 Hz. The maximum frequencies differed more, being 25 and 70 Hz. Therefore, muscle fibers of different types are exposed to differences in mean frequencies, but the overlap in discharge frequencies of slow and fast motor units is extensive making it unlikely that impulse frequency determines motor unit contraction time.

The tremor that occurs during stepping in some spinalized cats could alter the frequency of excitation and therefore be the cause of conversion of motor units from slow to fast twitch. However, this is characteristic of the LG as well, which showed no sign of changing its physiological or histochemical characteristics.

Tension. The tension produced within a muscle either actively or passively over a period of weeks may be a critical stimulus for the regulation of the rate of protein metabolism (23). It is not clear whether the type of protein, i.e. slow versus fast myosin, can also be altered by the chronic tension imposed on a muscle. There were several observations in these studies that could be related to this point. First, in each of the SOL muscles with a long CT, there were joint complications which resulted in a chronically plantar flexed position which presumably minimized the tension that could be imposed on the MG and particularly the SOL. In fact, the shorter CT in the SOL of the exercised cats probably reflected the preventive effect of exercise on dysarthrosis and therefore the plantar flexors were relatively normally loaded. If only the ankle remained hyperextended, no effect on the MG was seen because of its biarticular attachments. But in one cat the knee was fixed in flexion and the ankle in plantar flexion. The MG had a CT of 80ms. Also this cat demonstrated the least amount of movement. However, in this MG, atrophy was severe, particularly for the high alkaline ATPase. Therefore the long CT could have been due to a reduction in the functional cross section of fast twitch fibers rather than a conversion of fast to slow muscle fiber. Although these observations suggest that the passive and/or active load imposed on muscles over a period of weeks may be a factor in the regulation of protein metabolism, firing frequency cannot be discounted since the state of muscle stretch could have been manifested in the motoneuron firing patterns via the intact afferents.

The concept that a reduction in the tension imposed on the SOL by the fixed joint position can result in a shorter CT is also consistent with immobilization studies (13). For example, the shortened CT of the SOL muscle of immobilized muscle may be due to a reduction in the tension on the muscle. Assuming frequency is the controlling factor one would predict that the frequency of activation is increased in immobilized muscle but this has not been reported. The results of the present study on spinalized cats is also consistent with the hypothesis that the chronic tension imposed on a muscle may be an important regulator of protein metabolism.

Motoneuron and Synapse Properties. Although our data indicate that MG contractile properties in transected cats were unaffected by the exercise regime, motor axon conduction velocity (CV) in these exercised preparations was increased compared to unexercised controls suggesting that the treadmill training was not without effect. This increase is reminiscent of the findings of Walsh et al (25) who obtained similar results in muscles overloaded by tenotomy of a close synergist. This exercise induced CV increase appears to prevent the effects of transection which has previously been shown to attenuate the normal progressive increase of CV in kittens (22). Taken together these data indicate that motoneuron activity is required to ensure normal development of motoneuron properties, at least conduction velocity.

EPSPs in exercised and nonexercised preparations were within the normal range of variability seen in intact adults (10) or in preparations chronically transected at T_{13} as adults (19). Gallego et al (26) have demonstrated that disuse induced by a TTX cuff causes an increase in Ia-motoneuron EPSP amplitude. The present results suggest that an increase in use sufficient to increase motor axon conduction velocity (see above) does not affect EPSP amplitude. However, we cannot say which inputs to motoneurons achieved these changes; it is quite possible that under these exercised conditions total amount of the Ia fiber activity was not substantially altered. Thus it is not surprising that Ia-EPSPs might not change under these conditions.

References

1. Buller, A.J., Eccles, J.C., Eccles, R.M.: J. Physiol., Lond. 150, 417-439 (1960).

2. Sreter, F.A., Gergely, J., Luff, A.L.: Biochem. Biophy. Res. Commun. 56, 84-89 (1974).

3. Salmons, S., Vrbova, G.: J. Physiol. 210, 535-549 (1969).

4. Pette, D., Smith, M.E., Staudte, H.W., Vrbova, G.: Pflügers Arch. 338,257-272 (1973).

5. Lømo,T., Westgaard, R.H., Dahl, H.A.: Proc, R. Soc. B. 187, 99-103 (1974).

6. Salmons, S., Sreter, F.A.: Nature 263, 30-34 (1976).

7. Buller, A.J., Lewis, D.M.: J. Physiol. 176, 355-370 (1965).

8. Grillner, S.: Physiol. Rev. 55, 247-304 (1975).

9. Burke, R.E., Levine, D.N., Tsairis, P., Zajac, R.E.: J. Physiol. Lond. 234, 723-748 (1973).

10. Mendell, L.M., Henneman, E.: J. Neurophysiol. 34, 171-187 (1971).

11. Scott, J.G., Mendell, L.M.: J. Neurophysiol. 39, 679-692 (1976).

12. Nelson, S.G., Mendell, L.M.: J. Neurophysiol. 41, 778-787 (1978).

13. Edgerton, V.R., Barnard, R.J., Peter, J.B., Maier, A., Simpson, D.R.: Exp. Neurol. 46, 114-131 (1975).

14. Peter, J.B., Barnard, R.J., Edgerton, V.R., Gillespie, C.A., Stemple, K.E. Biochemistry. 11, 2627-2633 (1972).

15. Gardiner, P., Botterman, B., Eldred, E., Simpson, D., Edgerton, V.R.: Exp. Neurol. 62, 241-255 (1978).

16. Mann, W.S., Salafsky. B.J. Physiol. 208, 33-47 (1970).

17. Close, R. Physiol. Rev. 52, 129-197 (1972).

18. Smith, J.L., Smith, L.A., Dahms, L.L.: Soc. Neurosci. Abstracts. 5, 387 (1979).

19. Nelson, S.G., Mendell, L.M.: J. Neurophysiol. 42, 642-654 (1979).

20. Ariano, M.A., Armstrong, R.B., Edgerton, V.R.: J. Histochem. Cytochem. 21, 51-55 (1973).

21. Kernell, D.: Brain Res. 160, 159-162 (1979).

22. Gallego, R., Huizar, P., Kudo, N., Kuno, M.: J. Physiol. 281, 253-265 (1978).

23. Goldspink, D.F.: J. Physiol. Lond. 264, 267-282 (1977).

24. Spector, S., Gardiner, P.F., Roy, R.R., Edgerton, V.R.: Med. Sci. Sports Abstracts. 11, 114 (1979).

25. Walsh, J.V., Jr., Burke, R.E., Rymer, W.Z., Tsairis, P.: J. Neurophysiol. 41, 496-508 (1978).

26. Gallego, R., Kuno, M., Nunez, R., Snider, W.D.: J. Physiol.: 291, 191-205 (1979).

EFFECTS OF DISUSE BY LIMB IMMOBILIZATION ON DIFFERENT MUSCLE FIBER TYPES

F.W. Booth and M.J. Seider

Departments of Physiology and of Anesthesiology
University of Texas Medical School
Houston, Texas 77030 USA

G.R. Hugman

Baylor School of Medicine
Houston, Texas 77030 USA

Atrophy of the various types of skeletal muscles occurs in disuse, such as during limb immobilization (1,2). The purpose of this paper is to review our understanding of disuse atrophy. As will be demonstrated, many of the changes occurring in atrophying skeletal muscles can be explained, in part, by the duration of the half-lives of muscle proteins. For the sake of clarity, data presented in this review reflect experiments with animals' muscles whose limb lengths were fixed shorter than resting length during the limb immobilization.

The duration of time between immobilization of hind limbs and the first significant decline in the wet weight of a skeletal muscle of adult rats is shorter for fast twitch than slow twitch types of muscles. Since the percentage of the muscle that is water does not change during limb immobilization in rats (3,4), then wet weight is a valid index of protein content in these conditions. Significant declines in muscle wet weight begin between the 16th hour and the third day of limb immobilization in fast twitch muscles in various experiments (5), but they occur only at about the third day of limb immobilization in slow-twitch muscles (unpublished

Plasticity of Muscle

observation). At present, we have no consistent explanation for these variations. Another point to consider in the interpretation of these observations is that the first detectable decline in the wet weight of skeletal muscles in immobilized limbs of rats is not an accurate index of the time that muscle wasting actually begins because approximately ten percent of a muscle's weight must be lost before a significant difference from control values can be detected for an atrophying muscle (3). The amount of weight lost by a muscle is determined by both the rates of protein synthesis and degradation (10). Since five to ten percent of a muscle's protein turns over in a day in the adult rat (6), then it would take about one day for a weight loss of ten percent if protein synthesis stopped completely and the protein degradation rate was unchanged in the skeletal muscles within immobilized limbs. Thus, the initiation of muscular wasting must begin at least one day prior to the first detectable decline of a muscle's wet weight which occurs approximately one to three days post limb immobilization. These arguments are probably overly conservative. In reality, protein synthesis is not completely shut off in muscles of immobilized limbs (5). Consistent with less conservative arguments, the rate of protein degradation in these muscles does not in fact change for at least the first day of immobilization (24).

All proteins in animal cells undergo a continual process of degradation and replacement by the synthesis of new proteins (7). This process has been called turnover (8). When the level of proteins in muscles is unchanging or in steady state, this constant level is determined by both the rates of protein synthesis and of protein degradation (9). Thus, the steady state size of a muscle could be decreased during disuse atrophy by altering the rate of either protein synthesis or protein degradation, a point to be emphasized later. However, the time course of the decline in muscle size during disuse atrophy is determined solely by the first-order rate constant for protein degradation, which has been expressed as kd (10). Another term frequently used to describe protein degradation rate is half life, expressed as $t_{\frac{1}{2}}$ (8). These terms are related in the equation: $t_{\frac{1}{2}} = \ln 2/_{kd}$, where ln 2 is the natural logarithm of 2. Usage of the above formula has been applied to our experiments in the following manner. The time course of the decline in the wet weight of fast and slow-twitch muscles from larger, nonatrophied, steady-state size to their smaller, atrophied, steady-state size can be obtained by measuring muscle size at frequent time intervals during limb immobilization. This loss in the wet weight of skeletal muscles shows an exponential decay when plotted on a linear graph. Alternatively, a straight line results from plotting the following relationship on a semilogarithmic graph: the difference in muscular size between the apparent, final steady-state size after atrophy and muscular size at any time during atrophy (on logarithmic scale) versus the duration of the atrophy (on linear scale). The straight line thus obtained on the semi-logarithmic plot indicates that the loss of muscular mass can be approximated by a first-order rate constant. The $t_{\frac{1}{2}}$ of the decay in muscular size can therefore be procured from the semilogarithmic relationship either graphically or arithmetically. Using this relationship, we found that the time necessary for one-half of the final decline in the wet weight of either a fast-twitch or a slow-twitch muscle was five to six days in the immobilized limbs of adult rats (3). In control adult rats without hind-limb immobilization, the mean half life of proteins has been shown by use of the constant infusion procedure to be 13 days in the gastrocnemius muscle and 7 days in the soleus muscle (6). Thus, there is a greater increase in protein degradation rates in fast-twitch muscles than in slow-twitch muscles within immobilized limbs of adult rats when the control data by others (6) is compared to our atrophy data.

In addition to the exponential decay of muscle wet weight in immobilized limbs, we have observed that the loss of dry weight, total protein content, total cytochrome c content and total citrate synthase activity is exponential, or a first order process, in a fast-twitch muscle within an immobilized limb (3). We did not make these measurements in slow-twitch muscles. At this point in the discussion, it should be recalled that we previously indicated that the time course of the decay from the larger steady-state size after atrophy is determined solely by the first-order rate constant for degradation, or by the $t_{\frac{1}{2}}$ of the proteins. Since muscles are composed of many proteins, with each protein having its own distinctive half life, it is interesting that the algebraic sum of different half lives results in a single first-order approximation of the decay of muscle size during disuse atrophy caused by limb immobilization. However, it should be mentioned that not all proteins are lost at the same rate in skeletal muscles during disuse atrophy. We have shown that the time required for one-half of the final decrease in cytochrome c content to occur is shorter than that for one-half of the final decrease in total protein content in a fast-twitch muscle in an immobilized limb of the rat (3). Thus, cytochrome c concentration in fast-twitch muscle falls during the first four days of limb immobilization. Consistent with this finding of the rapid decline in a mitochondrial protein is our observation that significant decreases in succinic dehydrogenase activity per milligram of mitochondrial protein and decreases in state III respiration occur on the second day of hind-limb immobilization in fast-twitch skeletal muscle of the rat (11). In contrast to this relatively rapid decay of cytochrome c, as compared to the decay of total protein in fast-twitch skeletal muscle during disuse atrophy, is the failure of myoglobin content to change in atrophying fast-twitch muscle (12). Indeed, myoglobin concentration increases in fast-twitch muscle during disuse atrophy (3,12). The opposite response occurs in limb-immobilized slow-twitch muscle (12).

After the wet weight of muscles reached their new smaller, apparent steady states during limb immobilization, we calculated that the percentage of initial weight loss was greater in slow-twitch than in fast-twitch muscles (3). This suggests that protein synthesis rate declined more in slow-twitch than fast-twitch muscles during limb immobilization since rates of protein degradation changed less in slow-twitch muscles. Other investigators have also reported that slow-twitch muscles atrophy more than fast-twitch muscles during limb immobilization (4,13-18). Consistent with these observations are those reports which have indicated that slow-twitch muscle fibers degenerate during disuse atrophy (19-21). The same reports indicate minimal or no degeneration of fast-twitch muscle fibers in atrophying muscles. Since these observations suggested that slow-twitch muscle degenerated during atrophy, we undertook a study to determine whether the atrophy of slow-twitch muscle is reversible after prolonged limb immobilization.

After 90 days of hind-limb immobilization, the wet weight and protein content of a slow-twitch muscle returned more rapidly to normal, control values than did a fast-twitch muscle (22). This event is probably related to the half life of proteins and to usage. Since the half life of proteins in slow-twitch muscles is shorter than for fast-twitch muscles, then the time required for one-half of the final increase in muscular mass during regrowth from atrophy will be shorter in the slow-twitch muscle. Another factor is that tonic or slow-twitch muscle is probably used more than fast-twitch muscle during recovery. Despite the return of both the wet weight and the total protein content of slow-twitch muscle to control values by the 14th day of recovery after 90 days of limb immobilization, the recovery of the maximal isometric tension, which can by produced by a slow-twitch muscle, did not return to pre-atrophy values until the 120th day of recovery (22). During this recovery period, rats were permitted to have unrestricted cage exercise. At present, we have no explanation for the differential rate of recovery of wet weight and of maximal isometric tension in slow-twitch muscles. We also observed that in muscles regrowing after limb immobilization, the slow-twitch muscle was edematous during the first few days of recovery (unpublished observations). Others have previously reported that heavy usage of skeletal muscles are associated with edema (23). We speculate that atrophied slow-twitch muscles are heavily recruited in their function as an antigravity muscle during the first days after limb immobilization (23). Fast-twitch muscles were not studied.

After limbs have been immobilized for weeks, the contraction time of slow-twitch muscle becomes shorter, i.e., the contraction time for slow-twitch muscle approaches the value of contraction time observed in fast-twitch muscles (20,25-27). However, no change in the contraction time of a fast-twitch muscle occurs as a result of limb immobilization (20). In an attempt to determine some explanation for the shortened contraction time of slow-twitch muscle during immobilization, we directly counted every muscle fiber in a complete cross-section of the soleus muscle. A decline in the percentage of slow-twitch muscle fibers (and thus an increase in the percentage of fast-twitch muscle fibers) was observed in the soleus muscle from immobilized limbs as compared to controls (20). This alteration in the percentage of slow-twitch fibers was a result of a decrease in the total number of slow-twitch fibers and an increase in the total number of fast-twitch fibers (20). Maier et al. (27) also reported an increase in the percentage of fast-twitch muscle fibers in the soleus muscle of guinea pigs after four weeks of limb immobilization. The etiology of the change in the percentage of the muscle fiber types in the soleus muscle during limb immobilization is controversial. We believe this event is due to the preferential degeneration of slow-twitch muscle fibers in the immobilized limbs (20). Others have suggested that this alteration might be due to a transformation of fast-twitch muscle fibers from slow-twitch fibers (27). Indeed, the speeding of the contraction time of the soleus muscle in an immobilized limb may be due both to preferential degeneration of slow-twitch muscle fibers and to a transformation between the fiber types.

Numerous studies have investigated the metabolic capacity of muscles after limb immobilization. After ten days of immobilization, a reduction occurs in the capacities of homogenates, which have been prepared from either fast- or slow-twitch muscles, to oxidize any of the following substrates: pyruvate, β-hydroxybutyrate, palmitate, or glucose (28). A number of the enzymes in these metabolism pathways have also been studied. Prior to presenting these results, it is important to indicate that changes in enzyme activity during the decrease of a muscle's size will be dependent upon the relative rates of turnover for proteins. For example, the turnover of cytochrome c is relatively more rapid than for most muscle proteins (3). Thus, cytochrome c concentration in a fast-twitch muscle is significantly lower than in non-atrophied muscle on the fourth day of immobilization because it is being degraded more rapidly than most muscle proteins during early atrophy (Fig. 1).

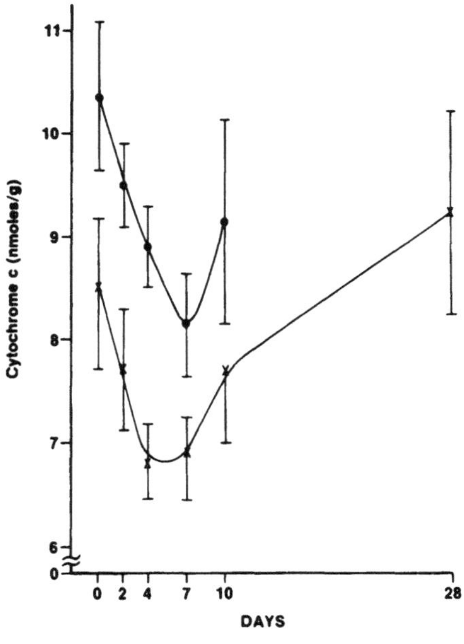

Fig. 1. Changes in the concentration of cytochrome c in the gastrocnemius (x——x) and the plantaris (●——●) muscles during hind-limb immobilization. (Published with permission from reference 3.)

Apparently, once cytochrome c reaches its new, atrophied steady state level, the majority of the muscle proteins, with their longer half-lives, continue to atrophy. Thus, after day four of limb immobilization, cytochrome c concentration begins to increase. By the 28th day of immobilization, cytochrome c concentration is now no different from control values. These principles demonstrated by cytochrome c also apply to enzyme activities. Thus, the activity of an enzyme during atrophy is dependent upon its relative rate of protein turnover. Therefore, great caution is needed in interpreting data on the activities of enzymes in muscles during disuse atrophy.

Edgerton et al. (17) observed no changes in the activities of phosphorylase, lactate dehydrogenase, α-glycerophosphate dehydrogenase, and succinate dehydrogenase in fast-twitch muscle after six months of limb immobilization in the *Galago senegalensis*. Their results suggest that cellular adjustments which occurred during the first few weeks of atrophy appear to return to control values after six months of disuse atrophy. However, after four weeks of limb immobilization, some alterations in enzyme activity are apparent. We have observed a decline in cytochrome oxidase activity in fast-twitch red, but not fast-twitch white or slow-twitch, muscles after four weeks of limb immobilization in the rat (29). Decreases in cytochrome oxidase activity have been reported to occur in fast-twitch muscles after only three days of limb immobilization (30).

An increase in tetanic tension per gram of muscle has been observed in fast-twitch, but not slow-twitch, muscles within immobilized limbs (27,31). Since a disproportionate loss of myofibrillar protein occurs in both fast- and slow-twitch muscles of immobilized limbs (27,32,33), concentration of contractile protein would not explain the increased tension per gram of muscle produced by limb immobilization. Further research for the cause of this increase in tetanic tension per gram of muscle in immobilized limbs is needed.

Earlier in this review, we mentioned that the size of a skeletal muscle is a result of both the rates of protein synthesis and of protein degradation. Goldspink has shown that the rates of protein synthesis declines in both fast- and slow-twitch muscles from rapidly growing, young rats when synthesis rates are measured during in vitro incubation (24). We have employed constant infusion techniques to measure in vivo rates of protein synthesis in adult, nongrowing rats. Using these techniques, we observed a 37% reduction in the fractional rate of protein synthesis in a fast-twitch muscle during the first six hours of limb immobilization (5). We have made preliminary observations during experimentation showing similar declines in the rate of protein synthesis in slow-twitch muscles in the first few hours of limb immobilization (unpublished observations).

Rates of protein degradation increase in slow-twitch muscles of rapidly growing rats after approximately one day of limb immobilization (24). However, no change in the rates of protein degradation has been noted in fast-twitch muscles of young rats (24). Previously in this paper, we presented indirect data that suggested an increase in the turnover of proteins in fast-twitch muscles from adult rats during limb immobilization. To our knowledge, no measurements of the rates of protein degradation exist in muscles from the immobilized limbs of adult rats. Whether the age of the rat affects the response of protein degradation rate in fast-twitch muscles to disuse atrophy awaits further investigation.

In summary, differences exist in the responses of fast- and slow-twitch muscles to disuse atrophy produced by limb immobilization. Many of the reported differences between these two types of skeletal muscle can be explained by their different rates of protein turnover. Also, our data show that the appearance of plasticity in skeletal muscle begins to occur very soon after changes in the level of contractile activity which are produced by limb immobilization.

This work was supported by grants from the National Institutes of Health AM 19393 and the National Aeronautics and Space Administration NAS 9-15388. Thanks are given to Ms. Carla Maywald and Ms. Homer Tolan for their assistance in preparing this manuscript.

References

1. Thomsen, P., Luco, J.V.: J. Neurophysiol. 7, 245-251 (1944).
2. Summers, T.B., Hines, H.M.: Arch. Phys. Med. Rehabil. 32, 142-145 (1951).
3. Booth, F.W.: J. Appl. Physiol. 43, 656-661 (1977).
4. Szöőr, Á., Boross, A., Hollósi, G., Szilágyi, T., Keszt-yűs, L.: Acta Biol. Acad. Sci. Hung. 28, 195-204 (1977).
5. Booth, F.W., Seider, M.J.: J. Appl. Physiol. In press (1980).
6. Waterlow, J.C., Garlick, P.J., Millward, D.J.: Protein Turnover in Mammalian Tissues and in the Whole Body, North Holland, Amsterdam 1978, p 534.
7. Goldberg, A.L., Dice, J.F.: Ann. Rev. Biochem. 43, 835-869 (1974).
8. Arias, I.M., Doyle, D. Schimke, R.T.: J. Biol. Chem. 244, 3303-3315 (1969).
9. Schimke, R.T., Doyle, D.: Ann. Rev. Biochem. 39, 929-976 (1970).
10. Schimke, R.T.: Meth. Enzymol. 40, 241-266 (1975).
11. Kreiger, D.A., Tate, C.A., Wood, J.M., Booth, F.W.: J. Appl. Physiol. In press (1980).
12. Takács, O., Sohár, I., Szilágyi, T., Guba, F.: Acta Biol. Sci. Hung. 28, 221-230 (1977).
13. Eccles, J.C.: J. Physiol. London 103, 253-266 (1944).
14. McMinn, R.M.H., Vrbova, G.: Q.J. Exp. Physiol. 49, 424-429 (1964).

15. Bergman, R.A., Afifi, A.K.: John Hopkins Med. J. 124, 119-131 (1969).

16. Mann, W.S., Salafsky, B.: J. Physiol. London 208, 33-47 (1970).

17. Edgerton, V.R., Barnard, R.J., Peter, J.B., Maier, A., Simpson, D.R.: Exp. Neurol. 46, 115-131 (1975).

18. Jansson, E., Sjodin, B., Tesch, P.: Acta Physiol. Scand. 104, 235-237 (1978).

19. McMinn, R.M.H., Vrbova, G.: Nature 195, 509 (1962).

20. Booth, F.W., Kelso, J.R.: Pflügers Arch. 342, 231-238 (1973).

21. Tomanek, R.J., Lund, D.D.: J. Anat. 118, 531-541 (1974).

22. Booth, F.W., Seider, M.J.: J. Appl. Physiol. 47, 435-439, 1979.

23. Bergstrom, J., Guarnieri, G., Holtman, E.: J. Appl. Physiol. 30, 122-125 (1971).

24. Goldspink, D.F.: J. Physiol. 264, 267-282 (1977).

25. Fischbach, G.D., Robbins, N.: J. Physiol. London 201, 305-320 (1969).

26. Mann, W.S., Salafsky, B.: J. Physiol. London 208, 33-47 (1970).

27. Maier, A., Crockett, J.L., Simpson, D.R., Saubert, C.W., Edgerton, V.R.: Amer. J. Physiol. 231, 1520-1526 (1976).

28. Rifenberick, D.H., Max, S.R.,: Amer. J. Physiol. 226, 295-297 (1974).

29. Booth, F.W., Kelso, J.R.: Can.J. Physiol. Pharmacol. 51, 679-681 (1973).

30. Rifenberick, D.H., Gamble, J.G., Max, S.R.: Amer. J. Physiol. 225, 1295-1299 (1973).

31. Fischer, E., Ramsey, V.W.: Amer. J. Physiol. 145, 571-582 (1946).

32. Helander, E.: Acta Physiol. Scand. 41, Suppl. 141, 1-99 (1957).

33. Takács, Ö., Sohár, I., Pelle, T., Guba, F., Szilágyi, T.: Acta Biol. Acad. Sci. Hung. 28, 213-219 (1977).

Section V. Effects of Chronic Stimulation

THE RESPONSE OF SKELETAL MUSCLE TO DIFFERENT PATTERNS OF USE - SOME NEW DEVELOPMENTS AND CONCEPTS

S. Salmons

Department of Anatomy, The Medical School, University of Birmingham, Birmingham, B15 2TJ, England.

Introduction

The plasticity of mammalian skeletal muscle is demonstrated to its most striking extent in its response to chronic low-frequency stimulation. Over a period of months a fast-twitch muscle subjected to this change in its normal pattern of use becomes indistinguishable from a slow-twitch muscle by a large number of physiological, histochemical and biochemical criteria. This transformation of fibre type takes place through an orderly sequence of changes rather than a single, synchronous process of change; as a result fibres at an intermediate stage of transformation exhibit a combination of properties not seen in any normal adult fibre type.

In terms of the proteins synthesised within the muscle cells, we can now point to three main groups of proteins which participate in the transformation process.

A. Proteins of the sarcoplasmic reticulum begin to change after only a few days of stimulation (1). These changes are paralleled by changes in the rate and extent of calcium uptake by preparations of fragmented sarcoplasmic reticulum (1,2) and are presumed responsible for the early reduction in contractile speed measured under isometric conditions (1,3,4).

B. After about 2 weeks of stimulation quantitative changes have taken place in enzymes of energy metabolism, representing an increase in the importance of oxidative pathways and a corresponding decrease in the use of the glycolytic pathway (4-6). These trends are also revealed by histochemical examination of transverse cryostat sections of the muscle (5-8). Together with an elaboration of capillary networks within the muscle (7-9) and a reduction in oxygen diffusion distances resulting from decreased fibre girth (4) these changes account for the remarkable resistance to fatigue shown by stimulated fast muscles (10-12).

C. Changes in both the heavy and light chains of myosin commence after 3-4 weeks of stimulation and reach completion between 12 and 20 weeks of stimulation (2,4,11,13-15). These changes are reflected in the specific activity and alkali stability of the calcium-activated myosin ATPase activity (2,11,13,14), and in the appearance of sections stained histochemically for the demonstration of myofibrillar ATPase (4,7) or by immunofluorescence techniques for the binding of antibodies specific to fast and slow muscle myosin (16). Changes in myosin ATPase activity would account for the continued reduction both in isometric contractile speed (3,4,11) and in the maximum speed of shortening measured under isotonic conditions (17).

These phenomena pose some interesting questions which will form the basis of the present discussion.

1. Can the influence of the motor nerve on the differentiation of fast and slow characteristics be explained entirely in terms of the pattern of impulse activity it imposes on the muscle?
2. What is the functional significance of muscle plasticity both during development and in the adult life of the animal?
3. What light can these phenomena shed on the relationship between structure and function in muscle?
4. What is the chain of events linking excitation of the muscle membrane to the synthesis of new muscle proteins, and what insight does it provide into the control of gene expression in muscle and, by extrapolation, in other tissues?

The neurotrophic influence

The existence of a neural influence on the differentiation of fast and slow muscle was first demonstrated in the cross-reinnervation experiments of Buller and his co-workers (18). While a chemotrophic explanation for this phenomenon was widely favoured, and indeed cannot yet be eliminated entirely, the bulk of experimental evidence now supports the alternative hypothesis (19) that the influence of the motor nerve resides in the nature of the impulse activity it delivers to the muscle. In framing this hypothesis it is sufficient to postulate a "slowing" influence of tonic activity, in the absence of which a muscle retains, or regains, fast characteristics (3,11). In this way it is possible to account both qualitatively and quantitatively for the changes consequent on cross-reinnervation (11, q.v. for a fuller discussion). Such an explanation nevertheless makes assumptions concerning the normal patterns reaching slow and fast muscles and their transposition after cross-reinnervation which would be worth testing in a rigorous way (20).

Functional significance of plasticity

Changes qualitatively similar to those brought about by chronic stimulation may be seen in the muscles of rats kept for 6 months under hypergravity conditions (21; Salmons & Oyama, in preparation). Thus the range of activity which can be generated by the nervous sytem is sufficient to elicit from a muscle the same type of changes as occur under the more artificial conditions of motor nerve stimulation. Exercise represents a less continuous stimulus to a muscle than hypergravity, and its effects, like those of intermittent stimulation (4), are more circumscribed. For example, while the response of muscle to endurance training involves extensive changes in energy metabolism (22), it has only exceptionally been reported to include changes in the type of myosin. Muscle therefore demonstrates its full capacity for adaptive change only when the deparature from normal use is both radical and sustained, but there is no good reason for supposing that the responses to stimulation (continuous or intermittent), hypergravity or endurance exercise are fundamentally different (see also 23).

Plasticity of muscle may therefore enable an animal to accommodate changes in the nature and level of its activity associated with seasonal changes of behaviour or longer term changes such as those associated with increase of body weight. Since the energy consumed and heat generated by muscle tissue are major factors in the energy budget of an animal, such a mechanism would well confer survival value.

A different perspective emerges from a consideration of the central nervous control of movement. There is now substantial evidence (24-26) that this operates through a hierarchy of firing thresholds of the motor neurones. Motor units recruited at low threshold are active more frequently, and for longer periods, than those recruited at high threshold. By responding adaptively to demand, the corresponding muscle fibres are able to acquire metabolic and contractile properties appropriate to their position in this hierarchy. In this way the neural and muscular elements of the locomotor system develop matching characteristics without the need for any predetermined pattern of connectivity between them. Viewed in this way the plasticity of muscle in the mature animal is little more than a carry-over into adult life of a mechanism whose primary value has been to reduce substantially the information needed to specify the growing organism.

The view was once prevalent that mammalian slow-twitch skeletal muscle was less differentiated than its fast-twitch counterpart. The adaptive response hypothesis calls for a reversal of this view, since genes coding for proteins specific to slow muscle would not be expressed until adult patterns of locomotor and postural activity had begun to emerge, up to which time only fast or foetal forms of muscle protein would be synthesised (27). Although direct evidence for this hypothesis is lacking, recent work on foetal and cultured muscle (28,29) confirms, in relation to the polymorphic forms of myosin, the predicted order of appearance of fast and slow muscle proteins.

Structural correlates of fibre type transformation

Recently we have been seeking evidence of changes at the ultrastructural level accompanying the transformation of fast muscle by chronic stimulation (27). Sections prepared from stimulated fast muscle showed an abundance of mitochondrial profiles and lipid droplets, structures seen only infrequently in control fast muscles. This accords well with the biochemical observation that the activities of enzymes of oxidative metabolism increase markedly as a result of stimulation (4-6). Measurement of the thickness of

Z-disc structure in stimulated muscle showed this to have dimensions characteristic of slow, and not fast, muscle. This was a significant finding, for Z-disc thickness is the ultrastructural parameter which discriminates most reliably between normal slow and fast muscle fibres (30). Evaluation of Z-disc thickness included measurement of the apparent overlap (which we termed Z-overlap) of filaments within the structure. A recent study, in which the Z-lattice was analysed by optical diffraction and reconstruction techniques indicates that the overlap is structurally significant and is based on a repeating unit of periodicity 38 nm (M.A. Goldstein, personal communication). In the light of this new evidence the figures we recorded for the Z-overlap acquire additional interest:

control fast muscle	39.6±1.8nm
stimulated fast muscle	76.6±4.1nm
control slow muscle	78.0±6.0nm

(all figures mean ± S.E.M.)

It is therefore possible that increases in the Z-overlap are generated in discrete steps rather than continuously, and we will be looking for evidence of this in our current studies (see below).

Comparison of changes in ultrastructural, physiological and biochemical parameters during fibre type transformation could provide further insight into the biochemical basis and functional significance of structural differences between fast and slow fibres. Such a comparison is now being undertaken in collaboration with Dr. B.R. Eisenberg (Rush Medical College, Chicago), using stereological techniques in order to analyse rigorously the ultrastructural changes in fast muscles stimulated for periods from 6 hours to 24 weeks. This study has now provided a time course for changes in mitochondrial content, Z-disc thickness and length of the luminal axis of the T-system. The mitochondrial volume fraction increased rapidly after about 12 days of stimulation and at 5 weeks had reached levels very much higher than those of control fast or slow muscles. The Z-disc width (measured in the conventional way in order to maintain randomness of sampling) increased to values characteristic of slow muscle in the period from 1½ to 2½ weeks following the onset of stimulation. The amount of T-system declined to the levels normally found in slow muscle in the period from 2 days to 2 weeks of stimulation. Associated elements of the sarcoplasmic reticulum were similarly reduced and it is therefore interesting to compare these observations with figures for the rate of calcium uptake by preparations of fragmented sarcoplasmic reticulum. Changes in calcium transport characteristics during the first few days (1) have a clear morphological counterpart in our data. However, the rate of calcium uptake does not decline to levels typical of slow muscle until some time after the morphological changes are complete (1,2). No doubt this extended time course reflects the progress of changes in the protein composition of the sarcoplasmic reticulum (1,31,32).

The control of gene expression in muscle

Since the response of fast muscle to continuous low-frequency stimulation involves qualitative as well as quantitative changes in protein synthesis it provides a useful experimental model for investigating regulatory mechanisms in muscle. Little is known of the sequence of events which starts with depolarisation of the muscle membrane and ends with the synthesis of new protein. The involvement of polyamines in the regulation of RNA and protein synthesis in a variety of other tissues prompted us to investigate changes in polyamine biosynthesis during stimulation of fast muscle for periods from 6 hours to 24 weeks. In these experiments stimulation was initiated via a subcutaneous switch at least 5 days post-operatively in order to dissociate the early effects of stimulation from those of surgery and anaesthesia. The study has been carried out by Miss C. Mastri, a postgraduate student with Dr. G.H. Thomas and myself. A preliminary account has been submitted for publication.

Polyamines are low molecular weight polycationic substances which are found in a variety of tissues, particularly those with a high biosynthetic activity. Their role *in vivo* is uncertain, since there is evidence to implicate them in almost every step from transcription to translation (33). Their biosynthesis involves two decarboxylating enzymes, ornithine decarboxylase (ODC) and S-adenosyl-L-methionine decarboxylase (SAM-D), both of which have exceptionally short half-lives and are readily induced by stimuli that enhance protein synthesis.

The concentrations of polyamines in muscle were very low compared with those found in most other tissues. In the rabbit the levels were somewhat higher in slow than in fast muscle, possibly reflecting a higher level of synthetic activity in the former. Chronic low-frequency stimulation of fast muscle eventually established polyamine levels typical of slow muscle, in keeping with the transformation of fibre type which has already been described in terms of other criteria. Of greater interest, however, were the striking phasic increases which occurred in the first week of stimulation. The polyamines putrescine, spermidine and spermine increased in succession, peak values being reached in 3 days for putrescine, 4 days for spermidine and over a period from 8 days to 3 weeks for spermine. The scale of the increases was approximately 50-fold for putrescine, 10-fold for spermidine and 2- to 3-fold for spermine. These changes were preceded by highly significant increases in the activity of the enzymes ODC and SAM-D, peak values being reached between 24 and 36 hours after the onset of stimulation.

The early occurrence, magnitude and transient character of the changes in polyamine biosynthesis which we have observed in stimulated muscle are consistent with an involvement of this pathway in the regulatory mechanism responsible for redifferentiation of the tissue. Further experiments are being undertaken in order to elucidate the possible relationship between these events.

Activity, gene-switching and muscle fibre types

Buller, Eccles and Eccles, in their seminal work on cross-innervation (18), found it difficult to accept an explanation for these phenomena based on a direct involvement of impulse activity. The problem they raised has not yet received an adequate answer and is worthy of reconsideration. How can one reconcile the inherent variability of motor neurone firing patterns with the sharp differentiation of muscle into fast and slow types?

More recent work has illustrated the need to modify one premise on which this reasoning was founded. The ability to discriminate between fibre types depends very much on the parameter being measured; most parameters show a normal range of variation and fibres typed according to one criterion may show a continuum of properties in another. For example Kugelberg (34) has determined microphotometrically the oxidative activity of individual fibres in a mixed muscle, and shown it to be distributed continuously, although it has been common for histochemists to classify such fibres into three, or even more, types. Again, Burke and his co-workers, by characterising motor units with a combination of physiological and histochemical techniques, have shown that considerable variation in contractile properties may be observed within a histochemically uniform population of fibres (35). Such considerations do not overcome the original problem; rather they restate it. Whatever mechanism is proposed for the generation of diversity within muscle fibres must account for continuous variation in some parameters and discontinuous variation in others. How can a continuously varying influence such as impulse activity produce two such dissimilar effects?

Figure 1 depicts the problem in two dimensional form, with "activity" as a single variable on the abcissa and an undefined measurement parameter on the ordinate.

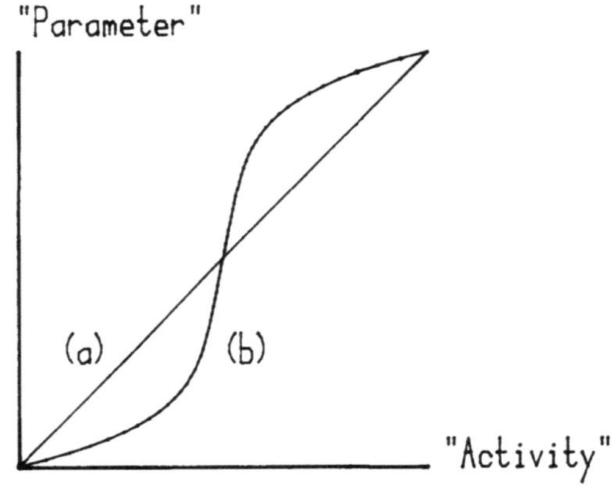

Fig. 1

Curve (a) represents a linear relationship in accordance with which continuous variations in activity will be reflected in continuous variations of the parameter. Curve (b) represents another continuous relationship with rather different properties; in this case the parameter is more likely to assume values in the upper or lower part of the range than in the intervening region, and movement between these extremes requires a substantial change in the level of activity.

Figure 2 takes the argument a step further.

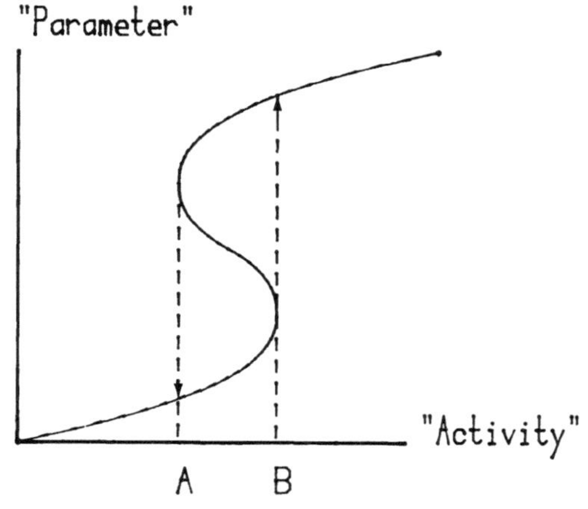

Fig. 2

The curve now represents a switching characteristic, with the parameter confined to one of two states. For the parameter to change from the upper to the lower state the activity must fall below a threshold level (A); likewise for a transition from the lower to the upper state activity must exceed another threshold level (B). The difference between the thresholds represents hysteresis, which imposes stability on the system by denying to the parameter values in the intermediate region.

The two-dimensional model is clearly a gross over-simplification. An adequate description of the properties of a muscle fibre would require a large number of parameters, each related differently to activity. The abcissa, too, should be elaborated in order to separate the influences of different aspects of the pattern of activity, such as the predominant frequency of impulses, and the way in which time is proportioned between rest and activity. In reality, therefore, the curve becomes a surface in multi-dimensional space, which may have a linear gradient in some directions and be folded in others. The properties of such systems have been described by Catastrophe Theory (36,37). One property worth emphasising is that a mere set of coordinates is not enough to define a point on the surface: it is also necessary to know the route by which the point was reached. In terms of muscle one is saying that the properties of the fibre depend not only on the current level of activity but also on the previous history of use.

The new feature which this model introduces, and one which is accessible to experimental test, is the concept of threshold. Do experiments involving changes in the level of activity provide any evidence of threshold behaviour? Certainly some observations can be interpreted in this way. For example fast muscles subjected to intermittent electrical stimulation, in which 8 hour periods of stimulation alternate with 16 hour periods of quiescence, undergo substantial changes in metabolism but show changes in myosin only after prolonged treatment (4). A threshold model for this situation is given in Figure 3.

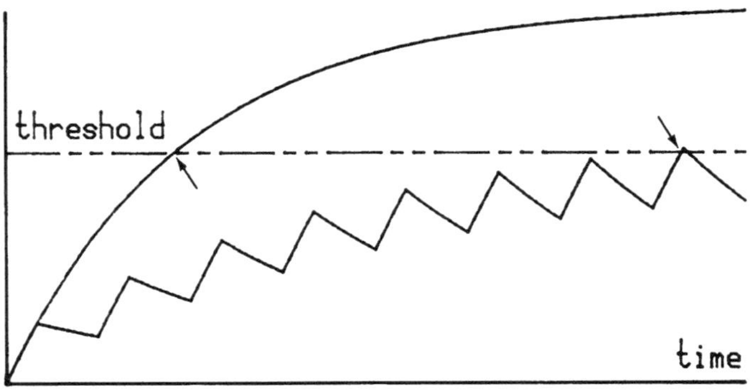

Fig. 3

The ordinate represents the intracellular level of a substance whose production is enhanced by an increase in contractile or metabolic activity. When this substance accumulates beyond a threshold value it triggers off a sequence of events the outcome of which is a change in the type of myosin synthesised by the cell. The upper curve denotes the response to continuous, and the lower curve to intermittent, stimulation. In the example given, the lower curve crosses the threshold after a delay. Another duty cycle could be chosen for which myosin synthesis would remain stable indefinitely under the intermittent regime.

It is not difficult to propose regulatory mechanisms which could produce such threshold behaviour. One of many possibilities is shown in Figure 4, where the messenger RNA transcribed from Gene A includes a sequence which suppresses the transcription of Gene B, and vice versa. Such a system is stable with either Gene A or Gene B active, but not both. A change of state is possible in circumstances which favour the expression of the inactive gene provided that these are maintained long enough for the threshold point to be crossed. Thereafter the system seeks its new stable state, a process which could take some weeks during which both gene products would be synthesised but in gradually changing proportions.

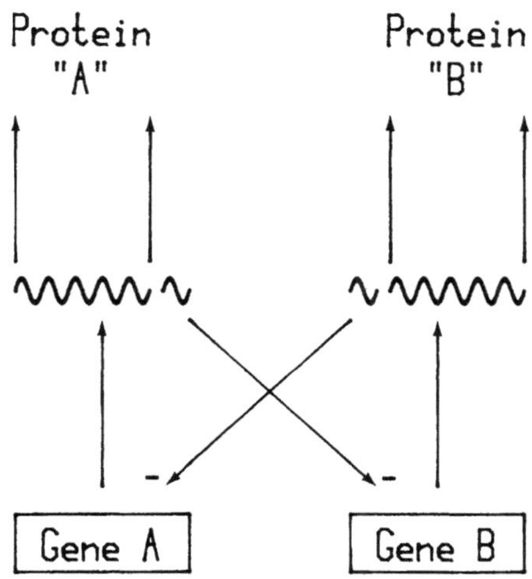

Fig. 4

Although such ideas are necessarily speculative in the current state of knowledge they nevertheless serve to show how discrete fibre types could be generated through a process of adaptive change to functional demand. Elucidation of the underlying mechanisms would represent a major step, not only in our understanding of muscle and its development, but in the broader context of cell biology.

Acknowledgement. The author gratefully acknowledges grant support from the Muscular Dystrophy Group of Great Britain and the Medical Research Council.

References

1. Heilmann, C., Pette, D.,: Eur. J. Biochem. 93, 437-446 (1979).

2. Sréter, F.A., Romanul, F.C.A., Salmons, S., Gergely, J.: Exploratory Concepts in Muscular Dystrophy II (Milhorat, A.T., ed.). Int. Congr. Ser. 333, 338-343, Excerpta Medica, Amsterdam (1974).

3. Salmons, S., Vrbová, G.: J. Physiol. 201, 535-549 (1969).

4. Pette, D., Müller, W., Leisner, E., Vrbová, G.: Pflügers Arch. 364, 103-112 (1976).

5. Pette, D., Smith, M.E., Staudte, H.W., Vrbová, G.: Pflügers Arch. 338, 257-272 (1973).

6. Pette, D., Ramirez, B.U., Müller, W., Simon, R., Exner, G.U., Hildebrand, R.: Pflügers Arch. 361, 1-7 (1975).

7. Romanul, F.C.A., Sréter, F.A., Salmons, S., Gergely, J.: Exploratory Concepts in Muscular Dystrophy II (Milhorat, A.T., ed.). Int. Congr. Ser. 333, 344-348, Excerpta Medica, Amsterdam (1974).

8. Brown, M.D., Cotter, M.A., Hudlická, O., Vrbová, G.: Pflügers Arch. 361, 241-250 (1976).

9. Hudlická, O., Myrhage, R.: J. Physiol. 258, 25-26P (1976).

10. Cooper, J., Hudlická, O.: J. Physiol. 263, 155-156P (1976).

11. Salmons, S., Sréter, F.A.: Nature 263, 30-34 (1976).

12. Hudlická, O., Brown, M., Cotter, M., Smith, M., Vrbová, G.: Pflügers Arch. 369, 141-149 (1977).

13. Sréter, F.A., Gergely, J., Salmons, S., Romanul, F.C.A.: Nature 241, 17-19 (1973).

14. Sréter, F.A., Elzinga, M., Salmons, S., Luff, A.R., Mabuchi, K.: FEBS Lett. 57, 107-111 (1975).

15. Pette, D., Schnez, U.: FEBS Lett. 83, 128-130 (1977).

16. Rubinstein, N., Mabuchi, K., Pepe, F., Salmons, S., Gergely, J., Sréter, F.A.: J. Cell Biol. 79, 252-261 (1978).

17. Al-Amoud, W.S., Buller, A.J., Pope, R.: Nature 244, 225-227 (1973).

18. Buller, A.J., Eccles, J.C., Eccles, R.M.: J. Physiol. 150, 417-439 (1960).

19. Vrbová, G.: J. Physiol. 169, 513-526 (1963).

20. Salmons, S.: A. Handbook of Biotelemetry and Radio Tracking (Amlaner, C.J., Jr., Macdonald, D.W., eds.) Pergamon Press, Oxford (in press).

21. Martin, W.D., Romond, E.M.: Exp. Neurol. 49, 758-771 (1975).

22. Holloszy, J.O., Booth, F.W.: Ann. Rev. Physiol. 38, 273-291 (1976).

23. Salmons, S., Henriksson, J.: Proc. IUPS Sympos. on Muscle, Szeged (1977) (in press).

24. Henneman, E., Somjen, G., Carpenter, D.: J. Neurophysiol. 28, 560-580 (1965).

25. Henneman, E., Somjen, G., Carpenter, D.: J. Neurophysiol. 28, 599-620 (1965).

26. Henneman, E., Clamann, H.P., Gillies, J.D., Skinner, R.D.: J. Neurophysiol. 37, 1338-1349 (1974).

27. Salmons, S., Gale, D.R., Sréter, F.A.: J. Anat. 127, 17-31 (1978).

28. Hoh, J.F.Y., Yeoh, G.P.S.: Nature 280, 321-323 (1979).

29. Rubinstein, N.A., Holtzer, H.: Nature 280, 323-325 (1979).

30. Eisenberg, B.R., Kuda, A.M.: J. Ultrastr. Res. 54, 76-88 (1976).

31. Ramirez, B.U., Pette, D.: FEBS Lett. 49, 188-190 (1974).

32. Pette, D., Heilmann, C.: Basic Res. Cardiol. 72, 247-253 (1977).

33. Cohen, S.S.: Introduction to the Polyamines, Prentice-Hall, New Jersey (1971).

34. Kugelberg, E.: Abstracts IVth Int. Congr. on Neuromuscular Diseases, Montreal (1978).

35. Burke, R.E., Tsairis, P.: Ann. N.Y. Acad. Sci. 228, 145-159 (1974).

36. Thom, R.: Structural Stability and Morphogenesis (Fowler, D., transl.) W.A. Benjamin, Inc. (1975).

37. Zeeman, E.C.: Sci. Am. 234, 65-83 (1976).

THE EFFECT OF LONG-TERM ELECTRICAL STIMULATION ON FUEL UPTAKE AND PERFORMANCE IN FAST SKELETAL MUSCLES

O. Hudlicka, K. R. Tyler and T. Aitman

Department of Physiology, University of Birmingham Medical School,
Birmingham B15 2TJ, England

Introduction

Long-term stimulation of fast skeletal muscles with a pattern of activity naturally occurring in nerves to slow muscles produces a decrease in the activity of glycolytic and increased activity of oxidative enzymes (1). Muscles stimulated in such way also have a prolonged contraction time (1, 2), higher capillary density (3, 4) and show increased resistance to fatigue (5, 6); they are in many respects similar to slow muscles. When fast muscles were stimulated with a pattern of frequency resembling that occurring in nerves to fast muscles (short bursts of tetani) using a much lower total number of stimuli (1440/hour) for 28 days, no changes were found in contraction times, capillary density or activity of succinate dehydrogenase (4); neither did we observe any changes with a similar pattern but a much higher number of stimuli (36,000/hour) after only four days of stimulation (4). After seven days, activity of succinate dehydrogenase was increased and muscle performance improved, but capillary density was not changed (7). These results are similar to findings of increased aerobic metabolism in endurance training (8, 9). Capillary density is increased after a relatively long period of training (10) but the increase in blood flow is smaller with similar performance than in control muscles (11). Fuel uptake shows a shift from a predominantly anaerobic to a predominantly aerobic metabolism with a greater utilisation of fatty acids (12); similar changes in fuel uptake were found in muscles stimulated at slow nerve frequency (6).

Both types of electrical stimulation activate all motor units at once,

Plasticity of Muscle

unlike most types of exercise with preferential recruitment of small motor units (13, 14). The aim of this study was to find out whether activation of all motor units and the total amount of activity, or the frequency pattern, are decisive in producing changes in capillary density, fuel uptake and muscle performance.

Methods

Experiments were performed on New Zealand Red rabbits of both sexes 2.5 to 3.5 kg body weight. Implantation of coiled multistranded stainless steel electrodes for chronic stimulation was performed under halothane anaesthesia and aseptic conditions as described previously (1, 4). The group of fast muscles innervated by the lateral popliteal (peroneal) nerve (tibialis anterior, TA, extensor digitorum longus, EDL and peroneal muscles) were stimulated for 8 hours a day for 4, 14 and 28 days using 0.3 msec stimulus width, with the voltage adjusted to give maximal contractions on palpation (usually between 2.5-5V) and using either continuous stimulation at 10 Hz or bursts of tetani at 40 Hz (each 5 sec duration 3 times per minute).

Final experiments were performed under pentobarbitone anaesthesia (35 mg/kg). Seven animals were used as controls and 4-5 animals were used in each stimulated group. Blood flow from the stimulated and control muscles was measured using a photoelectric drop-counter (15) inserted into the femoral vein after careful isolation of venous outflow (6). Oxygen, glucose and free fatty acids (FFA) uptake and lactate output were calculated from blood flow and A-V differences using sampling and methods for estimation described previously (6). Muscles were mounted for recording of isometric tension and stimulated via the cut motor nerve at 4 Hz for 10 min using 0.3 msec pulse width and supramaximal voltage. This was usually very close to the voltage used in each conscious animal on the day preceding the final experiment (indicating that all motor units were activated in conscious animals). Blood flow was recorded at rest and throughout the period of contraction; glucose and FFA consumption and lactate output were measured at rest and 2 and 10 min after the beginning of contractions. Oxygen consumption was estimated at rest and after 10 min of contractions;

this enabled us to measure oxygen content in blood samples within 10 min of withdrawal. Contraction times were recorded on a CRO.

At the end of each experiment muscles were quickly excised, weighed, and several blocks from each muscle were deepfrozen in isopentane precooled in liquid nitrogen for histochemical determination of succinate dehydrogenase and capillary density.

Results and Discussion

Previous results have shown that four days stimulation of fast muscles at a slow nerve frequency produced an increase in the capillary density with no changes in the activity of oxidative enzymes (1, 4) and no change in blood flow and fuel uptake (6). Stimulation at a fast nerve frequency did not affect the capillary density after seven days (7). We therefore compared the effect of the two different frequencies after a longer period of stimulation, 14 and 28 days.

Capillary density measured both as capillary/fibre ratio and as capillaries /mm^2 increased by 50% with both types of stimulation after 14 days and by 80% after 28 days (Table 1). There was a comparable increase in SDH activity. These changes were reflected in blood flow and oxygen consumption. Fig. 1 compares blood flow at rest and in the course of isometric contractions in control muscles and muscles stimulated for 14 and 28 days either at 10 Hz continuously or at 40 Hz intermittently.

Blood flow at rest was similar in all groups. The increase during isometric contractions was significantly higher in all chronically stimulated muscles, but no difference was found between the two patterns of stimulation. Oxygen consumption (Fig. 2) at rest was again similar in all groups with the exception of muscles stimulated at 10 Hz for 28 days where it was significantly higher. During muscle contractions, oxygen consumption in control muscles increased more than fivefold (from 0.54 ± 0.064 ml/100g. min at rest); it was twice as high in chronically than in control muscles. The increase was somewhat, but not significantly, greater in muscles

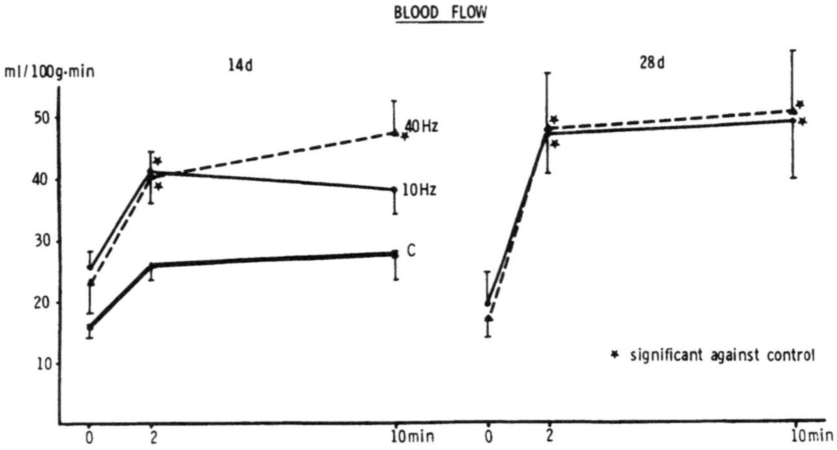

Fig. 1. Blood flow in rabbit fast skeletal muscles at rest (0 time on the abscissa) and during a 10 min period of isometric contractions at 4 Hz. C - blood flow in control muscles; solid line 10 Hz - muscles stimulated continuously at 10 Hz for 14 or 28 days. Interrupted line - muscles stimulated intermittently (5 sec three times/min) at 40 Hz. * denotes values significantly different from controls at $p < 0.05$.

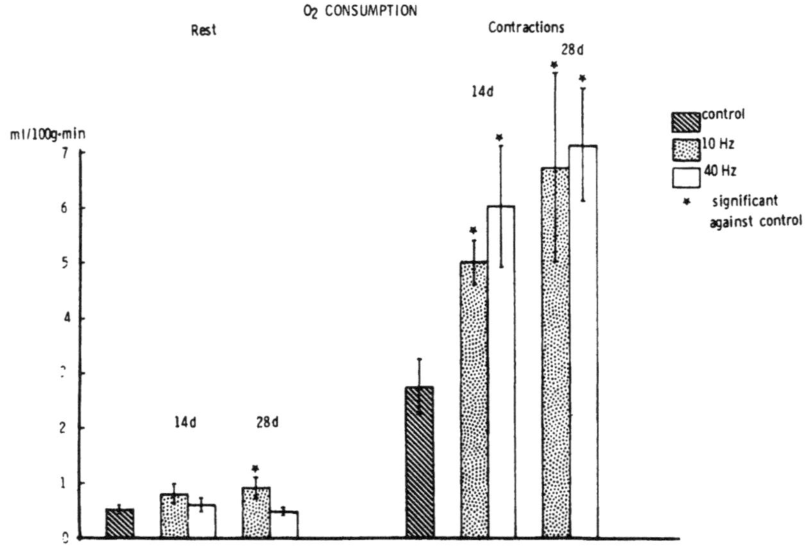

Fig. 2. Oxygen consumption in control and chronically stimulated muscles at rest and 10 min after the beginning of isometric contractions at 4 Hz. Significance at $p < 0.05$.

stimulated for 28 than in those stimulated for 14 days; again, there was no difference between muscles stimulated at different frequencies. The resting consumption of glucose was 7.0 ± 2.9 μmol/100g.min in control muscles and similar in all other groups. It increased during contractions to 28.1 ± 7.3 μmol/100g.min in control muscles and to the same extent in all stimulated muscles. Muscles stimulated for 28 days at 10 Hz had a greater uptake of FFA during contractions than control muscles (Fig. 3). No data are available for the group stimulated at 40 Hz; however, the activity of fatty acid activating enzyme, acylCoA synthetase in this group, was increased by 139 ± 24.7% as compared to the increase of 59 ± 12% in muscles stimulated at 10 Hz. This finding, together with the

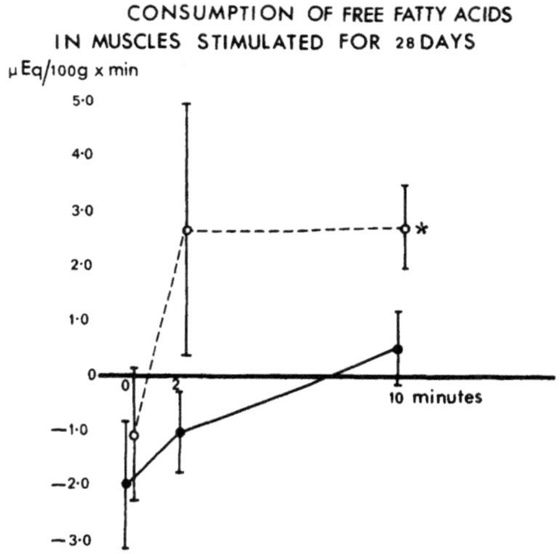

Fig. 3. Consumption of free fatty acids in muscles stimulated at 10 Hz for 28 days at rest (0 time) and during isometric contractions. Solid line - values in control muscles. Interrupted line - values in chronically stimulated muscles. Significance at $p < 0.05$.

increased oxygen consumption and SDH activity indicates a shift from predominantly anaerobic to predominantly aerobic metabolism in fast muscles under the influence of long-term electrical stimulation irrespective of frequency. The effect of frequency was however obvious in lactate output (Fig. 4) which was similar during muscle contractions in control muscles and muscles stimulated at 40 Hz but significantly lower in muscles stimulated at 10 Hz. This difference could be due to a smaller glycogen breakdown in the latter group where the activity of glycolytic enzymes (1) and muscle lactate concentration during contractions (16) were lower than in control muscles.

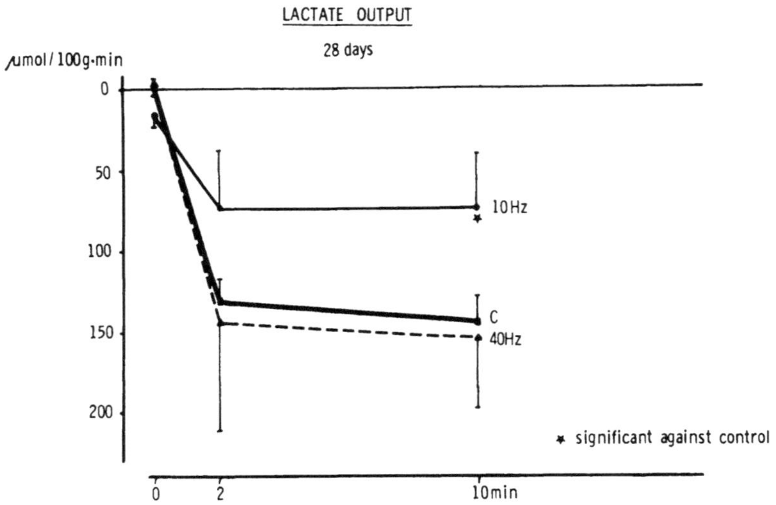

Fig. 4. Lactate output during isometric contractions at 4 Hz in control muscles (C) and muscles stimulated at 10 Hz (continuous line) and 40 Hz (interrupted line) for 28 days. 0 time resting values. Significance for $p < 0.05$.

Stimulation at different patterns of frequency thus changes the predominantly anaerobic metabolism in fast muscles towards predominantly aerobic metabolism found in slow or trained muscles if applied with a sufficient intensity. Electrically stimulated muscles, like slow or trained muscles, fatigue considerably less than control muscles (Table 1) and behave similarly with both patterns of stimulation. The peak tension developed during isometric contractions was similar in all groups, with the exception of muscles stimulated at 40 Hz for 28 days where it was somewhat higher.

The difference between endurance training and electrical stimulation is obvious with respect to muscle hypertrophy and contractile properties. Endurance training is known to produce fibre hypertrophy in muscles involved most heavily in any particular type of training (17). In contrast no fibre hypertrophy was found with electrical stimulation (4) and muscle weights did not change with either type of stimulation. Training does not affect contraction time (18), which was prolonged after only 4 days of stimulation at 10 Hz (1, 2, 4) and doubled after 28 days (1). Stimulation at 40 Hz did not affect contraction time within 4 days; it increased by 29.7 ± 9.9% after 14 days and by 39 ± 13.6% after 28 days.

Table 1

	Control	14 days 10 Hz	14 days 40 Hz	28 days 10 Hz	28 days 40 Hz
Capillaries/fibre EDL	1.20 ± 0.03	1.74 ± 0.03	1.80 ± 0.04	2.18 ± 0.03	2.07 ± 0.06
Peak twitch tension g/g	120 ± 14	141 ± 22	122 ± 9	117 ± 18	184 ± 42
% of peak twitch tension remaining after 10 min of 4 Hz stimulation	40.8 ± 9.1	63 ± 6.3	74.0 ± 8.05	77.3 ± 11.5	88.5 ± 5.6

There is a certain limitation to the use of the different frequencies of stimulation: while the 10 Hz is close enough to the frequency naturally occurring in nerves to slow muscles, 40 Hz produces only unfused tetanus in rabbit muscles and consequently does not mimic muscular contraction occurring with activation of fast motor units. However, it is the highest frequency tolerated by conscious animals in a chronic experiment. It is possible that some prolongation of the contraction time in animals stimulated at 40 Hz might be due to this factor.

The pattern of stimulation resembling that occurring in fast muscles produced similar changes in aerobic metabolism and fuel uptake as the pattern occurring in nerves to slow muscles, but only if the total number of stimuli was comparable and stimulation was carried on for a sufficiently long period. Riley & Allin (19) came to similar conclusions studying the effect of different frequencies on the activity of NADH diaphorase in cat tail muscles. So it seems that the total amount of activity rather than the frequency of firing is important for the changes in muscle metabolism, capillary density and muscle performance.

References

1. Pette, D., Smith, M. E., Staudte, H. W., Vrbova, G.: Pflügers Arch. 338, 257-272 (1973).

2. Salmons, S., Vrbova, G.: J. Physiol. 201, 535-549 (1969).

3. Cotter, M., Hudlicka, O., Vrbova, G.: Bibliotheca Anat. 11, 395-398 (1973).

4. Brown, M. D., Cotter, M. A., Hudlicka, O., Vrbova, G.: Pflügers Arch. 361, 241-250 (1976).

5. Pette, D., Ramirez, B. V., Müller, W., Simon, R., Exner, G. V., Hildebrand, R.: Pflügers Arch. 361, 1-7 (1975).

6. Hudlicka, O., Brown, M., Cotter, M., Smith, M., Vrbova, G.: Pflügers Arch. 369, 141-149 (1977).

7. Cotter, M., Hudlicka, O.: J. Physiol. 292, 20-21P (1979).

8. Holloszy, J. O., Booth, F. W., Winder, W. W., Fitts, R. H.: Biochemical adaptation of skeletal muscle to prolonged physical exercise. In 'Metabolic adaptation to prolonged physical exercise'. Ed. Howald, H., Poortmans, J. R. Birkhäuser, Basel, 438-447 (1975).

9. Saltin, B., Henriksson, J., Nygaard, E., Andersen, P.: Annals of the New York Academy of Sciences 301, 3-29 (1977).

10. Andersen, P., Henriksson, J.: J. Physiol. 270, 677-690 (1977).

11. Treumann, F., Schroeder, W.: Z. Kreislaufforsch. 57, 1024-1033 (1968).

12. Paul, P.: In 'Metabolic adaptation to prolonged physical exercise'. Ed. Howald, H., Poortmans, J. R. Birkhauser Verlag, Basel, 156-193 (1975).

13. Henneman, E., Olson, C.: J. Neurophysiol. 28, 581-598 (1965).

14. Gollnick, P. D., Piehl, K., Saltin, B.: J. Physiol. 241, 45-57 (1974).

15. Hilton, S. M., Lywood, D. W.: J. Physiol. 123, 64 (1954).

16. Minton, N.: B.Sc. Thesis, Birmingham University (1976).

17. Gollnick, P. D., Armstrong, R. B., Saubert, C. W., Piehl, K., Saltin, B.: J. Appl. Physiol. 33, 312-319 (1972).

18. Edgerton, V. R.: Amer. Zool. 18, 113-125 (1978).

19. Riley, D. A., Allin, E. F.: Exp. Neurol. 40, 391-413 (1973).

Acknowledgement

We would like to thank Dr. M. E. Smith for estimation of the activity of acyl CoA synthetase.

CHANGES INDUCED IN THE ENZYME ACTIVITY PATTERN BY ELECTRICAL STIMULATION OF FAST-TWITCH MUSCLE

A. Heilig and D. Pette
Fakultät für Biologie, Universität Konstanz,
D-7750 Konstanz, Germany

Chronic indirect stimulation of fast-twitch muscles with a frequency pattern resembling that of a slow motoneurone, has been shown to transform fast into slow-twitch muscle (1). This transformation involves a great number of different cellular systems, and characteristic time courses have been observed. Among the early changes are those of the enzyme activity pattern of energy metabolism (2), while changes in proteins of the contractile apparatus are found much later (3-10). Alterations of the enzyme activity pattern were previously observed at four days after onset of stimulation (2), but were not measured at earlier times. In those stimulations, periods of 8 hours were alternated with 16 hours rest. The induction stimulus for changes may have been suboptimal. This assumption is supported by the rapid transformation of the sarcoplasmic reticulum beginning 1-2 days after onset of stimulation (11-13). These early transformations were recorded in animals being stimulated 24 hours daily. It is of interest, therefore, whether changes in the enzyme activity pattern may be induced earlier than four days at prolonged daily periods of stimulation. The possibility of more rapid alterations was suggested by recent findings in our laboratory (14-16) on steady state turnover rates of enzymes of energy metabolism in muscle. Enzymes of glycogen metabolism and glycolysis as well as cytosolic and mitochondrial malate dehydrogenases revealed half-lives in the range between one and two days. Therefore, more rapid changes in enzyme levels may be expected to occur, especially if conditions of stimulation are optimal.

MATERIALS AND METHODS

Adult male rabbits of the strain "Weiße Wiener" were used. Implantation of electrodes and stimulation of the lateral popliteal nerve (frequency 10/sec, duration 0.15 msec, intensity 2-9 V) were performed as described previously (2). Stimulation lasted 12 hours per day with hourly alternating periods of stimulation and rest. At various days after onset of stimulation, animals were anesthetized and samples of stimulated (left) and control (right) extensor digitorum longus and tibialis anterior muscles were taken for biochemical analyses. For determinations of glycogen synthetase, probes were taken from the carefully exposed tibialis anterior muscle by rapid freeze-clamping. Thereafter, animals were killed and extensor digitorum longus muscles, as well as the remaining part of the tibialis anterior muscles, were excised, cooled on ice, minced in a micro meat grinder and mixed thoroughly. Activities of hexokinase (HK, EC 2.7.1.1), phosphofructokinase (PFK, EC 2.7.1.11), pyruvate kinase (PK, EC 2.7.1.40), glyceraldehydephosphate dehydrogenase (GAPDH, EC 1.2.1.12), fructose-1,6-diphosphatase (FDPase, EC 3.1.3.11), citrate synthase (CS, EC 4.1.3.7), 3-hydroxyacyl-CoA dehydrogenase (HAD, EC 1.1.1.35) were determined in supernatants of homogenates prepared from the pooled two muscles according to previously described methods (17,18). Succinate dehydrogenase (SDH, EC 1.3.99.1) and mitochondrial glycerolphosphate oxidase (GP-OX, EC 1.1.99.5) were measured in the resuspended sediments of the centrifuged homogenates (18). Cytochrome a was determined spectrophotometrically in sediments of the homogenates after high speed centrifugation (19,20). Glycogen synthetase (GS, EC 2.4.1.11) was measured in extracts of freeze clamped muscles according to Thomas et al. (21).

Electrophoresis of lactate dehydrogenase (LDH, EC 1.1.1.27) isoenzymes was performed in 6 % polyacrylamide gels in TRIS-glycine buffer at ph 8.4.

RESULTS AND DISCUSSION

The extent of changes induced by chronic 10 Hz stimulation in tibialis anterior muscle is illustrated by the activity patterns of selected enzymes of energy metabolism in Fig. 1.

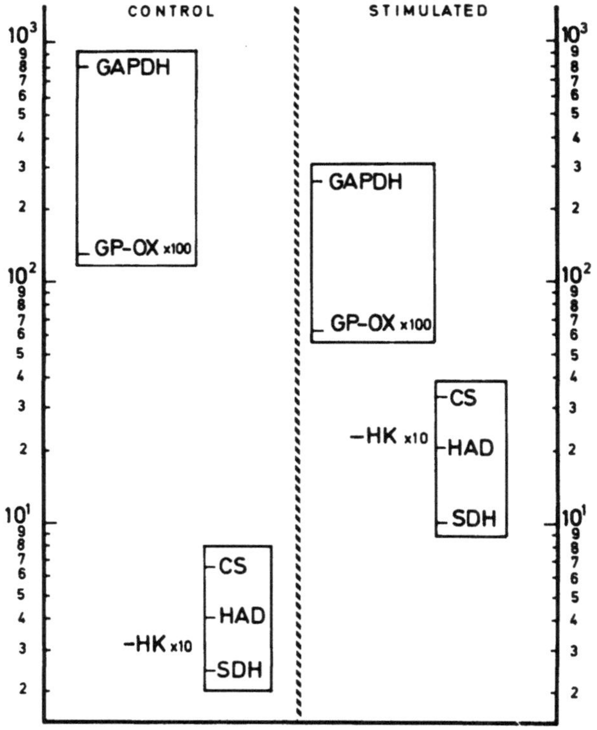

Fig. 1 Activity pattern of selected cytosolic and mitochondrial enzymes of energy metabolism in control and 21 d continuously stimulated (24 h daily) stimulated rabbit tibialis anterior muscle.

After 21 days of continuous stimulation (24 hours daily), the muscle resembles a slow-twitch "red" muscle by its profoundly rearranged enzyme activity pattern. As has been shown in previous studies (2,6,22), there is a decrease in glycolytic enzyme activities and a pronounced increase in activity levels of mitochondrial enzymes of aerobic substrate oxidation. These changes are illustrated in Fig. 1 by inverse shifts of glyceraldehydephosphate dehydrogenase (GAPDH) which decreases to about one third of the activity in the control, and of citrate synthase (CS), succinate dehydrogenase (SDH) and 3-hydroxyacyl-CoA dehydrogenase (HAD) which increased four to five fold. It is remarkable that cytosolic hexokinase (HK) behaved similarly to these mitochondrial enzymes and increased about 8-fold. Mitochondrial glycerolphosphate oxidase (GP-OX), on the other hand, decreased in parallel with GAPDH and therefore maintained its constant proportion with this enzyme of glycolysis (23). In agreement with our previous findings on cross-innervated muscle (24,25), induced changes in mitochondrial enzymes are not uniform. Augmented activity levels of CS, HAD and SDH suggest an increase in the surface area of the inner mitochondrial membrane, and probably in the cellular contents of mitochondria as well. Several authors have drawn similar conclusions on the basis of qualitative enzyme histochemistry (2,6,22,26-28). An increase in size and number of mitochondria is also suggested by qualitative electron microscopy, as illustrated in Fig. 2. The large shape and the abundance of mitochondria in the stimulated muscle is striking. Similar observations have recently been communicated by Salmons et al. (29). It is evident from Fig. 1 that the activity ratio of SDH and GP-OX, two membrane bound flavoproteins of the mitochondria, is profoundly shifted in the stimulated muscle. Its value is changed from 1.9 in normal to 16 in stimulated muscle. This might indicate a qualitative change of mitochondria in the stimulated muscle probably brought about by differential induction of various mitochondrial populations (e.g. subsarcolemmal and interfibrillary ones). This question remains open and will be answered by detailed morphometric analyses.

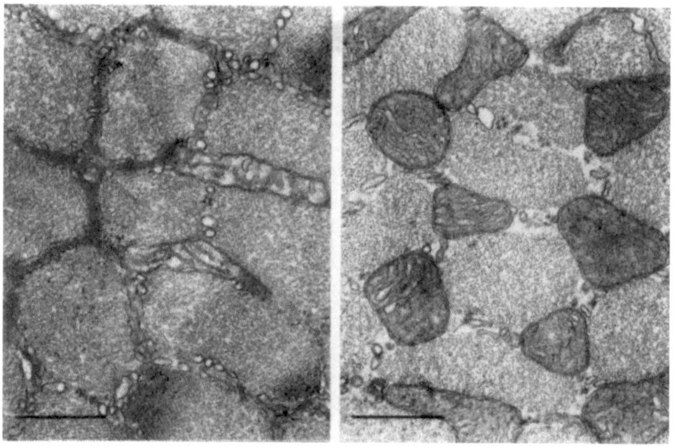

Fig. 2 Electronmicrographs of control (left) and 23 d continuously stimulated (right) rabbit tibialis anterior muscles. Electron microscopy was done by Dr. Werner Müller.

Changes in metabolic properties of stimulated muscles are illustrated not only by shifts in absolute enzyme activities but also by altered ratios of certain enzymes. The change in proportions of a more glycolytically oriented metabolism in normal fast-twitch muscle into that of a more aerobic oxidative one is depicted in Fig. 1. For example, the activity ratio GAPDH/CS is about 100 in control and 8 in stimulated muscle. These changes in the enzyme activity pattern can be related to the increased resistance to fatigue of the stimulated muscles (22,30).

A study of the development of metabolic changes suggests that increases as well as decreases in the activity levels of the various enzymes are initiated almost immediately after onset of stimulation. Time courses of the changes of various enzymes are depicted in Fig. 3 and Fig. 4. Confirming previous results (2,6), HK reveals the greatest changes. In the present experimental series with 12 hours daily stimulation, a 14-fold increase of this enzyme activity was observed after 13 days of stimulation. A pronounced increase by 50 % is seen after one day. Except for FDPase, all enzymes studied reveal small but significant changes in their activity levels at this time point and further increase with time. At 13 days of stimulation, PFK, PK and FDPase activities have decreased to about 40-50 % of the control values in the unstimulated muscles.

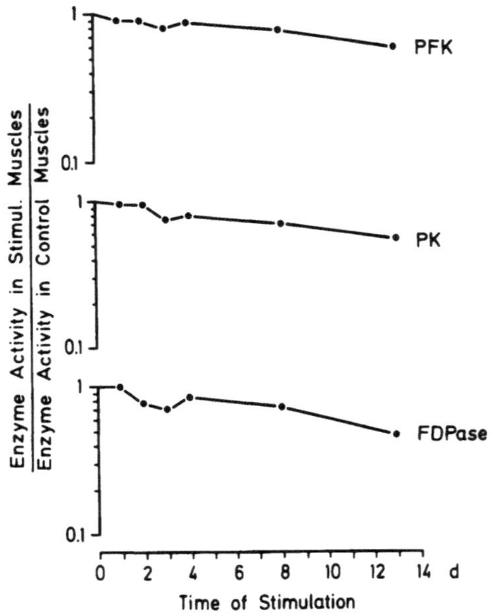

Fig. 3 Time course of changes in phosphofructokinase (PFK), pyruvate kinase (PK), and fructose-1,6-diphosphatase (FDPase) as induced by intermittent (12h daily) long term stimulation. Activities are expressed as ratios between those of stimulated muscles and those of the control contralateral muscles.

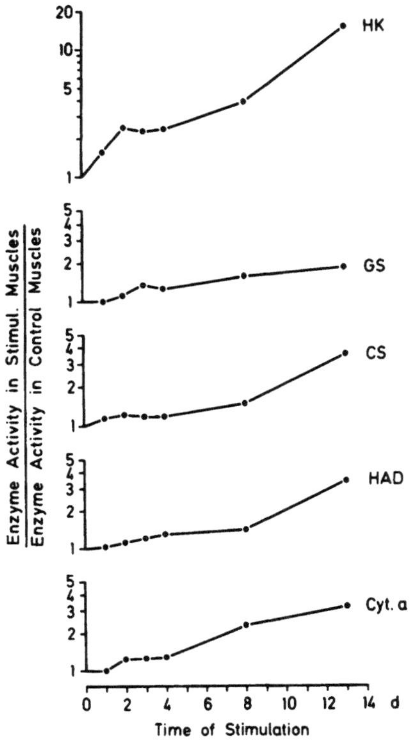

Fig. 4 Time course of changes in hexokinase (HK), glycogen synthetase (GS), citrate synthetase (CS), 3-hydroxy-acyl-CoA dehydrogenase (HAD) activities and in cytochrome a content as induced by intermittent long term stimulation (12 h daily). For further explanation see Fig. 3.

Glycogen synthetase (GS) does not follow the decrease of these enzymes, rather its activity increases moderately, such that at 13 days, its total activity is elevated by a factor of 1.8 in the stimulated muscle. The changes in the activity levels for CS, HAD and cytochrome a are similar in time course and extent. Small increases are observed at 1 day. Maximum changes amount to about 3.3-fold at 13 days of stimulation.

Table 1

ENZYME ACTIVITIES IN NORMAL SOLEUS, CONTROL AND 13 d STIMULATED EDL MUSCLES

	SOLEUS	NORMAL EDL	STIMUL.EDL
	Enzyme Activity (U / g w.wt.)		
Phosphofructokinase	12	78	46
Pyruvate kinase	94	580	333
Fructose-1,6-Pase	0.06	2.5	1.2
Glycogen Synthetase (D+I)	3.3	1.4	2.4
Hexokinase	2.4	0.4	6.1
Citrate synthetase	7.1	3.0	10
3-OH-Acyl-CoA dehydrogenase	7.5	2.1	6.7
Cytochrome a (nmol/g w.wt.)	8.3	3.7	11.3

Table 1 compares enzyme activities of control and 13 days stimulated extensor digitorum longus muscles with normal slow-twitch soleus muscle. Evidently, activities of PFK, PK and FDPase in the stimulated muscle have not yet reached the low levels of soleus muscle. On the other hand, CS and HAD activities in the stimulated muscle have already reached the range of soleus muscle. This holds also for cytochrome a, the level of which clearly exceeds that of soleus muscle. Hexokinase is found at about 2.5-fold higher activity level than in normal slow-twitch soleus muscle. It appears that the transformation of the enzyme activity pattern is complete after 13 days stimulation in the case of those enzymes which increase in activity levels. The transformation is, however, incomplete for those enzymes of which the activity decreases. In the case of HK it is obvious that the induced state may exceed that of normal slow-twitch muscle. Similar observations have been made for changes induced in the sarcoplasmic reticulum and in Ca-activated myosin ATPase (7,13).

Changes in the enzyme pattern are not restricted to activity levels, but include the isoenzyme pattern of lactate dehydrogenase (2). Fig. 5 shows electrophoreses stained for LDH activity in extracts of EDL muscles being stimulated for various

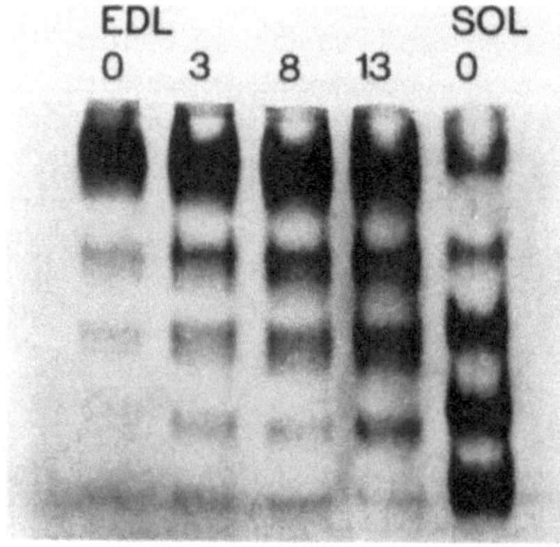

Fig. 5 Polyacrylamide gel electrophoresis of lactate dehydrogenase isoenzymes in normal soleus muscle, control, 3, 8 and 13 d stimulated extensor digitorum longus muscles.

periods and of normal EDL and soleus muscles. Shifts in the isoenzyme pattern concern a decrease in LDH-1 and an increase in the LDH hybrids LDH-2, LDH-3 and LDH-4, at 8 and 13 days of stimulation respectively. Slight changes are seen already after 3 days of stimulation. These alterations of the LDH isoenzyme pattern may be taken as a further evidence that gene expression in muscle may be modified by an altered pattern of neural input. It is evident, however, that the changes induced after 13 days of stimulation have not led to a complete transformation of the isoenzyme pattern as judged from that of normal soleus muscle.

Fig. 3 and Fig. 4 show that changes in enzyme activities of various cytosolic and mitochondrial enzymes of energy metabolism have been induced at a time course similar to that of the sarcoplasmic reticulum. These findings are in agreement with the demonstrated high turnover rates of enzymes of energy metabolism in muscle (14-16). The question may be raised whether changes are due to enhancements of synthesis and degradation. The results in Table 1 suggest that the transformation process is primarily brought about by a stimulation of synthetic activities. At steady state conditions, tissue levels of individual enzymes in muscle have been shown to result primarly from differences in apparent rates of synthesis, whereas degradation seems to be nonspecific (14-16). Applying these observations to the present findings of the induced state, all changes might be explained by simply assuming altered rates of synthesis at unchanged rates of degradation.

The time courses of enzyme induction shown in Fig. 4 suggest a biphasic process. It appears that there is a stronger increase in enzyme activities during the second week. It is tempting to speculate whether this biphasic time course expresses differences in translational and transcriptional activities. Transcription might be enhanced only after some delay due to a higher threshold.

A question which cannot be answered at present, is whether the observed increases in enzyme activity are brought about exclusively by increased synthetic rates. In the case of hexokinase which showed the highest increases, altered activity levels might also result from induced shifts of metabolic effectors. According to Beitner et al. (31,32) glucose-1,6-diphosphate acts as strong inhibitor of hexokinase isoenzymes I and II. A decrease in the level of this effector might partially explain the observed increase in hexokinase activity. This question is under study now.

References

1. Salmons, S., Vrbová, G.: J. Physiol. 201, 535-547 (1969).

2. Pette, D., Smith, M.E., Staudte, H.W., Vrbová, G.: Pflügers Arch. 338, 257-272 (1973).

3. Srêter, F.A., Gergely, J., Salmons, S., Romanul, F.: Nature New Biology 241, 17-19 (1973).

4. Srêter, F.A., Romanul, F., Salmons, S., Gergely, J.: Exploratory Concepts in Muscular Dystrophy II. Excerpta Medica Internatl. Congress Series No. 333, 338-343 (1973).

5. Srêter, F.A., Elzinga, M., Mabuchi, K., Salmons, S., Luff, A.R.: FEBS Lett. 57, 107-111 (1975).

6. Pette, D., Müller, W., Leisner, E., Vrbová, G.: Pflügers Arch. 364, 103-112 (1976).

7. Salmons, S., Srêter, F.A.: Nature 263, 30-34 (1976).

8. Pette, D., Schnez, U.: FEBS Lett. 83, 128-130 (1977).

9. Rubinstein, N., Mabuchi, K., Pepe, F., Salmons, S., Gergely, J., Srêter, F.A.: J. Cell Biology 79, 252-261 (1978).

10. Roy, R.K., Mabuchi, K., Sarkar, S., Mis, C., Srêter, F.A.: Biochem. Biophys. Res. Commun. 89, 181-187 (1979).

11. Ramirez, B.U., Pette. D.: FEBS Lett. 49, 188-190 (1974).

12. Pette, D., Heilmann, C.: Basic Res. Cardiol. 72, 247-253 (1977).

13. Heilmann, C., Pette, D.: Eur. J. Biochem. 93, 437-446 (1979).

14. Dölken, G., Pette, D.: Hoppe-Seyler's Z. Physiol. Chem. 355, 289-299 (1974).

15. Dölken, G., Leisner, E., Pette, D.: Eur. J. Biochem. 47, 333-342 (1974).

16. Illg, D., Pette, D.: Eur. J. Biochem. 97, 267-273 (1979).

17. Bass, A., Brdiczka, D., Eyer, P., Hofer, S., Pette, D.: Eur. J. Biochem. 10, 198-206 (1969).

18. Kubista, V., Kubistová, J., Pette, D.: Eur. J. Biochem. 18, 553-560 (1971).

19. Estabrook, R.W., Holowinky, A.: J. Cell Biol. 9, 19-28 (1961).

20. Chance, B.: In Methods of Enzymology IV (S.P. Colowick, N.O. Kaplan, eds.), pp. 273-329, Academic Press New York (1957).

21. Thomas, J.A., Schlender, K.K., Larner, J.: Anal. Biochem. 25, 486-499.

22. Pette, D., Ramirez, B.U., Müller, W., Simon, R., Exner, G.U., Hildebrand, R.: Pflügers Arch. 361, 1-7 (1975).

23. Pette, D., Bücher, Th.: Hoppe-Seyler's Z. Physiol. Chem. 331, 180 (1963).

24. Bücher, Th., Pette, D.: Verh. Dtsch. Ges. Inn. Med. 71, 104-124(1965).

25. Golisch, G., Pette, D., Pichlmaier, H.: Eur. J. Biochem. 16, 110-116 (1970).

26. Pette, D., Staudte, H.W., Vrbová, G.: Naturwissenschaften 8, 469-470 (1972).

27. Romanul, F., Srêter, F.A., Salmons, S., Gergely, J.: Exploratory Concepts in Muscular Dystrophy II. Excerpta Medica Internatl. Congress Series No. 333, 344-348 (1973).

28. Peckham, P.H., Mortimer, J.T., van der Meulen, J.P.: Brain Res. 50, 424-429 (1973).

29. Salmons, S., Gale, D.R., Srêter, F.A.: J. Anat. 127, 17-31 (1978).

30. Hudlická, O., Brown, M., Cotter, M., Smith, M., Vrbová, G.: Pflügers Arch. 369, 141-149 (1977).

31. Beitner, T., Haberman, S., Livni, L.: Biochim. Biophys. Acta 397, 355-369 (1975).

32. Beitner, R., Cohen, T.Y., Nordenberg, J., Haberman, S.: Biochim. Biophys. Acta 586, 266-277 (1979).

MOLECULAR TRANSFORMATIONS OF SARCOPLASMIC RETICULUM IN CHRONICALLY STIMULATED FAST-TWITCH MUSCLE

C. Heilmann and D. Pette
Fakultät für Biologie, Universität Konstanz, D-7750 Konstanz, Germany

Introduction

It has been shown that a white fast-twitch muscle can be progressively transformed into a red muscle with slow-twitch characteristics by means of superimposed electrical stimulation at a frequency resembling that normally received by a slow-twitch muscle (1). This phenomenon implies a process of molecular and functional transformations in several systems of cellular organization (2-15). This study deals with the changes in the sarcoplasmic reticulum of muscle which were continuously stimulated for different time periods.

Methods

Adult male rabbits (Weiße Wiener) were stimulated by implanted electrodes as has been described (4). Stimulation was 24 h per day with a frequency of 10 Hz for time periods up to 150 days. Sarcoplasmic reticulum (SR) was prepared as described (16) from stimulated and the contralateral unstimulated control tibialis anterior and extensor digitorum muscles. SR was also prepared from normal slow-twitch soleus muscles. The analysis of the SR included 1) gel electrophoresis in the presence of sodium dodecyl sulfate (SDS) of peptide patterns, 2) Ca^{2+} uptake characteristics, 3) phosphoprotein formation at various ionic conditions and in the presence of cyclic AMP and protein kinase. Biochemical methods have been described (16).

Results and Discussion

Peptide patterns of sarcoplasmic reticulum from normal and 75-day stimulated tibialis anterior muscle and from normal soleus muscle are shown in Fig.1.

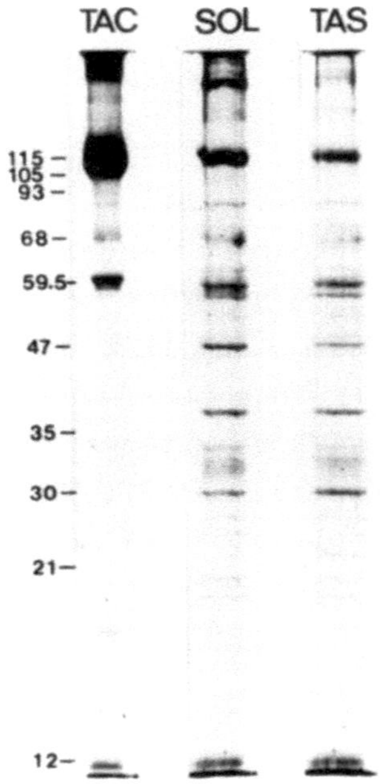

Fig. 1 Polyacrylamide gel electrophoreses in the presence of sodium dodecylsulphate of sarcoplasmic reticulum from control (TCA) and stimulated (TAS) tibialis anterior muscles and of normal soleus muscle (SOL). 10μg protein samples were applied to 12.5 % gels and electrophoreses were performed according to Laemmli (17).

It is seen that the intensity of the 115 000 peptide, which represents the Ca^{2+} transport ATPase (18), is markedly decreased by stimulation. This reduction is confirmed by the densitographs of electrophoreses from deoxycholate preextracted

SR shown in Fig. 2, peak A. Stimulation decreases also peptides with apparent molecular weights of 68 000 and 59 500 which represent calcium binding proteins. In fact, the 59 500 and 68 000 peptides can be hardly detected in sarcoplasmic reticulum from long-term stimulated muscle and soleus muscle. On

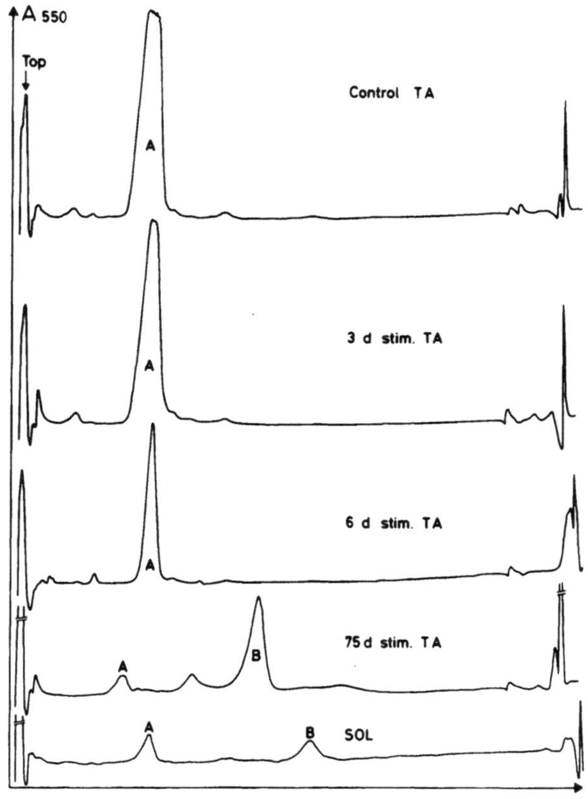

Fig. 2 Densitometric tracings at 550 nm of Coomassie Brilliant Blue-stained electrophoreses of deoxycholate extracted sarcoplasmic reticulum from control and stimulated tibialis anterior muscles (TA) and normal soleus muscle (SOL). Peak A corresponds to 115 000-Mr and peak B to 30 000-Mr.

the other hand, other bands are intensified or manifested themselves during stimulation. This particularly concerns a 105 000-Mr peptide (16) which was identified by immuno-electrophoresis as the protomer of glycogen phosphorylase (unpublished results). Bands in the region of 57 000-, 47 000-, 37 000- and 30 000-Mr and some other less prominent peptides are identified although the function of these peptides is unclear. A 30 000-Mr peptide has been suggested by Margreth et al. (19) to be an integral membrane component of sarcoplasmic reticulum in slow-twitch muscle. They showed this peptide to be resistent to extraction with cholate or deoxycholate. This finding is confirmed in Fig. 2 which shows densitometric evaluations of electrophoreses of deoxycholate-extracted sarcoplasmic reticulum from normal muscle and from control and stimulated tibialis anterior muscles. The 30 000-Mr compound (Fig. 2, peak B) is present in normal soleus muscle but not in unstimulated fast-twitch muscle. It is present, however, in SR of 75d stimulated tibialis anterior muscle at an even higher concentration than in soleus muscle. The results in Fig. 1 and Fig. 2 thus demonstrate a high degree of transformation of the peptide pattern.

Since the concentration of the Ca^{2+} transport ATPase is markedly reduced by stimulation (Fig. 1,2) one would expect lower Ca^{2+} transport activities of the respective sarcoplasmic reticulum. This is demonstrated in Fig. 3 in which time courses of Ca^{2+} uptake of sarcoplasmic reticulum from stimulated tibialis anterior muscles are compared to those of contralateral muscle and of normal soleus muscle. Both initial and total Ca^{2+} uptake are reduced from normal during the transformation process. The values, after 90 days of stimulation approach those of the normal soleus muscle. Changes in Ca^{2+} uptake may be correlated with the changes in ATPase activities (Table 1). Long-term stimulation induced a progressive reduction in Ca^{2+} dependent ATPase activities paralleled by an increase in Mg^{2+}-dependent ATPase activities. It is remarkable that the decrease in Ca^{2+}-dependent ATPase is almost 50% after only 2 days of stimulation.

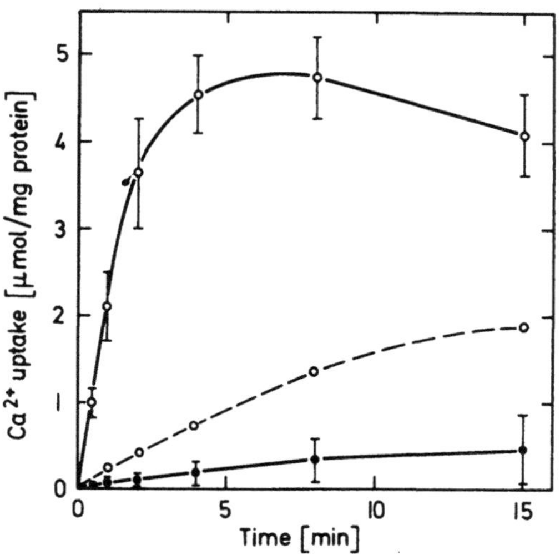

<u>Fig. 3</u> Time course of Ca^{2+} uptake in the presence of oxalate by sarcoplasmic reticulum from control (O—O) and 90 days stimulated (O---O) tibialis anterior, and normal soleus (●—●) muscles.

Table 1 ATPase activities in sarcoplasmic reticulum vesicles isolated from control and stimulated tibialis anterior muscles of rabbits subjected to different stimulation periods. Activities refer to measurements on control and stimulated muscles of single animals.

Days stimulated	ATPase activity			
	Mg^{2+}-dependent		Ca^{2+}-dependent	
	control	stimulated	control	stimulated
d		$\mu mol \times min^{-1} \times mg^{-1}$ protein		
0*	0.247	0.267	0.985	0.965
2	0.236	0.288	0.751	0.463
3	0.305	0.308	0.801	0.453
13	0.352	0.226	1.00	0.512
75	0.177	0.907	1.28	0.129
90	0.36	2.10	0.733	0.05
Soleus	1.67 ± 0.24 (n = 7)		0	

*unstimulated animal with implanted electrodes

Formation of the phosphorylated reaction intermediate of the Ca^{2+} transport ATPase is far more rapid than its hydrolysis. It appears therefore that the latter determines the rate of ATPase and Ca^{2+} transport activities (18). Time courses of phosphoprotein formation in sarcoplasmic reticulum from normal and 90 days stimulated fast-twitch and from normal slow-twitch muscle are illustrated in Fig. 4. Measurements were made at low ATP concentration and under different ionic conditions. These conditions make it possible to follow dephosphorylation while phosphoprotein formation is too fast to be followed. It is obvious that in the presence of Ca^{2+} and Mg^{2+}, dephosphorylation proceeds faster than with Ca^{2+} alone. This is because Mg^{2+} promotes the hydrolysis of the phosphoprotein intermediate (18). Measurements on sarcoplasmic reticulum of stimulated muscles show a lower phosphoprotein formation and

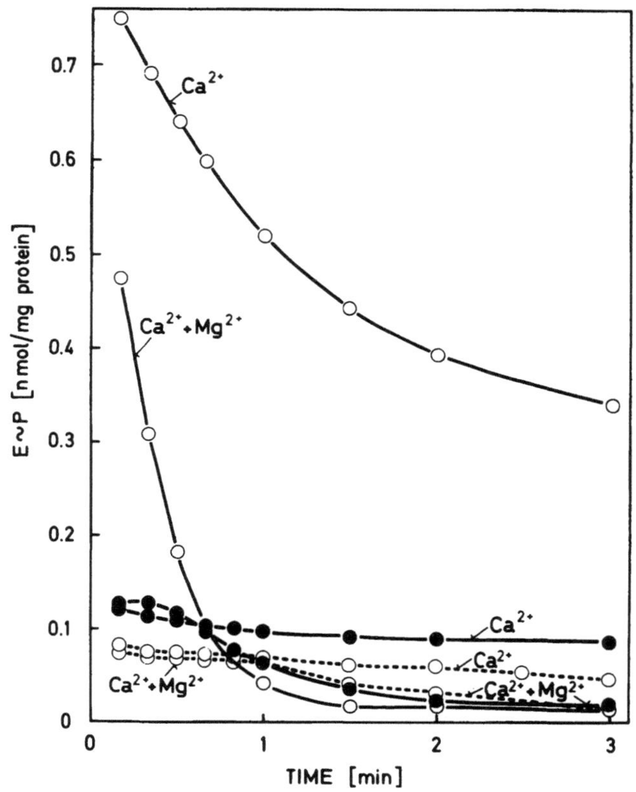

Fig. 4 Time course of ^{32}P incorporation from [γ-^{32}P] ATP into sarcoplasmic reticulum from control (○——○) and 90 days stimulated (○---○) tibialis anterior muscles and from normal soleus muscle (●——●). Incubations were performed at 0° C in the presence of 12 µm ATP, 100 mM KCl, 5 mM NaN_3, and 40 mM Hepes buffer (pH 6.9). 50 µM $CaCl_2$ or 50 µM $CaCl_2$ and 10 mM $MgCl_2$ were added.

a prolonged dephosphorylation as compared to SR from control muscles. Results from these measurements are summarized in Table 2. For all ionic conditions studied, the amount of phosphoprotein formed by the SR, decreases with length of stimulation. Moreover, dephosphorylation activity in the presence of Ca^{2+} and Mg^{2+} decreases as reflected by the increasing phosphoprotein half lives given in the right hand column of Table 2.

Table 2 ^{32}P incorporation into sarcoplasmic reticulum proteins of control and stimulated tibialis anterior muscles and soleus muscle of the rabbit at various ionic conditions and low ATP concentration (12 μM). In reactions where maximum phosphoprotein formation could not be measured directly, the respective values were obtained by extrapolating the reaction curves to zero time. Apparent half-lives of phosphoproteins formed in the presence of Ca^{2+} and Mg^{2+} were evaluated graphically from the initial linear part of reaction curves.

Ionic conditions	10 mM $MgCl_2$ + 1 mM EGTA	50 μM $CaCl_2$	50 μM $CaCl_2$ + 10 mM $MgCl_2$	
	^{32}P incorporation (pmol/mg protein)			$t_{1/2}$ (s)
Tibialis anterior				
control	56	794	794	15
11 d stimulated	40	263	316	43
90 d stimulated	23	87	79	95
150 d stimulated	20	90	94	85
Soleus	14	124	128	58

In the case of the 150 d stimulated muscle, the phosphoprotein half life exceeded that of normal slow-twitch muscle. Since the decay of phosphoprotein intermediate of the Ca^{2+} transport ATPase is probably the rate limiting step of ATPase and consequently of Ca^{2+} transport activity, it may be concluded that the lower ATPase and Ca^{2+} transport activities in sarcoplasmic reticulum from soleus muscle and stimulated muscle are at least partially due to the lower dephosphorylating activity of the Ca^{2+} transport ATPase.

Qualitative transformations are induced in the phosphoprotein pattern of sarcoplasmic reticulum by stimulation (14, 16). It has been shown by others (for review see 20) and ourselves (16, 21) that, apart from the phosphorylated intermediate of the Ca^{2+} transport ATPase, additional stable phosphoproteins are formed in sarcoplasmic reticulum preparations. Fig.5 shows electrophoretic patterns of phosphoproteins of sarcoplasmic reticulum from normal and stimulated fast-twitch tibialis anterior muscle and slow-twitch soleus muscle. Phosphorylation in the presence of Ca^{2+} and Mg^{2+} or with Ca^{2+} alone (autoradiographs c and d, Fig. 5) produces a prominent density with an apparent molecular weight of 115 000. This phosphopeptide represents the phosphorylated intermediate of the Ca^{2+} transport ATPase. The density of this band is considerably less in sarcoplasmic reticulum from slow-twitch as compared to fast-twitch muscle. A fainter band is seen with an apparent molecular weight of 68 000 in sarcoplasmic reticulum from both fast- and slow-twitch muscle. The function of this Ca^{2+} dependent phosphorylated peptide is unclear. It comigrated in SDS gel electrophoresis with a peptide which is eluted from DEAE-cellulose as a single peak at slightly lower ionic strength than calsequestrin. Therefore one may speculate that this peptide may be the so-called high affinity calcium binding protein, described by Ostwald and MacLennan (23). It is unlikely that this phosphopeptide is a cleavage product of the phosphorylated 115 000 Ca^{2+} transport protein, since its tryptic digestion gives 59 000- and 30 000-Mr phosphoproteins (24). These digestion products are demonstrated by electrophoresis (Fig.6c) or autoradiographs (Fig. 6a,b). Incubation for phosphorylation in the presence of Mg^{2+} and EGTA without Ca^{2+} gives no phosphorylated compounds (Fig. 6a). If Ca^{2+} is included and EGTA omitted, two phosphopeptides, of 59 000- and 28 000-Mr can be seen (Fig. 6b). Prolonged incubation of sarcoplasmic reticulum

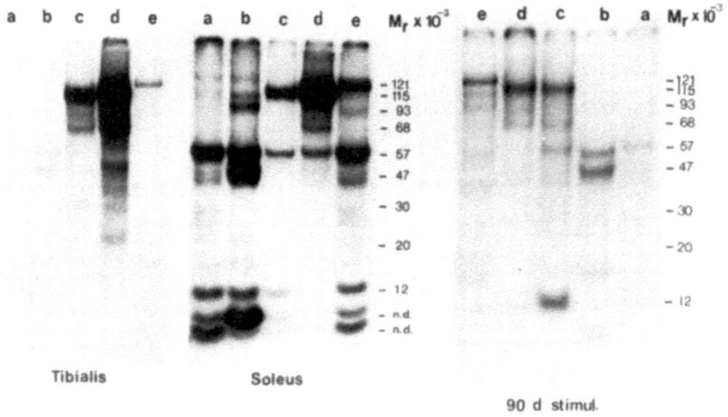

Fig. 5 Autoradiographs of SDS gel electrophoreses (22) phosphorylated sarcoplasmic reticulum of control and 90 days stimulated tibialis anterior muscles and of normal soleus muscle. Incubations lasted 20 s at 0° C in a 1 ml mixture of 35 mM Hepes (pH 6.9), 0.1 M KCL, 20 mM KF, 5 mM NaN_3, 12 μM $[\gamma\text{-}^{32}P]$ ATP and 0.5 mg vesicular protein.

Additions were:
- a) 10 mM $MgCl_2$, 2 mM EGTA
- b) 10 mM $MgCl_2$, 2 mM EGTA, 2μM cyclic AMP 0.1 mg skeletal muscle protein kinase (Peak II)
- c) 10 mM $MgCl_2$, 50 μM $CaCl_2$
- d) 10 mM $MgCl_2$, 2 mM EGTA, 0,11 mg phosphorylase b kinase

at 37° prior to phosphorylation does not increase the concentration of the 68 000-Mr peptide which argues against the action of endogenous proteases (Data not shown).

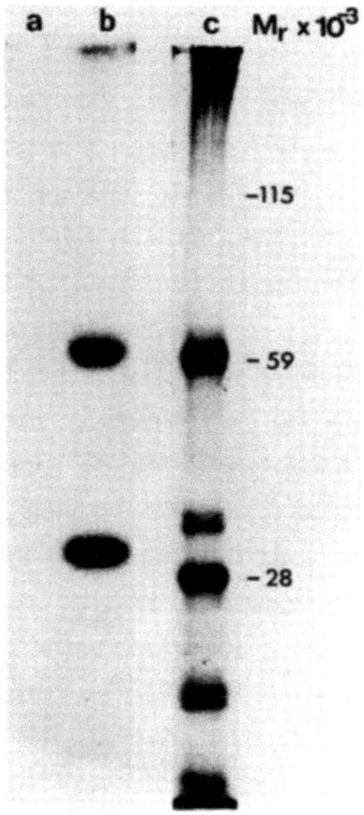

Fig. 6 SDS polyacrylamide gel electrophoresis according to Laemmli (17) of phosphorylated tryptic digests of sarcoplasmic reticulum from normal fast-twitch muscle (a and b autoradiograph, c stained for protein). Phosphorylation was performed as described in legend to Fig. 4. Additions were:

a) 10 mM $MgCl_2$, 2 mM EGTA,

b) 10 mM $MgCl_2$, 50 μM $CaCl_2$

Marked differences normally exist between fast and slow-twitch muscle SR if phosphorylation is performed in the absence of Ca^{2+} but in the presence of Mg^{2+} and EGTA (Fig. 5). Under these conditions practically no phosphorylated bands can be detected on autoradiographs from fast-twitch muscle SR. Slow-twitch muscle SR, however, gives phosphoproteins of 57 000-, 47 000-, 12 000-, 11 000- and 8 000-Mr. The identity of these phosphoproteins is unknown. It is possible that one of the low molecular weight phosphopeptides is identical with phospholamban (20). In a recent study by Le Peuch et al. (25) phospholamban has been shown to dissociate into monomers of 11 000-Mr in the presence of both ionic and nonionic detergents.

Studies on the stability of phosphoproteins formed in the presence of Mg^{2+} indicate the presence of stable esterphosphate bondages. Fig.7 shows time courses of phosphoprotein decay in the presence of hydroxylamine. The upper curves (Fig. 7A) represent measurements on sarcoplasmic reticulum from fast-twitch muscle which was phosphorylated in the presence of Ca^{2+}. A rapid hydroxylamine-induced decomposition of the intermediately formed aspartylphosphate of the Ca^{2} transport ATPase is observed. The lower part of this figure (Fig. 7B) shows the resistance to hydroxylamine of the Mg^{2+}-dependent phosphoproteins in sarcoplasmic reticulum from slow-twitch muscle. The respective autoradiographs are shown as inset. It is also shown that these compounds cannot be extracted by chloroform/methanol (Fig. 7B, inset labelled "Extr.") which excludes the possibility that one of these might be a lipid.

Phosphorylation experiments in the presence of c-AMP and exogenous protein kinase from skeletal muscle (peak II) suggest that some of these peptides might serve as substrates of an endogenous protein kinase or represent autophosphorylated subunits of a protein kinase itself (Fig. 8). Exogenous c-AMP dependent protein kinase markedly enhances phosphate incorpo-

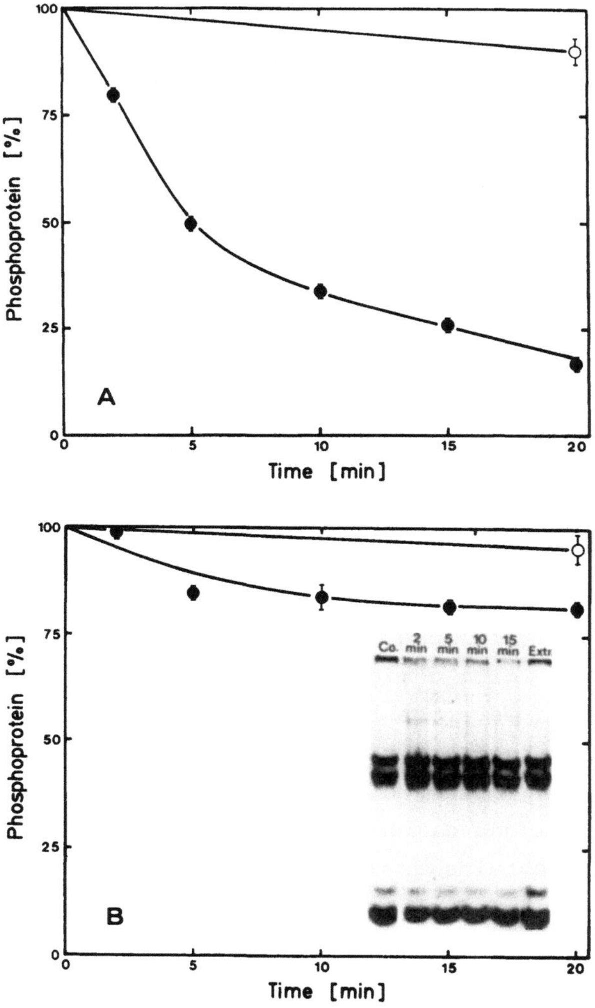

<u>Fig. 7</u> Effect of hydroxylamine on acid-denatured phosphoproteins of sarcoplasmic reticulum from fast-twitch(A) and slow-twitch muscle(B). Phosphorylation was performed as in Fig. 5b with the exception that EGTA was omitted and 50 μM $CaCl_2$ was added to the incubation mixture for sarcoplasmic reticulum from fast-twitch muscle. Precipitated and extensively washed protein

was suspended in sodium acetate buffer (pH 5.2) (○—○) or in 0.8 M hydroxylamine in sodium acetate buffer (pH 5.2) (●—●) and incubated for 20 min at 5° C. For extraction of lipids, phosphorylated vesicles were incubated 10 min with chloroform/methanol (2:1,v/v). Reactions were terminated with 10 % TCA. Inset: Radioautographs of SDS gel elctrophoreses (22) of control (Co.), hydroxylamine-treated(2-15 min) and chloroform/methanol-extracted (Ectr.) vesicles from slow-twitch muscle.

ration into peptides of 47 000- and 11 000-Mr (Fig. 5b, soleus). It may be that the 57 000- and 47 000-Mr phosphoproteins represent subunits of an endogenous c-AMP dependent protein kinase. The data shown in Fig. 8 support such a possibility. Incubation of purified c-AMP-dependent protein kinase from skeletal muscle (peak II) gives 57 000- and 47 000-Mr phosphoproteins which comigrate in SDS gel electrophoresis with those which were observed mainly in sarcoplasmic reticulum from slow-twitch muscle (Fig. 5a, b, soleus). Isoelectric focusing followed by electrophoresis in the presence of SDS in the second dimension shows that the 57 000- and 47 000-phosphoproteins from slow-twitch muscle SR and from purified c-AMP dependent protein kinase have similar, if not identical, isoelectric points and apparent molecular weights (Fig. 9). As is shown in Fig. 9, phosphoproteins of 57 000- and 47 000-Mr which are formed in the presence of Mg^{2+} and EGTA, are present also in a light vesicle fraction separated by sucrose density centrifugation from SR of fast-twitch muscle (21).

These phosphoproteins, however, are present at much higher concentrations in the SR of slow-twitch muscle. This suggestion is supported by findings obtained on the transformation of SR by chronic stimulation. Thus the autoradiographs of SR from a 90 days stimulated fast-twitch muscle presented in Fig. 5, show the same phosphoproteins formed in the presence of Mg^{2+} which are also found in SR of normal slow-twitch muscle.

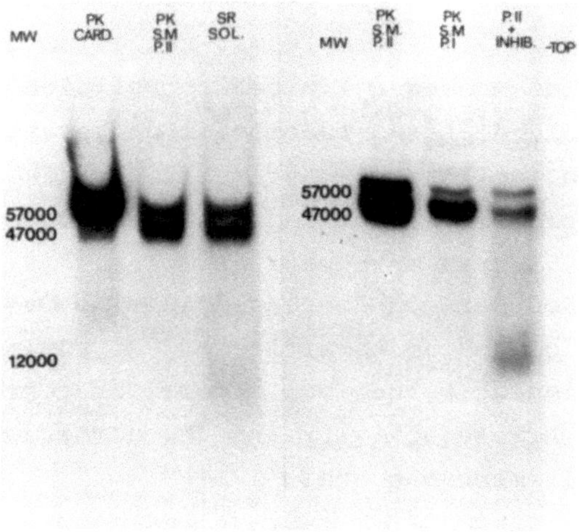

Fig. 8 Autoradiographs of SDS polyacrylamide gel electrophoreses (22) of cyclic AMP-dependent protein kinase from bovine cardiac muscle (PK CARD.), cyclic AMP-dependent protein kinase from skeletal muscle peak I (PK S.M. P.I) and peak II (PK S.M. P.II), sarcoplasmic reticulum from slow-twitch soleus muscle (SR SOL.) and heat-stable protein kinase inhibitor (INHIB.). Phosphorylations were performed in the presence of 10 mM $MgCl_2$, 2 mM EGTA, 2 µM c-AMP, 0.1 mg skeletal muscle protein kinase. Other conditions were as described in Fig. 5.

A final point concerns ultrastructural alterations of SR induced by continous electrical stimulation. The 7-9 nm intramembranous particles seen on concave freeze-fracture faces of isolated and native SR have been identified to be probably the morphological correlates of the Ca^{2+} transport ATPase molecule or an oligomer of it within the membrane (27, 28). Slow-twitch muscles from various species have been reported to have a lower content of SR and a lower number of intramembranous particles per unit area of the membrane as compared to fast-twitch muscles (29, 30). Combined biochemical and ultrastructural studies on sarcoplasmic reticulum during ontogenesis of muscles showed a limited correlation between particle density and molecular and functional differentiation of SR(31). Based on these implications, stimulation induced transformation of sarcoplasmic reticulum should be accompanied by a reduction in the intramembranous particles. Fig. 10 shows a direct correlation between decreased initial Ca^{2+} uptake and reduced density of 7-9 nm particles of SR from muscles subjected to various stimulation periods. A similar correlation exists between extra ATPase activity and the number of 7-9 nm particles per unit membrane area.

Our results show that sarcoplasmic reticulum of fast-twitch muscle can be transformed to resemble that of a slow-twitch muscle. This has been demonstrated as sequential changes of molecular, functional and ultrastructural properties of sarcoplasmic reticulum during long-term electrical stimulation. The qualitative nature of the transformation is emphasized by the appearance of compounds in the SR stimulated fast-twitch muscle which are specific of SR in slow-twitch muscle, as well as by the marked decrease in compounds characteristic of fast-twitch muscle SR. Changes in twitch-characteristic (10,11) occur as early as those reported here and previously (8, 14, 16) for the SR, whereas transformation of the myosin is observed much later (5). It is concluded, therefore, that the alterations in SR are primarily responsible for the early transformation of twitch-characteristic during chronic stimulation.

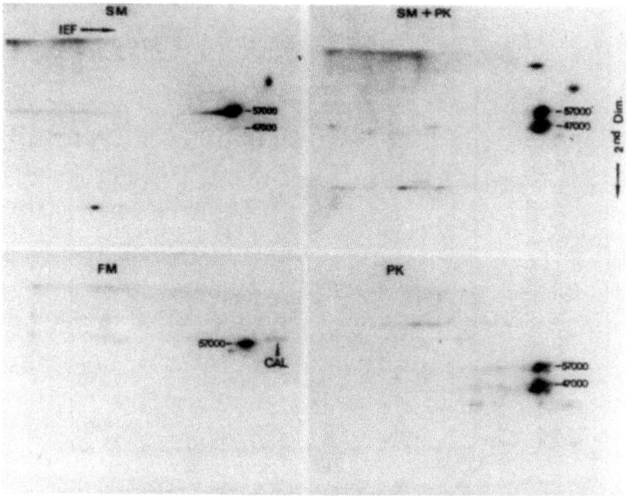

<u>Fig. 9</u> Autoradiographs of second dimension SDS polyacrylamide gel electrophoreses following isoelectric focusing, IEF, (26) of sarcoplasmic reticulum from slow-twitch (SM) and fast-twitch muscle(FM) and of cyclic AMP-dependent protein kinase (Peak II) from skeletal muscle. Phosphorylations of SM and of FM were made in the presence of 1 mM $MgCl_2$ and 2 mM EGTA; phosphorylation of SM+PK and of PK were made in the presence of 10 mM $MgCl_2$, 2 mM EGTA, 2 μM c-AMP, 0.1 mg skeletal muscle protein kinase. Other conditions were as described in Fig. 5.

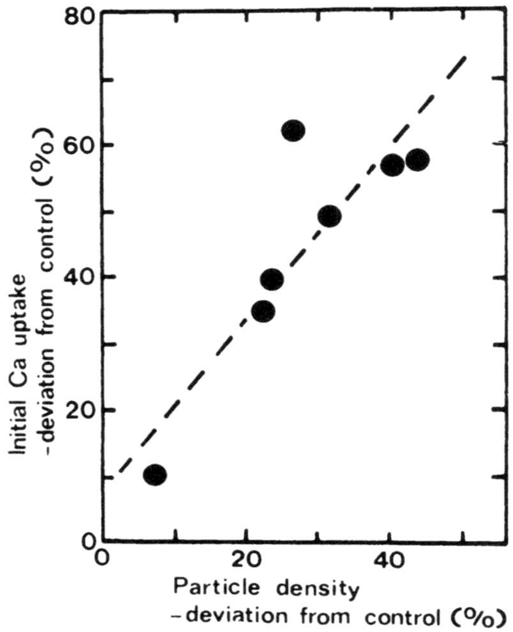

Fig. 10 Correlation between initial Ca^{2+} uptake (µmol x mg^{-1} x min^{-1}) and number of 7-9 nm particles per μm^2 of concave freeze-fracturing faces of sarcoplasmic reticulum from tibialis anterior muscles subjected to various times of stimulation. Electronmicroscopy was done by Dr. W.Müller.

Acknowledgements: This study was supported by grants of the Deutsche Forschungsgemeinschaft, Sonderforschungsbereich 138 "Biologische Grenzflächen und Spezifität" . Authors are grateful to Ms. U. Kutschka and Ms. E. Leisner for excellent technical assistence and to Dr. Patti Nemeth for reading the manuscript.

REFERENCES

1. Salmons, S., Vrbová, G.: J. Physiol. 201, 535-549 (1969).

2. Pette, D., Staudte, H.W.: Limiting Factors of Physical Performance, Inter. Symposium at Gravenbruch 1971 (Keul, J., ed) pp. 23-35, G. Thieme Publishers, Stuttgart (1973).

3. Pette, D., Staudte, H.W., Vrbová, G.: Naturwissenschaften 59, 469-470 (1972).

4. Pette, D., Smith, M.E., Staudte, H.W., Vrbová, G.: Pflügers Arch. 338, 257-272 (1973).

5. Sréter, F.A., Gergely, J., Salmons, S., Romanul, F.C.A.: Nature New Biol. 241, 17-19 (1973).

6. Sréter, F.A., Romanul, F.C.A., Salmons, S., Gergely, J.: Exploratory Concepts in Muscular Dystrophy II (Milhorat, A.T., ed) Int. Congr. Series No. 333, pp. 338-343, Excerpta Medica, Amsterdam (1974).

7. Romanul, F.C.A., Sréter, F.A., Salmons, S., Gergely, J.: (Milhorat, A.T., ed) Int. Congr. Series No. 333, pp. 344-348, Excerpta Medica, Amsterdam (1974).

8. Ramirez, B.U., Pette, D.: FEBS Lett. 49, 188-190 (1974).

9. Sréter, F.A., Elzinga, M., Mabuchi, K., Salmons, S., Luff, A.R.: FEBS Lett. 57, 107-110 (1975).

10. Pette, D., Ramirez, B.U., Müller, W., Simon, R., Exner, G.U. Hildebrand, R.: Pflügers Arch. 361, 1-7 (1975).

11. Pette, D., Müller, W., Leisner, E., Vrbová, G.: Pflügers Arch. 364, 103-112 (1976).

12. Salmons, S., Sréter, F.A.: Nature (Lond.) 263, 30-34 (1976)

13. Hudlická, O., Brown, M., Cotter, M., Smith, M.E., Vrbová, G.: Pflügers Arch. 369, 141-149 (1977).

14. Pette, D., Heilmann, C.: Bas. Res. Cardiol. 72, 247-253 (1977).

15. Buller, A.J., Pope, R.: Phil. Trans. Roy. Soc. Lond. B. 278, 295-305 (1977).

16. Heilmann, C., Pette, D.: Eur. J. Biochem. 93, 437-446 (1979).

17. Laemmli, U.K.: Nature (Lond.) 227, 680-685 (1970).

18. Hasselbach, W.: Topics in Current Chemistry 78, 1-56 (1979).

19. Margreth, A., Salviati, G., Mussini, I., Carraro, U.: in Exploratory Concepts in Muscular Dystrophy II (Milhorat, A.T., ed) Int. Congr. Series No. 33, pp. 406-415, Excerpta Medica, Amsterdam (1974).

20. Tada, M., Yamamoto, T., Tonomura, Y.: Physiol. Rev. 58, 1-79 (1978).

21. Heilmann, C., Brdiczka, D., Nickel, E., Pette, D.: Eur. J. Biochem. 81, 211-222 (1977).

22. Weber, K., Osborn, M.: J. Biol. Chem. 244, 4406-4412 (1969)

23. Ostwald, T.J., MacLennan, D.H.: J. Biol. Chem. 249, 974-979 (1974).

24. Stewart, P.S., MacLennan, D.H.: Ann. N. Y. Acad.Sci.264, 326-334 (1975)

25. Le Peuch, Ch.J., Haiech, J., Demaille, G.: Proc. of the 8th European Conference on Muscle and Motility at Heidelberg, pp. 58 (1979).

26. O'Farrell, P.: J. Biol. Chem. 250, 4007-4021 (1975).

27. Deamer, D.W., Baskin, R.J.: J. Cell Biol. 42, 296-307 (1969).

28. Packer, L., Mehard, C.W., Meissner, G., Zahler, W.L., Fleischer, S.: Biochim. Biophys. Acta 363, 159-181 (1974).

29. Van Winkle, W.B., Schwartz, A.: J. Cell Physiol. 97, 99-120 (1978).

30. Bray, D.F., Rayns, D.G.: J. Ultrastruct. Res. 57, 251-259 (1976).

31. Tillack, T.W., Boland, R., Martonosi, A.: J. Biol. Chem. 249, 624-633 (1974).

EFFECT OF CHRONIC STIMULATION ON CATION DISTRIBUTION AND MEMBRANE POTENTIAL IN FAST-TWITCH MUSCLES OF RABBIT

F.A. Sreter[1] and K. Mabuchi

Department of Muscle Research, Boston Biomedical Research Institute, Boston, Massachusetts, 02114

A. Kőver, I. Gesztelyi, Z. Nagy

Central Research Lab, University of Debrecen, Debrecen, Hungary

I. Furka

Department of Exp. Surgery, Medical School, University of Debrecen, Debrecen, Hungary

Introduction

During the past years it became firmly established that fast-twitch muscles are readily converted to the slow-twitch type by continuous indirect stimulation at 10 Hz, mimicking the impulse rate of slow-twitch fibers. The first report showed a slowing effect on the time course of contraction and relaxation (1). This was followed by the demonstration of biochemical changes upon chronic stimulation: decrease in the initial and total Ca^{2+}-uptake of the FSR; in the ATPase activities(2,3), in the light chain complement (4), in the light meromyosin paracrystal pattern (2); the N^{τ}-methylhistidine content of the myosin molecule (5); and a shift from glycolytic to oxidative enzymes (6,7), and more recently changes in the ratio of α and β subunits of tropomyosin to the value characteristics of slow-twitch fiber tropomyosin (8). Staining of thin sections with type-specific fluorescent antimyosins (9) as well as gel-electrophoretic analysis of myosin light chains in single fibers of stimulated muscle (10) indicated that the transformation is due to the conversion of pre-existing fast fibers to slow ones

[1]Exchange scientist in the Department of Biochemistry, Eotvos Lorand University, under the sponsorship of the National Academy of Science and the Hungarian Academy of Science.

rather than a replacement of the whole fast fiber population by new fibers, by switching off previously active genes and activating previously silent ones. The newly converted "slow-twitch" muscle will reconvert to the fast-twitch type within 4-6 weeks after cessation of stimulation.

The present experiments show extensive change in the resting membrane potential of the stimulated fast-twitch muscle fiber, concomitantly with a considerable shift in the content of intracellular sodium and potassium. Within 5 weeks of stimulation, a period during which the fast-twitch peroneus muscle showed an almost complete transformation, judging by histochemical methods, the membrane potential changed from the original −90 mV to more than −50 mV. The latter change, however, was entirely dynamic, since the original value was restored within 30 min after cessation of stimulation. This is in marked contrast to the transformation of various myofibrillar proteins and the properties of the sarcoplasmic reticulum where the reversibility usually takes several weeks or even months.

Methods

Chronic indirect stimulation of the left tibialis anterior (TA), extensor digitorum longus (EDL) and peronei muscles was carried out as described earlier (2). The surgical procedure was modified as follows: an approximately 4 cm incision was made just above the left knee joint and the superficial muscle fibers carefully separated to expose the common peroneal nerve. Another incision, approximately 3 cm in length, was made on the left lower quadrant of the abdominal wall and the miniature electric stimulator (11) placed subcutaneously into a pouch prepared under the skin. The capsule was immobilized by a few drops of Histoacryl-N-blau (B. Braun Melsungen A.G.) and the two flexible electrodes led subcutaneously to the left hind limb, where they were fixed near to the common peroneal nerve. By avoiding intraperitoneal implantation the surgical procedure was much faster and safer. After various lengths of stimulation the animals were anaesthetized by Nembutal (35 mg/kb i.v.) and the surface of the TA muscle was prepared for membrane potential measurements (12). The skin and the underlying fascia were drawn each side to form a pouch in which Tyrode solution was poured under a stereoscopic microscope. The solution was kept at 37° by infrared radiation. During membrane potential measurements the stimulation was temporarily suspended but reinstated at the end of the measurements. The blood supply to the kidney was then stopped by an intraperitoneal approach, the abdominal walls closed and [^{35}S]Na_2SO_4 (50 µCi) given intravenously. After 2 hours of equilibration the stimulation was stopped and the left side stimulated and the right side muscles (TA, EDL and peronei) were quickly removed. Simultaneously blood was obtained by heart puncture. The extracellular fluid volume and the cation content (K, Na, Ca and Mg) of the samples digested in nitric acid were determined as described earlier (13).

For membrane potential measurements a Keithley 615 digital electrometer and Ling-Gerard type glass microelectrodes (14) were used. The glass capillaries were filled with 2.5M KCl, and their resistance checked before use. Only capillaries with a resistance of less than 5M Ω (tip potential less than 5 mV) were used. In all cases membrane potential measurements were carried out in the deeper layers of the muscle to avoid possible erroneous results due to injury of the superficial layers. For statistical evaluations repeated measurements (usually at least 10) were taken and the time elapsed after interruption of stimulation recorded. When cation concentrations were measured the membrane potential values were calculated according to the Nernst equation.

Calcium uptake by sarcoplasmic reticulum (SR), quantitative actomyosin ATPase and histochemical fiber typing was carried out on cryostat serial sections as recently described (15). Briefly, Ca^{2+}-uptake was measured in 10 µm thin sections mounted on plastic coverslips in a medium containing 0.1M KCl, 5 mM $MgCl_2$, 25-25 mM Tris-maleate and Tris-oxalate, 5-5 mM ATP and NaN_3, 0.11 mM EGTA, 0.1 mM $^{45}CaCl_2$, pH 7.0, 37°C. Following incubation the coverslip was removed and the ^{45}Ca taken up by the SR determined in a liquid scintillation counter. The same cryostat sections were used for the quantitative determination of actomyosin ATPase activity by measuring the direct release of [^{32}P]P_i from [γ-^{32}P]ATP. In this procedure the liberated inorganic phosphate is trapped in sections as Ca-phosphate precipitate and quantification is carried out by radioactivity measurements. The assay medium contained 20 mM Na barbital, 5-5 mM $MgCl_2$ and NaN_3, 10 mM $CaCl_2$, 0.25 mM ouabain and 1.5 mM [γ-^{32}P]ATP. This histochemical actomyosin ATPase reaction was utilized also to distinguish between Type 1, Type 2A and 2B fibers in one step staining without any previous alkaline or acid preincubation (16).

Results

1) Ca^{2+}-uptake by SR, actomyosin ATPase and fiber type distribution in stimulated muscle. The effect of chronic stimulation showed the same pattern as that described by us earlier (2). The Ca^{2+}-uptake of the SR dropped considerably during the 5 weeks of stimulation reaching the values known to be present in the slow-twitch soleus muscle (Fig. 1). A similar decrease was noticeable in the actomyosin ATPase. It should be noted, however, that these results were obtained by the cryostat-thin-section method requiring a muscle sample of not more than 0.2 g. The advantage of this method is that the function of the sarcoplasmic reticulum can be evaluated in an intact muscle section without homogenization and differential centrifugation. Histochemistry (fiber typing, oxidative-glycolytic enzyme activity, etc.) and quantitative actomyosin ATPase can also be performed from the same block using serially cut sections.

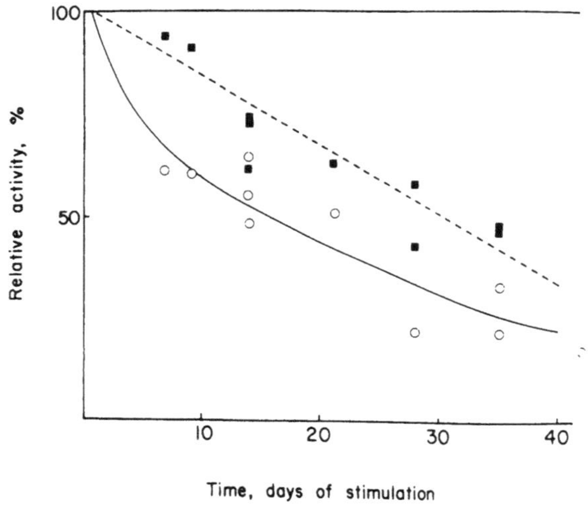

Fig. 1. Effect of chronic stimulation on the Ca^{2+}-uptake by sarcoplasmic reticulum ○ and actomyosin ATPase ■ of fast-twitch muscle. Normal values for unstimulated contralateral peroneus muscle. Ca^{2+}-uptake: 265±15 μmol Ca^{2+}/15 min/g protein; for actomyosin ATPase, 2.86±0.07 μmol P_i/min/g protein.

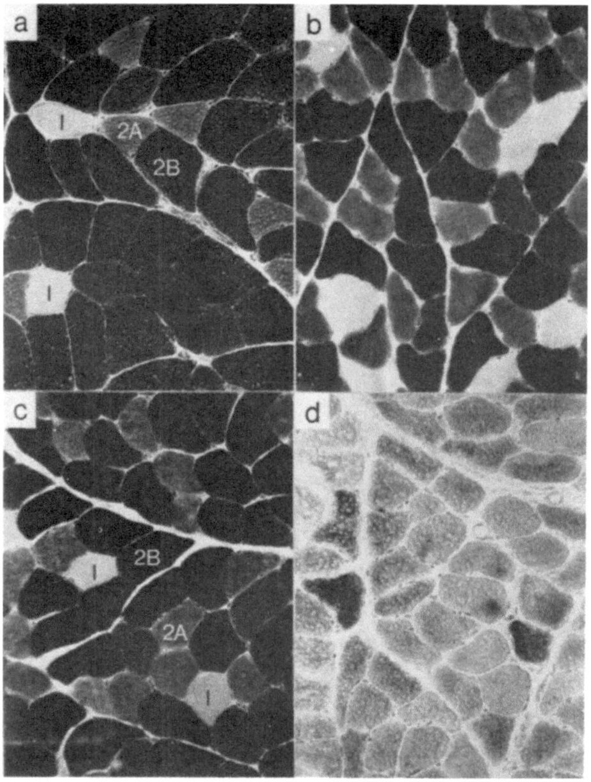

Fig. 2. Actomyosin ATPase staining of 9 and 35 days (b and d, respectively) stimulated peroneus muscle and their unstimulated counterparts (a and c, respectively).

Changes in the fiber type distribution during chronic stimulation for 9 days are illustrated in Fig. 2, showing sections of the stimulated left peroneus muscle (2b) and its unstimulated right counterpart stained for actomyosin ATPase (2a). In the control peroneus muscle (a) three types of fibers can be distinguished: light Type 1 fibers showing very negligible staining, dark Type 2B fibers and intermediate Type 2A fibers. The population of Type 1 and Type 2A fibers clearly increased after 9 days of stimulation (Fig. 2b). At this stage the total number of fibers in the peroneus muscle does not change but the average diameter of the individual fibers decreases. After 5 weeks of stimulation (Fig. 2d) the fast type fibers drastically decreased and almost all fibers showed the characteristics of Type 1 fibers. The contralateral unstimulated right peroneus showed the characteristics of the unstimulated control (Fig. 2a), but the number of Type 2A and B fibers increased, indicating that chronic stimulation of the left peroneus had some effect on the contralateral unstimulated muscle.

The actomyosin ATPase studies clearly indicate that the fast-twitch type actomyosin ATPase starts to transform to a slow-twitch type even as early as 9 days of chronic stimulation (Figs. 2a and b).

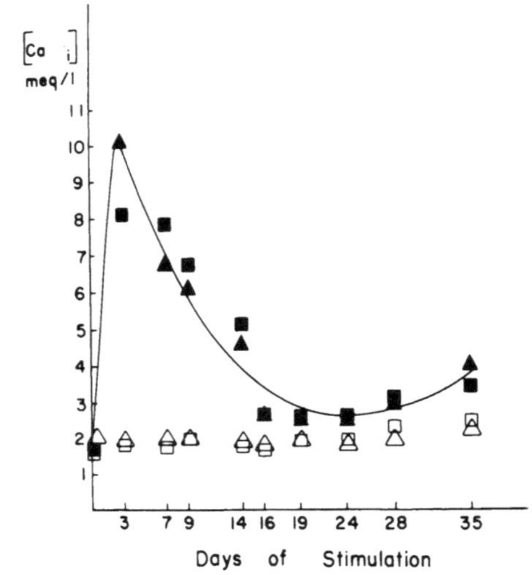

Fig.4. Effect of chronic low-frequency stimulation on the intracellular $[Ca^{2+}]$. For other details see legend to Fig. 1.

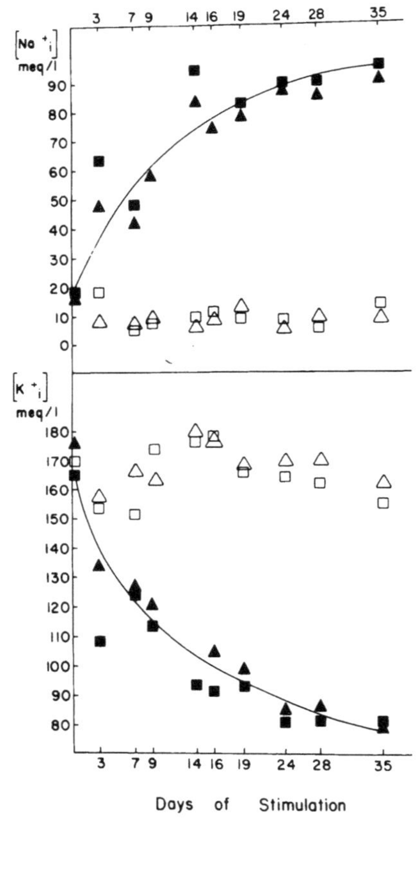

Fig. 3. Effect of chronic low-frequency stimulation on the intracellular $[Na^+]$ and $[K^+]$. Key: TA: △ , contralateral; ▲ , stimulated; EDL: □ , contralateral, ■ , stimulated.

2) Effect of chronic stimulation on the transmembrane distribution of sodium, potassium, calcium and magnesium in the stimulated muscles. Intracellular sodium content increases significantly in the stimulated muscle as a result of continuous stimulation, in good agreement with our earlier results (17). After 3 days of stimulation the intracellular sodium content increased from 8.2 and 13.7 to 47.6 and 63.9 mEq/l in the stimulated TA and EDL, respectively. Simultaneously the intracellular potassium content decreased from 156.3 and 163.5 to 115.6 and 111.5 mEq/l, respectively (Fig. 3). No significant change was present in the intracellular magnesium content but the intracellular calcium content after 3 days of stimulation increased by a factor of 4 to 5 in both muscles (Fig. 4). This increased value gradually decreased and after 2-3 weeks of stimulation showed the same results as those of contralateral muscles. This transitory increase in total intracellular Ca may be related to degenerative processes occuring during the early phase of transformation.

By continuation of chronic stimulation changes in the intracellular sodium and potassium content became even more pronounced and the maximal change developed after 5 weeks of stimulation.

The extracellular fluid compartment ($[^{35}S]Na_2SO_4$ space, not shown) slightly increased during stimulation; however, this change could not explain the large shift in sodium and potassium.

3) Effect of chronic stimulation on membrane potential of the TA muscle. Membrane potential measurements in normal (unstimulated) rabbit fast-twitch tibialis anterior and slow-twitch soleus muscles gave average values of −94 1.2 and −81.7 3.0 mV, respectively, indicating differences between these two types of muscles.

Fig. 5 illustrates membrane potential measurements carried out after 9-11 minutes following the interruption of stimulation. For comparison values calculated on the basis of intracellular potassium are also shown. These two sets of data indicate that even after 10 min following cessation of stimulation a repolarization takes place. This is shown more clearly in

Fig. 6, which illustrates several consecutive measurements in a muscle stimulated for 19 days. Note that the lowest values were always obtained in the first measurements and the repolarization was usually complete after 30 minutes. If these measured values were extrapolated to "zero" time the calculated value was always in good agreement with the result based on the actual concentration of intracellular K^+. The tendency of

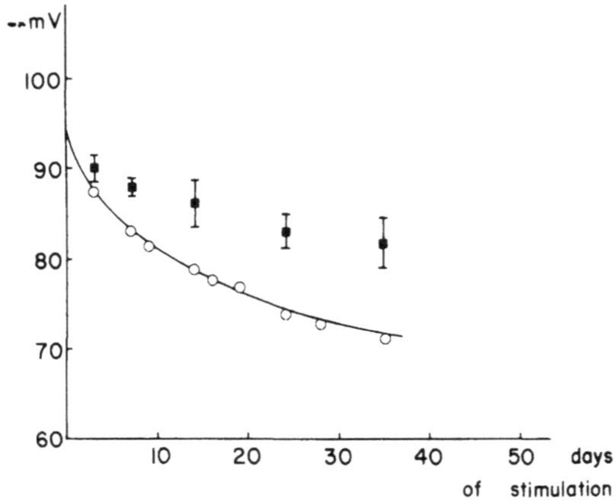

Fig. 5. Changes in measured (10 min after cessation of stimulation, ■) and calculated (on basis of K_o/K_i ratio, ○) membrane potential values on prolonged stimulation. Vertical brackets represent ±S.D., and are calculated using 8-10 membrane potential measurements. Abscissa indicates duration of chronic stimulation.

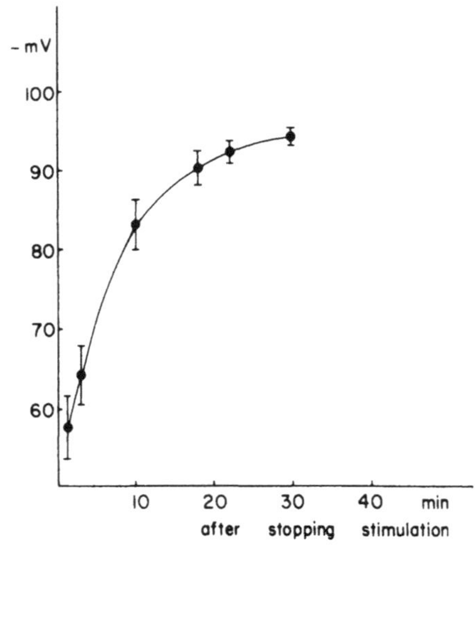

Fig. 6. Changes in measured membrane potential values after cessation of chronic stimulation (3 weeks of stimulation). Abscissa indicates time elapsed after cessation of stimulation.

repolarization also indicates that there is no membrane damage due to chronic stimulation, reconfirming our earlier data that the recovery process immediately takes place after stopping the low-frequency stimulation.

Discussion

Continuous low-frequency stimulation of a mammalian fast-twitch muscle at a frequency that imitates the impulse rate of a motoneuron supplying a slow-twitch muscle transforms the fast muscle into a slow type by a variety of physiological, biochemical and morphological criteria. The present experiments show extensive changes in the intracellular K^+/Na^+ content and the resting membrane potential values in chronically stimulated muscles. These changes induced by chronic stimulation are, however, readily restored within 30 min after the stimulation is stopped. This is in marked contrast to the reported biochemical changes where reversibility takes place only after a much longer period (7), e.g., in the case of the light chain pattern 6 weeks of recovery was necessary to obtain again the fast type light chain complement and to reach the level of myosin ATPase activity characteristic of fast-twitch muscle myosin. The slow reversibility of the changes in various myofibrillar proteins (myosin, tropomyosin, etc.) can be easily explained in terms of their turnover rate. It is also known that after a longer period of stimulation, e.g. 6-8 weeks, a fast muscle completely transformed to a slow one continues to show the characteristics of a slow muscle even after several days of rest. On the other hand, changes in the intracellular K^+/Na^+ content and the resting membrane potential values will be rapidly restored to their original values after the stimulation is stopped. The complete restoration of the original values also shows that no membrane damage was present in these chronically stimulated muscles.

It is well known that the intracellular potassium content and the resting membrane potential values of the slow-twitch muscles are lower than those of fast muscles (13) and in our experiments it should be noted that after cessation of stimulation the resting membrane potential values reaches the value known to represent fast muscle fibers, indicating a purely dynamic type of change during the stimulation time.

Summary

Rabbit muscles (tibialis anterior, extensor digitorum longus and peronei) were indirectly stimulated (8 Hz, continuously) for periods up to 5 weeks. Intracellular sodium, potassium, calcium and magnesium content were determined on excised muscles by atomic absorption spectroscopy and using $[^{35}S]Na_2SO_4$ administered i.v. prior to the excision of the muscles to correct for extracellular space. Membrane potential was measured with the use of intracellular microelectrodes. Transformation of muscles (fast → slow) was monitored by determination of changes in the Ca^{2+}-uptake by the SR and of actomyosin ATPase using sections cut in a cryostat. Both these parameters showed a complete transformation after 5 weeks of chronic stimulation. Intracellular K content decreased from 165 to 85 mEq/l and concomitantly the intracellular Na content increased from 15 to 90 mEq/l in the stimulated muscle. The membrane potential, measured immediately after cessation of stimulus, dropped from −94 mV to −50 mV. This change, however, was dynamic, since the original value was restored within 30 minutes after cessation of stimulation. A transitory increase in total intracellular Ca during the first 2 weeks of stimulation indicated some degenerative process during the early phase of transformation.

References

1. Salmons, S., Vrbova, G.: J. Physiol. (London) 201, 535-549 (1969).
2. Sreter, F.A., Romanul, F.C.A., Salmons, S., Gergely, J.: Exploratory Concepts in Muscular Dystrophy, ed. A. Milhorat, Excerpta Medica, Amsterdam, pp. 334-343, 1974.
3. Ramirez, B.U. and Pette, D.: FEBS Letters 49, 188-190 (1974).
4. Sreter, F.A., Salmons, S., Romanul, F., Gergely, J.: Nature (New Biol.) 241, 17-19 (1973).
5. Sreter, F.A., Elzinga, M., Mabuchi, K., Salmons, S., Luff, A.R.: FEBS Letters 57, 107-111 (1975).
6. Pette, D., Ramirez, B.U., Müller, W., Simon, R., Exner, G.V., Hildebrand, R.: Pflügers Arch. 361, 1-7 (1975).
7. Romanul, F.C.A., Sreter, F.A., Salmons, S., Gergely, J.: Exploratory Concepts in Muscular Dystrophy, ed. A. Milhorat, Excerpta Medica, Amsterdam, pp. 344-348, 1974.
8. Roy, R.K., Mabuchi, K., Sarkar, S., Mis, C., Sreter, F.A.: Biochem. Biophys. Res. Commun. 89, 181-187 (1979).
9. Rubinstein, N., Mabuchi, K., Pepe, F., Salmons, S., Gergely, J., Sreter, F.: J. Cell Biol. 79, 752-761 (1978).
10. Pette, D., Schnez, U.: FEBS Letters 83, 128-130 (1977).
11. Salmons, S.: J. Physiol. (London) 188, 13 (1967).
12. Haljamae, H., Jennische, E., Medegard, A.: Acta fysiol. scand. 101, 458-464 (1977).
13. Sreter, F.A., Woo, G.: Am. J. Physiol. 205, 1290-1294 (1963).
14. Ling, G., Gerard, R.W.: J. Cell. Comp. Physiol. 34, 383-396 (1949).
15. Mabuchi, K., Sreter, F.A.: Anal. Biochem. 86, 733-742 (1978).
16. Mabuchi, K., Sreter, F.A.: Muscle and Nerve, in press.
17. Sreter, F.A., Am. J. Physiol. 205, 1295-1298 (1963).

EFFECTS OF ELECTRICAL STIMULATION ON ACETYLCHOLINE RECEPTOR SYNTHESIS IN CULTURED MYOTUBES

Moshe A. Reis and Asher Shainberg
Department of Life Sciences, Bar-Ilan University, Ramat-Gan, Israel

An important example of muscle plasticity and one of its earliest manifestations is the process modulating the number of acetylcholine receptors (AChR) and their distribution on muscle cells. Embryonic muscle cells become highly sensitive to acetylcholine about the time mononucleated myoblasts fuse, to form myotubes. At this early stage AChR are distributed over the entire cell surface (1). Muscle innervation is accompanied by disappearance of many AChR and after synaptogenesis ACh sensitivity is restricted mainly to the neuromuscular junction. The process is reversible; deprivation of a matured muscle cell of its neuronal input causes the appearance of widespread extrajunctional chemosensitivity. Surgical denervation (2,3), blockage of axonal flow (4) or of axonal conduction (5,6), and neuromuscular transmission blockade by means of either pre- or post-synaptically-active neurotoxins (7,8), all bring about an increase of extrajunctional receptor levels. Reinnervation is followed by a gradual decrease of extrajunctional hypersensitivity, and AChR are once again confined to the end-plate area (9). Several aspects of these phenomena have been described, but the mechanism(s) underlying the regulation of AChR amount and distribution on muscle cells remain largely unknown. Any of the events in the sequence following acetylcholine release by the motoneuron could plausibly be a participant in the mechanism of AChR level regulation. These possible mediators include the release of neurotrophic factors, membrane depolarization, various ion fluxes, contractile activity, energy utilization or the formation of by-products, such as lactic acid, ADP, cyclic nucleotides, etc. Cessation of any of these activities after denervation could conceivably serve as a signal to trigger the changes of AChR levels.

In this chapter we wish to describe experiments in which AChR levels were either elevated or reduced by various treatments, with an emphasis on AChR level decrease caused by electrical stimulation (E.S.) of muscle cells grown in culture. A model is proposed in which intracellular calcium distribution plays a key role in the regulation of AChR synthesis. The suggested model might consistently explain many earlier reports concerning the action of various affectors of AChR level changes.

Embryonic muscle cells differentiated in vitro in the absence of nerve cells, are an excellent model system for the selective study of factors affecting AChR level regulation. The time course of AChR appearance in chick muscle cultures and receptor level alteration in response to different treatments is described in Fig. 1. Radioiodinated-α-bungarotoxin (^{125}I-α-BGT) binding shows that whereas tetrodotoxin (TTX) induces a >100% rise of AChR level, E.S. lowers it by 30-50%.

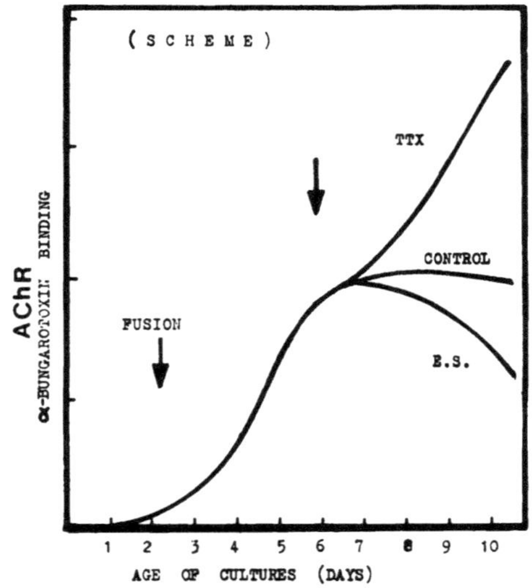

FIG. 1. Time course of AChR appearance and level changes in response to TTX or electrical stimulation. AChR levels were measured by counting bound radioactivity after exposing cultures to 6×10^{-8}M ^{125}I-α-BGT for 1 h followed by 6 washings of 10 min each. Spontaneously contracting 8-day-old chick muscle cultures were rendered paretic by addition of 0.5μg/ml TTX to the growth medium. Other cultures were stimulated through platinum electrodes immersed in the medium. 20V 2 msec square wave pulses were delivered 200 msec apart in 6-second trains, followed by a 54-second interval. Polarity of consecutive trains was alternated to prevent electrolyte polarization.

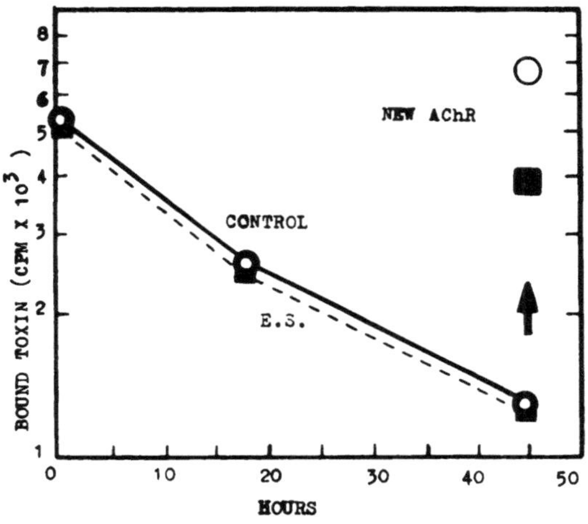

FIG. 2. Degradation and appearance of AChR during E.S. 8-day-old chick muscle cultures were first pulsed with ^{125}I-α-BGT. After washing away non-bound activity, the decrease of bound radioactivity was measured at various times during stimulation and compared with that of controls. Following termination of E.S., new AChR that appeared during the experiment were labeled by a second pulse of ^{125}I-α-BGT.

The amount of AChR in the cell membrane at any given time is determined by the relative rates of receptor synthesis, their incorporation into the plasma membrane and their degradation. Thus the reduction of the number of toxin binding sites on myotube membranes following E.S. may be achieved in several ways: a) an accelerated degradation of receptor molecules, b) a partial masking or inactivation of a steady-sized population, c) a reduced rate of receptor biosynthesis, and d) reduced incorporation into the cell membrane.

The possibility of accelerated degradation was examined by pulsing cultures with ^{125}I-α-BGT and then comparing the rate of decline of bound radioactivity on electrically-stimulated cultures with that of controls. Assuming this reflects the rate at which unlabeled AChR molecules undergo degradation (10), E.S. had no effect on the stability of AChR. Furthermore, labeling the same cultures with a second pulse of ^{125}I-α-BGT after the termination of E.S. revealed that this treatment diminished the population of new binding sites by about 40% as compared to controls (Fig. 2),

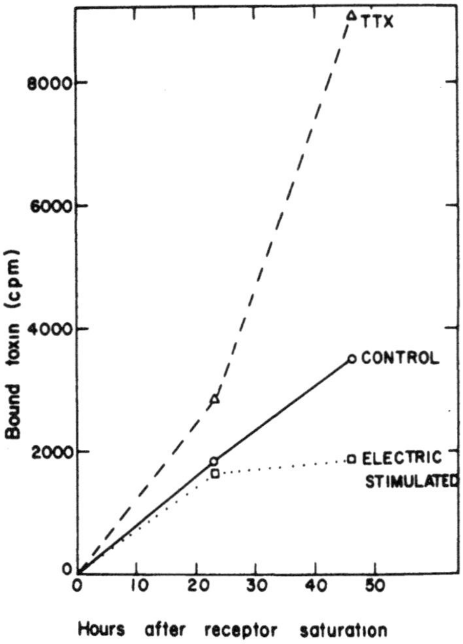

FIG. 3. Time course of AChR appearance during treatment with TTX or E.S. Preexisting receptors in 10-day-old chick muscle cultures were blocked with native α-BGT. New AChR that appeared in cultures grown in the presence of 0.5μg/ml TTX, or in cultures undergoing E.S. were measured by ^{125}I-α-BGT binding.

thus suggesting suppression of AChR biosynthesis.

The rates of appearance of new AChR in cultures pulsed with native (radio-inert) α-BGT before treatment with TTX or E.S. are shown in Fig. 3. Radiolabeled α-BGT binding after 48 h revealed that while TTX induced the appearance of ≈150% more AChR than in the controls, ≈50% less emerged in the electrically-stimulated cultures. This experiment confirms that level changes occur by modulation of the amounts of the newly-formed receptors appearing on the cell surface during the experiment. The extent of level lowering by E.S. is probably even more pronounced and the measured values should be regarded as minimal for two reasons: 1) 48 h after the first α-BGT pulse (>2 half lives) AChR levels are approaching their steady-state plateau; 2) a significant fraction (20-30%) of the new, unlabeled receptor population that emerged after the first α-BGT pulse appear at an

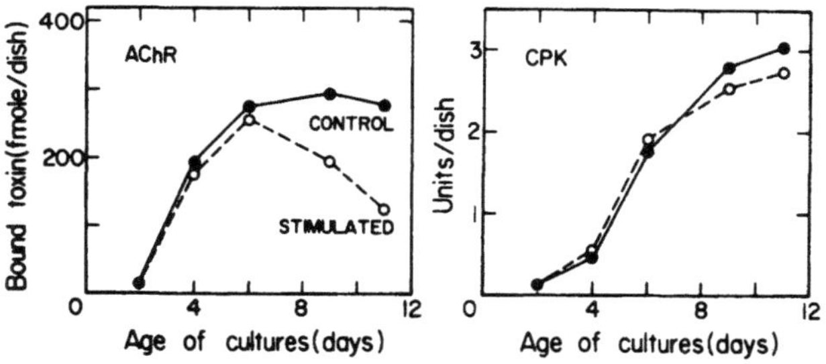

FIG. 4. The effect of E.S. on AChR and CPK in differentiating muscle cells. AChR and CPK were measured in chick muscle at the indicated times beginning on the second day in culture. CPK was assayed as described elsewhere (12).

accelerated rate following, and as a result of, the α-BGT binding reaction (11). Both phenomena are expected to partially obscure the extent of AChR level reduction.

To examine whether E.S. would inhibit the appearance of AChR during differentiation, muscle cells were stimulated beginning on the second day in culture (before fusion). No difference between electrically-stimulated cultures and controls was observed until the sixth day. From then on, E.S. selectively reduced AChR levels, whereas creatine phosphokinase (CPK) activity resembled that of the controls (Fig. 4). The onset of AChR level susceptibility to E.S. apparently coincides with the maturation of myotubes. Many of the cell properties emerging at this developmental stage could be responsible for this newly-acquired susceptibility. These include the beginning of contractile activity, establishment of membrane excitability, and the differentiation of some cellular organelles, e.g., the maturation of the sarcoplasmic reticulum (S.R.).

In an attempt to clarify whether mechanical activity or membrane depolarization played a role in AChR level regulation, cultures were stimulated in the presence of TTX. This toxin blocks action potential sodium channels (13), thus impairing membrane depolarization, and 0.5μg/ml in the growth medium suffices to abolish *spontaneous* contraction of cultured myotubes. However, under the conditions tested, the cells responded to E.S. and contracted vigorously, even in the presence of a 40-fold excess of TTX (20μg/ml). AChR levels measured following simultaneous treatment with 0.5μg/ml TTX and E.S. were about half way between those obtained by either of them alone (Fig. 5). We assume that E.S. can bring about myotube contraction in spite of the excess TTX present, by inducing intracellular Ca^{2+} mobilization. Thus, the fluctuating electrical field probably intervenes in the chain of events leading to contraction at a stage subsequent to membrane depolarization. Ca^{2+} may be transferred into the myofilamentous space from the transverse tubular system or released from an intracellular store such as the sarcoplasmic reticulum. Findings (14) that S.R. calcium is released by electrical fluctuations across this membrane strongly support the assumption that this process is of importance in enabling E.S.-induced muscle contraction, even when sarcolemmal excitability is impeded by TTX. Since AChR levels could be changed in the presence of TTX, it seems that plasma membrane depolarization *per se* is not an essential requirement for AChR level regulation.

At this stage the question arose whether mechanical activity (or its absence) played a mediatory role in the AChR regulation mechanism or whether contraction was merely concomitant with a more fundamental event, such as the redistribution of intracellular calcium ($[Ca^{2+}]_i$). The second possibility appears to be supported by findings (*) that the amount of AChR on cultured myotubes can be markedly increased (>200%) by growing them in Ca^{2+}-enriched growth medium (10mM instead of 1.8mM Ca^{2+}). Conversely, 5mM caffeine in the growth medium caused a significant decrease of AChR levels. It should be noted that both high extracellular Ca^{2+} ($[Ca^{2+}]_o$) and caffeine brought about these changes during continual contractile activity. Caffeine, which is known to affect the release of calcium from the S.R. into the myofilamentous space (15), brought about an AChR level reduction similar to the one caused by E.S. (Fig. 5). The $[Ca^{2+}]_i$ redistribution that follows caffeine treatment did not change the t ½ of AChR and a specific slowdown of receptor synthesis rate(s) is thus assumed.

Additional instances of the dependence of AChR levels on $[Ca^{2+}]_i$-distribution were demonstrated in experiments in which the amount of AChR was

(*) Birnbaum, Reis, Shainberg, submitted.

increased following treatment with different agents capable of increasing $[Ca^{2+}]_i$ or its accumulation in the S.R. As mentioned above, muscle cultures grown for 48 h in medium containing 10mM Ca^{2+}, bound about twice as much ^{125}I-α-BGT than controls grown in the regular (1.8mM Ca^{2+}) medium. The high external Ca^{2+} concentration is assumed to have increased diffusion of this ion into the cell followed by an increased sequestration into calcium-storing organelles in an attempt to maintain the steady state 10^{-7}M Ca^{2+} in the myoplasm. The S.R., mitochondria, calcium-binding proteins and the cell nucleus are all probable participants in this enhanced calcium sequestration. Marked increases of AChR amounts were observed after cultures grown in the presence of 1.8mM Ca^{2+} were treated with 20μg/ml ionophore 21873, 10mM $[Mg^{2+}]_o$, or sodium dantrolene (DaNa) (Fig. 5). Presumably at this concentration the ionophore promotes Ca^{2+} permeation (16), thus increasing $[Ca^{2+}]_i$ followed by acceleration of AChR synthesis. (At higher concentrations this compound is reported to raise $[Ca^{2+}]_i$ to levels that damage the cell through filament degradation (17).) Mg^{2+} ions are reported to stimulate Ca^{2+} uptake into the S.R. (18), and DaNa is believed to inhibit its release from this organelle (16,19). These experiments suggest that redistribution of intracellular calcium (by means of various agents acting in different ways) alters AChR levels. These levels appear to be proportional to S.R. calcium content. Experiments in which AChR-inducing agents were applied simultaneously, such as 10mM Ca^{2+} + TTX, or 10mM Ca^{2+} + 10mM Mg^{2+}, also support this assumption. Since the effects were not additive (Fig. 5), it is suggested that the different affectors all acted via a common 'saturable' stage. In other experiments, simultaneous treatments with converse effects, e.g., E.S. + TTX, E.S. + DaNa, E.S. + high $[Ca^{2+}]_o$ or caffeine + TTX, all resulted in intermediate AChR levels, between those obtained by either of the treatments alone (Fig. 5). This might be explained by contradicting actions on a common regulatory mediator.

The model suggesting that intracellular calcium distribution is a correlate and possible regulatory mediator of AChR synthesis rates may account for many phenomena. For instance, the dependence of E.S. effectiveness (as an affector of AChR level changes) on the temporal pattern of stimuli administration might be explained in terms of the time available for Ca^{2+}

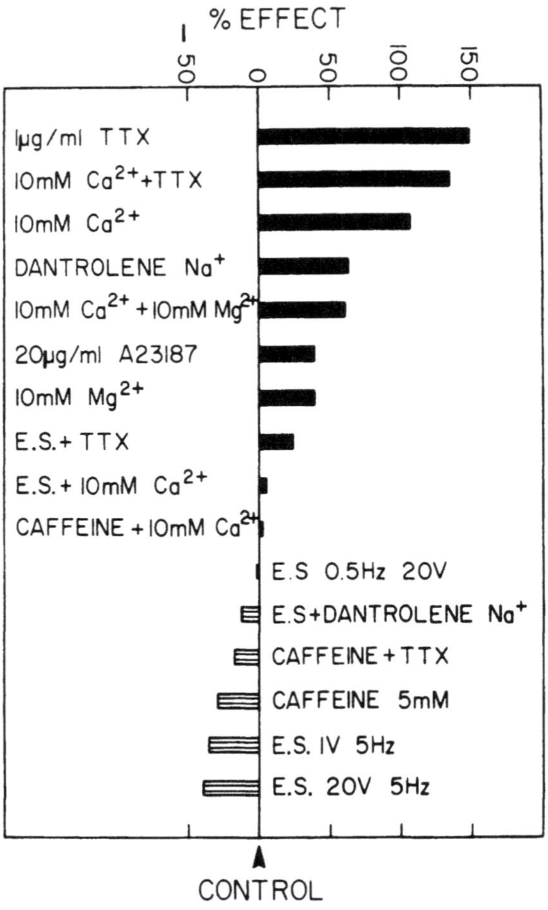

FIG. 5. Effects of various treatments on AChR levels (as compared to controls, Effect = 0). 8-10-day-old chick muscle cultures were given the indicated treatments for 48 h. AChR levels were measured by ^{125}I-α-BGT binding. Dantrolene Na^{+} was applied as a saturated solution in growth medium at 37°C. 0.5 Hz E.S. was continuous. 5 Hz stimulations were delivered as described in the legend to Fig. 1.

reuptake during intervals between consecutive pulses. Infrequent stimuli (*ca.* 0.5Hz) allow replenishment of calcium stores and, averaged over long periods, no deviation from normal calcium distribution evolves. Under these conditions no change of AChR level occurs (Fig. 5). Stimulation at higher frequency (*ca.* 5Hz) might affect a prolonged $[Ca^{2+}]_i$ redistribution by the combination of two additive effects. First, the interval between pulses arriving in rapid succession may not suffice to allow complete re-uptake of Ca^{2+} released by preceding stimuli and a progressive depletion of calcium stores ensues. Furthermore, residual-free Ca^{2+} is available for the facilitation of calcium release by subsequent pulses by an effect termed "Ca^{2+}-induced calcium release", thus increasing the efficiency of the process and exacerbating the deficit. The degree of AChR level changes following E.S. might be a function of the extent $[Ca^{2+}]_i$ distribution deviates from normal which, in turn, is determined by the pattern of the stimulation regimen. At even higher frequencies (*ca.* 50Hz) electrical stimulation damaged muscle cells. Since high doses of caffeine or ionophore 23187 had similar damaging effects on these cells, we assume that E.S., like caffeine or the ionophore, might cause an excessive free Ca^{2+} build-up, reaching concentrations that result in myofilament degradation (17).

The dependence of AChR level changes on the amount and temporal distribution of E.S. was pointed out by Lomo and Westgaard (20). They showed that longer intervals between trains of E.S. delayed and attenuated the decline of AChR levels. Furthermore, they found that the effectiveness of a given number of stimuli depended on the frequency at which they were delivered. The rate of AChR appearance after termination of E.S. indicated modulation of "fast" versus "slow" muscle characteristics by the pattern of stimulatory input. Earlier findings by Albuquerque et al. (21) that AChR levels depended on the type of innervation ("fast" or "slow") may also be an example of receptor level regulation by variation of stimulation pattern, possibly mediated by intracellular calcium.

Muscle disuse imposed by mechanical immobilization only caused a transient alteration of ACh sensitivity (22). This may be due to the continual stimulation by the intact motoneurons and restoration of the normal $[Ca^{2+}]_i$

balance. Powell and Friedman (23) showed another example of AChR biosynthesis rate dependence on excitatory signals (action potentials and possible calcium release) in the absence of contractile activity. Cultured myotubes from mice suffering from "muscular dysgenesis" generate normal action potentials, but lack the ability to contract. AChR levels on these myotubes were no higher than those on continuously-contracting control cultures from healthy mice.

The assumption that redistributed $[Ca^{2+}]_i$ is a regulatory mediator, rather than other events such as membrane depolarization, may both explain and moreover find strong support in experiments with sub-liminal E.S. In these experiments, low voltage E.S. (*ca.* 1 V) insufficient to elicit conraction, was highly effective in bringing about a decline of AChR levels (24) (Fig. 6). The fluctuating electrical field (unaided by action potentials, which would lead to contraction) is assumed to release stored calcium in decrements each smaller than the amount required to reach the contraction threshold concentration. If administered in an appropriate pattern, the sub-liminal E.S. could plausibly gradually drain calcium

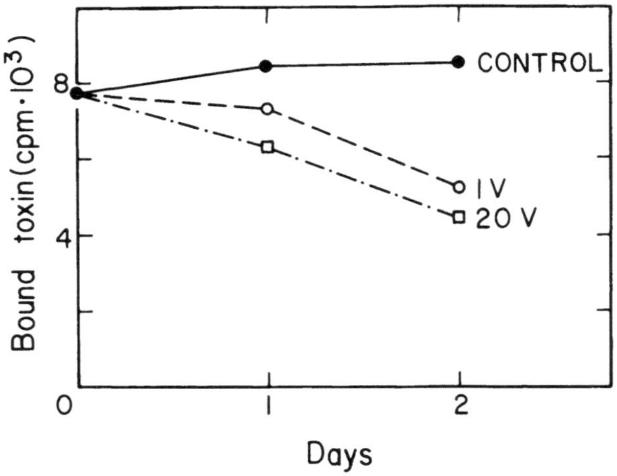

FIG. 6. Effects of E.S. at different voltages on AChR levels. 8-day-old chick muscle cultures were divided into 3 groups. One group was stimulated, as described in the legend to Fig. 1, at 20 V, which caused vigorous contraction. The second group was stimulated by 1 V pulses delivered in the same pattern as group 1 (contraction threshold was about 4 V). The third group served as controls.

stores, and although accompanied by no overt electromechanical activity, a marked drop of AChR levels ensues.

Gruner et al. (24) were first to demonstrate this phenomenon in their experiments with rat extensor digitorium longus muscles. The spread of post-denervation supersensitivity was restricted by sub-liminal E.S. They concluded that potential fluctuations of the membrane were responsible for inhibiting the appearance of extrajunctional ACh sensitivity. We have shown a similar effect of sub-liminal (1 V) E.S. on cultured myotubes. The amounts of both TTX-induced AChR and of receptors in untreated controls were specifically reduced (30-40%) after 48 h of this treatment (Figs.5,6).

A substantial body of data concerning AChR level suppression by E.S. has been gathered during the last decade. This and related subjects have been reviewed recently by Fambrough (25). A consistent scheme emerges when all those findings are considered together. About 18 h after E.S. of muscle cells is commenced, a reduction of extrajunctional AChR levels becomes noticeable. Concomitant contractile activity is not essential to the process (24) (Fig. 6), whereas the temporal pattern of stimulation is of prime importance (20) (Fig. 5). Specific suppression or acceleration (derepression ?) of *de novo* synthesis has been shown (10,26,27,28) to be the mode of AChR level regulation. The ability of electrical fluctuation to elicit calcium release from the S.R. (14), and the high correlation we found between $[Ca^{2+}]_i$ distribution and AChR levels, suggest that suppression of AChR synthesis rates by E.S. is mediated by $[Ca^{2+}]_i$ redistribution.

Calcium ions possess many qualities which make them highly suitable for regulatory messenger duties. Small variations in the absolute amount of this minor constituent of cytosolic ion composition result in large relative changes. This is especially so in the case of myoplasmic Ca^{2+}, maintained at a steady-state equilibrium concentration of 10^{-7}M. The high charge to surface ratio of this divalent ion sets very high energy barriers against permeation of phospholipid bilayers (29); thus, free diffusion across membranes is minimized. Transmembranal flow of Ca^{2+} is largely restricted to selective and controlled passages. Variations of muscular function (and incoming stimuli) could be expected to reflect on calcium distribution in a cell endowed with such an intricate system of calcium-manipulating organelles. Ca^{2+} ions could be envisaged to exert their effect on biosynthetic processes directly by forming divalent bridges that enable the close approach of polyanionic macromolecules otherwise kept apart by charge-charge repulsion forces. Indirectly, Ca^{2+} might affect the process through reaction with some Ca-activated mediator which, in turn, modulates the rate of receptor synthesis.

Insufficiency of currently available methods for the determination of $[Ca^{2+}]$ in various subcellular compartments, limits experimentation to a rather qualitative study of phenomena correlated to $[Ca^{2+}]$ in organelles for which selective pharmacological tools are known. It is for this reason that most of the data gathered to date concerns the S.R. However, the possibility that changes of the S.R. calcium content merely reflect calcium release or accumulation in some other relevant organelle (e.g., the nucleus), must not be overlooked.

It is interesting to examine whether the proposed model is applicable to other instances in which AChR levels are changed,and if denervation, hibernation and pharmacologically-induced paralysis, all cause redistribution of intracellular calcium.

Acknowledgments

We are grateful to Mrs. Ahuva Isak and Mrs. Tova Zinman for excellent technical assistance and to Mrs. Bluma Lederhendler for her help in preparing this manuscript. This work was supported by grants from the Muscular Dystrophy Association of America, and the U.S.-Israel Binational Science Foundation, Jerusalem, Israel.

References

1. Shainberg, A., Brik, H.: FEBS Lett. 88, 327-331 (1978).
2. Axelsson, J., Thesleff, S.: J. Physiol. 147, 178-193 (1959).
3. Miledi, R.: J. Physiol. 151, 1-23 (1960).
4. Albuquerque, E.X., Warnick, J.E., Sansone, F.M., Onur, R.: Ann. N.Y. Acad. Sci. 228, 224-243 (1974).

5. Pestronk, A., Drachman, D.B., Griffin, J.W.: Nature 260, 352-353 (1976).

6. Lavoie, P., Collier, A.B., Tenenhouse, A.: Nature 260, 349-350 (1976).

7. Thesleff, S.: J. Physiol. (Lond.) 151, 598-607 (1960).

8. Chang, C.C., Chuang, S.T., Huang, M.C.: J. Physiol. (Lond.) 250, 161-173 (1975).

9. Bennett, M.R., Pettigrew, A.G., Taylor, R.S.: J. Physiol. (Lond.) 230, 331-357 (1973).

10. Linden, D.C., Fambrough, D.M.: Neuroscience 4, 527-538 (1979).

11. Chiung Chang, C., Jai Su, M., Hsien Tung, L.: J. Physiol. 268, 449-465 (1977).

12. Shainberg, A., Yagil, G., Yaffe, D.: Develop. Biol. 25, 1-29 (1971).

13. Narahashi, T., Daguchi, T., Urakawa, N., Ohkubo, Y.: Am. J. Physiol. 198, 934-938 (1960).

14. Podolsky, R.J., Stephenson, E.W.: In "Human Pathogenesis of Muscular Dystrophies," editor L.P. Rowland, Excerpta Medica, Amsterdam, Oxford, pp. 626-632 (1977).

15. Weber, A., Hertz, R.: J. Gen. Physiol. 52, 750-759 (1968).

16. Desmedt, J.D., Hainaut, K.: Biochm. Pharmacol. 28, 957-964 (1979).

17. Duncan, C.J.: Experientia 34, 1531-1535 (1978).

18. Stephenson, E.W., Podolsky, R.J.: J. Gen. Physiol. 69, 1-16 (1977).

19. Ellis, K.O., Carpenter, J.F.: Arch. Phys. Med. Rehabil. 55, 362-369 (1974).

20. Lømo, T., Westgaard, R.H.: J. Physiol. (Lond.) 252, 603-626 (1975).

21. Albuquerque, E.X., McIsaac, R.J.: Exp. Neurol. 26, 183-202 (1970).

22. Fishbach, G.D., Robbins, N.: J. Neurophysiol. 34, 562-569 (1971).

23. Powell, J.H., Friedman, B.A.: J. Cell Biol. 75, 321a (1977).

24. Gruener, R., Baumbach, N., Coffee, D.: Nature 248, 68-69 (1974).

25. Fambrough, D.M.: Physiol. Rev. 59, 165-227 (1979).

26. Fambrough, D.M., Devreotes, P.N., Card, P.J.: In "Synapses," editors G.A. Cottrell and P.N.R. Usherwood, Glasgow, Blackie, pp. 202-236 (1977).

27. Reiness, C.G., Hall, Z.W.: Nature 268, 655-657 (1977).

28. Shainberg, A., Burstein, M.: Nature 264, 368-369 (1976).

29. Urry, D.W.: Ann. N.Y. Acad. Sci. 307, 3-27 (1978).

Section VI. Mechanisms of Hypertrophy and Atrophy

THE REGULATION OF PROTEIN TURNOVER BY ENDOCRINE AND NUTRITIONAL FACTORS

A.L. Goldberg
Department of Physiology, Harvard Medical School
Boston, Massachusetts, U.S.A.

After birth, the size of a skeletal muscle, i.e., its total protein content, is not fixed. Dramatic growth of skeletal and cardiac muscle occurs during postnatal development in response to a variety of hormonal signals. Even in the adult, reduced food intake, alterations in endocrine status (e.g. diabetes, Cushing's Syndrome, etc.), increased use or disuse will induce marked changes in muscle size. Whether a muscle undergoes growth or atrophy depends upon the net balance between the rates of protein synthesis and the rates of protein degradation (1-4) in this tissue. Protein balance in muscle is also of fundamental physiological importance in overall energy homeostasis, since this tissue contains most of the body's protein reserves. Thus, in fasting the mobilization of amino acids stored in muscle protein helps provide the organism with essential gluconeogenic precursors (5-8), and protein breakdown in skeletal muscle should be viewed as an initial rate-limiting step in gluconeogenesis.

This lecture will review recent findings from our laboratory concerning the influence of various hormones and food intake on protein breakdown in muscle. A major factor limiting progress in this area has been various technical problems involved in the measuring degradative rates of proteins (9). Such *in vivo* measurements are subject to a number of potential artefacts and have frequently engendered appreciable controversy. In recent years our laboratory has employed simple *in vitro* techniques to analyze rates of protein degradation and synthesis under carefully controlled conditions. These studies employed certain thin rat or mouse muscles that can be maintained *in vitro* for many hours in a good physiological state, the diaphragm, red soleus muscle, or pale extensor digitorum longus (10). The advantages of these methods have been discussed in detail elsewhere (9-12).

Rates of protein synthesis in the incubated mucles are determined by measuring rates of incorporation of [^{14}C]tyrosine or phenylalanine into muscle protein, after correcting for intracellular specific activity. Rates of protein degradation are measured by following the net release of tyrosine from cell protein (11). Tyrosine was chosen for such studies because sensitive fluorometric assays for tyrosine exist and because this amino acid is neither synthesized nor catabolized by muscle. Consequently, its production by isolated muscles must reflect net protein breakdown. (This consideration is an important one that has been often overlooked; for example, many investigators have incorrectly equated protein breakdown with the release from muscle of alanine and glutamine which are synthesized *de novo* in large amounts by muscle (7)). Absolute rates of protein degradation can be determined by measuring net tyrosine release in the presence of cycloheximide, an inhibitor of protein synthesis. Alternatively, absolute rates of protein catabolism can be calculated from simultaneous measurements of net protein balance and rates of protein synthesis (13).

Factors Affecting Protein Balance in Isolated Muscles

A number of factors known to be important for the growth of skeletal muscle has been shown by the use of such techniques to alter the overall rates of protein degradation as well as protein synthesis in isolated muscles (Table 1). Most results were obtained with rat skeletal muscles (diaphragm, soleus, or extensor digitorum longus) incubated *in vitro*, although similar findings seem also to apply to cardiac muscle. When skeletal muscles are incubated in unsupplemented Krebs-Ringer bicarbonate buffer, protein catabolism exceeds synthesis by several-fold (11,12). Because of the net protein breakdown, these

TABLE 1 Factors Affecting Protein Turnover in Isolated Rat Muscles

	Protein Synthesis	Protein Degradation
Glucose	—	↓
Insulin	↑	↓
Plasma Amino Acids	↑	↓
Leucine	↑	↓
Other Amino Acids	↑	↓
Passive Tension	—	↓
Repetitive Stimulation	—	↓

muscles show a net release of amino acids into the medium and thus resemble skeletal muscle during a short fast in vivo. Initially, we used these preparations to examine what factors normally present in vivo may be added to the incubation medium to improve overall nitrogen balance (Table 1).

Insulin is probably the most important physiological factor regulating overall protein balance in skeletal muscle (5). The rise in insulin levels after meals stimulates the net uptake of amino acids by skeletal muscle and their incorporation into protein. The fall in insulin in the fasting organism signals a net release of amino acids from muscle (5,6,8,14). In muscle and other tissues, insulin not only stimulates protein synthesis but also inhibits protein degradation (1,3,4,11,12, 15) apparently by an effect on lysosomal function (14). These two hormonal actions have complementary effects in producing a net accumulation of tissue protein. These effects on protein synthesis and degradation can be demonstrated in the absence of exogenous glucose or amino acids and thus are not secondary to insulin's well-known ability to stimulate nutrient transport (11).

Glucose by itself also inhibits protein degradation but does not affect overall protein synthesis (11). The effects of insulin and glucose in general, appear additive in improving overall protein balance in muscle. By contrast, free fatty acids or ketone bodies at physiological levels do not affect protein synthesis or degradation in skeletal muscle (11), even though these compounds are an excellent source of energy for this tissue.

The addition of plasma amino acids to the incubation medium retards protein breakdown further and stimulates protein synthesis, but most of the amino acids found in plasma are not necessary for such effects. Studies in our laboratory (1,11) and by others in skeletal (17,18) or cardiac muscle (3,4) have shown that leucine by itself promotes protein synthesis and inhibits protein breakdown (Table 2). No other plasma amino acid can influence protein turnover in these preparations.

TABLE 2 Stimulation of Protein Synthesis and Inhibition of Degradation in Rat Muscle by Leucine

	% Change In	
	Protein Degradation	Protein Synthesis
+ Plasma amino acids	-26 ± 2.9*	+23 ± 7.5*
Leucine	-25 ± 10.6*	+25 ± 8.3*
Isoleucine	- 8 ± 5.0	+14 ± 8.1
+ Other plasma amino acids	- 3 ± 3.1	+ 2 ± 6.6

*$p<.01$

Protein synthesis and degradation were measured in rat diaphragms incubated in the presence of Krebs-Ringer Bicarbonate buffer containing 10 mM glucose. Leucine or isoleucine were added at 0.5 mM and mixtures of plasma amino acids at concentrations found normally in rat plasma (24). Data are taken from Fulks et al.(11) and Tischler and Goldberg (submitted for publication).

Leucine also is the only amino acid that can provide significant energy for muscle (7,16,19) and in fasting, leucine, like fatty acids or ketone bodies, serves as an alternative substrate to glucose (7,16,19).

Our recent experiments (13) indicate that leucine's interesting inhibitory effects on protein balance in muscle are related to the rapid degradation of the branched-chain amino acids in skeletal and cardiac muscle (7). Thus if leucine transamination is blocked, leucine no longer inhibits protein breakdown. Furthermore, leucine does not affect rates of protein catabolism or synthesis in liver in which this amino acid is not rapidly degraded (14,15). Such effects on protein turnover in muscle are obtained with physiological concentrations of leucine (0.1-0.5 mM). We do not, however, know whether the normal variations in amino acid supply are of importance in regulating protein balance _in vivo_. However, intravenous administration of branched chain amino acids or leucine by itself, appear to be of therapeutic value in reducing the marked protein catabolism occurring in surgical patients (1).

All these factors together still do not prevent net protein catabolism in the isolated muscles. Unlike muscles _in vivo_, these isolated tissues do not perform work, and it has long been recognized that disuse or denervation leads to muscle atrophy. Earlier studies _in vivo_ and _in vitro_ indicate that this process is primarily a consequence of an enhancement of protein breakdown (20,21). Several years ago, Dr. Jeanne Li and I obtained evidence that _repeated contractions in vitro_ induced by electrical stimulation can retard protein breakdown by 15-25% in the isolated soleus or diaphragm (22,23). Under these same conditions, protein synthesis did not change in the muscles. This ability of contractile activity to retard protein breakdown probably is an important factor contributing to work-induced hypertrophy of muscle as well as denervation or disuse atrophy (1,22).

Using these _in vitro_ approaches, we also found that _passive stretch_ can mimic the effects of active contractions in decreasing protein catabolism (22,23). In fact, if the incubated muscles are provided with insulin, glucose, and amino acids and are maintained under some degree of tension, they are in a state of neutral or positive nitrogen balance. These findings with passive tension of muscle are of particular practical interest because of the evidence from several laboratories that passive tension of muscle may by itself induce compensatory hypertrophy (22). In addition, these effects may account for certain beneficial results obtained by physical therapy.

The exact mechanism by which contractile activity or muscle stretch influence protein balance is still unknown. Recent findings concerning these important phenomena will be discussed in the subsequent lectures by my former associates, Dr. Goldspink and Dr. Etlinger.

Pituitary and Thyroid Hormones

It has long been known that hypophysectomy of young animals prevents normal body growth and protein accumulation in skeletal muscles (24,25). Dr. Griffin has used these approaches to define further the influence of pituitary hormones on protein synthesis and degradation in muscle. As anticipated from earlier findings (24,25), protein synthesis was reduced following removal of the pituitary. In addition, however, rates of protein breakdown were much slower in muscles from hypophysectomized rats (Table 3). The fall in protein synthesis was an expected result of the lack of growth hormone (24,25), but the reduction in overall protein catabolism in muscle represents a previously undescribed effect of hypophysectomy (26-28).

In control rats growing normally, overall protein synthesis must exceed protein catabolism. The loss of the pituitary appeared to cause a greater decrease in protein synthesis than in protein degradation, such that these two processes were balanced, and no growth occurred. When the hypophysectomized animals were treated with

TABLE 3 Effects of Hypophysectomy and Thyroidectomy on Protein Turnover in Rat Skeletal Muscle

Physiological State	Change In Protein Synthesis	Protein Degradation	Net Effect
Hypophysectomized	-54%	-30%	Growth ceases
Thyroidectomized	-54%	-34%	Growth ceases

In experiment 1, protein synthesis or degradation was compared in solei from normal rats and hypophysectomized ones of similar size (70-80 g), four weeks after operation. During this period, the hypophysectomized rats did not grow. In experiment 2, a similar study was carried out on diaphragms of normal and thyroidectomized animals. Following the operation, the thyroidectomized rats grew, although at a decreased rate. To insure the absence of residual hormones, the experiment was performed four weeks after thyroidectomy, when growth had ceased (30). Muscles were incubated in Krebs-Ringer Bicarbonate buffer containing glucose and amino acids at 5x normal plasma levels.

growth hormone, overall body growth, including that of muscle, was reinitiated, and rates of protein synthesis in leg muscles increased (26,27). Treatment of hypophysectomized rats with growth hormones, however, did not alter rates of protein degradation in accord with our earlier findings (29).

The decrease in muscle protein breakdown after hypophysectomy appeared to result from the lack of thyroid hormones (22,26-28). Thyroidectomy like hypophysectomy also decreased protein breakdown (29) (Table 3), and treatment of the rats with either triiodothyronine (T_3) or thyroxine (T_4) increased the rates of protein degradation back to levels seen in control animals (Table 4). Upon treatment with normal or much higher doses of T_3, proteolysis increased in both dark and pale muscles after two days (27,30,31). A similar lag time is characteristic of many other responses induced by these hormones (32,33). Treatment of rats with T_3 or T_4 also stimulated protein synthesis in muscle (27,31), as has been shown previously for liver (33). In fact the thyroid hormones appeared only slightly less effective than growth hormone in stimulating this process in muscles. However, treatment of hypophysectomized rats with growth hormone was appreciably more effective in inducing growth than was T_3 or T_4 treatment, probably because the pituitary enhances protein synthesis without also increasing protein breakdown.

One puzzling and clinically important aspect of thyroid hormone action is that these agents have opposite effects on body growth when present in very high or normal levels. A minimal level of T_3 is essential for normal growth of muscle and other tissues, and lack of normal levels of this hormone leads to a failure of normal growth and a characteristic type of dwarfism, (i.e., cretinism). In contrast, excessive amounts of the thyroid hormones (e.g., in thyrotoxicosis) lead to a general loss of body weight and severe muscle wasting. Dr. Griffin succeeded in clarifying how high and low doses of thyroid hormones induce such different effects on muscle size (26,27, Table 4). Physiological levels of T_3 or T_4 must be inducing growth of

TABLE 4 Effect of Different Thyroxin Dosages on Protein Turnover in Skeletal Muscle

	Anabolic	Catabolic	% Change	p
	Protein Synthesis (nmoles TYR incorporated/mg muscle/2 hrs)			
Soleus	0.337 ± 0.039*	0.281 ± 0.010	NS	<0.1
EDL	0.204 ± 0.017*	0.165 ± 0.015	NS	<0.1
	Protein Degradation (nmoles TYR released/mg muscle/2 hrs)			
Soleus	0.229 ± 0.004*	0.391 ± 0.040	+ 71	<0.01
EDL	0.191 ± 0.009*	0.308 ± 0.020	+ 61	<0.001

*Significantly different from untreated hypophysectomized rats

Protein synthesis and degradation were measured from hypophysectomized rats that received an anabolic dose of T_4 (2.0 μg per day) for nine days or catabolic dose of T_4 (200 μg per day). During this period the groups receiving the anabolic dose gained 10.0 ± 0.9 g and the groups receiving the catabolic dose lost 6.5 ± 1.8 g (n=5 p<0.01). EDL stands for Extensor Digitorum Longus.

hypothyroid rats because they stimulate synthesis more than they promote breakdown. To determine the mechanism responsible for the muscle atrophy in hyperthyroidism, he compared the effects on muscle protein synthesis and breakdown of catabolic doses (200μg/rat/day), which cause the hypophysectomized animals to lose significant weight, and anabolic doses (2μg/rat/day) which cause a gain in weight and muscle mass. Rates of protein synthesis did not differ in the two groups. However, the rates of protein degradation were 50-75% greater in the group receiving the large dose of T_4 (Table 4). Thus, increased protein catabolism must be primarily responsible for the severe muscle wasting in hyperthyroidism.

To characterize further these effects of thyroid hormone on protein turnover in muscle, we used $^{14}CO_2$ or [^{3}H]leucine to label muscle proteins in the intact rats and then followed the disappearance of radioactive proteins from the muscle. These studies revealed an increase in breakdown of both soluble and myofibrillar proteins. In addition, measurements of 3-methylhistidine excretion in the urine in collaboration with Drs. Pamela Brown, Vernon Young and Hamish Munro at MIT also confirmed that T_3 and T_4 caused a marked increase in degradation of contractile proteins. We are presently attempting to define more precisely whether all cell proteins or only certain components turn over more rapidly under these conditions.

These isotopic experiments also indicated that treatment with T_3 stimulated proteolysis in liver, which like muscle underwent a net loss of weight. Under the same conditions, however, protein degradation did not rise in the brain and kidney, whose size was not affected by the hormones or in the heart, which actually hyperthrophied in response to thyroid treatment. Thus, the acceleration of protein breakdown in muscle and liver correlates with and can account for the marked atrophy in these tissues in hyperthyroidism. The basis for these selective effects on proteolysis in liver and muscle is unclear, especially since other tissues contain receptors for these hormones.

Since the effects of thyroid hormones on protein balance in muscle are long lasting, and have a long lag time, other factors (e.g., insulin or the level of contractile activity), probably have more important influences on these processes over shorter time periods. Nevertheless, changes in thyroid status probably influence the rates of protein breakdown under certain physiological conditions. For example, it is now well-established that in long term fasting of humans, protein breakdown in muscle seems to decrease, as evidenced by the reduction of total nitrogen excretion and amino acid release from muscles (6,8,). We have in fact observed similar effects in prolonged fasting of large rats (DeMartino & Goldberg, unpublished observations and 34). This inhibition of proteolysis in long term fasting seems to be an important adaptation to preserve body protein. Under such conditions, the need for continued gluconeogenesis from muscle protein is reduced, since the brain and other tissues of the fasting organism are adapted for use of ketone bodies (6,8).

Related unpublished studies (DeMartino & Goldberg, in preparation) argue strongly that decreased thyroid function plays an important role in this adaptation to prolonged food deprivation. In the fasted organism, the levels of T_3 are known to fall, and a less active hormone isomer, "reverse T_3", is produced in the liver from circulating T_4. In addition, the levels of thyroid hormone receptors decrease (35). Such effects may help explain the fall in muscle protein breakdown in prolonged fasting and would help preserve body musculature. Thus, we would like to suggest that decreased thyroid function may be of particular advantage to the organism during long term starvation in part by decreasing fuel needs and in part by retarding the loss of essential body proteins. In fact we have obtained evidence recently that the ability of rats to withstand food deprivation will increase dramatically if they are made hypothyroid (34 ; DeMartino & Goldberg, in preparation). These marked differences in weight loss and survival on fasting can not be explained by the rather small differences in oxyen consumption. Whatever their precise explanation, these findings indicate an important role for thyroid hormones in the regulation of protein metabolism in fasting and a coupling between the factors controlling muscle proteolysis and overall energy homeostasis.

Thyroid Hormones and Lysosomal Content

One obvious mechanism by which a hormone might stimulate protein degradation would be to increase the activity of intracellular proteases. The increase in proteolysis occurring in liver in the absence of insulin or amino acids is associated with a marked enlargement of lysosomes and autophagic vacuoles within the hepatocytes (14), but with no change in the content of the proteolytic enzymes. We have obtained evidence that the enhancement of protein degradation induced by thyroid hormones is associated with an increase in the content of the various lysosomal proteases (36). Treatment of hypothyroid animals with either catabolic or anabolic levels of T_3 or T_4 caused a 2-3 fold increase in the activity of cathepsin B and D (36, Table 5). This effect is not evident until two days after administration of the hormone, and thus correlates with the acceleration of protein breakdown. The regulation of lysosomal protease levels appears to be a physiological action of thyroid hormones, since after thyroidectomy (Table 3), the levels of these enzymes is approximately 50% of those seen in muscles and livers of control rats (36). These novel effects of the thyroid hormones are restricted to liver and skeletal muscle, the only tissues in which thyroid hormones stimulate overall proteolysis and tissue atrophy. Because of these correlations, we would like to suggest that the induction of these lysosomal enzymes is responsible for the acceleration of protein breakdown and for the resulting loss of liver and muscle weight in hyperthyroidism. However, direct proof that these changes in enzyme content are responsible for increased protein breakdown is lacking.

In addition to the proteases, the thyroid hormones influence the levels of a variety of other lysosomal enzymes including acid phosphatase, β-galactosidase, β-glucuronidase, DNAse and RNAse (36).

TABLE 5 Effect of Thyroidectomy and Treatment with Triiodothyronine on Lysosomal Enzyme Activities in Skeletal Muscle

Enzyme	Experiment I		Experiment II	
	Normal	Thyroidectomized	Hypophysectomized	T_3-Treated
	(Units/mg Protein x 10^3)			
Cathepsin B	1.49	0.42	0.98	1.97
Cathepsin D	3.44	1.06	2.47	3.60
Acid Phosphatase	1.80	1.20	2.63	4.38
N-Acetylglucosaminidase	2.90	1.10	0.49	0.95

Data taken from findings of DeMartino and Goldberg (39). Hypophysectomized animals were treated with 2.0 g/day of triiodothyronine for nine days prior to assay.

Treatment of hypophysectomized rats with these hormones increased the level of all eleven lysosomal hydrolases tested by 2-3 fold, and these changes occur synchronously. Thus, thyroid hormones either induced the production of more lysosomes or enhanced the enzyme content per lysosome. The ability of these hormones to influence the levels of apparently all lysosomal enzymes may also account for various other unexplained effects of thyroid disease (37). For example, two characteristic features of human hypothyroidism are the accumulation of mucopolysaccharides (i.e. myxedema) and the increase in cholesterol esters in serum. The reduced content of the rate-limiting lysosomal hydrolases that are rate-limiting in the catabolism of hyaluronic acid and cholesterol esters (37) may well contribute to the accumulation of these substances in hypothyroid patients.

These findings provide further evidence for a role of lysosomes in overall protein degradation in muscle and liver. As mentioned above, there is appreciable evidence that changes in protein breakdown induced by insulin, glucagon, or amino acids correlates with changes in lysosomal properties, such as lysosomal size or fragility (1,4,14). The ability of the thyroid hormones to affect lysosomal content appears to be a distinct hormonal mechanism which probably influences the maximal rate at which protein and other cell constituents may turn over in muscle and liver.

Glucocorticoids and the Response to Fasting

Like thyroid hormones, the glucocorticoids fall dramatically following hypophysectomy and can have large effects on muscle size. For example, large doses of the adrenal steroids, as occurs in Cushing's Syndrome, lead to a marked wasting of muscle and a loss of strength. In addition, the ability of the glucocorticoids at physiological levels to promote the net breakdown of muscle protein appears to be an important factor in the regulation of blood glucose in fasting. The classical studies of Long, Katzen and Fry (38) and others (39-41) demonstrated that adrenalectomized animals can not maintain blood glucose in fasting. When fasted, these animals show a large reduction in urea production, which indicates a failure in gluconeogenesis from endogenous protein reserves. The glucocorticoids thus appear to have important actions in facilitating the mobilization of amino acids from tissue proteins. Presumably such effects complement their other "permissive actions" in enhancing gluconeogenesis in liver and kidney (42,43).

The glucocorticoids also act in several ways to retard growth and to promote the release of amino acids from muscle. They decrease DNA (44) and protein synthesis in muscle (20, 45-49) and reduce amino acid uptake into this tissue (45). Although generally assumed to promote proteolysis, their effects on protein breakdown in muscle remain controversial. Early pulse-chase studies from this laboratory using intact rats suggested that cortisol stimulates protein breakdown in rat muscle, (20) in accord with earlier suggestions (50). More recently, Tomas *et al*. (48) found that administration of large amounts of corticosterone to rats enhanced excretion of 3-methylhistidine. However, several other groups have failed to observe a significant stimulation of proteolysis in muscle with either *in vitro* or *in vivo* approaches (47,51).

Because of these contradictory results Dr. Marc Tischler in my laboratory undertook to investigate systematically the mechanism by which glucocorticoids promote mobilization of amino acid from muscle proteins in fasting. Our studies employed muscles from young, growing rats (80-100g) in which food-deprivation normally leads to a marked decrease in protein synthesis and an enhancement of protein degradation within one-two days (Table 6). These changes in protein synthesis and degradation together are responsible for increased release of gluconeogenic precursors from muscle in fasting. When analogous experiments were performed with adrenalectomized animals, fasting produced little, if any, increase in the net release of amino

TABLE 6 Effect of Fasting on Protein Breakdown in Diaphragms from Normal and Adrenalectomized Rats

Animals	Fed	Fasted	% Change
	Net Protein Breakdown (nmol tyr/mg tissue/2 hrs)		
Experiment I			
Normal	.32 ± .01	.39 ± .02	23*
Adrenalectomized	.30 ± .01	.31 ± .01	--
+ Cortisol Treatment	.31 ± .02	.39 ± .02	26*
	Protein Breakdown		
Experiment II			
Normal	.42 ± .010	.52 ± .018	+24*
Adrenalectomized	.43 ± .009	.39 ± .014	- 9*
% Difference	---	-25*	

*p<.05

Net Protein Breakdown equals net protein balance (i.e. the difference between rates of protein synthesis and degradation). Diaphragms from normal and adrenalectomized rats in the fed and fasted (24 hr) state were incubated in Krebs-Ringer Bicarbonate buffer containing glucose. Cortisol or saline was injected for 2 days to fed (5 mg/100 g body wt) or fasted (1 mg/100 g body wt) animals 5 days after adrenalectomy. Other animals received saline injections and were sacrificed 24 hr after the last injection. Protein breakdown was the absolute rate of tyrosine release from muscle protein.

acids from the muscle (Table 6). In muscles from normal and adrenalectomized rats, protein synthesis was lower in the fasted than in the fed state (Table 6). However, protein breakdown in the muscles of the adrenalectomized rats did not increase in fasting as it did in muscles of normal rats (Table 6). On the contrary, fasting actually lowered rates of protein degradation in muscles of the adrenalectomized rats, for reasons that are not clear. In any case, net protein balance in the muscles of adrenalectomized rats did not change during fasting, and there was no enhancement of amino acid release from this tissue.

These simple experiments demonstrate a clear requirement for the glucocorticoids for the acceleration of protein breakdown during fasting. However, the reduction of protein synthesis after food deprivation (12) does not seem to require the adrenal hormones. Presumably this latter adaptation can be signalled solely by the fall in plasma levels of insulin (11). It is noteworthy that in the fed state the muscles of normal and adrenalectomized rats showed similar rates of protein degradation. In fasting, however, the lack of the adrenal resulted in a much lower rate of proteolysis in muscle (Table 6). When the fasting adrenalectomized animals were treated for two days with cortisol or corticosterone, rates of proteolysis increased in muscles from these rats, such that this process did not differ from that in muscles of fasting normal animals (53). In addition, after treatment with cortisol, the animals withstood fasting in a similar fashion to normal rats (Manuscripts in preparation).

Of particular interest was the finding that this acceleration of protein breakdown is also essential for the loss of muscle weight characteristic of the fasted state. Unless the adrenalectomized animals had received glucocorticoids, they showed little or no loss of muscle weight upon fasting. In contrast, the muscle content of normal rats decreased by 15-25% within two days or food deprivation. This requirement for glucocorticoids for the acceleration of muscle protein breakdown must be an important factor in the requirement of these steroids for increased gluconeogenesis and urea production on fasting.

In analogous experiments, when fed rats, either normal or adrenalectomized, were treated with cortisol or corticosterone, no increase in muscle protein breakdown was observed. On the contrary, high doses of cortisol decreased the rate of muscle proteolysis. In other words, these steroids had opposite effects in the fed and food-deprived animal. These findings suggest that some additional hormone dependent upon nutritional status, probably insulin can influence the muscle's response to the adrenal steroids. In fact Griffin and Wildenthal (54) also reported opposite effects of glucocorticoids on protein breakdown in heart muscle depending upon the presence of insulin. The biochemical explanation for the different effects of glucocorticoids in fed and fasted animals remains an interesting topic for future research. It is tempting to suggest that these contradictory effects are of importance to the organism in stress in assuring that the release of steroids does not cause an inappropriate breakdown of tissue protein when the organism is on a nutritionally adequate diet.

In related studies, we have demonstrated that glucocorticoids also regulate amino acid metabolism in skeletal muscle in a manner that complements their affects on proteolysis in fasting. Skeletal muscle synthesizes de novo and releases large amounts of glutamine and this process increases during fasting (7,53,55). Earlier studies by Chang and Goldberg (16) have demonstrated that most of this glutamine is synthesized from five amino acids generated by the net breakdown of protein in fasting. Specifically, aspartate, asparagine, isoleucine and valine are converted via the citric acid cycle into glutamate which is subsequently aminated into glutamine. Protein breakdown and the subsequent conversion of these amino acids into glutamine are related early steps in gluconeogenesis from muscle protein. Dr. Tischler has recently shown that muscles from fasting adrenalectomized rats have a reduced capacity to synthesize and release alanine and glutamine. Administration of glucocorticoids stimulates glutamine synthesis under the same conditions where it promotes net protein breakdown. This enhancement of glutamine production seems to depend upon induction of glutamine synthetase (Tischler & Goldberg, in preparation). It has long been known that cortisol can induce this enzyme in various tissues (56-59). Inhibitors of RNA synthesis can block this induction, and thereby also block the enhancement of glutamine production in muscle (Manuscript in preparation). Furthermore, these inhibitors also abolish the increase in net proteolysis. Such data is a further indication that the regulation of protein and amino acid metabolism are closely interrelated through physiologically important mechanisms that have not yet been elucidated.

SUMMARY

A variety of factors known to be important for muscle growth in vivo have been shown to retard rates of protein breakdown in rat skeletal muscles including insulin, glucose, leucine, and contractile activity. In addition overall rates of protein breakdown in muscle, like rates of protein synthesis are subject to precise regulation by pituitary hormones. Muscles from hypophysectomized rats show lower rates of protein synthesis and breakdown than those from normal controls. The lack of growth hormone is primarily responsible for the lower rates of protein synthesis, but this hormone does not affect protein catabolism. The reduced protein breakdown in muscle after hypophysectomy or thyroidectomy results from the lack of thyroid hormones. Treatment with T_3 or T_4 stimulates protein breakdown as well as synthesis in muscle after a lag time of two days. Under these conditions thyroid hormones also increase the content of lysosomal proteases and other acid hydrolases in muscle and liver, and this effect may be responsible for the acceleration of protein breakdown. This stimulation of proteolysis accounts for the decrease in weight of muscle and liver during hyperthyroidism.

During fasting, protein breakdown in muscle rises and protein synthesis falls to help provide the organism with amino acids for gluconeogenesis. In muscles of adrenalectomized rats, protein breakdown does not increase and may even decrease upon fasting. In these animals, muscles show little or no loss of weight after food deprivation unless treated with glucocorticoids. Injection of cortisol into the fasted adrenalectomized rats increased protein breakdown but had no such effects in the fed animals. Thus the net protein catabolism in muscle during fasting coincides with a stimulation of gluconeogenesis in liver; these complementary adaptations both require glucocorticoids.

ACKNOWLEDGEMENTS

This lecture has reviewed research of three former postdoctoral fellows, Dr. George Griffin, Dr. George DeMartino, and Dr. Marc Tischler, whose enthusiastic efforts are primarily responsible for any progress we have made. I am grateful to Ms. Robin Levine for her expert assistance in the preparation of this manuscript and wish to thank the National Institute of Neurological Diseases, and Stroke, the Muscular Dystrophy Association, and the National Institute for Heart, Lung, and Blood for their generous support of our research.

REFERENCES

1. Goldberg, A.L.: Diabetes 28, 18-24 (1979).
2. Goldberg, A.L., St. John, A.C.: Ann. Rev. Biochem. 45, 747-804 (1976).
3. Rannels, D.E., Kao, R., Morgan, H.E.: J. Biol. Chem. 250, 1694-1701 (1975).
4. Rannels, D.E., McKee, E.E., Morgan, H.E.: In Biochemical Actions of Hormones, Vol. IV, Litwack, G., ed., pp. 135-195, Academic Press, New York (1977).
5. Cahill, G.F., Aoki, T.T., Marliss, E.B.: In Handbook of Physiology-Endocrinology, Vol. 1, Greep, R.O., Astwood, E.B., eds., pp. 563-577, Am. Physiol. Soc., Washington (1972).

6. Felig, P.: Ann. Rev. Biochem. 44, 933-954 (1975).
7. Goldberg, A.L., Chang, T.W.: Fed. Proc. 37, 2301-2307 (1978).
8. Ruderman, N.B.: Ann. Rev. Med. 26, 245-258 (1975).
9. Goldberg, A.L., Dice, J.F.: Ann. Rev. Biochem. 43, 835-869 (1974).
10. Goldberg, A.L., Martel, S.M., Kushmerick, M.J.: Methods in Enzymol. XXXIX, 82-93 (1975).
11. Fulks, R., Li, J.B., Goldberg, A.L.: J. Biol. Chem. 250, 290-298 (1975).
12. Li, J.B., Goldberg, A.L.: Am. J. Physiol. 231, 441-448 (1976).
13. Tischler, M.E., Goldberg, A.L.: Submitted for publication (1979).
14. Mortimore, G.E., Neely, A.N.: In Intracellular Protein Turnover, Schimke, R.T., Katunuma, N., eds., pp. 265-279, Academic Press, New York (1975).
15. Mortimore, G.E., Mondon, C.F.: J. Biol. Chem. 245, 2375-2383 (1970).
16. Chang, T.W., Goldberg, A.L.: J. Biol. Chem. 253, 3685-3695 (1978).
17. Buse, M.G., Weigand, D.A.: Biochim. Biophys. Acta 475, 81-89 (1977).
18. Li, J.B., Jefferson, L.S.: Biochim. Biophys. Acta 544, 351-359 (1978).
19. Chang, T.W., Goldberg, A.L.: J. Biol. Chem. 253, 3696-3701 (1978).
20. Goldberg, A.L.: J. Biol. Chem. 244, 3223-3229 (1969).
21. Goldspink, D.F.: J. Physiol. 264, 283-296 (1977).
22. Goldberg, A.L., Etlinger, J.D., Goldspink, D.F., Jablecki, C.: Med. Sci. Sports 7, 248-261 (1975).
23. Goldberg, A.L., Jablecki, C.M., Li, J.B.: Ann. N.Y. Acad. Sci. 228, 190-201 (1974).
24. Kostyo, J.L., Isaksson, D.: In Review of Physiology: Reproductive Physiology II, Vol. 13, Greep, R.O., ed., pp. 255-274, University Press, Baltimore (1977).
25. Talwar, G.P., Pandian, M.R., Kumar, N., Hanjan, S.N.S., Saxena, R.K., Krishnary, R., Gupta, S.L.: Rec. Prog. Hormone Res. 31, 141-153 (1975).
26. Griffin, G.E., Goldberg, A.L.: J. Physiol. Comm. 54P-55P (1977).
27. Griffin, G.E., Goldberg, A.L.: J. Biol. Chem. (1980, In Press).
28. Flaim, K.E., Li, J.B., Jefferson, L.S.: Am. J. Physiol. 235, E231-E236 (1978).
29. Goldberg, A.L.: J. Biol. Chem. 244, 3217-3222 (1969).

30. Goldberg, A.L., Griffin, G.E., Dice, J.F.: In Pathogenesis of Human Muscular Dystrophies, Rowland, L.P., ed., pp. 376-385, Excerpta Medica, Amsterdam (1977).

31. Flaim, K.E., Li, J.B., Jefferson, L.S.: Am. J. Physiol. 234, E38-E43 (1978).

32. Baxter, J.D.: Pharmacol. Theor. Part D 2, 605-654 (1976).

33. Tata, J.R.: In Handbook of Physiology, Greep, R.O., Astwood, E.B., eds., pp. 469-478, Am. Physiol. Soc., Washington (1977).

34. Goldberg, A.L., DeMartino, G.N., Chang, T.W.: FEBS Symp. 42, 347-

35. Schussler, G.C., Orlando, J.: Science 199, 686-687 (1978).

36. DeMartino, G.N., Goldberg, A.L.: Proc. Nat. Acad. Sci. USA 75, 1369-1373 (1978).

37. DeMartino, G.N., Goldberg, A.L.: Submitted for publication.

38. Long, C.N.H., Katzin, B., Fry, E.G.: Endocrinology 26, 309-314

39. Engel, F.L.: Rec. Prog. Hormone Res. 6, 277-313 (1951).

40. Smith, O.K., Long, C.N.H.: Endocrinology 80, 561-566 (1967).

41. Steele, R.: In Handbook of Physiology-Endocrinology, Vol. VI, Greep, R.O., Astwood, E.B., eds., pp. 135-168, Am. Physiol. Soc., Washington (1975).

42. Exton, J.H., Mallette, L.E., Jefferson, L.S., Wong, E.H.A., Friedmann, N., Miller, T.B., Jr., Park, C.R.: Rec. Prog. Hormone Res. 26, 411-461 (1970).

43. Friedmann, N., Exton, J.H., Park, C.R.: Biochem. Biophys. Res. Commun. 29, 113 (1967).

44. Goldberg, A.L., Goldspink, D.F.: J. Physiol. 228, 310-317 (1975).

45. Kostyo, J.L., Richmond, A.F.: Endocrinology 79, 531-540 (1966).

46. Rannels, S.R., Rannels, D.E., Pegg, A.E., Jefferson, L.S.: Am. J. Physiol. 235, E134-E139 (1978).

47. Shoji, S., Pennington, R.J.T.: Mol. Cell Endocrinol. 6, 159-169 (1977).

48. Tomas, F.M., Munro, H.N., Young, V.R.: Biochem. J. 178, 139-146 (1979).

49. Wool, I.G., Weinshelbaum, E.I.: Am. J. Physiol. 197, 1089-1092 (1959).

50. Hoberman, M.D., Yale, J.: Biol. Med. 22, 341 (1950).

51. Millward, D.J., Garlick, P.J., Nnanyelugo, D.O., Waterlow, J.C.: Biochem. J. 156, 185-188 (1976).

52. Rannels, D.E., Pegg, A.E., Rannels, S.R., Jefferson, L.S.: Am. J. Physiol. 235, E126-E133 (1978).

53. Tischler, M.E.: Fed. Proc. 38, 283 (1979).

54. Griffin, D.E., Wildenthal, K.: Am. J. Physiol. 234, E306-E313 (1978).

55. Tischler, M.E., Chang, T.W., Goldberg, A.L.: J. Biol. Chem., in press.

56. Barnes, P.R., Youngberg, D., Kitos, P.A.: J. Cell Physiol. 77, 135-144 (1971).

57. Kulka, R.G., Cohen, H.: J. Biol. Chem. 248, 6738-6743 (1973).

58. Tiemeier, D.C., Milman, G.: J. Biol. Chem. 247, 5722-5727 (1972).

59. Tischler, M.E., Goldberg, A.L.: Submitted for publication (1980).

IN VITRO SKELETAL MUSCLE HYPERTROPHY AND Na PUMP ACTIVITY

Herman H. Vandenburgh and Seymour Kaufman
Laboratory of Neurochemistry, National Institute of Mental Health,
Bethesda, Maryland 20205

Introduction

The biological function of skeletal muscle is movement - either of the organism or of objects relative to the organism. Selective advantage would be high to an organism if it were capable of adapting the size and biochemical characteristics of its muscle tissue to the types of movement required of it. Thus has evolved a tissue which is able to enlarge when it is required to move heavy objects and which can increase its oxidative capacity when it is required to perform highly repetitive movements. The nervous system obviously plays a critical role, since it ultimately regulates the types of movement a muscle cell will undergo. Denervation normally leads to muscle atrophy, but this process can be slowed by immobilization of the muscle in a stretched position (1). In some situations, denervation can lead to an initial hypertrophy of the muscle tissue and increased rhythmic stretch appears to be involved (2). Compensatory hypertrophy of skeletal muscle following tenotomy of synergistic muscles also appears to be due to increased tension via stretch on the hypertrophying muscle (3). Mechanical stretch of skeletal muscle has been known for many years to stimulate its metabolic activity (4,5) but the biochemical mechanism by which stretch stimulates muscle growth and metabolism is poorly understood. We have developed an *in vitro* model system for mechanically stimulating muscle cell growth in cells which have never been innervated. With this system, we have been studying a variety of stretch-induced biochemical changes that appear to be related to the development of hypertrophy. Once the ability of stretch to alter muscle cell growth is known, the role played by the nervous system can be better defined.

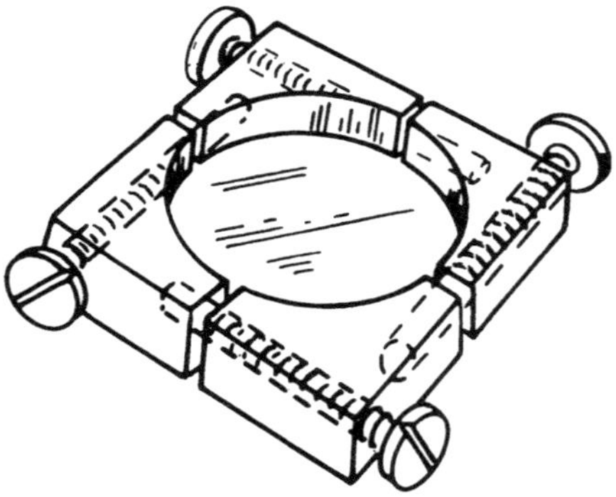

Figure 1. Diagramatic representation of the frame used to mechanically stretch muscle cells in culture.

In Vitro Skeletal Muscle Hypertrophy

Myogenic cells were isolated from 12 day old chicken embryos and grown in the well of a stretching frame (Figure 1) as described previously (6). The myoblasts began fusion to form multinucleated myotubes around 24 hours after plating, and by 72 hours, when the stretch experiments were initiated, well-developed, contractile myotubes were present. The muscle cells, attached to a collagen-coated elastic membrane in the central well of the stretching frame, could be mechanically stretched 5-20% by turning the calibrated screws. Longitudinal stretching of the myotubes was confirmed by examination with an inverted phase microscope. Several biochemical alterations, seen in the in vivo tenotomy model for compensatory hypertrophy, were studied with the in vitro system (compensatory hypertrophy of a skeletal muscle occurs in the tenotomy model when the tendons of synergistic muscles are cut and the remaining intact muscle is subjected to increased work (7)). These changes included: (a) increased long-term accumulation of the amino acid analog, α-aminoisobutyric acid (8); (b) increased incorporation of radioactive amino acids into total cellular protein (9,10) and into myosin heavy chains; and (c) increased accumulation of total cellular protein (9,10) and myosin heavy chains. The results of a number of experiments are summarized in Table I. The changes seen when cultured muscle cells are stretched 10.8% continuously for 18 hours are similar to those which occur in the tenotomy model. For example, in the tissue culture system, total cellular protein and content of myosin heavy chains increased 8.7 percent and 14.2 percent, respectively, after 18 hours of continuous stretch. In the in vivo tenotomy model, total muscle and myofibrillar nitrogen increased 16 percent and 9.5 percent, respectively, 4 days after the tenotomy operation (9). On the other hand, α-aminoisobutyric acid accumulation was increased to a smaller extent during 18 hours in the in vitro tissue culture system (30%) than during 15 hours in the in vivo tenotomy system (80%) (8). This difference may be due to the fact that the tenotomy model involves repeated stretch and relaxation while the in vitro system does not. It is known that the accumulation of α-aminoisobutyric acid is greater when organ-cultured muscle is stretched and electrically stimulated than when it is only stretched (11). We are currently developing an apparatus for repeated stretch and relaxation of cultured muscle cells to examine this possibility.

Table I. Effects of a Continuous 18 Hour 10.8% Stretch on Biochemical Changes in Muscle Cells

Biochemical Process	% Change	p^*
Accumulation of $[^3H]$α aminoisobutyric acid	+ 30%	< 0.01
Intracellular $[^{14}C]$radioactivity soluble in TCA	+ 5.8%	< 0.05
Incorporation of $[^{14}C]$radioactivity insoluble in TCA	+ 14.2%	< 0.02
Total protein	+ 8.7%	< 0.02
Incorporation of $[^{14}C]$radioactivity into myosin heavy chain	+ 16.3%	< 0.05
Myosin heavy chain	+ 14.7%	< 0.02

Seventy-two hour muscle cultures were stretched 10.8% and maintained in a stretched position for 18 hours in the presence of 5 μCi/ml$[^3H]$ AIB and 2 μCi /ml $[^{14}C]$leucine. Results are expressed as the mean percent change of 4 separate experiments with 18-20 cultures per experiment. TCA, trichloroacetic acid.
*Student's t-test for paired values.

In vivo skeletal muscle and muscle cell cultures contain a substantial number of non-myofiber cells (20-30% of total tissue nuclei in vivo (12) and a similar number in 72 hour muscle cultures). In the in vivo tenotomy model, DNA content increased 69% 4 days after tenotomy (9), predominantly due to proliferating interstitial fibroblasts (13). In vitro, stretch-induced muscle growth caused an increase in DNA content of 6.7% after 18 hours of continuous stretch (6). Since it has been found in the tenotomy model that certain of the biochemical alterations during stimulated growth (e.g., increased RNA synthesis) occurs predominantly in these fibroblast cells, rather than in the myofibers (14), it will be important to determine which of the stretch-induced biochemical changes in the tissue cultured cells are occurring in the myofibers and which are occurring in the fibroblasts. Fibroblast-free cultures can be used to examine this question. Quantitatively, the stretch-induced increase in content of myosin heavy chains of the cultured cells makes it unlikely that this increase is occurring solely in fibroblasts.

Mechanical Stretch and Protein Degradation

The accumulation of proteins in many cells, including skeletal muscle, is regulated by both their rates of synthesis and their rates of degradation (reviewed in 15). Denervation of skeletal muscle leads to muscle atrophy by decreased protein synthesis and increased protein degradation (16). Protein degradation is decreased in the tenotomy model for compensatory hypertrophy (17) and in stretched, organ-cultured adult skeletal muscle (18,19). This decrease in protein degradation in the tenotomy model compliments the increase in protein synthesis (9), leading to enhanced muscle growth. To determine the effect of stretch on protein degradation in the cultured muscle cells, two methods were used.

A. Intermittant Perfusion Technique

The technique developed by Bradley, Hayflick and Schimke (20) for measuring the rate of protein degradation in tissue cultured cells was adapted for skeletal muscle cells (21). Briefly, the cells were prelabeled for 20-24 hours with [^{14}C] phenylalanine and rinsed five times with basal medium (Eagles) containing 1 mM

Table II. Effect of Stretch on Protein Degradation in Muscle Cells

	A. ^{14}C labeled Protein Half-life (hours)	B. ^{3}H labeled Protein Half-life (hours)
Control	36.2 ± 2.4	31.4 ± 3.6
10.8% Stretch	48.9 ± 2.3*	40.5 ± 2.6**

48 Hour muscle cultures were labeled for 20-24 hours with 10 μCi/ml[2-^{3}H] phenylalanine and 2 μCi/ml[^{14}C]phenylalanine. Cultures were then rinsed five times with basal medium (Eagles) containing 1 mM phenylalanine, stretched 10.8% and the release of trichloroacetic acid soluble radioactivity followed for 10 hours into medium containing 5% horse serum, 5% embryo extract and 1 mM phenylalanine. Protein half-lives ($T_{1/2}$) were calculated from the equation $T_{1/2} = \frac{\ln 2}{k}$ where k = radioactive protein fraction degraded per hour. Half-lives are the mean ± S.E. of three cultures per group. Student's t-test for unpaired values: $^{*}p < .02$; $^{**}p < .05$.

excess unlabeled phenylalanine. Protein degradation was then measured by following at various times the release of trichloroacetic acid-soluble radioactivity into the degradation medium containing 1 mM phenylalanine. Control experiments with muscle cultures showed that [^{14}C] phenylalanine released by protein degradation effluxed rapidly from the cells into the medium, and that 1 mM excess phenylalanine in the medium inhibited reutilization of the radioactive amino acid. Phenylalanine was used as the amino acid for these studies since it is neither synthesized nor extensively degraded by skeletal muscle cells in culture. In some experiments leucine was used but we found that excess leucine in the medium can inhibit protein degradation under certain conditions (21).

Table II (column A) shows the results of a typical experiment performed in complete medium (5% serum, 5% embryo extract). The decrease in radioactivity remaining in cellular protein (% radioactivity remaining in protein = total initial radioactivity incorporated - trichloroacetic acid soluble radioactivity released into the medium ÷ total initial radioactivity incorporated) from two to ten hours after a 10.8%

stretch was used to calculate the half lives ($T_{1/2}$) of ^{14}C-labeled proteins. As can be seen, stretch led to an increase in $T_{1/2}$, i.e. stretch decreased the rate of protein degradation. From a number of experiments of this type, a 10.8% stretch was found to cause a 28% decrease in the rate of protein degradation. Similar results were obtained when the experiment was performed in basal medium containing no serum or embryo extract. Decreased protein degradation of the cultured muscle cells therefore contributes to the process of stretch-induced cell growth.

B. Transaminase Assay for Protein Degradation

Although the intermittant perfusion technique attempts to minimize the problem of reutilization, it cannot correct for an intracellular pool of radioactive amino acids released from proteins during degradation that does not exchange rapidly with the extracellular medium. A nonreutilizable radioactive amino acid would eliminate this possible source of error. Radioactivie arginine, labeled in the guanidine group, has been used for this purpose in liver cells (22), but this method works only in cells that contain the enzyme, arginase. We have developed a method based on the exchange reaction catalyzed by transaminases, a more widely distributed enzyme. The method is based on the fact that the hydrogen at the C-2 position of an amino acid is nonexchangeable when the amino acid is covalently linked in a polypeptide chain. In the free amino acid, this C-2 hydrogen becomes exchangeable with the medium due to the reaction catalyzed by the transaminase (23):

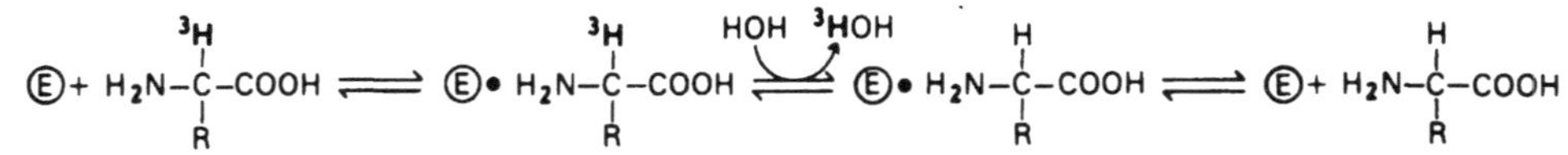

Thus, if an amino acid labeled with tritium in the C-2 position is incorporated into protein, release of the amino acid by protein degradation will make it available to the transaminase which will then catalyze the exchange reaction that results in the release of the tritium as ^3H_2O. Measuring protein degradation by the release of TCA soluble radioactivity will, using this label, be less subject to problems of reutilization of the radioactive label because ^3H_2O is effluxed rapidly from cells and is essentially not reutilized during the time period of these experiments.

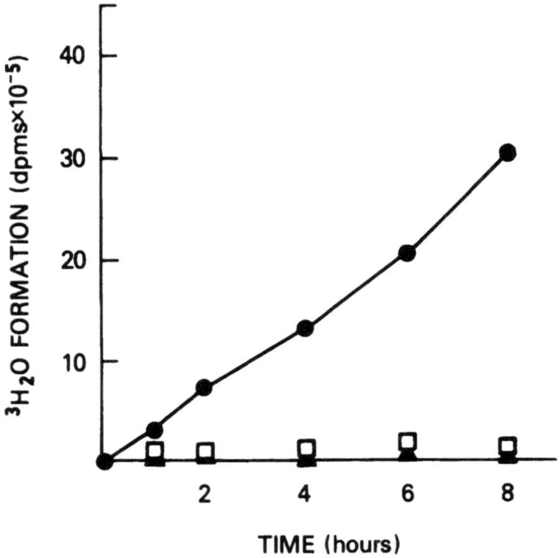

Figure 2. Transaminase activity in 72 hour muscle cultures incubated at 37° in basal medium (Eagle's), 10% horse serum containing 5 μCi/ml 2[^{3}H] phenylalanine. The rate of ^3H_2O formation was determined by collecting water samples by lyophilization of aliquots of the medium. The amount of radioactivity in the sample was measured by liquid scintillation counting. Cells (●), medium only (▲); and cells plus 1.7 mM aminooxyacetic acid (□).

Phenylalanine was synthesized with tritium exclusively at the C-2 position (24). Since skeletal muscle has been reported to neither synthesize nor degrade phenylalanine (25), it was necessary to first determine if muscle cells in culture have phenylalanine transaminase activity. This activity was measured by exposure of muscle cultures to [2-^{3}H] phenylalanine and following the formation of ^3H_2O with time. The results are shown in Figure 2. The formation of ^3H_2O was linear for at least 8 hours and was inhibited by 1.7 mM aminooxyacetic acid, a specific inhibitor of transaminase activity (26). No transaminase activity was present in the medium. The activity of the phenylalanine transaminase in the cultured cells was calculated to be 18.8 nmoles/hr/mg protein. Since radioactive amino acids are released from prelabeled proteins at a rate of 15 pmoles/hr/mg protein, the transaminase activity should be adequate for removing a major fraction of the tritium from the [2-^{3}H] phenylalanine that is released during degradation. This conclusion is based on the assumption that the released [2-^{3}H] phenylalanine is not compartmentalized from the enzyme. We have shown that [2-^{3}H] phenylalanine taken up from the medium is accessible to the enzyme (Figure 2) and that at least some of the tritium released from proteins during protein degradation is accessible to the transaminase as measured by the formation of $^{3}H_2O$ during protein degradation (Vandenburgh and Kaufman, unpublished observation). It is difficult to determine whether any fraction of the released amino acid is inaccessible to the transaminase because it was found that the rate of efflux into the medium of the ^{3}H-labeled amino acid released from prelabeled proteins in the cultured cells is rapid when compared to the rate of transaminase activity. Therefore, a large fraction (> 80%) of the radioactivity effluxed from the cells before it was acted upon by the enzyme. Nonetheless, these results indicate that the method should be useful in minimizing the reutilization of any [2-^{3}H] phenylalanine that remains intracellularly or that reenters the cells from the medium.

Muscle cell proteins were labeled with [2-^{3}H] phenylalanine for 20-24 hours and the rate of protein degradation followed as described for the intermittant perfusion technique. The results are shown in column B of Table II. In the experiment described in Table II, the same cultures were labelled with both [^{14}C] phenylalanine and [2-^{3}H] phenylalanine so that the two methods of measuring protein degradation could be compared. Protein half-lives calculated from the rate of release of tritium were 15-20% lower than those calculated from the rate of release of ^{14}C radioactivity, an indication that some reutilization of amino acid still occurred with the intermittent perfusion technique. With either radioactive label, stretch of the cells decreased the rate of protein degradation to a similar extent (35% ^{14}C labeled proteins, 29% ^{3}H labeled proteins). Thus, it appears that the _in vitro_ stretch-induced model for hypertrophy is also similar to the tenotomy model in that stimulation of muscle cell growth involves a decrease in the rate of protein degradation.

In contrast to the decreased protein degradation found in the cultured muscle stretch system, in stretched, organ-cultured muscle (18, 19) and in the tenotomy model (17), others have found increased protein degradation with _in vivo_ immobilization of rat (27) or fowl (28) muscle in a stretched position for several days. In both of these studies, compensatory hypertrophy of the muscles was shown to be due to a greater increase in the rate of protein synthesis than of the rate of degradation. These different results could be due to: (a) the different methods used to induce muscle cell growth; (b) the different methods used for assaying the rate of protein degradation; or (c) measurement of protein degradation at different time periods after the initial stimulus. For example, the methods used for measuring protein degradation in the above studies could be measuring predominantly the degradation of a different class of proteins compared to those being measured in the present study. In this regard, it is known that the degradation of short and long-lived proteins are regulated by different mechanisms in many cell types (29,30), including muscle cells in culture (21), Alternately, the different results could be due to the time period at which protein degradation was measured. We have found that the stretch-induced decrease in protein degration in the cultured muscle cells is greater at time points early after the stretch stimulus than at later time points (21). In the two in vivo stretch studies where protein degradation was found to be increased after several days, an early decrease in protein degradation apparently occurred (27,31). Thus, the effect of stretch on protein degradation in muscle cells may be biphasic, with an early decrease, followed by a more prolonged increase. In the tenotomy model, where repeated stretch and relaxation occurs rather than immobilized stretch, a prolonged decrease in protein degradation is found (17).

Mechanical Stretch and Na Pump Activity

Stretch of adult skeletal muscle is known to stimulate heat production (32) and helps to maintain a muscle in an excitable state in vitro for a longer period of time (33). The close relationship between thermogenesis in muscle (34), and the maintenance of resting membrane potential (35) to the activity of the Na pump (Na^+-K^+ ATPase) indicated that stretch may stimulate Na pump activity. Indeed, direct support in adult muscle for Na pump activation on stretch has been obtained (36). This stimulation leads to the maintenance of a higher K^+/Na^+ ratio in the cells (36). Because of the postulated regulatory role of K^+ and Na^+ levels in cell growth (37,38), and in amino acid transport (39), we measured the effect of stretch on Na pump activity in the cultured muscle cells and its relationship to several of the biochemical reactions found to be stimulated in the stretched cells.

Na pump activity in 72 hour cultured muscle cells was found to be sensitive to inhibition by the cardiac glycoside, ouabain. When measured by the uptake of

Table III. Ouabain Sensitivity of[^{3}H] AIB Accumulation and [^{14}C] Leucine Incorporation in Muscle Cells

	Time (hours)	No Ouabain	1 mM Ouabain	% Inhibition
[^{3}H] AIB Accumulation (dpms x 10^{-2}/μg protein)	1	6.24 ± 0.23	2.01 ± 0.06**	68
	2	7.04 ± 0.40	1.75 ± 0.15**	75
TCA Insoluble Radioactivity (dpms x 10^{-2}/μg protein)	1	3.00 ± 0.05	2.07 ± 0.04*	31
	2	5.68 ± 0.29	3.23 ± 0.26*	43

72 Hour muscle cultures were preincubated at 37° in basal medium (Eagles), 5% serum, 5% embryo extract and 1 mM ouabain for 30 minutes. 5 μCi/ml[^{3}H] AIB and 1 μCi/ml[^{14}C]leucine was then added to each culture and incubated at 37° for 1-2 hours before processing for[^{3}H] AIB accumulation and[^{14}C]leucine incorporation into TCA insoluble material. Each value is the mean ± S.E. of three cultures. Student's t-test for paired values: $^{*}p < .01$; $^{**}p < .001$.

$^{86}Rb^{+}$, a K^{+} tracer used to measure Na pump activity, 1 mM ouabain inhibited $^{86}Rb^{+}$ uptake 54% within 60 minutes of incubation. Table III shows the effect of 1 mM ouabain on the accumulation of α aminoisobutyric acid and the incorporation of [^{14}C] leucine into protein in these cells. This concentration of ouabain caused a dramatic decrease within one hour in both of these reactions. α-Aminoisobutyric acid accumulation and incorporation of[^{14}C] leucine were inhibited 68% and 31%, respectively. This inhibitory response of ouabain on aminoisobutyric acid accumulation in cultured muscle cells has been found by others (40). Stretch of the muscle cells was found to stimulate the uptake of $^{86}Rb^{+}$ (Figure 3A) without having a significant effect on the rate of efflux of $^{86}Rb^{+}$ from the cells (Figure 3B). This stretch-induced stimulation of Na pump activity occurred within one hour of stretch and is maintained for at least 6 hours when the cells are held in a stretched position. Ouabain is able to inhibit most of this stretch-stimulated activity (Table IV). In a number of different experiments, a small ouabain-insensitive stimulation of $^{86}Rb^{+}$ uptake is usually seen. This result is similar to that seen

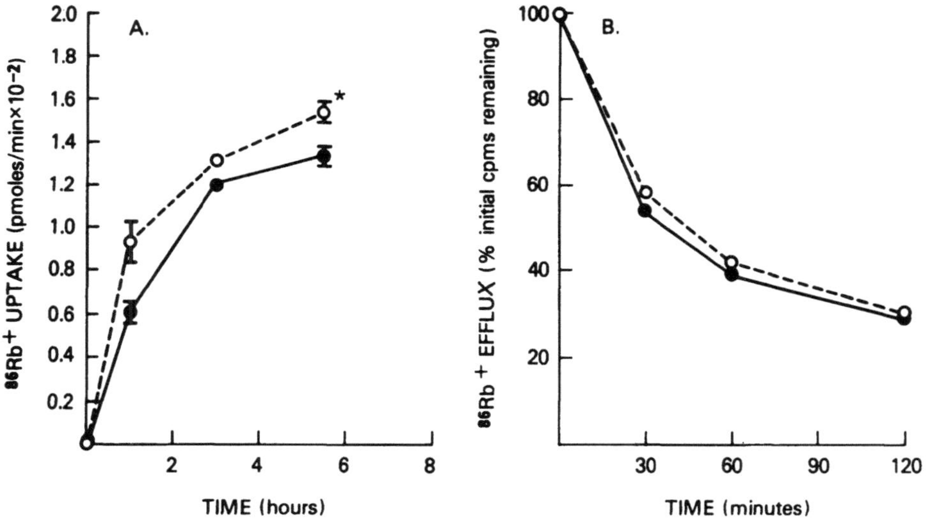

Figure 3. Uptake and efflux of $^{86}Rb^{+}$ in muscle cells stretched 10.8%. A. Cells were stretched 10.8% at 0 time in medium containing 5% serum and 5% embryo extract. 2.5 μCi/ml $^{86}Rb^{+}$ and 0.1 μmole/ml RbCl was added for a 15 minute period before processing the cells for total intracellular $^{86}Rb^{+}$. B. Cells were preincubated for 30 minutes with 2.5 μCi/ml $^{86}Rb^{+}$ and 0.1 μmole/ml RbCl in medium containing 5% serum, 5% embryo extract, rinsed rapidly, stretched 10.8%, and the intracellular amount of $^{86}Rb^{+}$ determined at several time points after stretch. Values are the mean ± S.E. of 3 to 5 cultures. *$p < .01$, Student's t-test for paired values.

with serum stimulation of $^{86}Rb^{+}$ uptake and cell division in quiescent 3T3 fibroblasts where both a ouabain-sensitive and ouabain-insensitive stimulation of $^{86}Rb^{+}$ uptake occurred (41).

As mentioned earlier, proliferation of interstitial fibroblasts is a characteristic of the in vivo tenotomy model for hypertrophy. Since some of the biochemical changes that occur in the hypertrophied muscle in this model have been localized predominatly to these non-muscle cells (14), it was of interest to identify the site of the stimulated Na pump activity. Initial results indicate that stretch stimulates $^{86}Rb^{+}$ uptake in cultures containing only myotubes (Vandenburgh and Kaufman,

Table IV. Ouabain Sensitivity of Stretch-Induced Stimulation of $^{86}Rb^{+}$ Uptake by Muscle Cells

	No ouabain (pmoles/min · mg protein)	1 mM ouabain (pmoles/min · mg protein)
Control cultures	181.6 ± 5.9	83.4 ± 4.7
10.8% Stretched cultures	235.4 ± 10.0*	97.0 ± 6.0

Muscle cells were stretched 10.8% at 72 hours in medium containing 5% serum, 5% embryo extract. 1 mM ouabain was added to the medium 15 minutes after stretch, 2.5 μCi/ml $^{86}Rb^{+}$ and 0.1 μmole/ml RbCl was added 45 minutes after stretch, and the cells were processed for intracellular $^{86}Rb^{+}$ at one hour after stretch. Values are the mean ± S.E. of 5 cultures per group. *$p < .01$, Students t-test for paired values.

unpublished observations). Although we have not yet determined whether $^{86}Rb^{+}$ uptake into fibroblasts is also stimulated by stretch in our in vitro system, it seems likely that it may be. It has been found, for example, that tension-induced stimulation of fibroblast proliferation occurs (42) and Na pump activation has been implicated as a possible early event in cellular proliferation (reviewed in 43). Thus, a common, early event in both myofiber hypertrophy and fibroblast proliferation in response to stretch may occur. Stimulated fibroblasts, which are capable of mitosis, would subsequently proliferate whereas postmitotic myofibers would only enlarge in size.

In summary, stretch of embryonic muscle cells which have never been innervated leads to cell growth by increased amino acid accumulation and incorporation into proteins, and by decreased protein degradation. Activity of the Na pump in these cells appears to be related to amino acid accumulation and incorporation into proteins. Stretch-induced activation of the Na pump may be an early and possibly critical event in stimulating these biochemical changes and in inducing the growth of muscle cells.

References

1. Goldspink, D.F.: Biochem. J. 174, 595-602 (1978).
2. Sola, D.M., Martin, A.W.: Am. J. Physiol. 172, 325-332 (1953).
3. Schiaffino, S., Hanzlikova, V.: Experientia 26, 152-153 (1970).
4. Gotschlick, E.: Arch. Ges. Physiol. 56: 355 (1894).
5. Feng, T.P.: J. Physiol. 74, 441-454 (1932).
6. Vandenburgh, H.H., Kaufman, S.: Science 203, 265-268 (1979).
7. Goldberg, A.L.: Am. J. Physiol. 213, 1193-1181 (1967).
8. Goldberg, A.L., Goodman, H.M.: Am. J. Physiol. 216, 1111-1115 (1969).
9. Hamosh, M., Lesch, M., Baron, J., Kaufman, S.: Science 157, 935-937 (1967)
10. Goldberg, A.L.: J. Cell Biol. 36, 653-658 (1968).
11. Goldberg, A.L., Etlinger, J.D., Goldspink, D.F., Jablecki, C.: Med. Sci. Sports 7, 248-261 (1975).
12. Enesco, M., Puddy, D.: Am. J. Anat. 114, 235-244 (1964).
13. Schiaffino, S., Bormioli, S.P., Aloisi, N.: Virchows Arch. Abt. B. Zellpath. 11, 268-273 (1972).
14. Jablecki, C.K., Heuser, J.E., Kaufman, S.: J. Cell Biol. 57, 743-759 (1973).
15. Goldberg, A.L., St. John, A.C.: Ann. Rev. Biochem. 45, 747-803 (1976).
16. Goldberg, A.L.: J. Biol. Chem. 244, 3223-3229 (1969).
17. Goldberg, A.L.: J. Biol. Chem. 244, 3217-3222 (1969).
18. Goldberg, A.L., Jablecki, C., Li, J.B.: Ann. N.Y. Acad. Sci. 228, 190-201 (1974).
19. Kameyama, T., Etlinger, J.D.: Nature 279, 344-346 (1979).
20. Bradley, M.O., Hayflick, L., Schimke, R.T.: J. Biol. Chem. 251, 3521-3529 (1976).
21. Vandenburgh, H.H., Kaufman, S.: submitted for publication.
22. Swick, R.W.: J. Biol. Chem. 231, 751-764 (1958).
23. Hilton, M.A., Barnes, F.W., Henry, S.S., Enns, T.: J. Biol. Chem. 209, 743-754 (1954).
24. Brundish, D.E., Elliott, D.F., Wade, R.: J. Labelled Compounds 7, 473-493 (1971).
25. Manchester, K.L.: Biochim. Biophys. Acta 100, 295-297 (1965).
26. Wallach, D.P.: Biochem. Pharmacol. 5, 323 (1961).
27. Goldspink, D.F.: J. Physiol. (London) 264, 267-282 (1977).
28. Laurent, G.J., Sparrow, M.P., Millward, D.J.: Biochem. J. 176, 407-417 (1978).

29. Hershko, A., Mamont, P., Shields, R., Tomkins, G.M.: Nature New Biol. 232, 206-211 (L971).
30. Hopgood, M.F., Clark, M.G., Ballard, F.J.: Biochem. J. 164, 399-407 (1977).
31. Laurent, G.J., Sparrow, M.P.: Growth 41, 249-262 (1977).
32. Feng, T.P.: J. Physiol. 74, 441-454 (1932).
33. Weiss, P.: Am. J. Physiol. 106, 156-169 (1933).
34. Himms-Hagen, J.: Ann. Rev. Physiol. 38, 315 (1976).
35. Baker, P.F.: Metabolic Pathways (ed. Hokin, L.E.) 6, 243-268, 3rd edition, Academic Press, New York, 1972.
36. Rapaport, S.I., Bidinger, J.M.: Am. J. Physiol. 226, 452-457 (1974).
37. Lubin, M.: Nature 213, 451-453 (1967).
38. Ledbetter, M.S., Lubin, M.: Exp. Cell. Res. 105, 223-236 (1977).
39. Schultz, S.G., Curran, P.F.: Physiol. Rev. 50, 637-718 (1970).
40. Grove, B.K., Stockdale, F.E.: Develop. Biol. 66, 142-150 (1978).
41. Rozengurt, E., Heppel, L.A.: Proc. Natl. Acad. Sci. USA 72, 4492-4495 (1975).
42. Curtis, A.S.G., Seehar, G.M.: Nature 274, 52-53 (1978).
43. Kaplan, J.G.: Ann. Rev. Physiol. 40, 19-41 (1978).

THE EFFECT OF DISUSE ON PROTEIN PATTERNS IN FAST AND SLOW TWITCH MUSCLES

F. Guba, Ő. Takács and Z. Kiss
Institute of Biochemistry, University of Medical Sciences, 6701 Szeged, P.O. Box 415, Hungary
Á. Szöör and T. Szilágyi
Institute of Physiology and Pathophysiology, Univ. Med. Sci., 4012 Debrecen, Hungary

Introduction

It is well established that the motor activity pattern regulates physiological and molecular processes in skeletal muscle (1,5,6,7). The biological equilibrium of the physiological and metabolic properties of skeletal muscles with different functions can be changed by a wide variety of experimental conditions: by transection of the medulla (4), by tenotomy (9), by denervation (19) and by different forms of immobilization, such as plaster casts (2) and articular fixation (11). These manipulations lead to muscular atrophy, alteration in the composition of protein compartments, as well as tissue damage (20) and ultrastructural changes (14,29). These changes suggest that the motor activity pattern regulates molecular processes to yield the optimum protein profile for a particular type of function.

Plaster cast immobilization, being non-invasive and reversible, appears to be the most convenient method for producing reduced activity of muscles. The effects of immobilization on muscle depend upon the position in which the muscle is held. Recently Goldspink (13) demonstrated that rat muscles fixed in the stretched position show increases in weight and in the number

Plasticity of Muscle

of sarcomeres per fibre. In general, immobilization of muscle leads to atrophy, which is reflected in a decrease in muscle weight (23), and in changes of the RNA/DNA ratio and of other parameters (15). It appeared reasonable to investigate the changes occurring in various compartments of the muscle cell as a result of immobilization. Disuse of the muscles leads to a gradual reduction in the contractile proteins (26). This specific protein fraction - which can be termed "luxury" protein - is degraded to a greater extent in the slow muscles than in the fast ones as was shown earlier in our laboratory. The proteins of the sarcoplasm perform many different functions and are extremely heterogeneous. It is obvious, therefore, that disuse will exert different effects on the equilibrium of protein synthesis and degradation. There are several proteins, for example myoglobin and myokinase (27), which display marked differences in the direction of their change, and others such as glycolytic enzymes, which change in the same direction in both muscle types. In general, from the experiments summarized above, one can conclude that immobilization leads to a differentiation of the muscle cell and that mammalian skeletal muscle is capable of adapting to deficiencies in function.

To obtain further information about disused muscle, systematic and detailed investigations were performed. The following main questions were examined: 1. The nucleic acid content. 2. Activities of RNA-ase and RNA-ase inhibitor. 3. Mechanical properties of glycerol-treated muscle fibres. 4. Alteration in the myofibrillar syneresis. 5. Submolecular changes in the structural proteins.

Methods

In all experiments described here, rabbits were used as experimental animals. Experimental conditions of plaster cast fixation were as described previously (23). RNA and DNA were determined according to Hollősi et al. (15). One part of the muscle sample was used for measuring the RNA-ase and RNA-ase inhibitor (for details see ref. 16). For the investigation of the mechanical properties of muscle fibres, the samples were treated according to Szöör and Boross (24). Isometric tension was recorded with the use of a transducer and potentiometric recorder. Structural proteins were isolated and purified as described in our previous communication (26). In the syneresis experiments, the purified myofibrillar suspension was treated as described by Ebashi (8). Quantitative analysis of the composition of myofribrillar proteins was performed after separation of the solubilized samples in a detergent solution (1% sodium dodecyl sulphate, 1% 2-mercaptoethanol and 10% sucrose) on 7.5 and 10% polyacrylamide gels with the procedure of Weber and Osborn (31), modified by Talbot and Yphantis (28). The quantitative evaluation was made with a Kipp and Zonen densitometer. For higher resolution and more accurate characterization of structural proteins, the myofibrillar pellet was treated by the method of O'Farrel (18) and separated with two-dimensional electrophoresis.

Results

The DNA concentration compared to the control showed a threefold increase by the 28th day of immobilization in the case of the soleus muscle (Fig. 1). However the DNA content of the whole soleus muscle whose weight decreases according to published data (26) remained nearly constant. In the gastroc--nemius muscle the DNA concentration did not increase to such an extent as in the soleus muscle, but nearly constant values were also obtained for the DNA content of the whole muscle, because in this case the decrease of weight was also less. In the soleus muscle the RNA concentration was nearly constant (Fig. 2).

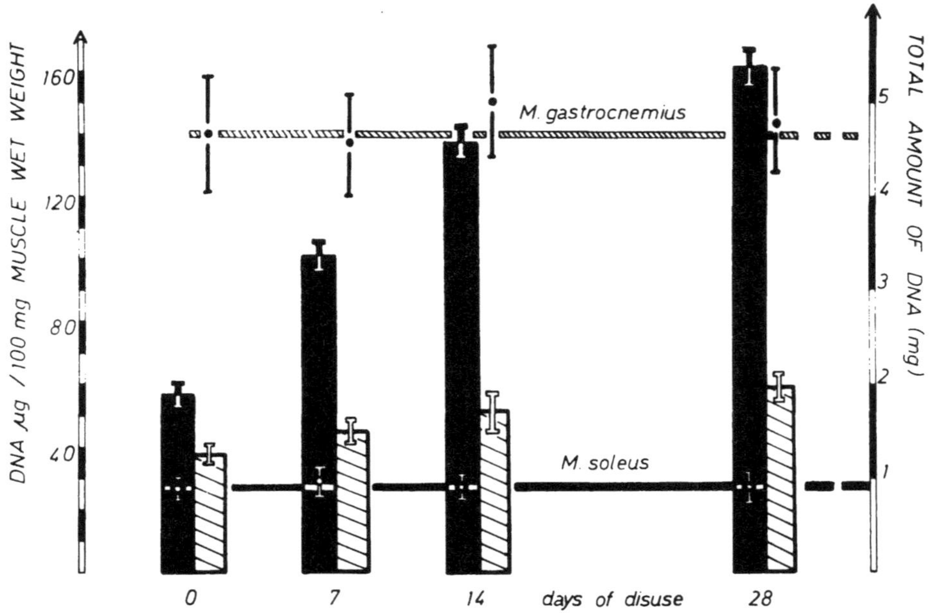

Fig. 1. DNA content and total amount of DNA in soleus (black columns) and gastrocnemius (dashed columns) muscles.

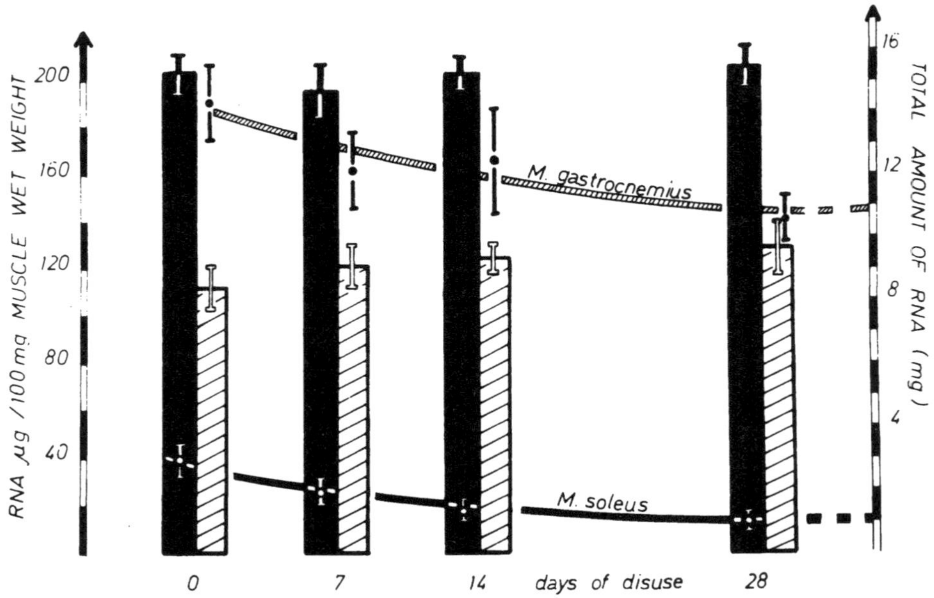

Fig. 2. RNA content and total amount of RNA in soleus (black columns) and gastrocnemius (dashed columns) muscles.

The RNA content of the whole muscle showed a slight decreasing tendency. In the gastrocnemius muscle, an insignificant increase in RNA concentration was encountered, while the total RNA content decreased.

With the supposition that the slight decrease in RNA content was due to an increased RNA-ase activity, the non-specific alkaline RNA-ase activity and the activity of the RNA-ase inhibitor regulating its function were investigated. The RNA-ase activity was measured in the particulate fraction, and that of the RNA-ase inhibitor in the cytosol fraction. As a result, we found that the RNA-ase activity increased with increasing immobilization time (Fig. 3 A). The activity of RNA-ase inhibitor in the cytosol fraction decreased with progressing immobilization (Fig. 3 B).

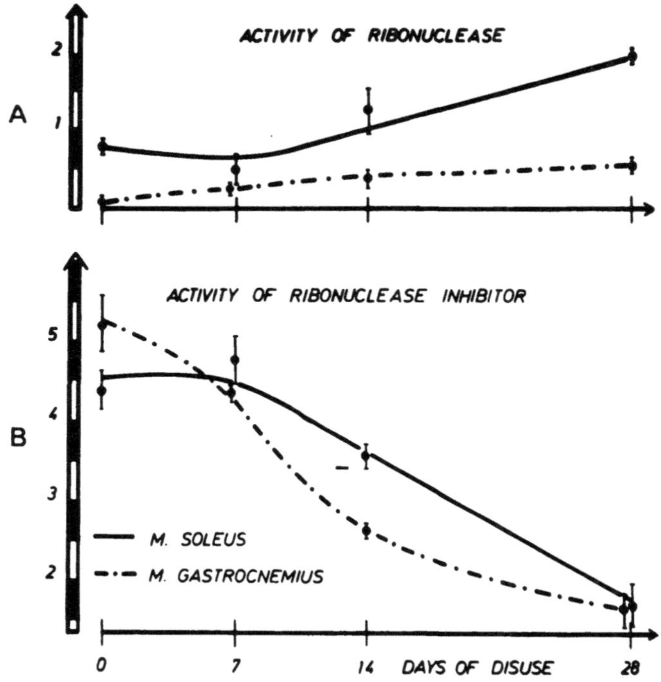

Fig. 3. Activities of RNA-ase (units of activity/mg protein) and RNA-ase inhibitor (units of inhibitor activity/mg protein) in the course of four weeks' immobilization.

This result is in accordance with the definitions given by Kraft (17) concerning the balance of RNA-ase and RNA-ase inhibitor.

Investigation of the mechanical parameters of the muscles gave the following results. Figure 4 shows the ATP-induced tension of glycerol-treated gastrocnemius fibres. One can see that 28 days of disuse causes the contraction force ($g \times cm^{-1}$) and the speed of tension development ($g \times cm^{-1} \times sec^{-1}$) to decrease significantly. The syneresis of myofibrils prepared from immobilized gastrocnemius muscle was compared with that of controls (Fig. 5). It can be seen that there was less increase in the turbidity of the myofibrillar suspension prepared from the immobilized muscle. Immobilization did not change the temporal course of syneresis.
The protein patterns of the myofibrils prepared from the gastrocnemius muscles of intact and immobilized rabbits are presented in Fig. 6.

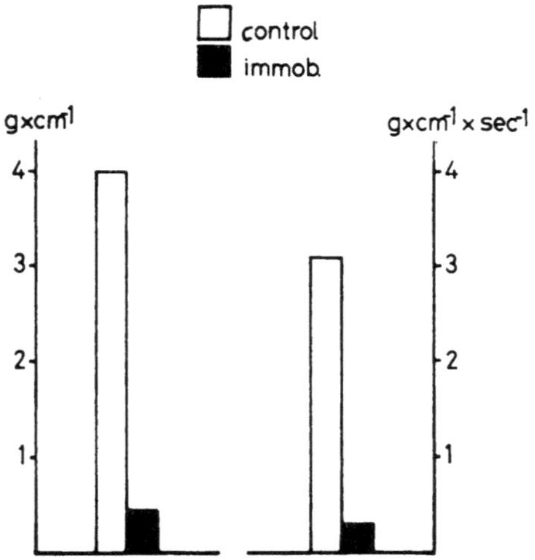

Fig. 4. Changes in the contraction force ($g \times cm^{-1}$) and in the speed of shortening ($g \times cm^{-1} \times sec^{-1}$) of immobilized soleus mucle (representative result).

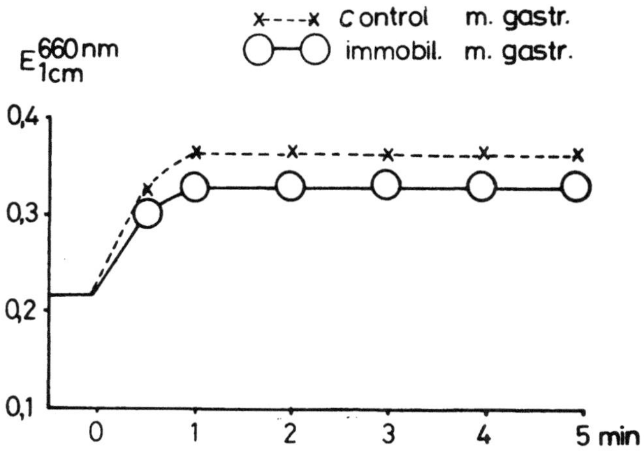

Fig. 5. Effect of disuse on the syneresis of myofibrils prepared from gastrocnemius muscle.

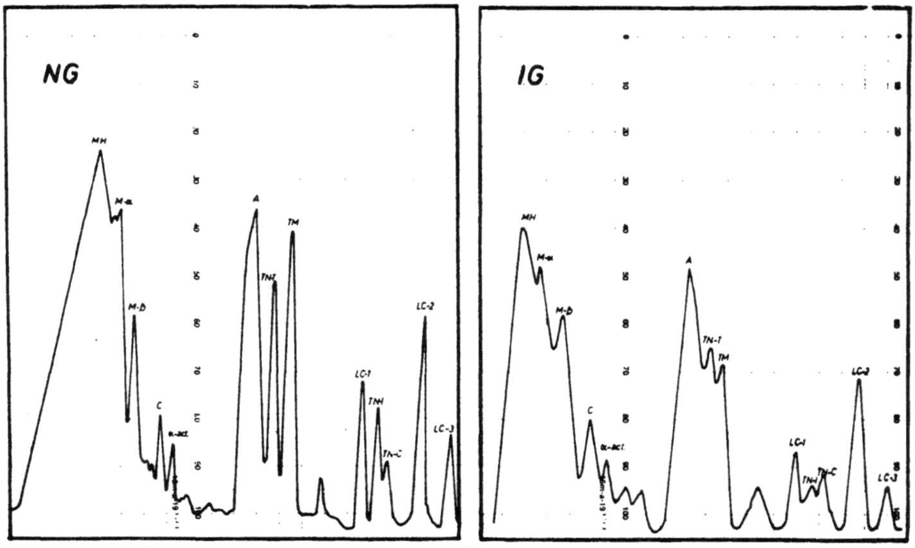

Fig. 6. Densitometric scans of normal (NG) and four weeks immobilized (IG) gastrocnemius myofibrillar proteins.

The major myofibrillar components of the gastrocnemius muscle, which is made up mainly of fast fibres, are as follows: myosin heavy chain (MHC), actin (A), tropomyosin (TM), troponins (TN), and myosin light chains (LC). The LC-1:LC-2: LC-3 ratio is 1:2:1. The myosin/actin weight ratio is 5.0, while that of the thick and thin filament-forming components is 2.45. The changes brought about by immobilization are depicted in Fig. 7.

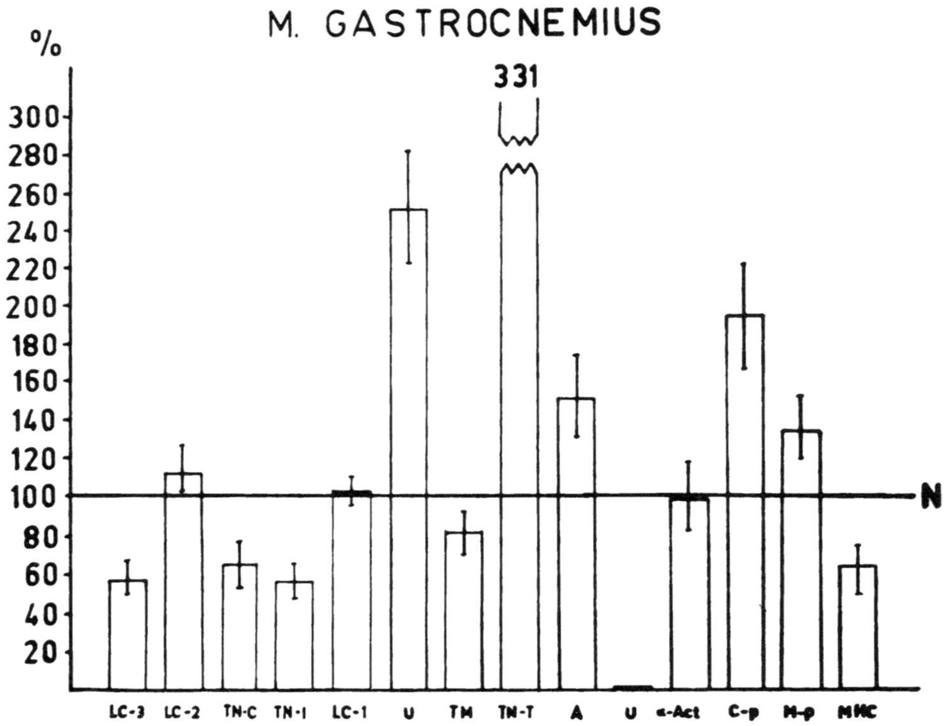

Fig. 7. Relative changes in the composition of immobilized gastrocnemius myofibrillar proteins taking the normal (N) values as 100%.

Disuse of the limb for 28 days resulted in a relative decrease in some of the individual components and in an increase of other components. In regards to the low molecular weight peptides, there was a significant decrease in the LC-3 component of myosin, as well as in the calcium-binding (TN-C) and inhibitor (TN-I) components of troponin. Of the high molecular weight components, only the myosin heavy chain showed a decrease, and it was of about 37% relative to its normal value. The data obtained on the composition of the myofibrillar protein content of the soleus muscle is shown in Fig. 8.

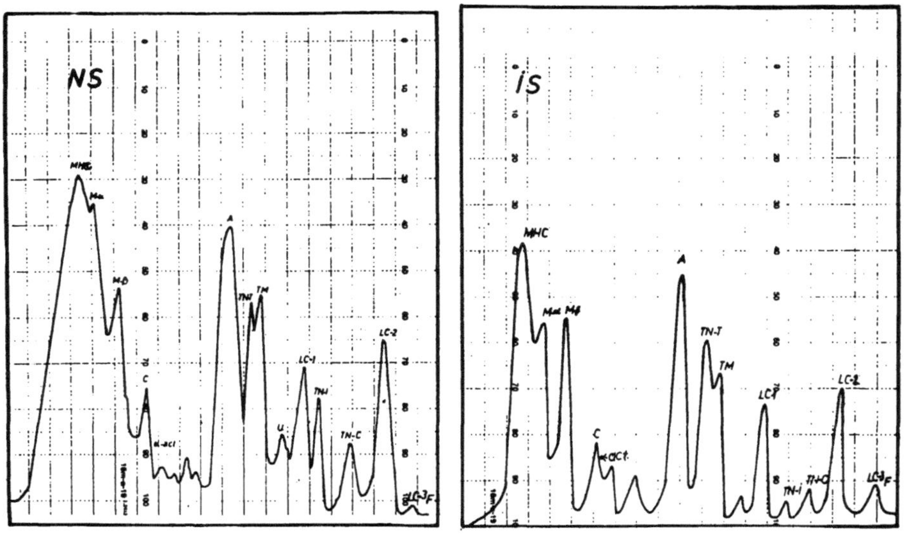

Fig. 8. Densitometric scans from normal (NS) and four weeks immobilized (IS) soleus myofibrillar proteins.

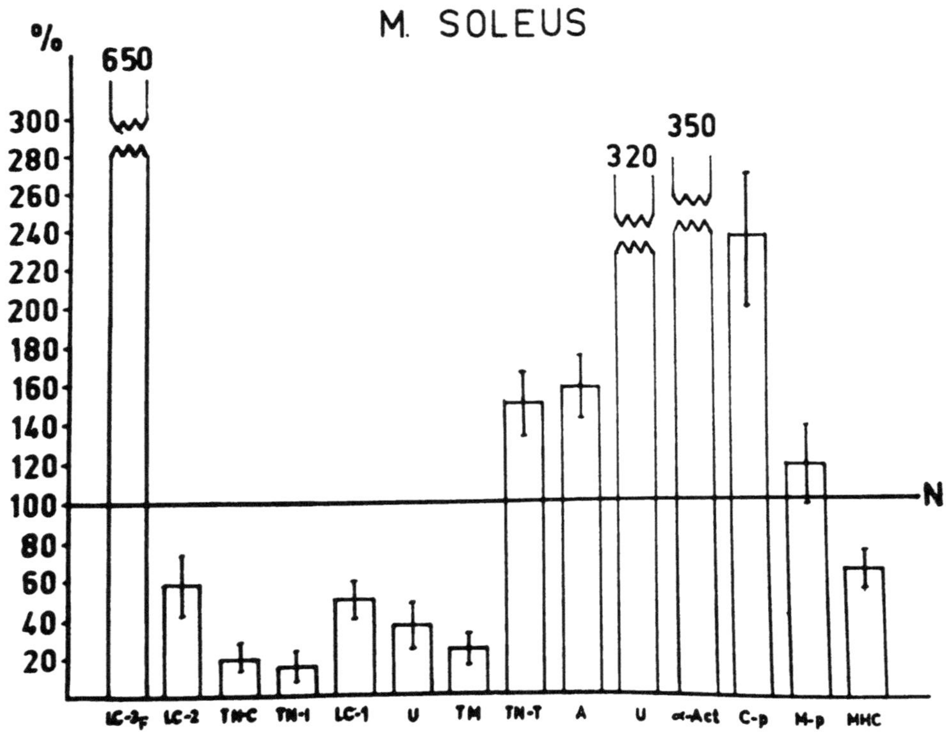

Fig. 9. Relative changes in the composition of immobilized soleus myofibrillar proteins taking the normal (N) values as 100%.

The relative changes in the disused soleus muscle myofibrillar proteins are illustrated in Fig. 9.

One of the more notable changes was that in the immobilized soleus muscle there was an increase in the amount of the fast light chain with a molecular weight of 15,000 dalton ($LC\text{-}3_F$), which can be detected in the normal soleus muscle only in very small amounts.

For a better understanding of these results, one has to take into account the findings of our previous investigations (23). Since the loss in weight of the gastrocnemius and soleus muscles after four weeks of immobilization are, on the average,

35 and 55%, respectively, one can calculate the total amounts of the individual myofibrillar components remaining at the end of the experimental period. The calculated data are shown in Table I.

Table I.

Protein	Normal G. mg	Immob.G. mg	D %	Normal S. mg	Immob.S. mg	D
Myosin	896	307	-65	113	22	-80
Actin	263	100	-62	33	11	-65
Holotroponin	130	84	-36	16	7	-58
a./ TN-C	32	11	-67	6	1	-80
b./ TN-I	39	13	- 6	5	1	-79
c./ TN-T	59	58	- 2	6	5	-16
Tropomyosin	148	71	-52	18	9	-69
α-actinin	23	10	-55	2	2	0
C-protein	28	25	-10	9	5	-43
M-protein	119	97	-18	15	7	-51

Total amounts of contractile proteins in normal and immobilized gastrocnemius and soleus muscles, calculated from the average muscle weight and the absolute amounts of individual components as measured by quantitative densitometry.

It is evident that the individual components are degraded to different extents. TN-T seems to be very stable, for its content was not altered in the gastrocnemius muscle, and only a 16% decrease was found in the soleus muscle. The quantities of C and M proteins decreased to a moderate extent. One can conclude that the muscle protein alteration resulting from disuse affects the amounts of contractile proteins to different degrees: it seems that TN-T is the most stable, and myosin, the most sensitive.

On two-dimensional gels obtained for normal (Fig. 10 A) and one-week-immobilized (Fig. 10 B) gastrocnemius, the main myofibrillar components can be identified.

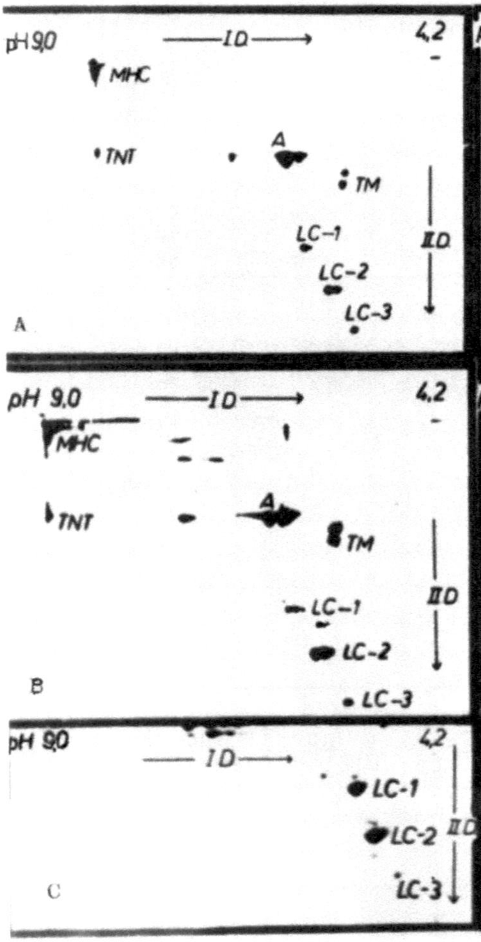

Fig. 10

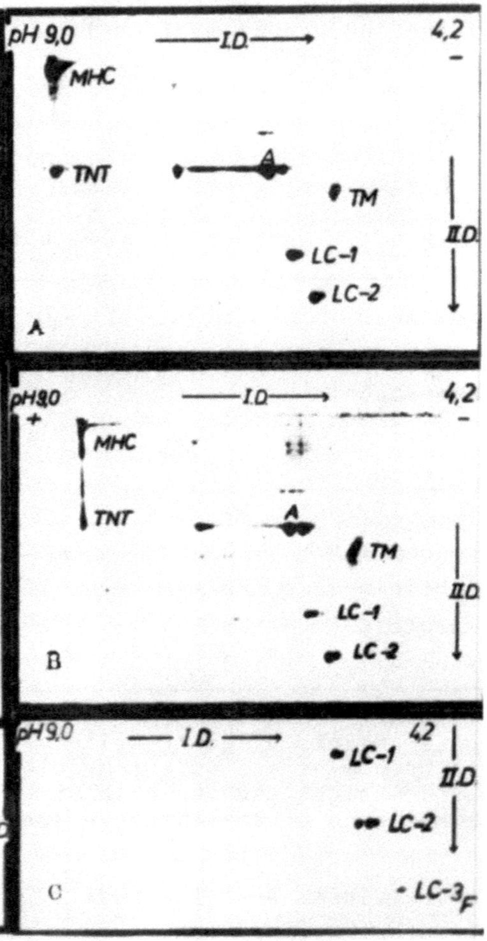

Fig. 11

Fig. 10. Two-dimensional gel electrophoresis of myofibrillar proteins from: A normal, B 1-week, C 4-weeks immobilized gastrocnemius.

Fig. 11. Two-dimensional gel electrophoresis of myofibrillar proteins from: A normal, B 1-week, C 4-weeks immobilized soleus.

After one week of immobilization, a double spot can be observed between LC-1 and LC-2, such that LC-1 and LC-2 appear in the form of a triplet. Figure 10.C shows the light chain region from four-weeks-immobilized gastrocnemius muscle. LC-3 in the disused muscle has decreased. After 4 weeks of immobilization, LC-2 shows up in one spot, probably as a fully dephosphorylated peptide. Besides the normal fast myosin light chains, at least four spots can be observed, with molecular weights of 17,000 - 27,000 dalton. These can be considered as slow myosin light chains or proteolytic products. These peptides are absent from normal proteinograms, so their appearance could be due to the altered protein metabolism.

In the proteinogram of the normal soleus muscle (Fig. 11.A) the components of slow myofibrils can be detected. Only two slow light chains are present. After one week of immobilization (Fig. 11.B), LC-2 and LC-1 appear in polymorphic forms and some faint spots can be detected corresponding to molecular weights of 16,000-35,000 dalton. In the light chain region of four-week-immobilized slow muscle one can observe a considerable amount of the fast LC-3 component (Fig. 11.C).

These results suggest that, due to disuse, a polymorphism of the normal components occurs, which suggests the possibility of transformation of muscle cells.

Discussion

Comparison of the increase of DNA concentration with the decrease of weight encountered during immobilization reveals constant values for the DNA content. This experimental finding indicates that the genetic reservoir is not impaired with the decrease of proteins during immobilization. Thus, no appreciable cellular death seems to occur during the "no use" period.

Our investigation suggests that, due to the lack of activity, the characteristics of muscle cells, as such, fade away and a widespread genetic reprogramming takes place in order to meet the requirements of the metabolically optimum balance of the cell.
From a comparison of the submolecular changes one can calculate that the subunit structure of the fast muscle myosin changes characteristically due to experimental immobilization. The change is manifested in the shift of the ratio of the light chains from 1:2:1 to 1:2:0.5. In contrast, in the slow muscle, the relative amount of peptides with higher molecular weight decreases, and the LC-3 peptide characteristic of the fast muscle appears and displays an increasing trend as a function of time. According to recent investigations (see J.F.Y. Hoh and A. Weeds: this Symposium), the light chain pattern of fast fibres corresponds to three types of myosin isoenzymes: homodimeric LC-1 and LC-3, and heterodimeric LC-1 + LC-3.

Our investigations indicate that the lack of function due, for example, to denervation (25) manifests itself, apart from the decrease in the myosin content of the cells, primarily in the decrease of the homodimeric LC-3 isoenzyme, and partly that of the heterodimeric LC-1 + LC-3 isoenzyme. The loss of myosin therefore is due to the proteolytic disintegration of the latter isomyosin molecules. In the case of the soleus muscle, immobilization results in a gradual decrease in the amount of myosin, without any essential change in the relative distribution of the slow light chains. We emphasize the fact that even in the first week of immobilization there is an increase in the amount of the fast LC-3 subunit, which is present only in trace amounts in the normal soleus muscle. This suggests that disuse impairs the re-synthesis of the slow LC-1 and LC-2 isoenzymes, and at the same time the rate of synthesis of the homodimeric LC-3 fast isoenzyme is enhanced. Consequently, the slow muscle takes up the properties of the fast one to a certain degree. This hypothesis seems to be in agreement with observations of Booth and Kelso (3) and the physiological and biochemical investigations of Fischbach *et al*.(12),Vrbová(30), Salmons and Vrbová (21) and Edgerton *et al*.(10).

Summary

Changes in the compositions of the myofibrillar proteins in the soleus and gastrocnemius muscles of rabbit hind limbs have been studied in the course of 1-4 weeks of immobilization by plaster casts. The following main results have been obtained. The DNA content of immobilized muscles does not change. The RNA content shows a decreasing tendency during immobilization. The RNA-ase activity increases with increasing immobilization time. The activity of RNA-ase inhibitor in the cytosol fraction decreased with progressing immobilization. The myofibrillar protein content and the amount of myosin within the cells show significant decreases. The ratio of the light chain peptides of the fast muscle myosin changes from 1:2:1 to 1:2:0.5 as a result of 4 weeks of disuse. The LC-1:LC-2 ratio of slow myosin does not change considerably, whereas the amount of fast LC-3 peptide, hardly detectable in the normal soleus muscle, increases markedly. The amount of tropomyosin decreases significantly in both muscles. The submolecular composition of troponin changes, mostly in the slow muscle; TN-C and TN-I decrease significantly, whereas there is an increase in TN-T.

References

1. Bárány, M., Close, R.J.: J. Physiol. (Lond) 213, 455-474 (1971).
2. Booth, F.W., Kelso, J.R.: J. Appl. Physiol. 34, 404-406 (1973).
3. Booth, F.W., Kelso, J.R.: Pflügers Arch. ges. Physiol. 342, 231-238 (1973).
4. Brooks, J.E.: Arch. Neurol. 22, 27-30 (1970).
5. Buller, A.J., Eccles, J.C., Eccles, R.M.: J. Physiol. (Lond.) 150, 417-439 (1960).
6. Close, R.: Nature 206, 831-832 (1965).
7. Close, R.: J. Physiol. (Lond.) 204, 331-346 (1969).
8. Ebashi, S.: J. Biochem. 50, 236 (1961).
9. Eccles, J.C.: J. Physiol. 103, 253-270 (1944).
10. Edgerton, V.W., Barnard, R.J., Peter, J.B., Maier, A., Simpson, D.: Exp. Neurol. 46, 115-131 (1975).
11. Ferguson, A.B., Vuhn, L., Ward, L.: J. Bone Joint Surg. 39A, 583-624 (1957).
12. Fischbach, G.D., Robbins, N.: J. Physiol. (Lond.) 201, 305-320 (1969).
13. Goldspink, D.F.: J. Anat. 127, 459-468 (1978).
14. Guba, F., G. Mészáros, M. and Takács, Ö.: Acta biol. med. germ. 36, 1605-1619 (1977).
15. Hollósi, G., Takács, Ö., Guba, F., Szöör, A., Szilágyi, T. Acta biol. Acad. Sci. Hung. 28, 205-214 (1977).
16. Kiss, Z., and Guba, F.: FEBS Lett., in press (1979).
17. Kraft, N., and Shortman, K.: Biochim.Biophys. Acta 217, 164 (1970).
18. O'Farrel, P.H. (1975) High resolution two dimensional electrophoresis of proteins. J. Biol. Chem. 250, 4007-4019.
19. Pellegrino, C., Franzini, C.: J. Cell. Biol. 17, 327-343 (1963).
20. Ralston, H.J., Feinstein, B., Inman, V.T.: Fed. Proc. 11, 217-238 (1952).
21. Salmons, S., Vrbová, G.: J. Physiol. (Lond.) 201, 535-549 (1969).
22. Sréter, F.A., Gergely, J., Salmons, S., Romanul, F.: Nature New Biol. 241, 17-18 (1973).

23. Szöör, Á., Boross, A., Hollósi, G., Szilágyi, T., Keszthyüs, L.: Acta biol. Acad. Sci. Hung. 28, 195-204 (1977).

24. Szöör, Á., Boross, A.: in press.

25. Takács, Ö., Mészáros, M., Guba, F.: In Symposium on the muscle, Symposia Biologica Hungarica Vol. 17, Akadémiai Kiadó, Budapest (1974).

26. Takács, Ö., Sohár, I., Pelle, T., Guba, F. and Szilágyi, T.: Acta biol. Acad. Sci. Hung. 28, 213-219 (1977).

27. Takács, Ö., Sohár, I., Szilágyi, T., and Guba, F.: Acta biol. Acad. Sci. Hung. 28, 221-230 (1977).

28. Talbot, D.N. and Yphantis, D.A.: Analyt. Biochem. 44, 246-253 (1971).

29. Tomanek, R.T. and Lund, D.: J. Anat. (Lond.) 118, 531-541 (1974).

30. Vrbová, G.: J. Physiol. (Lond.) 169, 513-526 (1963).

31. Weber, K., Osborn, M.: J. Biol. Chem. 244, 4406-4410 (1969).

THE INFLUENCE OF CONTRACTILE ACTIVITY AND THE NERVE SUPPLY ON MUSCLE SIZE AND PROTEIN TURNOVER

D.F. Goldspink

Department of Physiology, Medical Biology Centre, The Queen's University, Belfast, Northern Ireland.

Introduction

Skeletal muscle undergoes a greater post-natal growth than most other body tissues (1). Further, its ability to adapt by changing its size in response to different physiological demands is well known. For example, increased work loads induce compensatory growth, whereas diminished activity leads to either a slowing of growth or to atrophy. Such factors may be quite selective affecting only a single muscle, or a small group of functionally-related muscles. In addition, other factors, eg, varying concentrations of nutrients or hormones within the circulation, can exert more generalized effects on the size of the body's musculature.

Since protein represents a large proportion (16-18%) of the muscle mass, its accumulation is reflected as muscle growth, while its loss induces a state of atrophy. Such changes in the intracellular protein content are determined by an imbalance between the rates of protein synthesis and protein breakdown. Any further divergence of these rates induces greater muscle growth. In contrast, a narrowing or reversal of the imbalance (ie, breakdown exceeding synthesis) creates a situation of slower growth or atrophy. Our present understanding concerning these two processes differs markedly. Although extensive information is available on the mechanisms involved in the manufacture of new proteins (2), little is known of the processes concerned with protein degradation (3). Nonetheless, both the processes of protein synthesis and protein breakdown appear to be under physiological control, and some of the factors known to influence protein turnover will be described below, with particular reference being paid to the influence of the nerve supply and contractile activity.

Plasticity of Muscle

The Relationship between Contractile Activity and Protein Turnover

A. Inactivity induced by immobilization

The extensor (eg, soleus) and flexor (eg, extensor digitorum longus; EDL) muscles normally involved in the movement of the ankle joint can be effectively immobilized by placing a plaster of Paris cast around the ankle (4). The particular length at which these leg muscles are immobilized depends upon the anatomical location of the muscles within the limb and the chosen restraining position of the ankle. In the studies described here the ankle of one hind limb was immobilized in one of two extreme positions, ie, with the foot fully extended (ie, plantar flexed) or dorsi flexed. In the former position the soleus is held in a fully shortened state. In order to produce the same state in the EDL, the dorsi flexed position of immobilization has to be employed. Both the EDL and soleus muscles were rendered largely inactive with the sarcomeres held in this shortened state (5). Precise quantitation of the remaining activity, if any, in these immobilized muscles is not currently available, but is under investigation. Nonetheless, the drastic reduction (5), or total loss, of activity causes the immobilized EDL to grow less (Figure 1, curve B) and the soleus muscle to undergo atrophy (Figure 2, curve B). In contrast, the respective unrestrained control tissues continue to grow throughout the experimental period (curve A): a point serving to emphasize the fact that young rats (approx. 50-60 g) were used throughout all of these studies. Other investigators studying the same muscles in the mouse (7,8) and cat (9) have found immobilization in this shortened position to induce a rapid and marked loss (up to 30%) of sarcomeres in series. It is, therefore, highly probable that the muscle atrophy (relative or true), and concomitant loss of protein, observed in our studies (Figures 1 & 2) arises from a decrease in longitudinal growth as well as a possible decrease in fibre diameters. It should be noted that immobilization in the shortened position (as used in this paper) is not to be confused with immobilization in a fully lengthened position, which produces the opposite

Figure 1. The wet wt. of the EDL after various treatments.

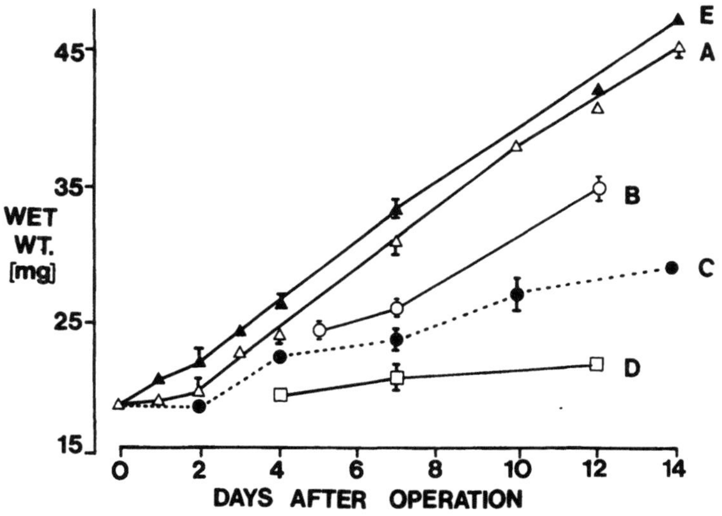

EDL muscles (A; controls) were either immobilized in a shortened (B) or lengthened (E) state (4), denervated (C), or denervated and immobilized (ie, shortened; D). Values are means ± S.E.M. of six muscles.

Figure 2. Immobilization and the return of activity to the soleus muscle.

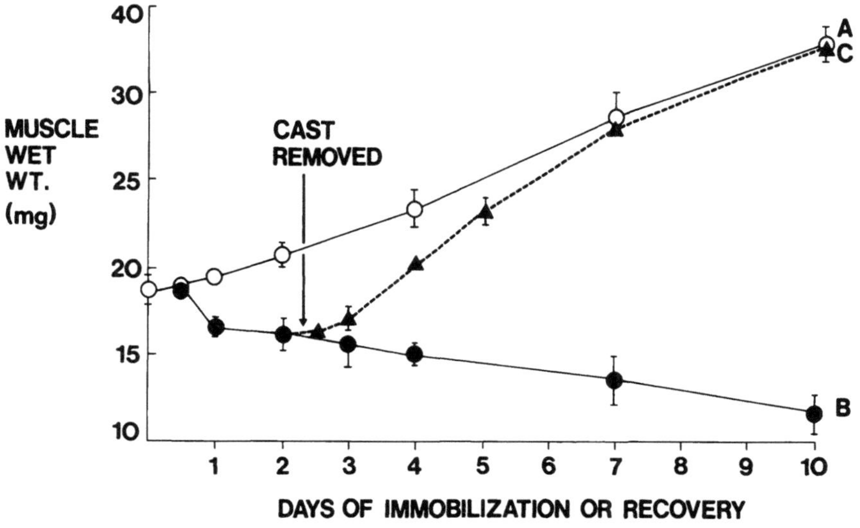

Soleus muscles (unrestrained controls: A) were permanently immobilized in a shortened state (B), or allowed to regain normal activity (C).

situation of additional growth (see Figure 1, curve E) and increased sarcomere numbers (7,8).

The net loss of intracellular protein, and corresponding muscle wasting described above was then investigated with respect to the possible changes in the regulatory processes, ie, protein synthesis and protein breakdown. As exemplified by the soleus muscle (Figure 3), the adjustment of the protein content after being rendered inactive correlated with complementary changes in both protein synthesis and protein degradation. Protein synthesis rapidly declined within the immobilized soleus (curve A), compared with its unrestrained control. In addition, after a delayed onset, the rate of protein breakdown was also found to be higher in the immobilized muscle (curve B). The early decline (ie, after 6 h) in the rates of protein synthesis within the immobilized muscle occurred in advance of any detectable fall in the RNA concentration. This possibly suggests that the initial effect of inactivity is exerted directly on the translation process, with a probable lowering of the rate of polypeptide chain initiation (2). At later times of immobilization however (eg, 7 days), RNA was lost at a faster rate than the general loss of muscle mass (4). Because approximately 80% of the total RNA is ribosomal, the immobilized soleus must acquire a lower capacity for making new proteins, since it must inevitably contain fewer ribosomes.

It is evident from our studies that immobilization of the EDL and soleus muscles to produce a state of inactivity did not induce the same degree of atrophy in the two muscle types. Although the directional changes in protein synthesis and protein breakdown were the same in both muscles after 2 days of immobilization (Figure 4A), these effects on protein turnover were more pronounced in the soleus muscle. A correspondingly greater atrophy was therefore observed in the soleus muscle (Figure 4A). These findings, when based on a relationship between activity and protein turnover, are as might be expected. From electromyographic studies (5) it is known that the soleus muscle exhibits fairly continuous activity in performing its role in the maintenance of body posture. In contrast, the EDL is much less frequently recruited (5), mainly during the locomotion of the animal. Hence the overall level of activity in these two muscles

Figure 3. Changes in protein turnover in immobilized and recovering muscles

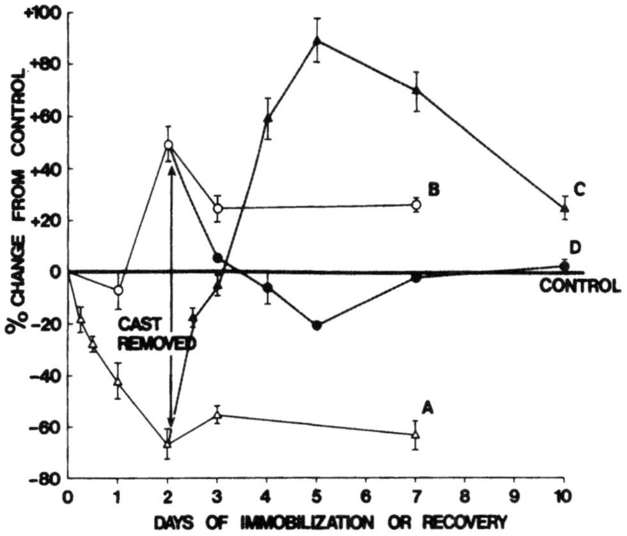

Average rates of protein synthesis (ie, A & C) and protein breakdown (ie, B & D) were measured in vitro (6,25) in immobilized (A,B) and recovering (C, D) soleus muscles. Rates are expressed as percent changes from controls.

Figure 4(A & B). Changes in the EDL and soleus after immobilization or starvation.

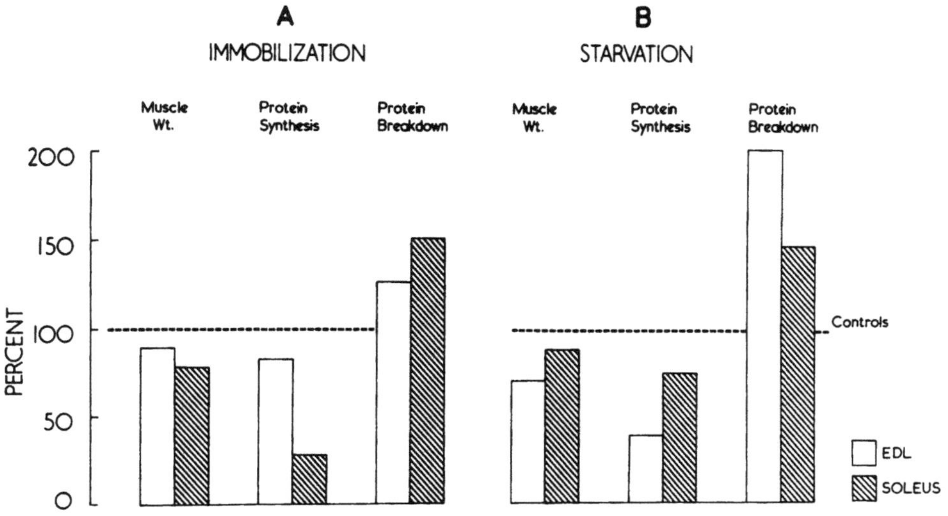

Measurements made 48 h after either immobilization (A), or starvation (B), are expressed as a percent of their unrestrained (A), or fed (B), controls.

is very different and this is correspondingly reflected in their differing basal rates of protein turnover. That is, both higher levels of activity and rates of protein turnover are found in the soleus, compared with the EDL (10,16). Since the normal soleus possesses a higher initial level of activity than the EDL, it follows that of the two muscles the former should be the more affected when rendered inactive.

B. The return of activity to formerly immobilized muscles.

Several methods (eg, tenotomy, denervation, dorsal root or spinal cord section) have been used over many years to induce muscle inactivity. However, most of these procedures are quite traumatic and irreversible. In contrast, the simple immobilization procedure described here is readily reversed by removing the plaster cast. Since there has been no interference with the nerve supply, normal activity is quickly restored to the formerly immobilized tissue. For example, the return of the normal shortening activity to the 2 day immobilized soleus rapidly restored the recovering tissue (curve C) to the same size as that of its unrestrained control (Figure 2, curve A). In so doing, two aspects of growth have probably taken place: (i) reversal of the true atrophy and loss of sarcomeres in series, which occurred during the period of immobilization, and (ii) a phase of 'catch-up' growth to compensate for the continued growth of the unrestrained, control soleus (curve A) during the 2 days the experimental muscle was immobilized. Further confirming the relationship between contractile activity and protein turnover (Figure 3), was the finding of a stimulation of protein synthesis (curve C) and an inhibition of protein breakdown (curve D) as the muscle once again operated throughout its normal functional range. After the recovering muscle had been fully restored to the same size as that of its control (Figure 2, curve A), the rates of protein synthesis and breakdown became indistinguishable in both the formerly immobilized and control muscles (Figure 3). The fall in the content of RNA and DNA after immobilization was also restored with the return of normal activity (11).

The Influence of the Nerve Supply on Protein Turnover.

A long standing experimental method of inducing inactivity in skeletal muscle has been to section the nerve supply to that tissue (12). In adult animals this usually leads to a true atrophy of the muscle (12,13,14), while in young animals some denervated muscles simply exhibit diminished growth (15,16). In certain situations, eg, the denervated hemidiaphragm of the rat (17,18,19), or anterior latissimus dorsi of the fowl (20), transient increases in size above that of their innervated controls were initially found.

In the case of the denervated soleus muscle of young rats (Figure 5), an initial atrophy was followed by a later phase of renewed growth (curve B). This latter accumulation of protein and growth has been attributed to the appearance of anabolic trends, which are probably caused by an increased stretching of this denervated tissue (see below). Although several studies have attempted to define the effects of denervation on protein turnover, most have been confined to measurements on the protein synthetic side of the regulatory processes. These studies have, however, left a confusing and often contradictory account of the events post denervation. This situation partly arises from the study of many different animal species and the examination of changes at single time points only after nerve section. These events may be summarized as follows. Most measurements performed only a few hours, or 1 or 2 days after denervation suggest a decreased rate of protein synthesis in the denervated tissue (13, 21,22). In contrast, measurements taken after longer durations, ie, days or weeks, describe the opposite situation, with increased rates of synthesis in the disused muscles (13,23,24). Most of these findings are consistent with the reversal of changes in protein synthesis (Figure 6, curve A) in our own detailed study on the denervated soleus muscle of the rat. The initial decrease in protein synthesis (Figure 6, curve A) correlates with the initial atrophy of the denervated tissue (Figure 5, curve B) when compared with the growing, innervated control (curve A). However, the subsequent accumulation of protein and hence growth (Figure 5, curve B) was in keeping with the delayed stimulation of protein synthesis in the denervated soleus. This reversal of changes in protein synthesis was

Figure 5. The wet wt. of the soleus after various operative procedures.

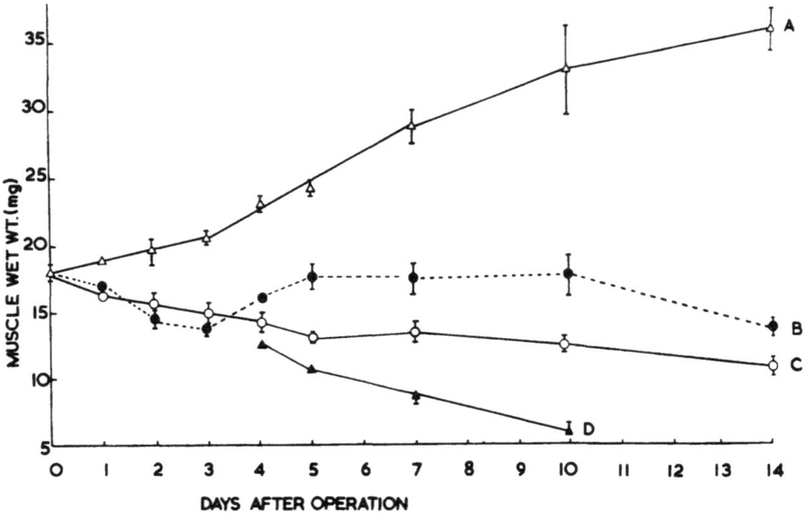

Soleus muscles (A; controls) were either denervated (B), immobilized (C) or, denervated and immobilized (D).

Figure 6. Changes in protein turnover in inactive soleus muscles.

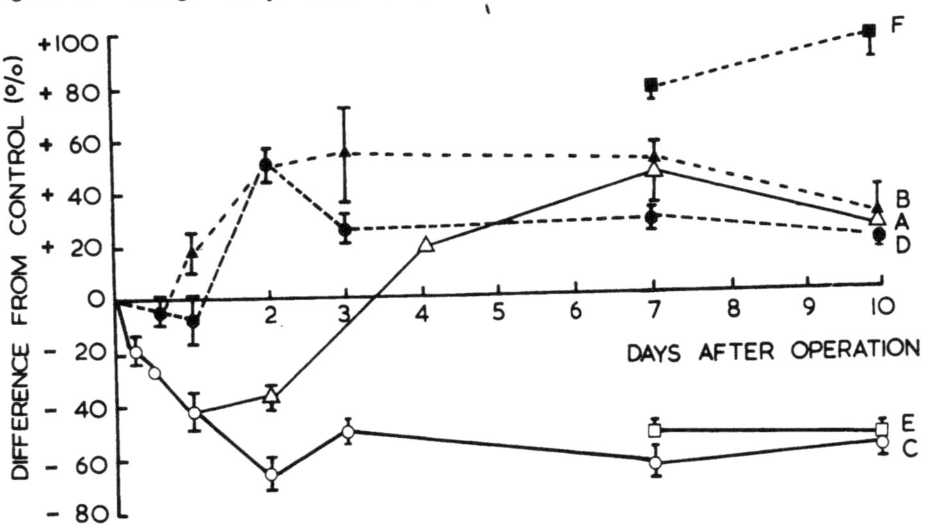

Protein synthesis (ie, A,C & E) and breakdown (ie, B,D & F) were measured in denervated (A,B), immobilized (C,D) or, denervated and immobilized (E,F) muscles. Values are percent changes from respective controls.

found to be even more pronounced in the denervated EDL muscle of young rats (16). The latter unexpected stimulation of the synthesis of both sarcoplasmic and myofibrillar proteins in the denervated leg muscles (16) probably arises from an increased stretching of these tissues by the altered gait of the animal. Stretching denervated (25,26), as well as innervated (4,27), muscles has been shown to stimulate protein synthesis, hence the probable growth of the denervated leg muscles (25), the hemi-diaphragm (17,18,19) and wing muscles of the fowl (20).

Increased rates of protein catabolism (Figure 6, curve B) in the denervated soleus were also consistent with the initial atrophy of this tissue (Figure 5, curve B). This maintained higher rate of breakdown (Figure 6, curve B) was then subsequently opposed by the increased synthetic rate (curve A). Other investigators have either measured (22), or calculated (19), similarly higher rates of protein degradation following nerve section, and presumably this is associated with higher activities of intracellular proteinases (14,28).

A Comparison of the Effects of Immobilization and Denervation on Protein Turnover.

For comparative purposes the changes in protein turnover after immobilization (Figure 3) have been superimposed on those of the denervated muscle (Figure 6). It can be seen that these two methods of rendering the same muscle inactive are very similar with respect to the induced changes in protein breakdown (curves B & D) and the early changes in protein synthesis (curves A & C). The most striking difference was the presence of the stimulation of protein synthesis in the denervated soleus (curve A) but its absence from the immobilized muscle (curve C). This difference was reflected in the lower, overall wasting of the denervated muscle (Figure 5, curve B), compared with the immobilized tissue (curve C).

Immobilization of the denervated ankle in the plantar flexed position prevents any stretching of the soleus (ie, it is held in a fully shortened state) through a change in the animal's gait. The consequence of combining these two procedures (Figure 6) was such that the anabolic trends (16), including the stimulation of protein synthesis (curve A), were abolished (curve E). The removal of passive stretch from the denervated EDL has also been shown to cause a significant decrease in the rate of protein synthesis (25). In addition to suppressing protein synthesis, the removal of stretch from the denervated soleus induced a further increase in the rate of protein degradation (curve F). The two combined changes in synthesis (curve E) and breakdown (curve F), probably serve to explain the even greater atrophy of the soleus (Figure 5, curve D), upon combining these two operative procedures, compared with their separate use (curves B and C). Hence the denervated soleus, now without the influence of stretch (Figure 5, curve D), underwent a greater atrophy than that induced by immobilization alone (curve C).

In morphological terms denervation causes a reduction in fibre diameters (29), but not in sarcomeres in series (30). Conversely, the principal changes after immobilization of the innervated muscle in a shortened position are concerned with longitudinal growth (7,8,9,30). Presumably the combined denervated and immobilized soleus loses sarcomeres both transversely and in series, thus accounting for its even greater atrophy (Figure 5, curve D). The procedures of denervation (31) or immobilization (5) produce quite different responses in the muscle's electrical and mechanical properties. Hence, on the basis of morphological, electrophysiological and biochemical evidence, it may be concluded that immobilization and denervation are not equivalent, even though they do render muscles inactive. Denervation itself leads to a complicated sequence of events caused by depriving the muscle of its electrical input and possibly trophic chemical substances. The full extent of the atrophy which should be observed may be masked by other factors, eg, passive stretch (25) or spontaneous fibrillation, which probably give rise to opposing anabolic trends. Immobilization in the fully shortened state does not interfere with the nerve supply, neither does it incur the problems of stretch. It is known that activity is drastically reduced when a muscle is immobilized at a slightly shortened length (5). However, it is not yet certain if any residual activity remains when the sarcomeres are fully shortened. This latter question should be answered shortly.

Possible Relationships between Contractility and the Responsiveness of Muscles to Hormones

A. Changes in protein turnover during starvation

Tonically active muscles (eg, the soleus) may be more resistant than phasic muscles (eg, the EDL) to changes in their immediate environment. For example, the soleus muscle of young rats undergoes less atrophy (12%) than the EDL (30%) when the body's musculature is subjected to a strongly catabolic environment as created by 2 days of starvation. The resultant inhibition of protein synthesis and enhancement of protein breakdown was more marked in the EDL, hence the more pronounced atrophy of this muscle (Figure 4B).

In subsequent experiments, we found that the level of activity appeared to influence the degree of responsiveness of a muscle to certain circulating hormones. In particular we examined the effects of the glucocorticoid hormones and insulin on both normally active and inactive muscles. The plasma concentrations of these two hormones are known to change appreciably in the starved state.

B. Muscle activity and its effect on the responsiveness to glucocorticoids

Although the 4 days of treatment with exogenous cortisone (Figure 7) caused significant decreases in both the rates of protein synthesis and protein breakdown, the normally active soleus fully maintained its protein content (ie, these changes effectively cancelled each other out). However, when the same muscle was rendered inactive throughout this period by immobilization, the resultant decrease in the protein content (40%) was further reduced (to 55%) upon exposure to the glucocorticoid (Figure 7). This additional atrophy produced by the hormone correlated with a further suppression (to 65%) of the already lowered rate of protein synthesis (46%) within the inactive muscle. In addition, the increased rate of protein breakdown (18%) caused by immobilization was further increased (to 61%) upon exposure to the steroid hormone (Figure 7).

Figure 7. Effects of exogenous cortisone on the normal and immobilized soleus.

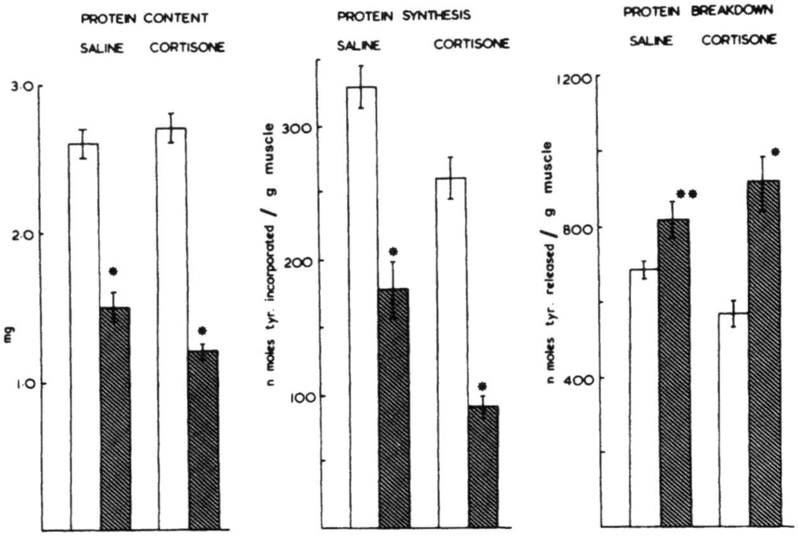

In each rat the soleus of one leg was immobilized. Thereafter, saline (ie, control group) or cortisone acetate (2.5 mg/Kg/day) was injected for 4 days. ☐ normal and ▨ immobilized muscles; *P < 0.005 and ** P < 0.025.

Figure 8. Insulin's action on protein synthesis of normal and denervated EDL.

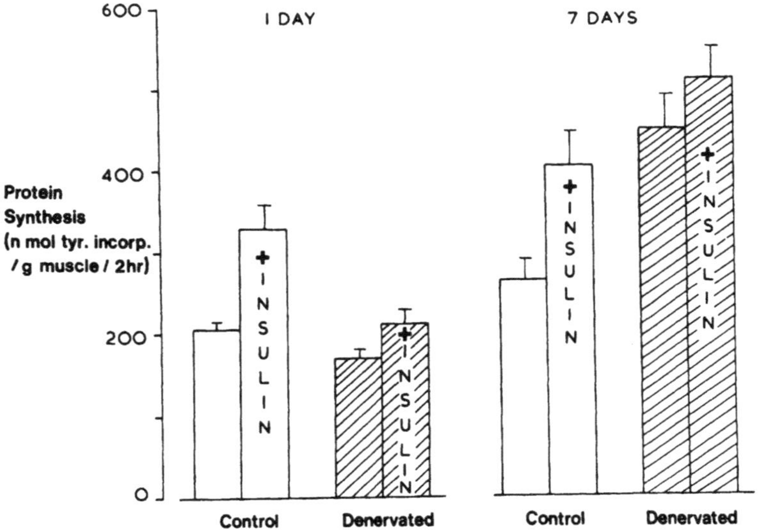

Protein synthesis was measured in vitro in control and 1 or 7 day denervated EDL muscles in the presence or absence of insulin (1 mU/ml).

The normally high level of activity in the soleus probably protects this muscle from the full catabolic effects of the adrenal steroids. This conclusion is further supported by the findings that the more phasic muscles, like the EDL, are more severely affected by either the oversecretion or exogenous administration of glucocorticoid hormones (1,32,33).

C. Muscle activity and its effect on the responsiveness to insulin

Earlier investigators using the denervated diaphragm (34,35) concluded that this inactive muscle became insensitive to the anabolic effect of insulin. However, in these earlier in vitro studies protein synthesis was measured without any allowance for possible changes in the specific activity of the intracellular amino acid pools. In addition, the concentrations of insulin used were far in excess of normal plasma levels. Recently we (McGeown & Goldspink - unpublished) undertook some similar experiments, but using physiological concentrations of insulin (1) and a suitable method for measuring protein synthesis with appropriate corrections for changes in pool sizes (6,25). In the presence of insulin we found protein synthesis increased between 50 to 55% in the innervated EDL muscles (Figure 8). Such an anabolic action of insulin is now well established (1). Although the 1 day denervated EDL also responded significantly (24%) to the addition of insulin, protein synthesis in the 7 day denervated tissue was not enhanced by the hormone (Figure 8). Hence, in agreement with the earlier studies (34,35) the denervated EDL appeared to lose its sensitivity to insulin, at least with respect to the hormone's action on protein synthesis. It is, however, worthy of note that insulin's inhibitory influence on protein breakdown remained unchanged after denervating the EDL (1).

In conclusion, the muscle atrophy induced by either immobilization or denervation is probably enhanced by alterations in the responsiveness of the inactive tissue to circulating hormones. As illustrated here, inactive muscles probably acquire an increased responsiveness to catabolic hormones like the glucocorticoids, and a decreased responsiveness to anabolic hormones such as insulin.

Acknowledgements

This work was made possible by the skilful assistance of Patricia C. Verner, P. Anderson, Josephine A. McGrath and J.G. McGeown. The author also acknowledges the financial support of the Fund for Research into Crippling Diseases and the Medical Research Council.

References

1. Goldspink, D.F.: The Development and Specialization of Skeletal Muscle, S.E.B. Seminar Series (ed. Goldspink, D.F.), Chapter 6, Cambridge University Press, Cambridge 1980.
2. Waterlow, J.C., Garlick, P.J., Millward, D.J. (eds.): Protein Turnover in Mammalian Tissues and in the Whole Body. Chapter 1, North Holland, Amsterdam 1978.
3. Goldberg, A.L., St.John, A.C.: Ann. Rev. Biochem. 45, 747-803 (1976).
4. Goldspink, D.F.: J. Physiol. (London) 264, 267-282 (1977).
5. Fischbach, G.D., Robbins, N.: J. Physiol. 201, 305-320 (1969).
6. Fulks, R.M., Li, J.B., Goldberg, A.L.: J. Biol. Chem. 250, 290-298 (1975).
7. Williams, P.E., Goldspink, G.: J. Anat. 116, 45-55 (1973).
8. Williams, P.E., Goldspink, G.: J. Anat. 122, 455-465 (1976).
9. Tabary, J.C., Tabary, C., Tardieu, C., Tardieu, G., Goldspink, G.: J. Physiol. 224, 231-244 (1972).
10. Goldberg, A.L.: Nature, 216, 1219-1220 (1967).
11. Goldspink, D.F.: J. Physiol. 264, 283-296 (1977).
12. Gutmann, E. (ed.): The Denervated Muscle. Czechoslovakian Academy of Science, Prague 1962.
13. Goldspink, D.F.: Comp. Biochem. Physiol. 61B, 37-41 (1978).
14. Goldspink, D.F., Harris, J.B., Park, D.C., Parson, M.E., Pennington, R.J.: Int. J. Biochem. 2, 427-433 (1971).
15. Lewis, D.M.: Nature, New Biol. 241, 285-286 (1973).
16. Goldspink, D.F.: Biochem. J. 156, 71-80 (1976).
17. Sola, O.M., Martin, A.W.: Am. J. Physiol. 172, 324-332 (1953).
18. Manchester, K.L., Harris, E.J.: Biochem. J. 108, 177-183 (1968).
19. Turner, L.V., Garlick, P.J.: Biochim. Biophys. Acta. 349, 109-113 (1974).
20. Sola, O.M., Christensen, D.L., Martin, A.W.: Exptal. Neurol. 41, 76-100 (1973).

21. Dreyfus, J.C.: Rev. Fr. Etud. Clin. Biol. 21, 461-466 (1967).

22. Goldberg, A.L., Jablecki, C., Li, J.B.: Ann. N.Y. Acad. Sci. 228, 190-201 (1974).

23. Ferdman, D.L.: The Effect of Use and Disuse on Neuromuscular Functions (ed. Gutmann, E. & Hnik, P.) 407-412, Elsevier Publishing Co., Amsterdam 1963.

24. Pater, J.L., Krohn, R.R.: Proc. Soc. Exp. Biol. Med. 125, 476-481 (1967).

25. Goldspink, D.F.: Biochem. J. 174, 595-602 (1978).

26. Buresova, M., Gutmann, E., Klicpera, M.: Experientia 25, 144-145 (1969).

27. Laurent, G.J., Sparrow, M.P., Millward, D.J.: Biochem. J. 176, 407-417 (1978).

28. Pennington, R.J.T.: Proteinases in Mammalian Cells and Tissues (ed. Barrett, A.J.) 515-543. North-Holland, Amsterdam 1977.

29. Muscatello, V., Margreth, A., Aloisi, M.: J. Cell Biol. 27, 1-22, (1965).

30. Goldspink, G., Tabary, C., Tabary, J.C., Tardieu, C., Tardieu, G.: J. Physiol. 236, 733-742 (1974).

31. Lewis, D.M.: J. Physiol. 222, 51-75 (1972).

32. McGrath, J.A., Goldspink, D.F.: Ir. J. med. Sci. 147, 328 (1978).

33. Goldberg, A.L., Goodman, H.M.: J. Physiol. 200, 667-675 (1969).

34. Harris, E.J., Manchester, K.L.: Biochem. J. 101, 135-145 (1966).

35. Buse, M.G., McMaster, J., Buse, J.: Metabolism 14, 1220-1232 (1965).

CALCIUM AND STRETCH-DEPENDENT REGULATION OF PROTEIN TURNOVER AND MYOFIBRILLAR DISASSEMBLY IN MUSCLE

J.D. Etlinger, T. Kameyama, K. Toner, D. van der Westhuyzen and K. Matsumoto

Department of Anatomy and Cell Biology, State University of New York, Downstate Medical Center, 450 Clarkson Avenue, Brooklyn, New York 11203

Abstract

We have investigated the roles of Ca^{++} and tension in regulating protein degradation and synthesis in rat muscles incubated <u>in vitro</u>. Proteolysis was stimulated by addition of Ca^{++} to the incubation medium and was further enhanced by the Ca^{++} ionophore A23187 or by depolarization with elevated K^+. K^+-induced degradation was specifically inhibited by dantrolene which uncouples Ca^{++} release from excitation. Protein synthesis was also stimulated by Ca^{++} and the increased synthesis could be partially inhibited by actinomycin D. Fixing the length of the muscle during incubation decreased proteolysis when compared to muscles incubated in a free state; an effect opposite to Ca^{++}. In contrast, tension stimulated synthesis in the presence of elevated Ca^{++}. The overall effect of Ca^{++} in the absence of applied tension appears to favor a net catabolic state i.e. the increase in synthesis is smaller than the increase in degradation. However, in the presence of tension Ca^{++} produces affects which may relate to hypertrophy since Ca^{++}-dependent degradation is reduced or reversed while synthesis is further stimulated.

We have also examined the role of Ca^{++} and tension in myofibrillar disassembly. Gentle treatment of isolated myofibrils with an ATP-containing relaxing solution results in the release of a small fraction of myofilaments. Ca^{++} increases the size of the releasible filament fraction whereas stretch decreases the amount of releasible myofilaments. Leupeptin, an inhibitor of Ca^{++}-dependent protease inhibits protein

Plasticity of Muscle

degradation in ionophore-treated muscles and also reduces myofilament release. Thus, changes in filament release parallel rates of proteolysis. Digestion of isolated myofibrils with Ca^{++}-dependent protease also increases the quantity of releasible myofilaments. Therefore, releasible filaments may be intermediates in myofibrillar degradation as well as assembly and these processes like overall protein synthesis and degradation appear coordinately regulated by independent effects of Ca^{++} and tension.

Introduction

Large and rapid changes in protein content occur during hypertrophy and atrophy of cardiac and skeletal muscle (1-3). In addition, the protein composition of a muscle fiber can change markedly in response to the imposed activity pattern (this symposium). Since muscle protein normally undergoes continuous intracellular turnover, alteration in protein levels can occur through changes in degradative and/or synthetic rates (3). Thus, muscle can, in principle, grow as a result of decreased protein degradation as well as increased synthesis. Protein degradation plays an essential role in producing atrophy as well as permitting the replacement of one form of a protein, e.g. myosin by a different isozyme. However, despite the importance of protein degradation in muscle plasticity, little is known about the relevant cellular processes.

To investigate the mechanisms responsible for controlling protein degradation as well as synthesis, we have focused on the role of Ca^{++} for several reasons. First, Ca^{++} plays a central role in contractility and thus could serve to directly couple muscle work-load and activity patterns to adaptive changes in size or composition. Second, evidence already exists for ability of Ca^{++} to couple energy metabolism to activity (6,4). Third, certain hereditary dystrophies involve increases in protein degradation, protein synthesis and cellular uptake of Ca^{++} (5). Finally, there is now evidence that response to several hormones may, in part, be Ca^{++} dependent (4,6). Our results indicate that Ca^{++} can stimulate protein degradation as well as protein synthesis under certain conditions.

However, such effects by themselves can not account for muscle growth. Maintainance of tension also appears necessary so as to oppose the stimulatory effect of Ca^{++} on degradation and to further enhance synthesis, thus favoring an anabolic state.

Regulation of Protein Degradation and Synthesis by Calcium

We have utilized several methods to examine the effect of increased intracellular Ca^{++} levels on rates of overall protein degradation and synthesis in muscles incubated *in vitro*. Paired rat soleus, extensor digitorum longus (EDL), diaphragm and left atria were incubated *in vitro* in defined media under conditions in which the muscles maintain linear rates of protein synthesis and degradation for at least several hours (7,8). Protein degradation was measured as the amount of tyrosine released from protein into the muscle pools and medium in the presence of cycloheximide, an inhibitor of protein synthesis (5,7,8). The first approach utilized to alter Ca^{++} levels was the addition of the Ca^{++} ionophore A23187 to the incubation medium. Initially ionophore was added to muscles incubated for 2 hours in a free state so that no tension was maintained in either the presence or absence of shortening (Fig.1-no stretch). Under such conditions protein degradation was markedly increased when A23187 concentration was raised to 10 µg/ml. Similar stimulation of protein degradation was seen when A23187 was added to the EDL and bisected left atria (not shown). When muscles were incubated in medium prepared without Ca^{++}, the ionophore produced no significant increase in protein degradation indicating that movement of Ca^{++} into the cells is probably required (5). Production of tension in the rat soleus in response to A23187 also showed an absolute requirement for extracellular Ca^{++} (9) unlike results obtained with certain other muscle preparations (10,11).

A second method used to elevate Ca^{++} involved increasing K^{+} in the medium to depolarize the plasma membrane, in turn, leading to Ca^{++} release from the sarcoplasmic reticulum. High K^{+} increased protein degradation in the soleus and EDL by about 55% (Table 1) an effect similar to that produced by A23187 (Fig.1). To further define this effect, we tested dantrolene

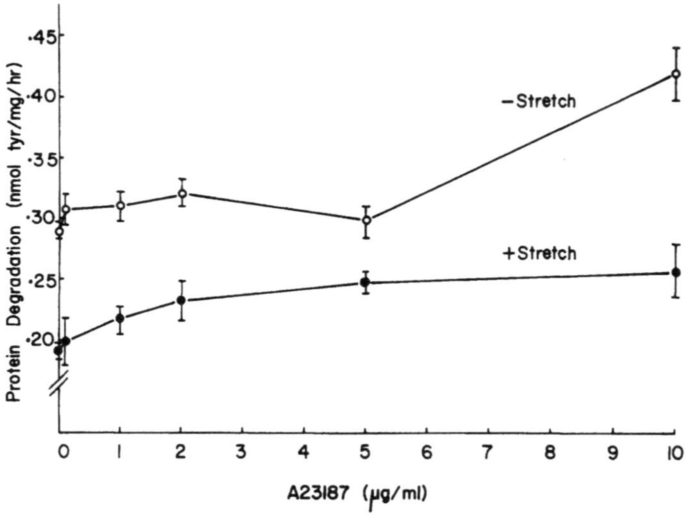

Figure 1. Effect of Ca^{++} on protein degradation. Normal female rats (40-55 gm) were killed by cervical dislocation and paired soleus muscles were immediately removed with tendons intact and weighed. Each muscle was pre-incubated in the presence or absence of A23187 (Eli Lilly) as indicated for 1 hour at 37°C. The incubation medium (3.0ml) also contained Krebs-Ringer bicarbonate solution, (containing 2.5mM Ca^{++}), saturated with 95% O_2 - 5% CO_2 and supplemented with 10 mM glucose, 5 X plasma levels of the branched chain amino acids (valine, isoleucine and leucine), .1 unit/ml of insulin and 0.5mM cycloheximide (5,7). A23187 was dissolved in absolute ethanol and the final concentration of ethanol in all incubations was adjusted to .5% v/v. After one hour, the muscles were blotted and transferred to fresh medium of the same composition. Protein degradation, measured during the second hour, was proportional to tyrosine appearing in the medium during that period. Tyrosine was measured flurometically as previously described (5,7). Muscles were incubated under tension (+ stretch) by fixing the length of the tissues at 110% of rest length with pinning through the tendons into small plastic supports. Other muscles (- stretch) were incubated in a flaccid state and would shorten against no load. Results are expressed as the mean value ± the standard error of the mean N=6.

Table 1
Effect of K^+ Depolarization and Dantrolene on Protein Degradation

Addition	Protein Degradation (nmol tyr/mg/hr)	Change %
A. EDL		
None	.183 ± .004	
High K^+	.287 ± .013	+56 (P < .0005)
High K^+	.299 ± .013	
High K^+, Dantrolene	.229 ± .012	-23 (P < .0005)
None	.169 ± .010	
Dantrolene	.152 ± .004	-10 (NS)
B. Soleus		
None	.246 ± .020	
High K^+	.381 ± .035	+55 (P < .005)
High K^+	.320 ± .018	
High K^+, Dantrolene	.288 ± .016	-10 (P < .05)
None	.222 ± .012	
Dantrolene	.211 ± .011	- 5 (NS)

Table 1. Measurement of protein degradation was carried out as described in Figure 1. High K^+ medium was prepared by substituting 100mM KCl for 100mM NaCl in the supplemented KRB medium (Fig. 1). Dantrolene was dissolved in NaOH and added at a final concentration of 10µg/ml. Muscles were preincubated for 1 hour with the indicated additions before measuring degradation as in Fig. 1. P values were calculated for paired differences using the Student's t-test.

sodium on protein degradation. The mode of action of this muscle relaxant is to inhibit Ca^{++} release by interfering with excitation-contraction coupling (12). Treatment of muscles with dantrolene significantly inhibited protein degradation in the depolarized soleus and EDL preparations (Table 1). In contrast, dantrolene failed to significantly affect basal proteolytic rates in individual experiments, although a small amount of inhibition was consistantly noted.

Finally, we have also examined the effect of Ca^{++} supply itself on protein degradation. Treatment of muscles for 2 hours in the usual medium (Fig.1) but prepared without Ca^{++} produced a significant decrease in proteolysis (20-35%). Thus, in light of the similar results obtained with A23187 and high K^{+}, it appears likely that removal of Ca^{++} from the medium leads to lower intracellular Ca^{++} concentrations.

Similar experiments were also performed to determine the effect of Ca^{++} on protein synthesis in these muscles. To measure protein synthesis rates, the incorporation of ^{14}C-tyrosine into total protein was corrected for the specific activity of the intracellular tyrosine pool (5,7). As was noted for protein degradation in muscles incubated without tension, a significant stimulation could be observed at high ionophore concentration (Fig.2). Increased synthesis could be inhibited partially, (50-70%), by addition of actinomycin D (G. Silver and J. Etlinger in preparation), thus indicating the involvement of some class of RNA synthesis in the Ca^{++}-dependent stimulation of protein synthesis. However, unlike the consistent stimulatory effect of A23187 on degradation, the stimulation of synthesis was highly variable sometimes even showing significant inhibition in certain groups of animals. This may reflect variable Ca^{++} sensitivity of the fibers and a highly sensitive mechanism which may respond in a positive manner over a narrow range of Ca^{++} concentration. More recently we found that the presence of Ca^{++} itself reproducibly enhanced protein synthesis over rates obtained in Ca^{++}-free medium (not shown). In fact, because of the more uniform results obtained by simply varying Ca^{++} in the incubation medium we are presently using this approach to further analyze the Ca^{++}-dependent stimulation of protein synthesis. Such results appear related to the Ca^{++}-dependent enhancement of certain enzyme activities in cultured muscle (4), a process also inhibited by actinomycin D.

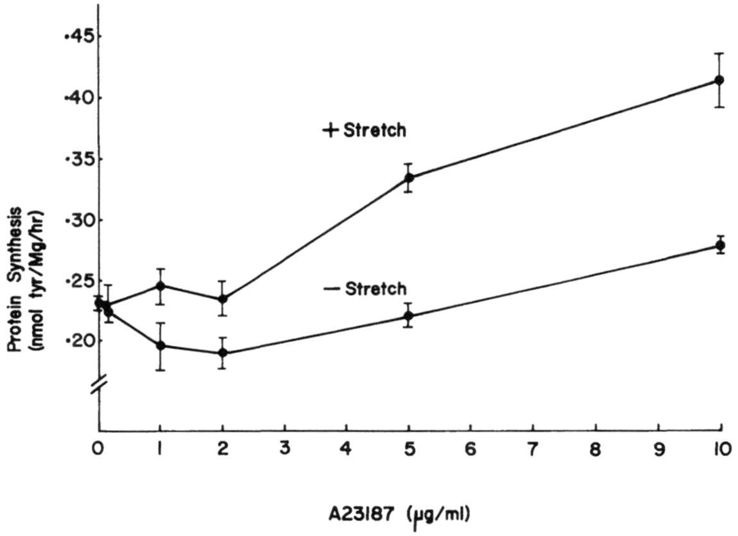

Figure 2. Effect of Ca^{++} on protein synthesis. Muscles were incubated as described in Figure 1 but without cycloheximide and with 0.5mM ^{14}C-tyrosine (Schwartz-Mann), .06 μCi/ml. Determination of synthesis was carried out by correcting rates of ^{14}C-tyrosine incorporation into acid-precipitable protein using the specific activity of the intracellular tyrosine pool (7).

Influence of Tension on Protein Turnover

The studies described above demonstrate that Ca^{++} can stimulate protein degradation as well as synthesisin muscles which are not permitted to develop tension. In fact, these muscles shorten against no load during the experiment. Since several previous studies (see 1 for review and 13-15), have suggested that passive stretch or activity can influence synthesis and/or degradation we examined the effects of active as well as passive tension in relationship to the action of Ca^{++} on protein turnover. Maintaining the muscle in a stretched vs. a free state decreased proteolysis approximately 50% (Fig.1) in accord with previous results (1). The ability of stretch to decrease protein degradation could be seen over a wide range of lengths. In fact, simply fixing the muscle at rest length (defined as the length at which passive tension is first detected), resulted in a similar decrease in protein catabolism (16). Interestingly, a 6-10% stimulation of degradation was seen when stretched muscles, (110% X Lo), were compared to control muscles maintained at rest length (16). The importance of this opposite but smaller stretch effect on degradation is unclear. The major effect of tension in decreasing degradation, however, appears opposite to the effect of Ca^{++}.

To further study the relationship between the tension and Ca^{++} effects, we examined the ability of tension to affect proteolysis in the presence of varying concentrations of Ca^{++} ionophore (Fig.1). Tension in the presence of ionophore included an increment of active tension in addition to the passive tension resulting from 10% stretch (16). Maintainance of tension resulted in a reduction of proteolysis at all concentrations of ionophore. In fact, in the presence of levels of A23187 which stimulated degradation in free muscles, stretch had an equal or greater ability to reduce proteolysis (Fig.1). Thus, the major effect of active or passive tension on protein degradation appears independent to that of Ca^{++}.

We have also examined the effect of tension on protein synthesis. When muscles were maintained in a stretched state and compared to flaccid muscles, protein synthesis was unchanged for periods up to 2 hours as seen previously (1). However, maintainance of tension in the presence of

A23187 stimulated synthesis above any effect due to increased Ca^{++} alone. Thus, tension appears to potentiate the effect of Ca^{++} on synthesis in contrast to its opposing effect on Ca^{++}-dependent degradation. All of the effects described above apply to average protein in the cells and it should be emphasized that no conclusion can necessarily be drawn about the pattern of turnover of individual proteins with regard to the influence of tension and Ca^{++}.

Coordinate Regulation of Protein Degradation and Myofilament Release

Although we have not yet examined the degradation or synthesis of specific muscle proteins,we have, however, examined certain steps in myofibrillar disassembly which may proceed degradation of the constituent proteins to amino acid. Several years ago Etlinger et al. observed that gentle treatment of isolated myofibrils with an ATP-containing relaxing solution resulted in the separation of only a small amount of thick and thin myofilaments from a residual population of intact skeletal myofibrils (17,18). Repeated extraction of the myofibrils with relaxing solution did not appreciably increase the amount of releasible myofilament protein. Incorporation of radioactive amino acid into protein _in vivo_ demonstrated that the specific activity of the easily releasible myofilaments was several fold greater than that of the residual myofibrils, although with time, the specific activities became similar (18). Analysis of the composition of these myofilaments revealed that they were devoid of M-line, Z-line and certain high molecular weight protein of the myofibril (17). These results were later confirmed in the heart (19). More recently, we established methods for the isolation of myofibrils and releasible myofilaments from the small rat cardiac and skeletal muscles following pulse-labeling _in vitro_ (20-22). Such experiments indicated that myosin heavy chain and actin in releasible myofilaments are highly labeled after brief exposure to radioactive precursor but rapidly mix with the residual myofibrils in the presence of cycloheximide, an inhibitor of protein synthesis. Various arguments suggested that the releasible myofilaments may originate from the surface of individual myofibrils consistent with the peripheral localization of newly synthesized myofibrillar protein as determined by electron microscopic autoradiography (29).

Other observations prompted us to investigate whether the releasible protein fraction also contains myofilaments destined for degradation. For example, the size of this fraction is greater than the total amount of protein synthesized during even several hours and often varies between experiments. Therefore, we examined whether the amount of releasible myofilament would increase when rates of protein degradation were elevated. Pre-treatment of the muscles with Ca^{++} ionophore for 2 hours increased the amount of releasible myofilament; an effect which paralleled the increase in rates of proteolysis in these tissues (Table 2). In contrast, stretch decreased the amount of releasible myofilament in parallel with reduced proteolysis (Table 2).

Such observations suggest that release of certain myofilaments may require initial limited proteolytic cleavage(s) and that the loosening of myofilaments may be an initial step in the degradation of the myofibril. To test this idea, we examined the effect of leupeptin, a non-toxic protease inhibitor, on protein degradation and myofilament release. Leupeptin has been shown to enter cells and retard muscle degeneration (23), and proteolysis (24) without inhibiting protein synthesis or affecting other cellular reactions. We found that leupeptin inhibits protein degradation in ionophore-treated muscles (5) and also decreases the amount of releasible myosin and actin filaments (Fig.3). The reduction of myofilament release by leupeptin must occur primarily during the muscle incubation and not during subsequent fractionation procedures as a result of residual leupeptin adhering to the outside of cells. This follows from the observation that inhibition of filament release can be seen independently of the presence of protease inhibitors added during fractionation of the tissues (21,22).

Leupeptin is known to inhibit at least two proteases found in muscle. These are cathepsin B-1, a lysosomal enzyme, and soluble Ca^{++}-activated neutral protease (24-27). Although it is difficult to prove which, if any,

Table 2

Effect of Tension and Ca^{++} on Myofilament Release

Treatment	Releasible Myofilament (% Myosin Heavy Chain)	Change
No stretch	.49 ± .02	
Stretch	.38 ± .02	-22 ($P < .01$)
Stretch	.31 ± .03	
Stretch + A23187	.46 ± .02	+48 ($P < .01$)

Muscles were incubated in the presence or absence of applied tension as described in Fig. 1. After 2 hours, 8 soleus muscles were pooled and myofibrils and releasible filaments prepared by a modification (22) of earlier methods (17,18). The finely minced tissues were dispersed in a Dounce homogenizer followed by purification of myofibrils as previously described (18,17). Filaments were released from myofibrils by suspension in Mg ATP-containing relaxing solution at 4°C (18) with 10 passages through a Pasteur pipette. The suspension was immediately layered over relaxing solution containing 20% glycerol and centrifuged for 5 min x 1500xg. The pellet contained residual myofibrils. The supernatant was collected and recentrifuged through the glycerol solution. The final supernatant contained the released myofilaments. Proteins in the filaments and myofibrils were separated by SDS-gel electrophoresis and stained with Coomasie blue. Quantitation of the releasible protein was performed by slicing out the myosin heavy chain band followed by extraction of the dye with pyridine and measurement of absorbance at 605nM as previously described (33).

of these proteases is important in protein degradation in the intact cells we have examined the possible role of the Ca^{++}-dependent protease in myofibrillar disassembly for several reasons. This protease(s) can quickly remove Z- and probably M-lines from myofibrils before substantially degrading other proteins (26), thus, this action is consistent with the intact structure of the released filaments. Secondly, the Ca^{++}- dependency of this enzyme would permit a simple explanation of the ability of increased intracellular Ca^{++} to stimulate proteolysis. In muscle Ca^{++} interacts with an ATP-dependent proteolytic system (J. Etlinger and M. Glucksman, in preparation). This soluble degradative activity is analogous to the ATP-dependent proteolytic system first described in reticulocytes (28) and is probably responsible for the apparent energy-requirement of proteolysis in all cells (3,28). When myofibrils were digested with Ca^{++}-dependent protease for increasing periods of time followed by separation of the releasible myofilaments, it was found that the amount of releasible myosin and actin increased with digestion (Fig.4). Such effects were Ca^{++}-dependent and inhibitable by leupeptin (Fig.4). Thus, at least one intracellular protease is capable of modifying myofibrils in a manner which reflects changes occurring in the intact muscles.

Implications

Our studies have shown that Ca^{++} stimulates protein degradation as well as synthesis in several rat muscles. However, the relative magnitudes of these two effects depends strongly on whether tension is maintained during the experiment. In the absence of any active or passive tension Ca^{++} primarily stimulates protein degradation, especially at high Ca^{++} levels, and thus, increases the net negative nitrogen balance. Such large increases in degradation accompanied by smaller increases in synthesis resemble both denervation atrophy (14) and several hereditary dystrophies (5). In the case of the dystrophies existent stretch may be insufficient to balance the catabolic effects of high cellular Ca^{++} uptake known to

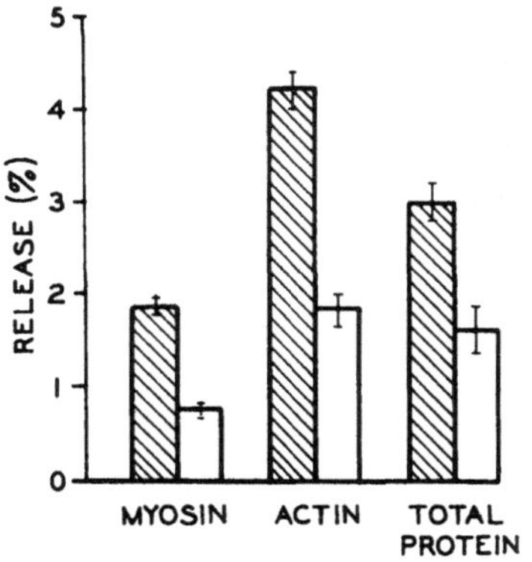

Figure 3. Effect of leupeptin on myofilament release. Hemidiaphragms were incubated in the presence (□) or absence (▧) of leupeptin (50μm) for 3 hrs. Myofibrils and releasible myofilaments were then prepared as described for Table 2 and the percentage of releasible myosin heavy chains, actin and total releasible myofilament protein was determined. Leupeptin was a gift from the US-Japan Cooperative Cancer Research Program.

occur in certain diseased muscles (5). Alternatively protein degradation and synthesis may fail to respond properly to normal stretch as a result of some unknown biochemical defect. The acceleration of degradation produced in our experiments differs significantly from the effects of the calcium ionophore A23187 which have been described in different muscle systems (10). These other studies describe a rapid destruction of cellular structure and loss of resting potentials. In contrast, our experiments show no visible loss of myofibrillar structure or ability to develop maximum twitch tension after two hours of exposure to A23187 (16). Probably the proteolysis which we describe proceeds massive cytoplasmic destruction which occurs at even higher Ca^{++} levels.

The observation that tension is required to oppose the net catabolic tendency of Ca^{++} (Fig.1) agrees with early _in vivo_ studies (32). In these studies, denervation appeared to block atrophy resulting from tenotomy suggesting that in the absence of tension activity causes an increase in net protein catabolism. In fact, our results suggest that Ca^{++} in the presence of maintained tension may even approximate situations which resemble muscle growth. It is interesting that while work-induced growth

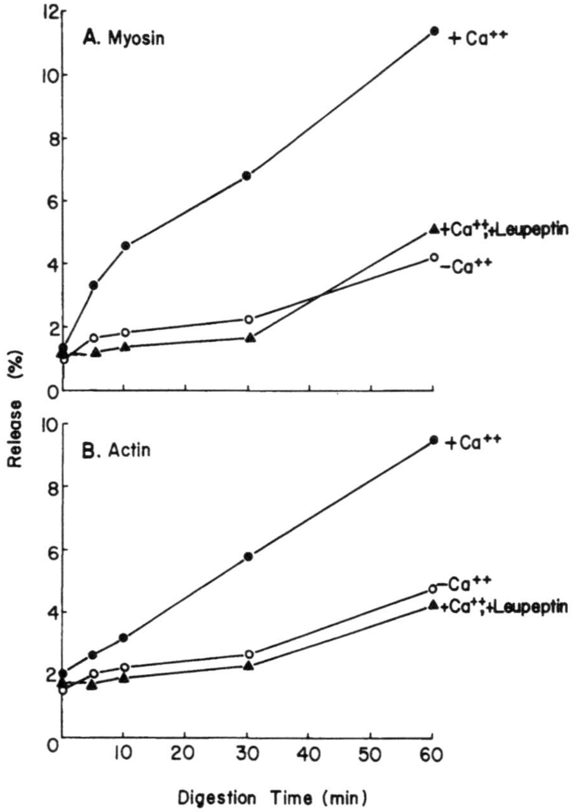

Figure 4. Effect of Ca^{++}-dependent protease on myofilament release. Myofibrils were prepared from rat soleus muscles as described in Table 2. Myofibrils were digested with Ca^{++}-dependent protease prepared from rabbit skeletal muscle as described previously (26). The protease fraction was partially purified as described except activity was eluted with a single 0.5M KCl step elution from DEAE-Sephacel. Digestion was performed in either Ca^{++} (2mM), - Ca^{++} (+2mM EGTA) and/or in the presence of leupeptin (10µM) as indicated. Myofilament release was performed as described in Table 2.

of muscle involves increased rates of protein synthesis (1), degradation rates have been reported to increase (29,30), decrease (1,2), and remain unchanged (31). Although such apparently conflicting results may be due to methodological differences, a more interesting possibility is that protein breakdown may increase as well as decrease depending on the particular phase or extent of hypertrophy. Furthermore, whether degradation rates appear to increase or decrease may depend on the precise tension or activity state of the control muscle. When compared to a control muscle maintained with only passive tension, Ca^{++} mainly stimulates synthesis while producing a smaller increase in degradation (Fig.1,2). However when the control is a muscle incubated without any applied tension, Ca^{++} together with stretch results in rates of degradation which are decreased or unchanged while synthesis is stimulated. It is not clear why degradation might increase under certain conditions of growth and decrease under others. Possibly, when it is of advantage to obtain maximal growth rates, the synergistic action of increased synthesis and decreased degradation would be desirable. On the other hand, when it is important to replace one myosin isozyme by another during hypertrophy (K. Schwartz et. al. this symposium), increased degradation and synthesis may be favored even though growth would not be as rapid.

Finally, there are several important questions which arise from our observations and those of others. First, can the particular frequency or pattern of Ca^{++} fluctuations determine which genes are expressed within a myofiber? Second, are there effects of Ca^{++} below the contraction threshholds and can Ca^{++} mediate certain hormonal effects on protein metabolism? Last, what agent(s) mediate the effects of stretch or tension which oppose the Ca^{++} effects? An attractive candidate is a cyclic nucleotide. Possibly tension or stretch applied to the cell surface activates membrane enzymes e.g. cyclase activities. These questions are currently being investigated in our laboratory.

References

1. Goldberg, A.L., Etlinger, J.D., Goldspink, D.F., Jablecki, C.: Med and Sci. in Sports. 7, 248 - 261 (1975).

2. Zak, R., Martin, A.F., Reddy, M.K., Rabinowitz, M.: Circulation Res. 38-I-145- 1-52 (1976).

3. Goldberg, A.L., St. John, A.C.: Ann. Rev. Biochem. 45, 747-803 (1976).

4. Schudt, C., Pette, D.: Adv. Enzyme Regulation. 16, 121-139 (1978).

5. Kameyama, T., Etlinger, J.D.: Nature 279, 344-346 (1979).

6. Rasmussen, H., Goodman, D.: Physiological Rev. 57, 421-510 (1977).

7. Fulks, R.M., Li, J.B., Goldberg, A.L. : J. Biol Chem. 250, 290-298 (1975).

8. Li, J.B., Goldberg, A.L.: Amer. J. of Physiol. 231, 441-448 (1976).

9. Grinstein, S., Erlij, D.J.: Membrane Biol. 29, 313-328 (1976).

10. Statham, H.E., Duncan, C.J., Smith, J.L.: Cell Tissue Res. 173, 193-209 (1976).

11. Desmedt, J.E., Hainaut, K.: J. Physiol. 257, 87-107 (1976).

12. Ellis, K.O., Bryant, S.H.: Arch. Exptl. Pathol, Pharmakol. 27, 107-113 (1972).

13. Peterson, M.D., Lesch, M.: Circ. Res. 31, 317-327 (1972).

14. Goldspink, D.F.: Biochem. J. 174, 595-602 (1978).

15. Vandenburgh, H., Kaufman, S.: Science. 203, 265-268 (1979).

16. Etlinger, J.D., Kameyama, T., Erlij, D., Matsumoto, K.: (Submitted for publication).

17. Etlinger, J.D., Zak, R., Fischman, D.A.: J. Cell Biol. 68, 123-141 (1976).

18. Etlinger, J.D., Zak, R., Fischman, D.A., Rabinowitz, M.: Nature. 255, 259-261 (1975).

19. Sims, J.M., Patzer, B., Kumudaralli-Reddy, Martin, A.F., Rabinowitz, M., Zak, R.: Recent Advances on Studies on Cardiac Structure and Metabolism. 12, 19-28 (1978).

20. Van der Westhuyzen, D.R., Matsumoto, K., Etlinger, J.D.: Circulation 58, II-73 (1978).

21. Van der Westhuyzen, D.R., Matusmoto, K., Etlinger, J.D.: J. Cell Biol. 79, 325a (1978).

22. Van der Westhuyzen, D.R., Matsumoto, K., Etlinger, J.D.: (Submitted for publication).

23. Stracher, A., McGowan, E.B., Shafiq, S.A.: Science. 250, 50-53 (1978).

24. Libby, P., Goldberg, A.L. : Science. 199, 534-536 (1978).

25. Toyo-oka, T., Shimizu, T., Masaki, T.: Biochem. Biophys. Res. Comm. 82, 484-491 (1978).

26. Reddy, M.K., Etlinger, J.D., Rabinowitz, M., Fischman, D.A., Zak, R.: J. Biol. Chem. 250, 4278-4284 (1975).

27. Dayton, W.R., Reville, W.J., Goll, D.E., Stromer, M.H.: Biochemistry 15, 2159-2167 (1976).

28. Etlinger, J.D., Goldberg, A.L.: Proc. Natl. Acad. Sci. USA 74, 54-58 (1977).

29. Morkin, E., Kimata, S., Skillman, J.J.: Circ. Res. 30, 690-702 (1978).

30. Laurent, G.J., Sparrow, M.P., Millward, D.J.: Biochem. J. 176, 407-417 (1978).

31. Kao, R.L., Rannels, D.E., Whitman, V., Morgan, H.E.: Recent Advances on Studies on Cardiac Structure and Metabolism. 12, 105-113 (1978).

32. McMinn, R.M.H., Vrbova, G.,: Nature. 195, 509 (1962).

33. Fenner, C., Traut, R.R., Mason, D.T., Wikman-Coffelt, J.W.; Analytical Biochem. 63, 595-602, (1975).

Acknowledgements

We wish to thank K. Kameyama for excellent assistance with the experiments and L. Abott and R. Cintron for expert help in the preparation of the manuscript. These studies were supported by grants from NIH (HL2197002), the Muscular Dystrophy Association and the Commonwealth Foundation. D.v.d.W. held a U.S.P.H.S. Fogerty Fellowship during these studies.

MYOSIN POLYMORPHISM, CELLULAR HETEROGENEITY AND PLASTICITY OF CARDIAC MUSCLE

S. Schiaffino, L. Gorza, S. Pierobon Bormioli and S. Sartore
National Research Council Unit for Muscle Biology and Physiopathology, Institute of General Pathology, University of Padova, 35100 Padova, Italy.

Cardiac muscle is composed of different types of muscle cells. We have previously shown that atrial, ventricular and Purkinje cells of the chicken heart contain antigenically and structurally different forms of myosin (1,2). Different types of cardiac muscle cells have now been identified in the mammalian heart using immunofluorescence procedures with specific antimyosin antibodies. We show here that subpopulations of muscle cells are present in the ventricular myocardium of the rabbit and that their relative proportions change in the course of postnatal development and under the influence of thyroid hormone.

1. Atrial and Ventricular Myosins in the Mammalian Heart

Antigenic differences between atrial and ventricular myosins were demonstrated in several mammalian species using a variety of antimyosin antisera. Like in the chicken, mammalian ventricular myosin was found to display antigenic similarities with slow skeletal myosin, whereas atrial myosin was cross-reactive with fast skeletal myosin. However, atrial and ventricular myosins also showed common antigenic determinants. This is clearly illustrated by the immunological analysis of beef atrial and ventricular myosins. Antisera were prepared in rabbits against beef atrial and ventricular myosins, using procedures which have in part been reported (1) and will be described in greater detail elsewhere. Specific IgG were isolated by affinity chromatography on the respective insolubilized immunogens and used for indirect or direct immunofluorescence. When applied to cryostat sections of the beef heart anti-atrial myosin was found to stain both atrial and ventricular muscle cells: however, after absorption with ventricular myosin it stained selectively atrial cells. Similar results were obtained with anti-ventricular myosin, which initially stained both ventricular and atrial cells and after absorption with atrial myosin showed a specific reactivity with ventricular cells. These results indicate that beef atrial and ventricular myosins are structurally distinct molecules with a number of common antigenic determinants. Cross-absorption permits to isolate antibodies directed to determinants unique to each myosin. Biochemical evidence for differences between atrial and ventricular myosin in mammals has been previously reported (3,4). These differences in myosin composition are probably related to differences in contractile properties, the maximum velocity of shortening of atrial muscle being significantly higher than that of ventricular muscle (5). However, immunofluorescence studies indicate that the situation is more complex in that a number of muscle cells in the atrial myocardium are stained by cross-absorbed anti-ventricular myosin (anti-bV) and a number of cells in the ventricular myocardium are stained by cross-absorbed anti-atrial myosin (anti-bA). The degree of cellular heterogeneity revealed by anti-bA in the ventricles varies in different species and is especially evident in the rabbit heart.

2. Myosin Polymorphism and Cellular Heterogeneity in the Ventricular Myocardium of the Rabbit

As shown in Figure 1 anti-bA stains all atrial muscle cells

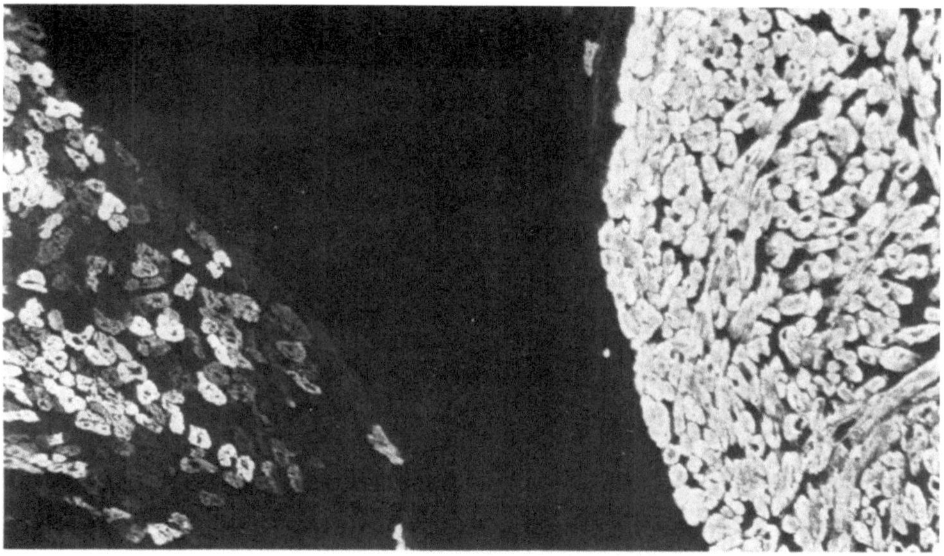

Figure 1. Adult rabbit.Cryostat section through a block composed of atrial (right) and ventricular (left) myocardium, stained with anti-bA (indirect immunofluorescence). x 150

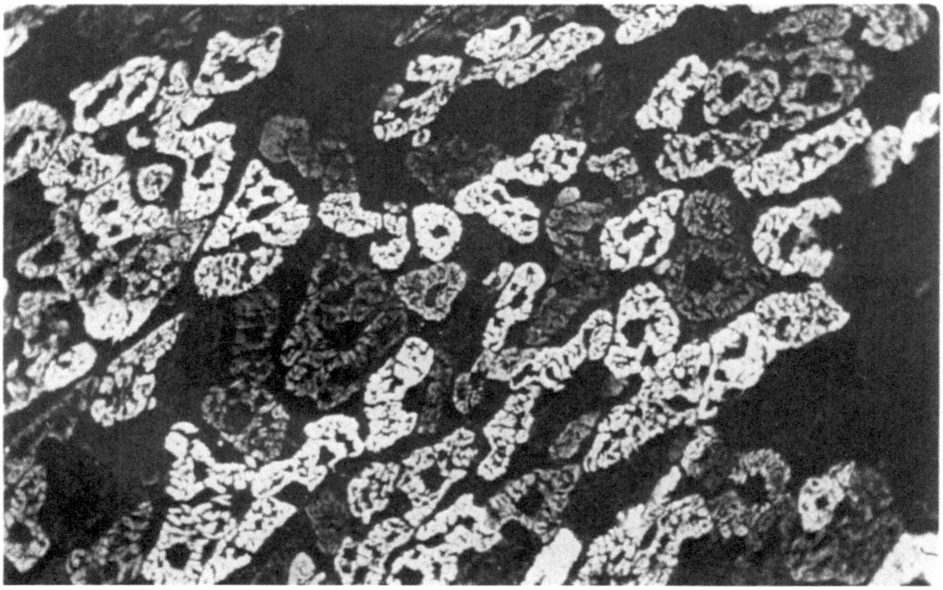

Figure 2. Higher magnification of ventricular myocardium stained with anti-bA. x 360

and a proportion of ventricular muscle cells in the rabbit heart. Figure 2 illustrates the amazing cellular heterogeneity revealed by this antimyosin antiserum in the ventricular myocardium. Completely unstained muscle cells are seen side by side with brightly stained cells or cells showing varying degrees of reactivity: the resulting picture reminds the familiar mosaic pattern of skeletal muscle fibers in a mixed muscle. In longitudinal sections reactive cells show a clear transverse striation with positively stained A bands. They are distributed with variable frequency in different areas of the ventricles: extensive study of serial sections will be necessary to determine whether there is a constant pattern of distribution of the different fibers and to work out a precise distribution map.

These findings indicate that there are different myosins in the rabbit ventricles. Direct evidence for the presence of two distinct myosins has been recently obtained using insolubilized anti-bA to fractionate rabbit ventricular myosin by affinity chromatography. Two-dimensional electrophoresis of the tryptic peptide maps showed that the myosin fraction retained by the immunoabsorbent has different heavy chains when compared with the unretained fraction: in contrast light chains appeared identical in the two myosins (6). A plausible interpretation of immunofluorescence findings based on the existence of two ventricular myosins is that the retained type of myosin is predominantly segregated in the brightly stained cells, the unretained type of myosin in the unstained cells and mixtures of the two or hybrid forms in cells displaying intermediate reactivity.

3. Myosin Changes and Diversification of Ventricular Cells during Postnatal Development

Diversification of ventricular cells occurs during postnatal

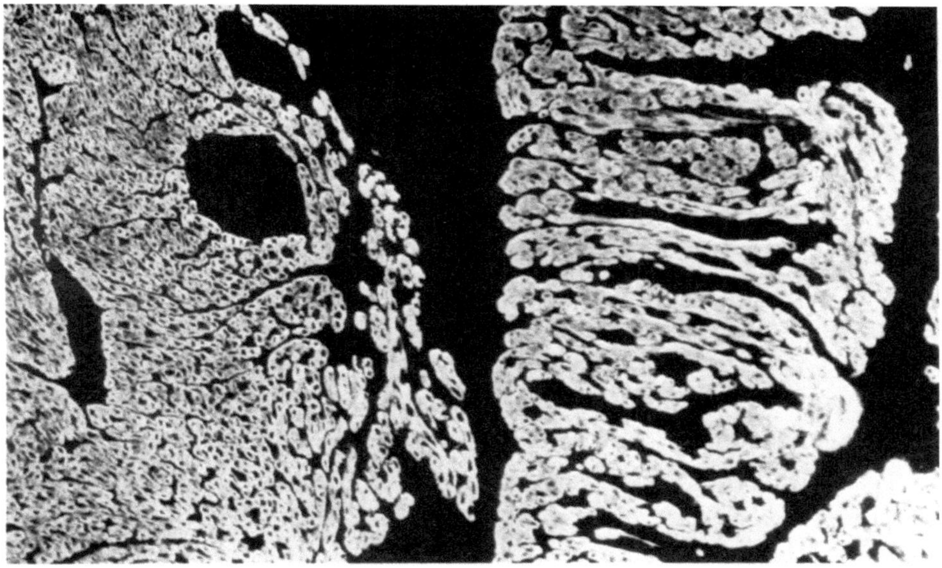

Figure 3. One-week-old rabbit. Cryostat section through a block composed of atrial (right) and ventricular (left) myocardium, stained with anti-bA. x 150

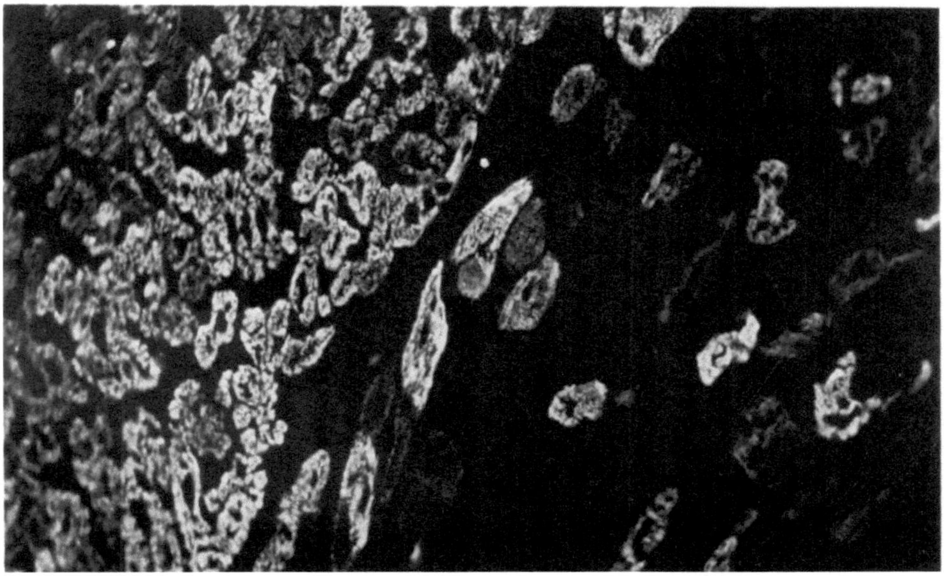

Figure 4. Composite block of ventricular myocardium from one--month-old (left) and adult (right) rabbit, stained with anti--bA. x 360

development. In one-week-old rabbits all ventricular cells are brightly stained by anti-bA, the intensity of the reaction being similar to that of atrial muscle cells (Fig. 3). In one--month-old rabbits the intensity of the reaction becomes lower in ventricular compared to atrial muscle cells and initial signs of diversification are present (Fig. 4). At subsequent stages cellular heterogeneity becomes more and more evident. It appears that during postnatal development there is a progressive transformation of a proportion of ventricular cells with substitution of one type of myosin with another type; other ventricular cells seem to retain the type of myosin present in developing cardiac muscle. A comparative biochemical analysis of adult and neonatal myosins in the rabbit heart is now in progress.

4. The Effect of Thyroid Hormone

The factors responsible for the diversification of ventricular myosins and cell types in the ventricular myocardium of the rabbit are yet to be determined. In view of the growing evidence for the synthesis of a new type of myosin in the cardiac muscle of thyrotoxic animals (for a review see 7) we have analized the changes induced by thyroid hormone in the reactivity of rabbit ventricular cells with anti-bA. Adult rabbits were injected with L-thyroxine (200 μg/kg) daily for 2 weeks, a treatment known to induce myosin changes in the rabbit heart (8). As shown in Figures 5 and 6 the ventricular myocardium from thyrotoxic rabbits showed no evidence of heterogeneity, all cells being brightly stained by anti-bA. The differential response of atrial and ventricular tissue to anti-bA was no longer seen after thyroxine treatment and in this respect the reactivity of ventricular cells became similar to that of the newborn rabbit. It will be of interest to compare myosins from thyrotoxic and newborn rabbits in order to establish whether immunological cross-reactivity reflects structural identity.

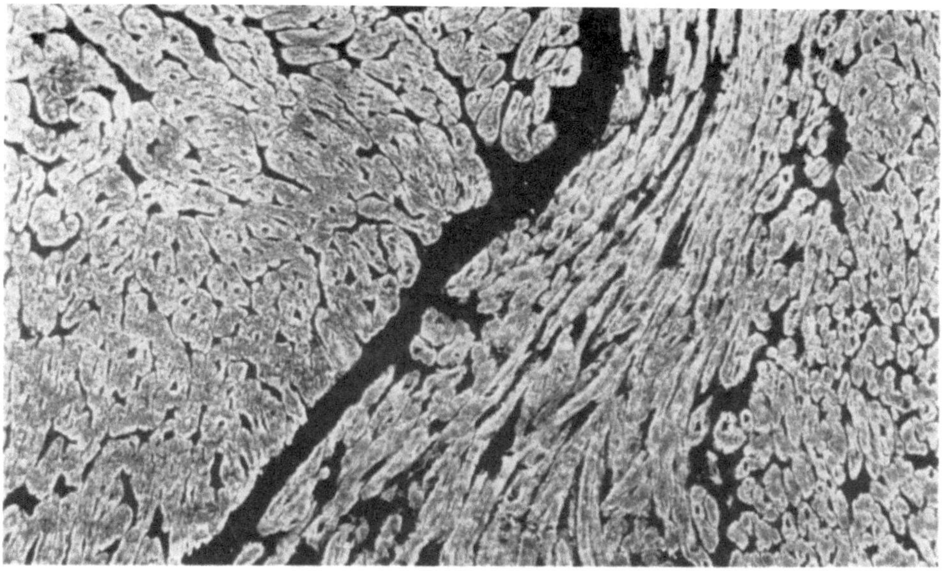

Figure 5. Thyrotoxic rabbit. Cryostat section through a block composed of atrial (right) and ventricular (left) myocardium, stained with anti-bA. x 150

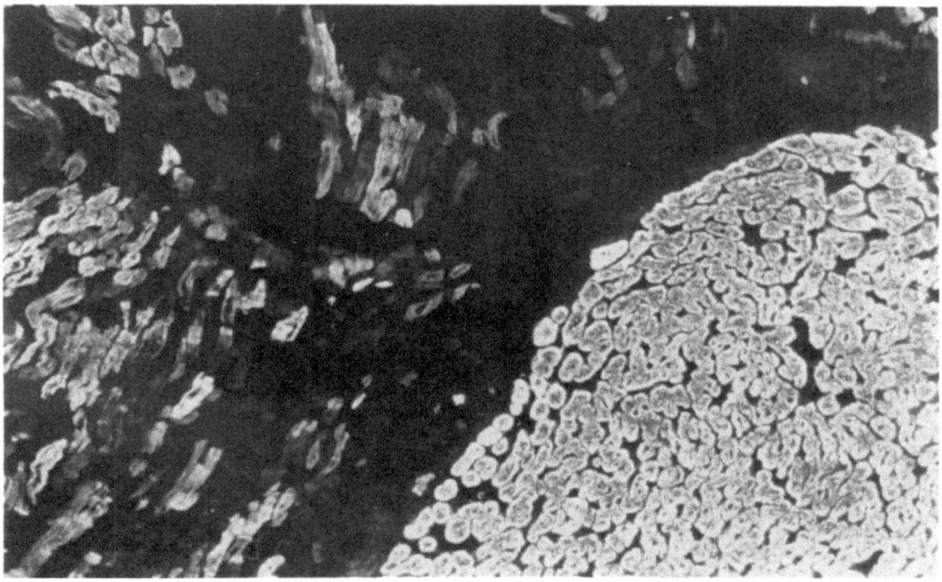

Figure 6. Composite block of ventricular myocardium from thyrotoxic (right) and normal adult (left) rabbit, stained with anti-bA. x 150

5. Conclusions and Open Questions

Cardiac muscle, like most skeletal muscles, is a mixed muscle and can undergo plastic changes involving transformation of muscle cell types during development and under hormonal influence. Functional overload, such as induced by systemic hypertension, can also modify the "immunohistochemical profile" of ventricular myocardium in the rat (in preparation). The results presented here concerning rabbit ventricular myocardium are essentially in agreement with those previously reported by Hoh et al. (9) in an important study on rat myocardium based on gel electrophoresis of intact myosins. Myosin polymorphism and myosin changes in ventricular myocardium have thus been demonstrated using two completely different methodological approaches. Immunofluorescence studies provide one further piece of information, namely that different myosins are often segregated into different cells. This finding raises new problems concerning the function of the various cardiac muscle cells and the factors responsible for their differentiation. Though thyroxine administration can induce myosin changes and cell type changes in ventricular myocardium it is not clear how thyroid hormone under physiological conditions could determine differences among neighbouring muscle cells. Other factors must be implicated in inducing and maintaining cell diversity in cardiac muscle: sympathetic innervation could be a candidate for such a role.

The functional significance of cell heterogeneity in the ventricular myocardium is not clear. In particular, it remains to be determined what relevance muscle cell heterogeneity may have with respect to the functional adaptations by which the heart can quickly modulate its contractile properties in response to altered circulatory demands. It is generally assumed that whereas both the force and the speed of contraction of skeletal muscle can be modulated by the recruitment of a variable number and type of motor units, "this type of control is not possible in the heart, which is functionally a syncytium" (10). However, the demonstration of muscle cell heterogeneity in the myocardium stimulates a reconsideration of this dogma and makes one wonder whether recruitment or activation of various cell types might be operative in cardiac muscle as well. Adaptive changes in myocardial contractility under different conditions might be explained by selective recruitment. The changes induced by catecholamines in the contractile properties of the myocardium could be a case in point.

References

1. Sartore, S., Pierobon Bormioli, S., Schiaffino, S.: Immunohistochemical evidence for myosin polymorphism in the chicken heart. Nature 274, 82-83 (1978).

2. Dalla Libera, L., Sartore, S., Schiaffino, S.: Comparative analysis of chicken atrial and ventricular myosins. Biochim. Biophys. Acta, in press.

3. Long, L., Fabian, F., Mason, D.T., Wikman-Coffelt, J.: A new cardiac myosin characterized from canine atria. Biochem. Biophys. Res. Commun. 76, 626-635 (1977).

4. Flink, I.L., Rader, J., Banerjee, S.K., Morkin, E.: Atrial and ventricular cardiac myosins contain different heavy chain species. FEBS Lett. 94, 125-130 (1978).

5. Urthaler, F., Walker, A.A., Hefner, L.L., James, T.N.: Comparison of contractile performance of canine atrial and ventricular muscles. Circ. Res. 37, 762-771 (1975).

6. Sartore, S., Dalla Libera, L., Schiaffino, S.: Fractionation of rabbit ventricular myosin by affinity chromatography with insolubilized antimyosin antibodies. FEBS Lett., in press.

7. Morkin, E.: Stimulation of cardiac myosin adenosine triphosphatase in thyrotoxicosis. Circ.Res. 44, 1-7 (1979).

8. Flink, I.L., Morkin, E.: Evidence for a new cardiac myosin species in thyrotoxic rabbit. FEBS Lett. 81, 391-394 (1977)

9. Hoh, J.F.Y., McGrath, P.A., Hale, P.T.: Electrophoretic analysis of multiple forms of rat cardiac myosin: effects of hypophysectomy and thyroxine replacement. J. Mol. Cell. Cardiol. 10, 1053-1076 (1978).

10. Katz, A.M.: Physiology of the heart, Raven Press, New York 1977.

MYOSIN IN CHRONIC CARDIAC OVERLOAD

K. Schwartz, A.M. Lompré, A. D'Albis*, G. Lacombe, N.V. Thiem and B. Swynghedauw.

U 127 INSERM, Hop. Lariboisière, Paris, *Lab. de Biologie Physico-chimique, Univ. Paris-Sud, Orsay, France.

Introduction

Subjecting the heart to a chronic mechanical overload causes the myocardium to hypertrophy. In this hypertrophied heart, the contractile activity of the cardiac muscle is depressed (rev. in 1). Most authors agree that this mechanical alteration is accompanied by a decrease of the Ca^{2+}-dependent ATPase activity of myosin (rev. in 2). Several isoenzymes of myosin exhibiting different Ca^{2+}-dependent ATPase activities exist not only within the different muscles of the same animal species (3), but also within the same muscle (4-8). Various authors hypothetized therefore that the alterations observed during mechanical overload could be due, at least in part, to the synthesis in the heart of a different isoenzyme. Though extensive work has been devoted to this problem, evidence for a different molecule has not yet been clearly presented. Several observations, however, suggested subtle changes in the molecule (9-11).

Our strategy was first to demonstrate that the decrease in ATPase activity was not an artifact, and then to study myosin structure using two different analytical approaches, known both for their ability to detect small structural changes.

A. Animal model

Left ventricular hypertrophy was induced in male Wistar rats, 180-200 g, either by constriction of the abdominal aorta (12) or by induction of aortic insufficiency (13) or by combining both procedures (13). At various times after surgery (35-141 days), rats were killed by a blow on the head and hearts were rapidly excised and weighed (Vw). Depending on the experiment, either both ventricles, or right and left ventricles, or endocardium and epicardium were used. All samples were frozen in liquid nitrogen within 3 to 5 min. For each animal, theoritical ventricular weight (Th w) was calculated as in (14), according to the following regression curve, established in 152 controls : Th w (g) = $[(2.02 \times \text{body wt})] + 197 \times 10^{-3}$. The % of hypertrophy was equal to (Vw - Th w) x 100/Th w. Animals exhibiting 60 to 100% cardiac hypertrophy by any of the above procedures were selected for this study. Enzymatic, immunological and electrophoretic tests were performed on myosins from the same hypertrophied hearts and from hearts of sham-operated animals respectively. Myosins were prepared by minor modifications of the procedure described by Offer et al (15, 16).

B. Enzymatic analyses

Steady-state Ca^{2+}-dependent ATPase activities of myosins from hypertrophied hearts were significantly depressed, whilst the K^{+}-activated ATPase remained unchanged (Table I). This is in full agreement with earlier observations in rats (14) and in other animal species (9-11). The number of ATP moles cleaved per myosin molecule was essentially the same for control and hypertrophied hearts (Table I). As shown in (17, 19) this ruled out the possible presence of denatured molecules in the preparation and clearly indicated that the decrease in ATPase activity was not an artifact.

Table I

	Sham-operated	Hypertrophy
Phosphate burst size mol ATP x mol myosin^{-1}	1.64	1.56
Steady-state activity n mol Pi x mg myosin^{-1} x min^{-1}		
Ca^{2+}-ATPase	1114	750*
K^{+} -ATPase	644	692

*$p < 0.001$

Table I. Myosin ATPase in control and hypertrophied tissue. Early phosphate burst size was determined manually essentially according to Taylor and Weeds (17), under stoichiometric conditions (18). Steady-state activities were measured as in (19).

C. Immunological studies

A double immunological approach was used, by preparing antibodies to either native or sodium-dodecyl-sulfate (SDS) denatured heavy-meromyosins extracted from normal rat cardiac ventricles. Antigen-antibody interactions were analysed by a very discriminant technique, micro-complement fixation, MCF (20). The methods used for the production and characterization of antisera and the protocol for micro-complement fixation reactions have all been described (14, 20-22). Fig. 1 and 2 show typical experiments of complement fixation by myosins from sham-operated and hypertrophied rat hearts. With both types of antisera, anti-native HMM and anti-SDS denatured HMM, the amount of complement fixed by the hypertrophied heart was decreased as compared to the control, in agreement with our previous observations (14). The difference was more pronounced with SDS-denatured myosins, confirming that the use of antibodies to SDS denatured HMM is a discriminant means

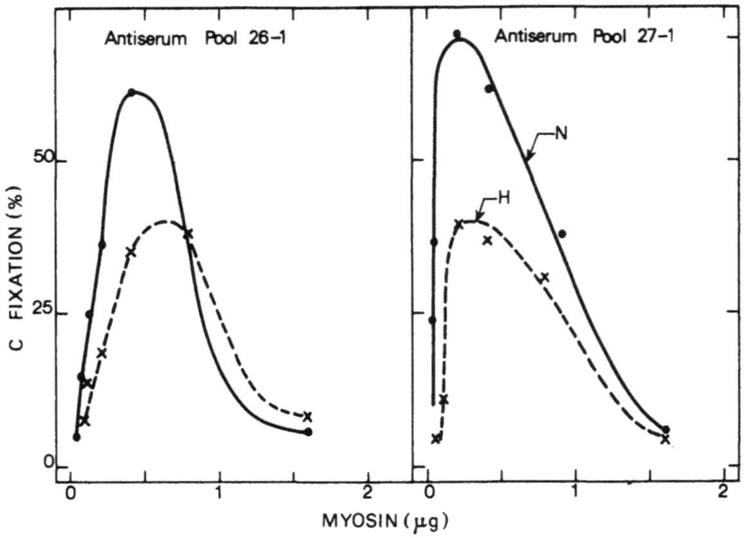

Fig. 1. Micro-complement (MC) fixations with antisera to native normal rat cardiac heavy-meromyosin (HMM) and native myosins obtained from the ventricles of normal (N) and hypertrophied (H) rat hearts (80% hypertrophy).

Fig. 2. MC fixations with an antiserum to normal rat ventricular HMM denatured myosins (22). 100% hypertrophy Maximum amounts of C fixed by SDS-myosins with various antiserum dilutions.

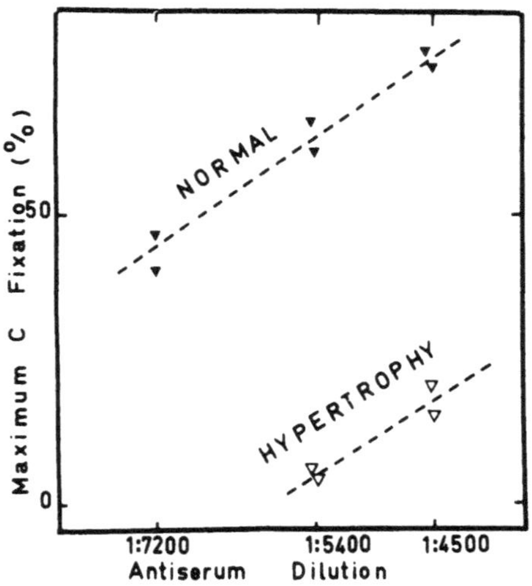

to probe structural differences between myosin isoenzymes (K. Schwartz et al submitted). Numerous observations on globular and monomeric proteins have shown that vertical shifts of the MCF curves reflect differences in primary structure (23), and the few available data indicate that such a relation also is true for myosin (21, 24). Our results thus indicated that a different population of myosin molecules was present within the hypertrophied heart.

D. Electrophoresis of native myosins

Conditions permitting the electrophoresis of myosins in their native state have been set up independently by Hoh et al (6) and by one of us (25). The method has allowed the identification of distinct species of myosins, not only in various

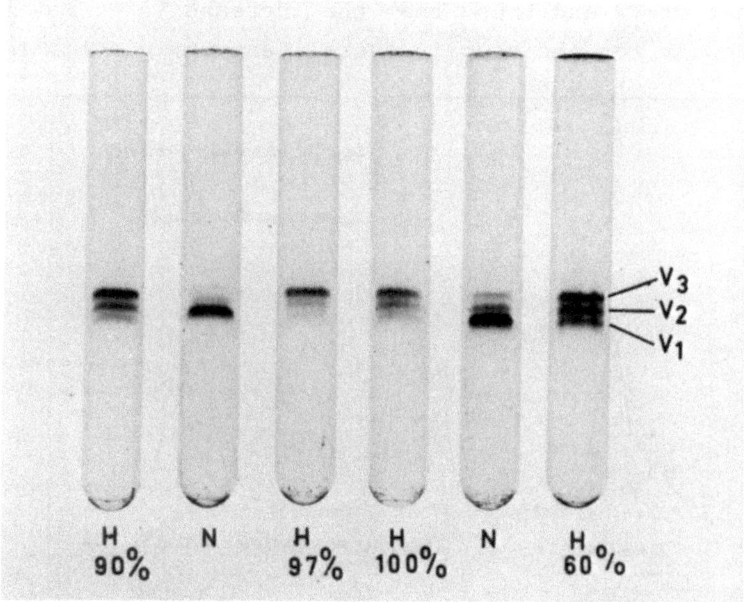

Fig. 3. Electrophoretic patterns for native myosins from various hearts. Electrophoresis was performed as in (7), and similar results were obtained in a slightly different medium (8).

types of muscles (6, 25) but also in the same muscle (7, 8). Typical electrophoretic patterns for native myosins from normal and hypertrophied hearts are shown in Fig. 3. Normal rat cardiac myosin, as already described by Hoh et al (7), migrates as three bands, V_1, V_2 and V_3, V_1 being the predominant species. A redistribution of these isoenzymes was observed in the hypertrophied hearts, with a predominance of V_3 in the largest (~100% hypertrophy) and an intermediary pattern in the less affected hearts (60% hypertrophy)(16). The same redistribution was observed on myocytes isolated from an hypertrophied heart according to Cutilletta et al (26), showing that non-muscle cells did not interfere. The Ca^{2+}-activated ATPase activity of individual components were measured in gels by methods similar to those described by Hoh et al (6, 8). It is apparent from the absorbance profiles shown in Fig. 4 that the specific activity of V_3 is lower than that of V_1 and thus, that the increase in V_3 may therefore account for the drop in overall enzymatic activity

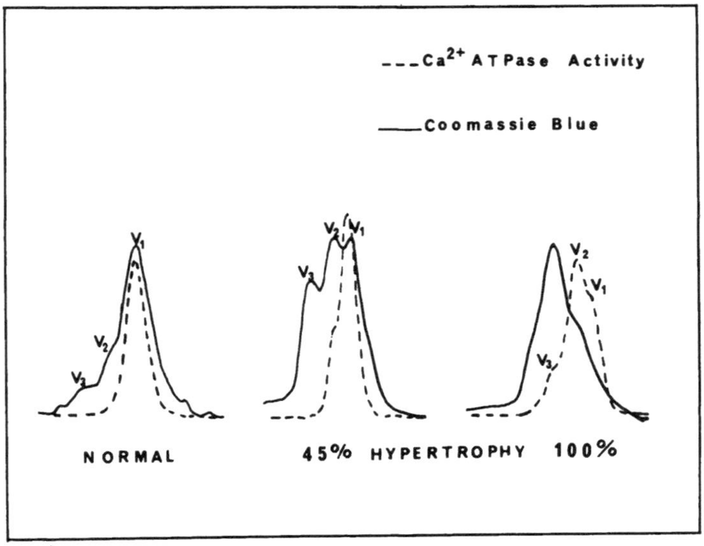

Fig. 4. Comparison of the absorbance profiles of myosin components after ATPase and protein stainings. Gel electrophoresis was run in non-dissociating conditions (7).

Fig. 5. Absorbance profiles after protein staining of gels in non-dissociating conditions (7) of endocardial and epicardial myosins. Crude tissue extracts prepared with Guba-Straub buffer were run in these experiments.

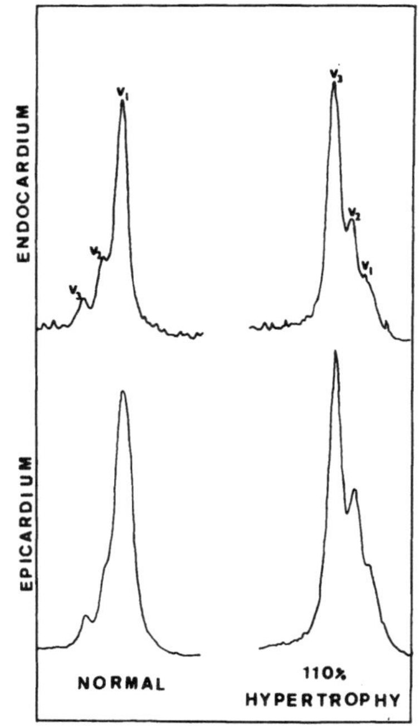

Fig. 6 Absorbance profiles after protein staining in non-dissociating conditions of myosins from right and left ventricles. Crude tissue extracts.

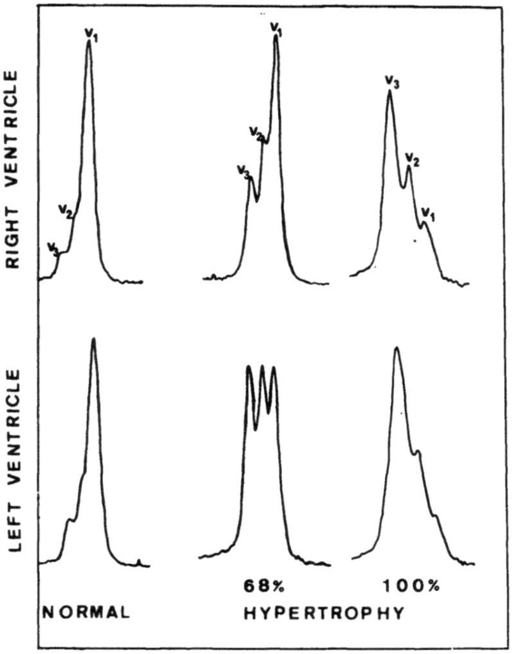

(Table I). Analysis of samples extracted from the endocardium and epicardium of left ventricles showed no difference in pattern to that of the whole ventricle, both in normal and hypertrophied hearts (Fig. 5). There was no difference either between right and left ventricles (Fig. 6) in control and large hypertrophies. In contrast, in the intermediate hypertrophies, the isoenzymic shift was more pronounced in the left ventricle (Fig. 6). In view of the difference in functional demand on the left and right sides of the heart at this stage, such a difference is not surprising.

E. Conclusions and open questions

The studies reported here have demonstrated that an increase in cardiac work produced by mechanical overloading in rats induces the preferential synthesis of a cardiac myosin isoenzyme exhibiting a lower ATPase activity and characterized by specific immunological and electrophoretic properties. In view of the close correlation found in different types of skeletal muscles (27) and in hearts of different animal species (28) between contractile element shortening capacity and myosin ATPase, this isoenzymic shift may account for the reduced shortening speed of this hypertrophied cardiac muscle (1). On the contrary, thyroid hormone increases cardiac performance (rev. in 27) and induces the synthesis of the cardiac isoenzyme of high enzymatic activity, V_1 (7). Adaptation in cardiac performance seems thus to be mediated by a change in myosin species, a process which can be related to the adaptive responses of the skeletal muscles to new functional requirements. However, the isoenzymic shift we describe cannot entirely account for the increase in time-to-peak tension described by Hamrell et al (28). It might be postulated that, as already described in other models (29), internal Ca^{2+}-movements are also altered, but this remains to be demonstrated.

Acknowledgments

We acknowledge the excellent technical assistance of P. Bouveret, J. Bercovici and C. Pantaloni, and thank M. Albaret for the secretarial work. Grants from the D.G.R.S.T. (77.7. 0308), I.N.S.E.R.M. (76.5.188.5) and C.N.R.S. (480).

References

1. Maughan, D., Low, E., Litten, R., Brayden, J., Alpert, N.: Circulat. Res. 44, 279-287 (1979).
2. Swynghedauw, B., Léger, J.J., Schwartz, K.: J. Mol. Cell. Card. 8, 915-924 (1976).
3. Katz, A.M.: Physiol. Rev. 50, 63-158 (1970).
4. Gröschel-Stewart, U., Doniach, D.: Immunology 17, 991-994 (1969).
5. Gauthier, J., Lowey, S.: J. Cell. Biol. 74, 760-779 (1977).
6. Hoh, J.F.Y., Mc Grath, P.A., White, R.I.: Biochemical J. 157, 87-95 (1976).
7. Hoh, J.F.Y., Mc Grath, P.A., Hale, P.T.: J. Mol. Cell. Card. 10, 1053-1076 (1978).
8. D'Albis, A., Pantaloni, C., Bechet, J.J.: Europ. J. Biochem. 99, 261-272 (1979).
9. Katagiri, T., Morkin, E.: Biochim. Biophys. Acta 342, 262-274 (1974).
10. Wikman-Coffelt, J., Walsh, R., Fenner, C., Kamiyama, T., Salel, A., Mason, D.T.: J. Mol. Cell. Card. 8, 263-270 (1976).
11. Shiverick, K.T., Hamrell, B.B., Alpert, N.R.: J. Mol. Cell. Card. 8, 837-852 (1976).
12. Cutilletta, A.F., Rudnik, M., Zak, R.: J. Mol. Cell. Card. 10, 677-687 (1978).
13. Jouannot, P. Gourdier, B., Courtalon, A., Hatt, P.Y.: Path. Biol. 21, 623-627 (1973).
14. Schwartz, K., Bouveret, P., Bercovici, J., Swynghedauw, B.: FEBS Letters 93, 137-140 (1978).
15. Offer, G., Moos, C., Starr, R.: J. Mol. Biol. 74, 653-676 (1973).
16. Lompré, A.M., Schwartz, K., d'Albis, A., Lacombe, G.,

Van Thiem, N., Swynghedauw, B.: Nature in the press.

17. Taylor, R.S., Weeds, A.G.: Biochem. J. 159, 301-315 (1976)
18. Taylor, E.W. Biochemistry 16, 732-740 (1977).
19. Van Thiem, N., Lacombe, G., Swynghedauw, B.: Eur. J. Biochem. 91, 243-248 (1978).
20. Levine, L. in Handbook of Experimental Immunology, Second edition, Weir, D.M. ed., Blackwell Scientific, Oxford 1973.
21. Schwartz, K., Bouveret, P., Sebag, C., Swynghedauw, B.: Biochim. Biophys. Acta 495, 24-36 (1977).
22. Lompré, A.M., Bouveret, P., Léger, Joc. Schwartz, K.: J. Immunol. Methods 28, 143-148 (1979).
23. Prager, E.M., Wilson, A.C.: J. Biol. Chem. 246, 5978-5989 (1971).
24. Whalen, R., Schwartz, K., Bouveret, P., Sell, S., Gros, F. : Proc. Natl. Acad. Sci, in the press.
25. D'Albis, A., Gratzer, W.B.: FEBS Lett. 29, 292-296 (1973).
26. Cutilletta, A.F., Aumont, M.C., Nag, A.C., Zak, M.: J. Mol. Cell. Card. 9, 399-407 (1977).
27. Morkin, E.: Circulat. Res. 44, 1-7 (1979).
28. Hamrell, B.B., Alpert, N.R.: Circulat. Res. 40, 20-25 (1977).
29. Heilmann, C., Pette, D.: International Symposium, Experimental cardiac hypertrophy and heart failure Tübingen, April 5-7 (1979): Basic Res. Cardiol., in press.

Section VII. Effects of Thyroid Hormones

EFFECTS OF THYROID HORMONES ON DIFFERENT TYPES OF SKELETAL MUSCLE

W. Winder

Division of Biochemistry, Physiology, and Pharmacology
University of South Dakota School of Medicine
Vermillion, South Dakota 57069

R. Fitts

Department of Biology, Marquette University
Milwaukee, Wisconsin 53233

J. Holloszy, K. Kaiser, and M. Brooke

Departments of Preventive Medicine and Neurology
Washington University School of Medicine, St. Louis, MO 63110

Fiber Types in Rodents

Three major types of skeletal muscle fibers exist in the rodent: fast-twitch white fibers have a low respiratory capacity, a high glycogenolytic capacity, and high myosin ATPase activity; fast-twitch red fibers have a high respiratory capacity, a high glycolytic capacity, and high myosin ATPase activity; and slow-twitch red fibers have a moderately high respiratory capacity, a low glycogenolytic capacity, and a low myosin ATPase activity (1-3). In the rat, relatively homogeneous regions of muscle composed of primarily one fiber type may be used to study responses of the different fiber types to physiological stimuli. The soleus of the rat, for example, is composed predominantly of slow-twitch red fibers. The deep portion of the quadriceps nearest the bone is made up of fast-twitch red fibers, whereas the superficial region of the quadriceps is composed of fast-twitch white fibers (1). These three fiber types do not always respond equally to the same apparent physiological stimulus. Levels of glycolytic enzyme activities increase slightly in soleus muscle of rats in response to 12 weeks of treadmill running, whereas the same enzymes decrease in fast-twitch red muscle from the quadriceps (2).

Effect of Thyroid Hormones on Skeletal Muscle Mitochondria

Tata, *et al.* (4) and Gustafsson, *et al.* (5) previously reported that intermittant treatment of normal rats with "close to physiological" amounts of thyroxine (T_4) or triiodothyronine (T_3) increased the rate of weight gain, increased the resting metabolic rate, and increased the respiratory capacity of hindlimb muscle mitochondria. Electron micrographs of "hindlimb muscles" also showed increased numbers of mitochondrial profiles in thyroxine treated rats (5).

These results stimulated our interest in the use of thyroid hormone treatment as a tool for studying the regulation of mitochondrial biogenesis in muscle. In our initial experiments (6) we were unable to detect increases in mitochondrial marker enzymes in fast twitch muscles (gastrocnemius, quadriceps) in response to dosage and treatment schedules described by earlier workers (4,5). We found that only after prolonged treatment with relatively high doses of T_3 and T_4 could we detect increases in mitochondrial content of hindlimb skeletal muscle. In our hands, a six week treatment with near-maximal tolerable doses of T_3 and T_4 produces only a 30% increase in cytochrome c and a 48% increase in citrate synthase in gastrocnemius muscle (6).

Effect of Thyroid Hormones on Different Fiber Types

Nolte, *et al.* (7) and Kubista, *et al.* (8) previously reported wide disparities in responsiveness of different muscles to thyroid hormones, particularly with respect to mitochondrial α-glycerophosphate dehydrogenase which increased to a much greater extent in the soleus (composed of slow-twitch red fibers) than in the rectus femoris (composed predominantly of fast-twitch red and white fibers) in rats. In light of these findings and our observation of the difference in response pattern of the three muscle fiber types to treadmill running, we further characterized the responses of these three fiber types to a 12 week treatment with excess thyroid hormones (9).

Thyroid hormones were given in powdered food (3 mg T_4 + 1 mg T_3 per Kg powdered Purina rat chow). Controls were food restricted slightly so that they gained weight at the same rate as the treated animals. The effectiveness of this treatment in producing a hypermetabolic state is evidenced by increased food intake, increased heart weight, and a 13-fold increase in liver mitochondrial α-glycerophosphate dehydrogenase activity (Table 1).

Table 1. Evidence of thyrotoxicosis in rats given thyroid hormones in food for 12 weeks

	Controls	$T_3 + T_4$ Treated
Food intake (g/day)	32 ± 1	44 ± 1 *
Heart weight (g)	0.92 ± 0.01	1.81 ± 0.05 *
Liver mitochondrial α-GPDH (ul/g/min)	16 ± 2	218 ± 20 *

* $p<0.001$ (Data from reference 9)

Table 2. Effect of long-term thyrotoxicosis on mitochondrial enzymes of skeletal muscle

	% Change from Control Values		
	Red Quad	White Quad	Soleus
Citrate Synthase	+ 44*	+ 42*	+ 162*
Cytochrome c	+ 44*	+ 36	+ 140*
Cytochrome oxidase	+ 37*	+ 27	+ 89*
Mitochondrial α-GPDH	+ 76*	- 8	+ 170*
3-Hydroxybutyrate DH	+ 73*	Not measurable	+ 69*

*Control vs. thyroid hormone treated, 12 weeks, $p<0.01$
(Data calculated from reference 9)

Activities of several mitochondrial enzymes were examined in the three different muscle types (Table 2). Citrate synthase, cytochrome c, and cytochrome oxidase increased in skeletal muscle in response to 12 weeks treatment with T_3 and T_4. Particularly interesting is the observation that these enzymes are much more responsive to thyrotoxicosis in the slow-twitch red type of muscle (soleus) than in the fast-twitch white or fast-twitch red types of muscle. Approximately 2.5 fold increases in citrate synthase and cytochrome c occurred in soleus muscle, whereas the red and white portions of the quadriceps showed only 27-44% increases in these enzymes in response to the same treatment. The same holds true for cytochrome oxidase, with soleus muscle being much more responsive to thyroid hormones than either fast-twitch red or fast-twitch white muscle.

3-Hydroxybutyrate dehydrogenase and mitochondrial α-glycerophosphate dehydrogenase show somewhat different patterns of response. Relative to initial concentration, 3-hydroxybutyrate dehydrogenase activity increased approximately 70% in fast-twitch red as well as in the soleus. The activity of this enzyme was too low to measure in white muscle. Mitochondrial α-glycerophosphate dehydrogenase increased 76% in the fast-twitch red portion of the quadriceps, was not significantly affected in white muscle, and increased 170% in the soleus muscle (Table 2).

Why Do Fiber Types Respond Differently?

It is not possible from available data to determine the reason for the difference in responsiveness of the three different types of fibers to thyroid hormone treatment. It is conceivable that T_3 and T_4 have no effect on the muscle per se, but instead influence the activity of the motoneurons supplying the muscle. It is well established that increasing the contractile activity of muscle either by artificial stimulation of motoneurons with implanted electrodes (10) or by use of treadmill exercise (11) causes an increase in several mitochondrial enzymes. Many of the same qualitative changes in oxidative enzyme levels occur in response to thyrotoxicosis as do in response to endurance exercise training (9). Soleus muscle fibers are innervated by smaller, more excitable neurons than are the fast-twitch fibers (12). It is conceivable that neurons innervating slow-twitch fibers are more sensitive to T_3 or T_4 than are neurons innervating fast-twitch fibers. If this is true, denervation should abolish the effect of thyroid hormones on soleus mitochondrial enzymes.

To test this possibility, the right soleus nerve was cut near to the point where the nerve enters the muscle, taking care to avoid disturbing the blood supply. Half the rats were then given thyroid hormones in their food for 10 days. Although the results are difficult to interpret because of the atrophy which occurs rapidly after denervation, it does seem clear that thyroid hormones do have an effect on mitochondria of

Table 3. Effect of denervation on the response of soleus muscle to thyroid hormones (10 days treatment)

	Soleus Wt. (mg)		Soleus Citrate Synthase (μmoles/g/min)	
	Denervated	Normal	Denervated	Normal
Controls	49 ± 3	77 ± 2	15 ± 1	19 ± 1
T_3T_4 Treated	41 ± 3	72 ± 3	29 ± 1*	37 ± 1*

*Controls vs. T_3T_4 Treated, $p < 0.001$. (Data from reference 13).

denervated soleus muscle (Table 3, 13). Citrate synthase activity was two fold higher in the denervated soleus of thyroid hormone treated animals compared to the denervated soleus of controls. This data implies therefore a direct action of thyroid hormones on muscle rather than an action mediated by the nervous system.

We should point out that the effective concentrations of T_3 or T_4 within the fibers may not be identical for all fiber types due to differences in blood supply or rate of uptake. Janssen, et al. (14) reported that in euthyroid rats the concentration of T_3 in "red" muscle (consisting of a combination of slow and fast-twitch red muscles) is 25% greater than in white muscle. To our knowledge, no data is available on T_3 concentrations in the specific muscles we used.

One of the proposed theories for the mechanism of those actions of T_3 requiring protein synthesis is that T_3 binds to nucleoproteins, thereby altering transcription rates of specific genes and consequently increasing rates of synthesis of corresponding proteins. With few exceptions, tissues which are less responsive to thyroid hormones apparently have fewer T_3 binding sites in the nucleus (15). The maximal physiological response is reached when all binding sites are occupied (15). It is possible therefore that the slow-twitch red type of fiber contains a greater quantity of T_3 binding sites in the nuclei than do other types of muscle fibers, thereby rendering it more sensitive to T_3. We are not aware of any information on quantitation of T_3 receptors in the different types of skeletal muscle.

Time Course of the Response in Soleus Muscle

The time course of the increase in citrate synthase, cytochrome c, and cytochrome oxidase activity in soleus muscle in response to thyroid hormones (30 ug T_4 + 10 ug T_3/100 g body weight/day) has been studied (13). The half-time of the increase in levels of these enzymes to a higher steady state was approximately 11-15 days. After cessation of treatment, these enzymes returned to the lower steady state control level with a half-time of 5-8 days (13). Thus, the half-time for the induction process tends to be longer than the half-time for the decay process. This would suggest that a reduced rate of degradation of these enzymes is responsible for the higher steady state levels in response to thyrotoxicosis (cf 16). This interpretation must be viewed with caution,however. Studies of the effects of thyrotoxicosis on cytochrome c turnover in the liver suggest that the degradation rate is increased (17). Another possible explanation is that there was a progressive increase in the stimulus for protein synthesis during the course of treatment, even though the daily dose of T_3 and T_4 was held constant. An increase in the magnitude of intracellular stimulus during the course of treatment would prolong the approach of mitochondrial enzyme levels to a new steady state, thus resulting in a longer half-time for the process.

Does Thyrotoxicosis Cause a Change in Fiber Type in Soleus?

The enzymes citrate synthase, cytochrome oxidase, cytochrome c, and α-glycerophosphate dehydrogenase increase in the soleus to levels higher than normally seen in fast-twitch red muscle in response to thyroid hormone treatment. The activity of 3-hydroxybutyrate dehydrogenase is 2.7 fold higher in soleus muscle than in fast-twitch red muscle in control animals. If slow-twitch red fibers of the soleus were being converted to fast-twitch red fibers, we might expect to see a reduction in the 3-hydroxybutyrate dehydrogenase activity. This did not occur. We actually observed an increase in the activity of this enzyme in response to thyrotoxicosis (9). Thus, the soleus becomes more like a fast-twitch red muscle with respect to some, but not all mitochondrial oxidative enzymes.

Fitts, et al (18) have studied the contractile properties of soleus muscle in response to 6-8 weeks of treatment with excess thyroid hormones. Marked changes were observed in isometric contractile properties of the soleus, thus confirming the work of Gold, et al (19). The soleus becomes more like fast-twitch muscle in terms of shorter isometric contraction and relaxation times and higher tetanic fusion frequency. The isotonic contractile properties were not altered by thyroid hormone treatment, however. The maximum shortening velocity (V_{max}) of the soleus in the thyrotoxic rats was the same as in normal rats.

Fitts, et al (18) also found that the protein content of the sarcoplasmic reticulum (SR) fraction from the soleus was increased two-fold by 6-8 weeks treatment with excess thyroid hormones (18). The calcium stimulated ATPase activity and the initial rate of calcium uptake by the SR were also increased two-fold when results were expressed per mg protein. These results provide evidence that the total amount of SR and the calcium transport activity of SR are increased in soleus muscle in the thyrotoxic state. These adaptations in the SR likely allow more rapid removal of calcium from the sarcoplasm after excitation-contraction and are probably responsible for the faster contraction and relaxation times in the isometric condition.

Histochemical fiber typing was done on soleus muscles from rats given excess thyroid hormones for 6-8 weeks using the pH sensitivity of ATPase to classify the fiber types (20). The fiber distribution of control muscles was 85% type 1 (slow-twitch red), 12% type 2A (fast-twitch red), and 3% type 2C (undifferentiated fetal fibers). Thyrotoxic animals showed a markedly different distribution: 40% type 1, 26% type 2A and 34% type 2C. Thus, there appears to be a transition from type 1 to type 2 fibers as a result of thyroid hormone treatment. This confirms the findings of Ianuzzo, et al (21) who also found an increase in the number of type 2 fibers in thyrotoxic animals. These observations appear to be inconsistent with our data on isotonic contractile properties wherein no change in V_{max} was observed in the soleus in response to 6-8 weeks treatment with thyroid hormones. Fibers which demonstrate alkali stability for the histochemical ATPase stain are generally considered to be fast-twitch fibers. A positive correlation between actin-activated myosin ATPase activity and the V_{max} for different types of muscles was observed by Barany (22). This apparent inconsistency in our data may be explained by the observation that 57% of the type 2 fibers in the soleus of thyrotoxic rats are type 2C fibers. Although the properties of these fibers are not well-characterized in the mature animal, their histochemical characteristics are similar to undifferentiated fetal fibers which have a shortening velocity similar to slow muscle (23). These "undifferentiated" fibers which predominate in fetal muscle, exhibit a slow shortening velocity despite the alkali stable histochemical ATPase (24).

Thus, it seems clear that major changes in the properties of the slow-twitch red fibers of the soleus occur in response to prolonged thyrotoxicosis. Large increases occur in mitochondrial oxidative enzymes. The SR increases and isometric contraction and relaxation times are markedly reduced. This does not appear to be a simple conversion of slow-twitch red fibers to fast-twitch red fibers however. Some of the characteristic properties of the slow-twitch red fibers are retained, including high 3-hydroxybutyrate dehydrogenase activity and low maximal shortening velocity. These observations demonstrate clearly the potential which adult muscle retains for change and adaptability in response to physiological stimuli. Experimental thyrotoxicosis thus provides a very interesting and useful model for studying factors which control phenotypic expression in adult muscle.

References

1. Baldwin, K.M., Klinkerfuss, G.H., Terjung, R.L., Mole, P.A., Holloszy, J.O.: Am. J. Physiol. 222, 373-378 (1972).
2. Baldwin, K.M., Winder, W.W., Terjung, R.L., Holloszy,J.O.: Am. J. Physiol. 225, 962-966 (1973).
3. Barnard, R.J., Edgerton, V.R., Furukawa, T., Peter, J.B.: Am. J. Physiol. 220, 410-414 (1971).
4. Tata, J.R., Ernster, L., Lindberg, O., Arrhenius, E., Pedersen, S., Hedman, R.: Biochem.J. 86, 408-428 (1963).
5. Gustafsson, R., Tata, J.R., Lindberg, O., and Ernster, L.: J. Cell. Biol. 26, 555-570 (1965).
6. Winder, W.W., Baldwin, K.M., Terjung, R.L., Holloszy,J.O.: Am. J. Physiol. 228, 1341-1345 (1975).
7. Nolte, J., Pette, D., Bachmaier, B., Kiefhaber, P., Schneider, H., Scriba, P.C.: Eur. J. Clin. Invest. 2, 141-149 (1972).
8. Kubista, V., Kubistova, J., Pette, D.: Eur. J. Biochem. 18, 553-560 (1971).

9. Winder, W.W., Holloszy, J.O.: Am. J. Physiol. 232, C180-C184 (1977).

10. Pette, D., Smith, M.E., Smith, H.W., Staudte, H.W., Vrbova, G.: Pflügers Arch. 338, 257-272 (1973).

11. Holloszy, J.O., Booth, F.W.: Ann. Rev. Physiol. 38, 213-291 (1976).

12. Henneman, E.: Medical Physiology, Mosby, St. Louis 1974, p. 617-650.

13. Winder, W.W.: Am. J. Physiol. 236, C132-C138 (1979).

14. Janssen, J.W., Hardeveld, C. van, Kassenaar, A.A.H.: Acta Endocrinol. 90, 81-89 (1979).

15. Oppenheimer, J.H.: New Engl. J. Med. 292, 1063-1068 (1975).

16. Schimke, R.T.: Mammalian Protein Metabolism, Academic, New York 1970, p. 177-228.

17. Booth, F.W., Holloszy, J.O.: Arch. Biochem. Biophys. 167, 674-677 (1975).

18. Fitts, R.H., Winder, W.W., Brooke, M.H., Kaiser, K.K., Holloszy, J.O.: Am. J. Physiol. (In Press).

19. Gold, H.K., Spann, J.F., Branuwald, E.: J. Clin. Invest. 49, 849-854 (1970).

20. Brooke, M.H., Kaiser, K.K.: Arch. Neurol. 23, 369-379 (1970).

21. Ianuzzo, D., Patel, P., Chen, V., O'Brien, P., Williams, C.: Nature 270, 74-76 (1977).

22. Barany, M.: J. Gen. Physiol. 50, 197-218 (1967).

23. Close, R.: J. Physiol. 173, 74-95 (1964).

24. Brooke, M.H., Williamson, E., Kaiser, K.K.: Arch. Neurol. 25, 360-366 (1971).

A POSSIBLE THYROIDAL TROPHIC INFLUENCE ON FAST AND SLOW SKELETAL MUSCLE MYOSIN

C.D. Ianuzzo, P. Patel, V. Chen, and P. O'Brien
Departments of Biology and Physical Education, York University, Toronto
Ontario, Canada M3J 1P3

Introduction

Several factors have been identified that influence the differentiated character of skeletal muscle fibres. These include neural factors, physiological demand, and myopathies. Experimental manipulation of neural factors using cross-innervation and long-term electrical stimulation appear to produce the most extensive interconversion in the differentiated characteristics of slow and fast skeletal muscle fibre types (1-8). The change that occurs in the myosin phenotype is of considerable interest as fast and slow myosins are qualitatively distinct proteins and result from the expression of different genes. Recently we presented findings (9,10) indicating a possible thyroidal trophic influence on skeletal muscle myosin. The magnitudes of these changes compare favorably with those of cross-innervation studies (2,3,6). This paper describes additional studies confirming and extending our original experiments.

Methods

Euthyroid (EU), hypothyroid, (TX) and hyperthyroid male Sprague-Dawley albino rats with initial body weights of 180-200 grams were used in these experiments. Hypothyroidism was produced by surgical thyroidectomy (Hormone Assay, Chicago, IL, U.S.A.). Hyperthyroidism was induced in hypothyroid (TX+T3) and euthyroid (EU+T3) rats by subcutaneous injections of sodium 3,5,3' -tri-iodo-L-thyronine (T3) dissolved in 0.5 mM NaOH at a dose of 300 $\mu g.kg^{-1}$ body weight on alternate days for six weeks. Control animals were sham injected with 0.5 mM NaOH. Rats were maintained as previously described (11).

Following the six week treatment period, animals were exsanguinated while under sodium pentobarbital anesthesia (60 mg·kg^{-1}, ip.). Heart ventricles, soleus and plantaris muscles were excised _in toto_, blotted, cleaned of visible fat and connective tissue, and weighed. The thyroid conditions were confirmed by slower than normal total body growth rates, differences in heart to body weight ratios, and changes in Vmax activities of succinate dehydrogenase (EC 1.2.99.1) and mitochondrial alpha-glycerol phosphate dehydrogenase (GPDH, EC 1.1.99.5) of liver and GPDH for three of the muscles studied (11) (Table 1).

The mid-portions of the soleus and plantaris muscles and a sample from the left ventricle at the base of the papillary muscle were frozen in 2-methyl-butane cooled in liquid nitrogen. Ten micra thick sections were stained for myofibrillar adenosine triphosphatase (ATPase) at pH 9.40 (12) following pre-incubation at either pH 4.35 or 10.30. Approximately 300 fibres were counted from eight randomly selected locations equally distributed within each section. Fibres were classified as alkali-labile (slow-twitch) or alkali-stable (fast-twitch) based on the intensity of the stain.

Muscles used for the isolation of myosin were stored in glycerol at -20°C until used. The preparation of myosin was accomplished using the method of Bārāny and Close (6) or by the following procedure. Muscles were coarsely minced and rinsed in 40 mM KCL at pH 7.0, then homogenized in 10 ml of 10 mM KCl and extracted in Guba-Straub solution containing 2 mM dithiothreitol, 2 mM EDTA, 20 mM Mg-ATP at pH 6.6 for 10 min. Samples were centrifuged at 150,000 g for 1 hr. $(NH_4)_2 SO_4$ was added to the supernatant fluid and the precipitate from the 34-45% fraction was collected by centrifugation at 20,000 g for 10 min. The myosin pellet was resuspended in 600 mM KCl and dialyzed 12-14 hr at 4°C versus 0.5 mM dithiothreitol 1mM EDTA 10 mM KCl and 1mM Tris-HCl at pH 7.2. The myosin precipitate was collected and dissolved in 500 mM KCl and 50 mM Tris at pH 7.0.

TABLE 1 Selected characteristics of different muscle types from rats with altered thyroid conditions

	Euthyroid	Hypothyroid	Hyperthyroid
Final body weight (g)	399 ± 6 (50)	287 ± 8 (46)[b]	334 ± 7 (44)[b]
Liver			
SDH activity	21.9 ± 0.6 (30)	16.3 ± 0.6 (37)[b]	28.9 ± 1.8 (27)[b]
GPDH activity	3.5 ± 0.2 (32)	2.4 ± 0.2 (36)[b]	8.6 ± 0.5 (28)[b]
Soleus muscle			
Wet weight (mg)	137 ± 3 (28)	120 ± 4 (28)[a]	119 ± 2 (16)[a]
Protein $(mg \cdot g^{-1})$	185 ± 9 (16)	188 ± 13 (13)	186 ± 6 (14)
Myosin yield $(mg \cdot g^{-1})$	31 ± 2 (17)	32 ± 2 (17)	31 ± 2 (15)
GPDH activity	1.5 ± 0.04(18)	1.1 ± 0.02(14)[b]	2.6 ± 0.10(13)
Plantaris muscle			
Wet weight (mg)	357 ± 8 (19)	267 ± 8 (13)[b]	268 ± 9 (16)[b]
Protein $(mg \cdot g^{-1})$	194 ± 10 (16)	199 ± 12 (13)	195 ± 11 (14)
Myosin yield $(mg \cdot g^{-1})$	37 ± 2 (32)	35 ± 2 (39)	37 ± 2 (29)
GPDH activity	1.8 ± 0.04(17)	1.5 ± 0.02(13)[b]	2.2 ± 0.07(10)[b]
Cardiac muscle			
Wet weight (mg)	1031 ± 14 (50)	684 ± 13 (46)[b]	1219 ± 16 (44)[b]
Protein $(mg \cdot g^{-1})$	192 ± 9 (14)	190 ± 9 (13)	201 ± 11 (14)
Myosin yield $(mg \cdot g^{-1})$	23 ± 2 (4)	22 ± 2 (4)	24 ± 1 (4)
GPDH activity	1.9 ± 0.03(17)	1.7 ± 0.05(14)[b]	3.3 ± 0.07(14)[b]

Values are mean ± SE. Number in parenthesis is number per group.
Succinate dehydrogenase (SDH) and glycerol phosphate dehydrogenase (GPDH) activities are in umoles$\cdot$min$^{-1} \cdot$g wet weight^{-1}.
[a,b] Indicates values are statistically significant from euthyroid at $P<0.01$ and 0.001, respectively.

The Ca^{2+} -activated myosin ATPase (EC 3.4.1.3) activity was determined at different pHs as described by Bárány and Close (6). Inorganic phosphate liberated by the myosin ATPase reaction was determined by the method of Rockstein and Herron (13). Protein was measured by the method of Lowry et al. (14) with bovine serum albumin as standard. Sodium dodecyl sulfate (SDS) polyacrylamide gel electrophoresis was accomplished on 1.5 mm slabs using 4% (stacking gel) and 12% (separating gel) acrylamide as described by Laemmli (15).

Results

Histochemical observations indicate an altered thyroid status can determine the proportion of alkali-stable and alkali-labile fibres in rat skeletal muscle (Table 2 and Fig. 1). The adductor longus and soleus muscles, which in the EU condition contains 15-16% fast fibres, had a two-fold increase in the percentage of alkali-stable fibres following six weeks of T3 treatment. The opposite occurred in the TX condition. Both of these muscles had significantly less alkali-stable fibres than the EU muscles. The soleus was comprised totally of alkali-labile fibres in seven of the 10 muscles examined. There was a 40% difference in alkali-stable fibres between the hypothyroid and hyperthyroid soleus muscles.

The EU rat diaphragm and plantaris muscles are comprised predominantly of alkali-stable fibres (Table 2). The changes in the percentage of fibre types in these fast muscles compare favorably to those of the slow muscles, with the exception of the hyperthyroid plantaris. The lesser change in the hyperthyroid plantaris may be expected since it normally consists of 90% alkali-stable fibres.

Fibres with intermediate staining intensity were observed in all skeletal muscles following both pH 4.35 and pH 10.30 pre-incubation (Fig. 2). At times, these fibres were observed either at pH 4.35 or pH 10.30 and at other times at both pHs.

The myofibrillar ATPase stain had a lesser intensity for the TX cardiac

TABLE 2 Percent alkali-stable fibres in muscles of rats with altered thyroid status.

Muscle	Euthyroid	Hypothyroid	Hyperthyroid
Adductor Longus	15.2 ± 1.6 (7)	5.6 ± 1.5 (5)[a]	29.4 ± 0.8 (5)[a]
Soleus	16.3 ± 0.7(10)	0.2 ± 0.1(10)[b]	40.4 ± 1.4(10)[b]
Diaphragm	68.1 ± 1.2 (5)	58.2 ± 0.9 (5)[a]	81.6 ± 2.1 (4)[a]
Plantaris	90.4 ± 0.6(13)	78.6 ± 0.6(13)[a]	96.4 ± 0.6(11)[a]

Values are mean ± SE. Numbers in parenthesis are number per group.
[a,b] Indicates values are statistically significant compared to euthyroid at $P<0.01$ and 0.001, respectively.

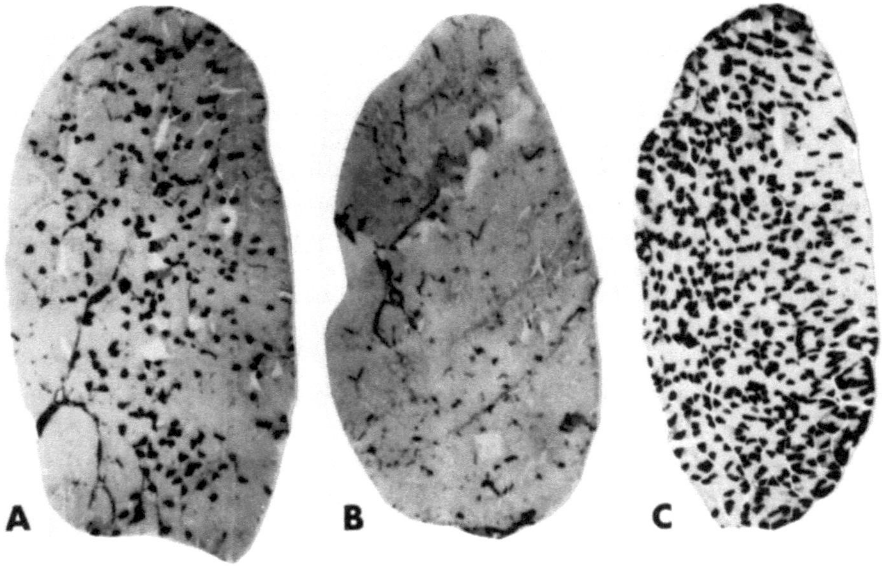

Fig. 1 Transverse sections from the soleus muscle of (A) euthyroid, (B) hypothyroid and (C) hyperthyroid rats stained for myofibrillar ATPase (pH 10.30).

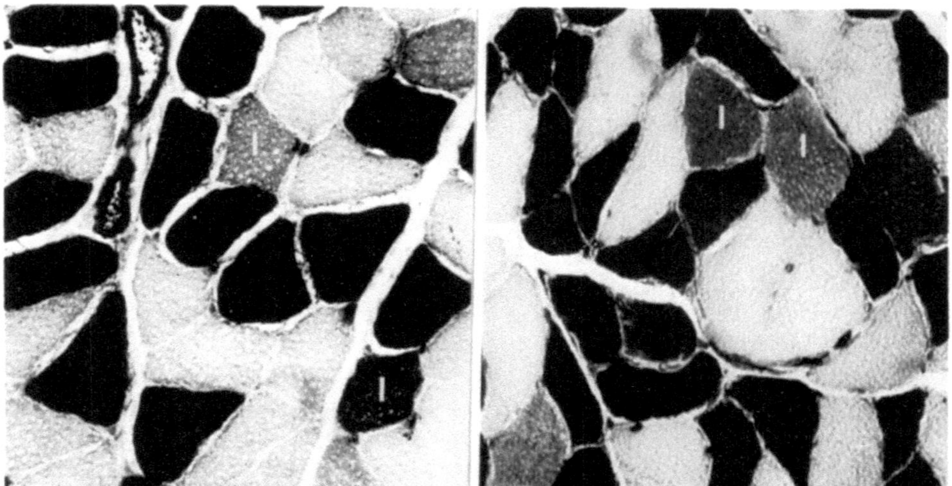

Fig. 2. A. Fibres with intermediate (I) staining intensity (pH 10.30). B. Fibres with intermediate staining intensity (pH 4.35).

muscle than for the EU. But no difference in staining intensity was observed between the hyperthyroid and EU hearts using routine histochemical procedures.

Biochemical determination of Ca^{2+}-activated myosin ATPase was performed at pH 7.4 and 10.0 with myosin isolated from soleus, plantaris and cardiac muscles of EU, TX, EU+T3 and TX+T3 rats. Myosin ATPase activity for the soleus muscle was significantly lower (-27%) and higher (67%) for the TX and hyperthyroid (EU+T3 and TX+T3) groups, respectively, than the EU at pH 7.4 (Fig. 3). At pH 10.0 the myosin ATPase from the EU+T3 and TX+T3 soleus had a two-fold activation compared to activities at pH 7.4, which is a characteristic response of fast muscle myosin, whereas, the EU and TX myosin activity declined by 50%. This resulted in an approximate eight-fold difference between the EU and TX compared to the EU+T3 and TX+T3 enzyme activities at pH 10.0. A pH profile of the soleus muscle is shown in Fig. 4. In EU soleus muscle two peaks in myosin ATPase occur, one in the acidic range at approximately pH 5.5 and the other in the alkaline range at about pH 9.8-10.0. The enzyme activity at the alkaline peak best differentiated the thyroid effects on myosin ATPase.

Myosin ATPase activity of the plantaris muscle was the least affected of the three muscles studied. The specific activity of EU plantaris muscle myosin was twice that of the EU soleus. The only significant difference in ATPase activity from that of the EU was for the TX+T3 myosin at pH 10.0 (Fig. 5).

Myosin ATPase from EU rat cardiac muscle had an activity that was 2.5-fold greater than the fast plantaris muscle (Fig. 6). It exhibited a three-fold activation at alkali pH, as does fast muscle myosin. The EU and TX+T3 cardiac myosin ATPase activities were similar throughout the pH range. The activity of the TX cardiac myosin was considerably less than normal. At pH 7.0 and 9.4 it had only 35% and 5%, respectively, the activity of the EU cardiac myosin.

SDS gel electrophoresis was performed to determine whether the changes in histochemical fibre types and myosin ATPase activity were accompanied by changes in myosin light chains. Changes in the light chain patterns were observed for both the soleus and plantaris muscles (Fig. 7).

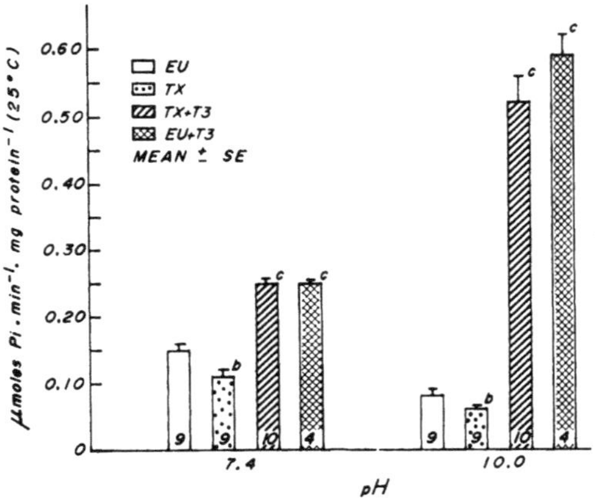

Fig.3 Ca^{2+}-stimulated myosin ATPase activity from soleus muscle. Number in each bar is n/group.

[b,c]Significant from euthyroid at P<0.01 and 0.001, respectively.

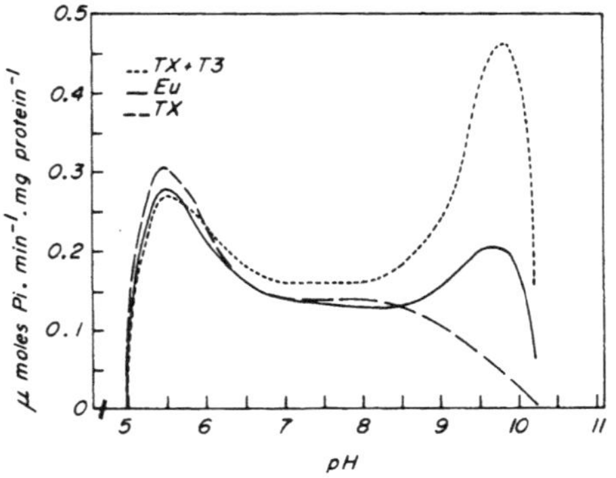

Fig.4 pH profile of Ca^{2+}-stimulated myosin ATPase activity (25°C) from soleus muscle.

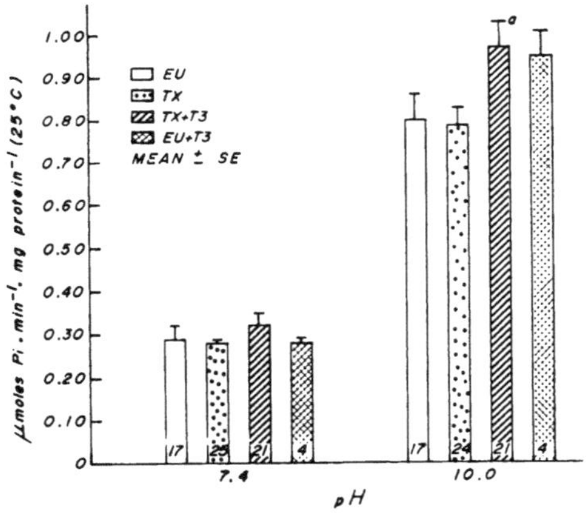

Fig.5 Ca^{2+}-stimulated myosin ATPase activity from plantaris muscle. Number in each bar is n/group.

[a]Significant from euthyroid at P<0.05.

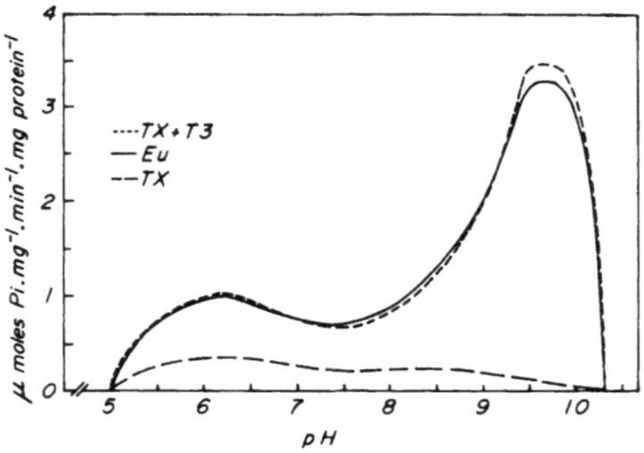

Fig.6 pH profile of Ca^{2+}-stimulated myosin ATPase activity (25°C) from cardiac muscle.

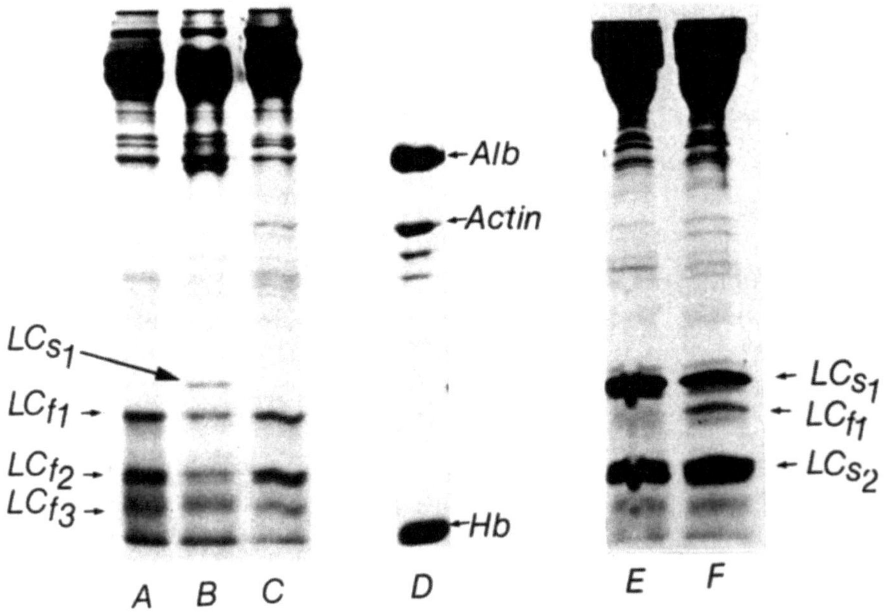

Fig. 7 Electrophoretogram of myosin light chains using 50 ug myosin protein in 12 % acrylamide from euthyroid (A), hypothyroid (B) and (C) hyperthyroid plantaris muscles. Proteins of known molecular weight, i.e., albumin (68,000), actin (42,000), and hemoglobin (15,500) are shown in D. Myosin light chains from hypothyroid and hyperthyroid soleus muscles are shown in E and F, respectively. (A 100 ug of protein was applied to these gels.)

The gel electrophoretogram for the TX+T3 soleus myosin had an increased staining intensity for the first light chain of fast myosin (LCf_1). The TX plantaris muscle had an increased staining intensity for the first light chain of slow myosin (LCs_1).

Discussion

An increase in cardiac myosin ATPase activity with thyrotoxicosis has been shown to occur in some commonly used experimental animals (18,19,20), but it has not been observed in the rat (16,17). The findings of this study substantiate previous findings on rat cardiac muscle myosin. The lack of effects on rat cardiac ATPase has been explained by the inherently high specific enzyme activity in the euthyroid state, which was believed to be already maximally activated (17). Hoh et al. (21,22) have electrophoretically identified three isoenzymes (V_1, V_2, V_3) with different $Ca^{2\pm}$ activated ATPase activities in rat ventricles. These isoenzymes are present in approximately equal proportions in adult euthyroid cardiac muscle. Hypothyroidism results in an increased proportion of V_3, which has the lowest ATPase activity, whereas, hyperthyroidism induced the V_1 form. This indicates that some change in ATPase activity should be observed in hyperthyroid, as well as, hypothyroid rat cardiac ventricles. It should be mentioned, that originally we observed an increase in ATPase activity in hyperthyroid rat ventricles using the myosin isolation procedure of Bárány and Close (6). However, using the myosin isolation procedure described in Methods, differences between the euthyroid and hyperthyroid cardiac ATPase were not observed. This disparity may have resulted from the less pure cardiac myosin obtained from the Bárány and Close procedure, as indicated by gel electrophoresis and the reddish tinge in the myosin pellet. An added benefit of the procedure described in Methods, is that myosin could be isolated in less time (i.e. one day versus two-and-a-half days), which could in part explain the two-fold greater specific activity of myosin ATPase using this method.

The histochemical and biochemical findings presented in this paper demonstrate that alterations in thyroid status can determine the proportion of fast and slow twitch fibres in rat skeletal muscle. In previous investigations (16,17), in which both cardiac and skeletal muscle myosin were studied following experimentally imposed dysthyreosis, changes in ATPase activity were found for cardiac but not for skeletal muscle. This led to the conclusion that the effect of thyroid hormones was specific to cardiac myosin. The lack of change reported in previous studies can probably be attributed to: (a) the shorter treatment times considering skeletal muscle myosin has a longer half life than cardiac myosin, (b) the type of skeletal muscle studied, and (c) the pH at which the ATPase activity was determined.

The most significant aspect of these findings is that alterations in thyroid status result in an interconversion of slow and fast twitch skeletal muscle fibres. It is well established that fast and slow myosin are qualitatively distinct proteins (3,4,6). The regulation of the myosin phenotype is thought to be determined primarily by neural factors, i.e., neurotrophic chemicals and/or neuronal impulse activity patterns (1-5,6,8). When comparing the magnitude of change in Ca^{2+} activated myosin ATPase activity between cross-innervation and thyroid alterations they appear similar. For example, cross-innervation of the soleus with a nerve from fast muscle for 6-15 months produced a 35-68% increase in ATPase activity at pHs ranging from 7.4-8.0 (2,3,6). In the present study a 66% increase occurred in the soleus at pH 7.4 after only six weeks of hyperthyroidism. In a previous study (9) we observed a 35% increase at pH 7.4. At pH 10.0, Bárány and Close (6) observed a 225% activation in myosin ATPase from cross-innervated soleus muscle relative to its activity at pH 7.4. At pH 10.0, myosin ATPase from the EU+T3 and TX+T3 soleus was activated 235% and 208%, respectively. These comparisons indicate that hyperthyroidism brings about similar changes in myosin ATPase in less time than cross-innervation.

Electrical stimulation, which produces the best results when converting fast muscle to slow, generally results in a more complete transformation than occurs in dysthyroidism. However, there are findings that are comparable. A study (8) using continuous electrical stimulation for about 40 days found an 11-17% decline in fast-twitch fibres in three of the four muscles examined (one muscle was atypical). This change compares favorably with the 12% decrease in fast fibres in the hypothyroid plantaris muscle. Another study (4) observed decreases of 28 to 36% in Ca^{2+} stimulated myosin ATPase activity following four weeks of continuous stimulation of fast muscles. In comparison, hypothyroidism resulted in a 27% decrease in soleus myosin ATPase activity and hyperthyroidism a 67% increase in activity.

It may be questioned whether some of these effects result from the parathyroidectomy that occurs with surgical thyroidectomy. This seems unlikely since both euthyroid and the thyroidectomized animals made hyperthyroid by T3 administration showed similar effects. Further evidence that hypoparathyroidism is not involved in bringing about these differences are the comparable percent changes observed in muscle fibre types with chemically (methyl-thiouracil) induced hypothyroidism (personal communication, C. Williams, Loughborough, England) with those of surgical thyroidectomy (9). Another concern is whether the myocardial changes may in part result from an altered functional demand that accompanies thyroid dysfunction (23). This suggestion is supported by the myosin ATPase activity of different mammals being relative to their resting heart rate (24) and that changes in myocardial ATPase activity are induced by endurance training (25). Also changes in functional demand of the diaphragm may be expected to occur, however, we found no differences in respiratory rate and minute ventilation between the altered thyroid and euthyroid animals (unpublished data). This may also indicate that impulse traffic to the diaphragm has not been altered.

The changes observed in myosin light chains in the thyroid altered muscles are similar to those that occur with cross-innervation (7) and chronic electrical stimulation (4,5,8). This is consistent with a hypothesis that there is a common trophic mechanism for neural and thyroidal induced changes in skeletal muscle myosin phenotypes. However, at present it would be premature and scientifically unwise to assume a common mechanism for neural and thyroidal induced changes of skeletal muscle myosin polymorphs. At least four possible mechanisms by which the thyroid state may alter the phenotypic isozyme of skeletal muscle myosin can be suggested: (a) changes in neural trophic substances, (b) alteration in impulse activity pattern, (c) direct induction of new gene expression in the muscle cell by thyroid hormone, or (d) imbalances in other hormones that induce this change.
(Supported by Canada NSERC grant 0404).

References

1. Buller, A.J., Eccles, J.C., Eccles, R.M.: J. Physiol. (Lond.) 150, 417-439 (1960).

2. Buller, A.J., Mommaerts, W.F.H.M., Seraydarian, K.: J. Physiol. (Lond.) 205, 581-597 (1969).

3. Jean, D.H., Guth, L., Albers, R.W.: Exptl. Neurol. 38, 458-471 (1973).

4. Sréter, F.A., Gergely, J., Salmons, S., Romanul, F.: Nature 241, 17-19 (1973).

5. Salmons, S., Sréter, F.A.: Nature 253, 30-34 (1976).

6. Bárány, M., Close, R.I.: J. Physiol. (Lond.) 213, 445-474 (1971).

7. Weeds, A.G., Trentham, D.R., Kean, C.J.C., Buller, A.J.: Nature 247, 135-139 (1974).

8. Pette, D., Müller, W., Leisner, E., Vrbová, G.: Pflügers Arch. 364, 103-112 (1976).

9. Ianuzzo, D., Patel, P., Chen, V., O'Brien, P., Williams, C.: Nature 270, 74-76 (1977).

10. Patel, P., Chen, V., Gollnick, P., Ianuzzo, D.: Fed. Proc. 37, 375 (1978).

11. Keens, T.G., Chen, V., Patel, P., O'Brien, P., Levison, H., Ianuzzo, C.D.: J. Appl. Physiol., 44, 905-908 (1978).

12. Padykula, H.A., Herman, E.: J. Histochem. Cytochem. 3, 170-195,(1955).

13. Rockstein, M., Herron, R.W.: Anal. Chem. 23, 1500-1501 (1951).

14. Lowry, O.H., Rosebrough, N.J., Farr, A.L., Randall, R.J.: J. Biol. Chem. 193, 265-275 (1951).

15. Laemmli, U.: Nature 227. 680-685 (1970).

16. Rovetto, M.J., Hjalmarson, A.C., Morgan, H.E., Barrett, H.J., Goldstein, R.A.: Circ. Res. 31, 397-409 (1972).

17. Yazaki, Y., Raben, M.S.: Circ. Res. 36, 208-215 (1975).

18. Morkin, E.: Circ. Res. 44, 1-7 (1979).

19. Goodkind, M.J., Dambach, J.E., Thyrum, P.T., Luchi, R.J.: Amer. J. Physiol. 226, 66-72 (1974).

20. Conway, J., Heazlitt, R.A., Fowler, N.O., Gabel, M., Green, S.: J. Mol. Cell. Card. 8, 39-51 (1976).

21. Hoh, J., McGrath, P., Hales, P.: J. Mol. Cell Card. 10, 1053-1076 (1978).

22. Hoh, J., Yeoh, G., Thomas, M., Higginbottom, L.: FEBS Lett. 97,330-335 (1979).

23. Buccino, R.A., Spann Jr., J.F., Pool, P.E., Sonnenblick, E.H., Braunwald, E.: J. Clin. Invest. 46, 1669-1682 (1967).

24. Delcayre, C., Swyngedauw , B.: Pflügers Arch. 355, 39-47 (1975).

25. Bhan, A.K., Scheur, J.: Amer. J. Physiol. 228, 1178-1182 (1975).

A NEURALLY MEDIATED EFFECT OF THYROID HORMONE DEFICIENCY ON SLOW-TWITCH SKELETAL MUSCLE?

M. A. Johnson, F. L. Mastaglia and A. Montgomery

Muscular Dystrophy Group Research Laboratories,
Newcastle General Hospital, Westgate Road,
Newcastle upon Tyne, NE4 6BE, England.

B. Pope and A. G. Weeds

MRC Laboratory of Molecular Biology,
University Medical School, Hills Road,
Cambridge, CB2 2QH, England.

Introduction

The occurrence of structural and physiological changes in skeletal muscle as a result of prolonged hypothyroidism has been well documented. It has been shown in man (1-3) and in the experimental animal (4-5) that both the time taken to reach peak tension and the time to half relaxation is prolonged in evoked muscle responses. The mechanism of this slowing of muscle contraction remains unclear. The results of histochemical studies have suggested that it may be related to changes in the muscle fibre types. Thus, an increase in the proportion of Type I (slow-twitch oxidative fibres) has been found in patients with long-standing hypothyroidism (6). In the rat soleus muscle, thyroidectomy has been shown to lead to an increase in the proportion of Type I fibres with a concomitant increase in the alkali lability of myosin ATPase (7). On the basis of these findings Ianuzzo and his colleagues have suggested that thyroid status plays a part in determining the phenotypic properties of skeletal muscle myosin. The aim of our study was to elucidate the mechanism whereby thyroid hormone deficiency leads to a change in the metabolic and physiological properties of skeletal muscle and, in particular, to determine whether these changes result from a direct action of the hypothyroid state on muscle or from a neurally mediated effect. Our findings suggest that the latter is probably the case.

Physiological and Histochemical Studies

Male Wistar rats aged 8 weeks were used since it is known that the developmental changes in the muscle twitch characteristics are complete by this age (8). A total thyroidectomy was performed under sodium pentobarbitone anaesthesia (50 mg/kg) on two groups of rats which were subsequently examined 12 weeks after thyroidectomy. The first group of animals was used to study the isometric twitch characteristics and histochemical fibre type profiles of the fast-twitch extensor digitorum longus (EDL) and slow-twitch soleus (SOL) muscles (8). The second group of animals was used for studies on the myosin light chain patterns of SOL and EDL; in addition the fibre type proportions of all the solei were calculated and the isometric twitch characteristics of some of the soleus muscles were also measured. Terminal serum thyroxine and calcium levels were measured in each of the thyroidectomised and control animals. It was found that the mean thyroxine level of thyroidectomised animals was about 25% of that of the control group. There was no significant difference between the mean serum calcium level of the control and thyroidectomised groups. Thyroidectomised animals failed to grow and their mean weight at the time of the terminal experiment was 54% of that of the control animals. The weights of the EDL and SOL muscles were lower in thyroidectomised animals but were not reduced relative to body weight.

The in vivo isometric twitch characteristics of the EDL and SOL were recorded using previously described techniques (9). The isometric studies showed a pronounced slowing of the contraction of the soleus muscle (Table 1, Figure 1), while the twitch characteristics of the EDL remained essentially normal. The changes in the soleus

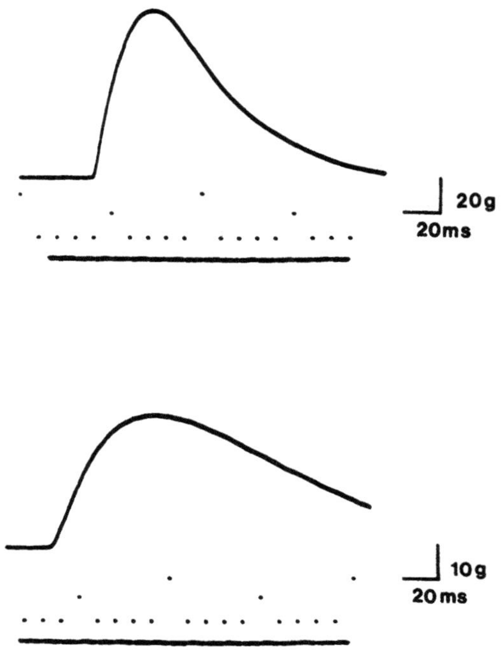

Fig. 1 In vivo isometric myograms recorded at 36°C from normal (upper trace) and hypothyroid (lower trace) soleus muscles.

Table 1

	Control	Hypothyroid
Serum T_4 nmole/l	51.6 ± 4.62(24)	13.7 ± 1.63(24)*
Serum Ca^{++} mmole/l	2.52 ± 0.32(22)	2.59 ± 0.81(20)
Final body weight (g)	525.7 ± 12.3(24)	284.0 ± 11.9(24)*
Soleus weight (mg):	214 ± 4.74(48)	115.6 ± 4.60(48)*
Muscle weight (mg): Animal weight (g) ratio	0.404 ± 0.008(48)	0.411 ± 0.008(48)
Soleus contraction time (ms)	34.5 ± 0.99(8)	62.7 ± 2.87(8)
% Type I fibres in soleus	78.3 ± 3.0(10)	98.7 ± 0.78(10)

*Statistically significant, p = 0.05 using Student's t test.

comprised a slowing of both the contraction and relaxation phases of the isometric twitch myogram, a reduction in the maximum rate of

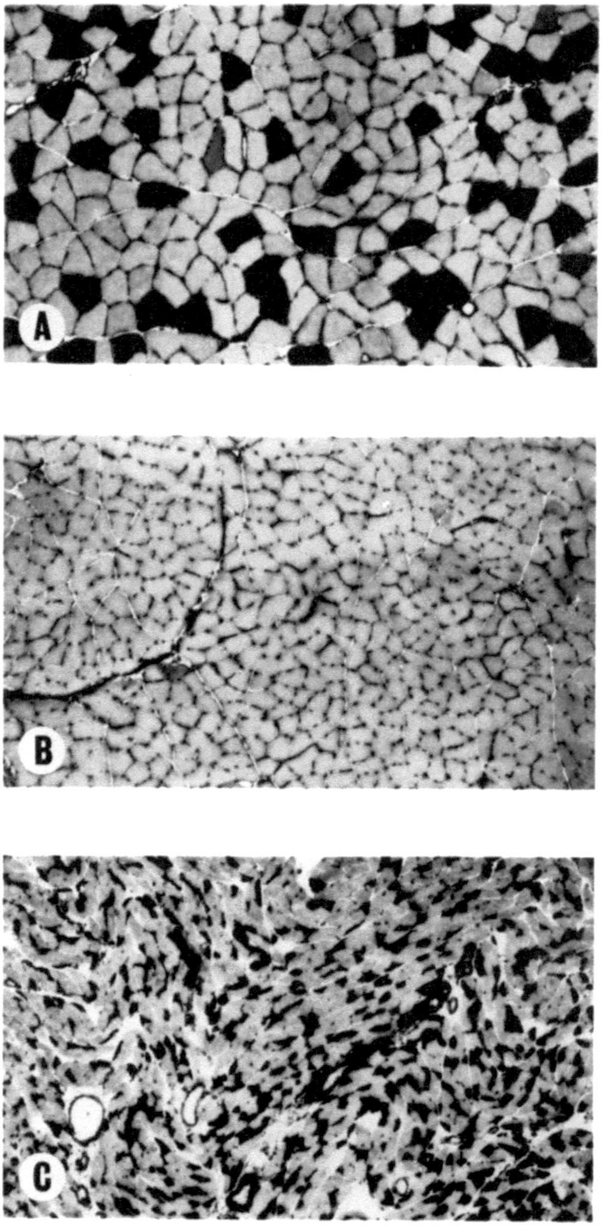

Fig. 2 (A) Normal rat SOL muscle, (B) Hypothyroid rat SOL muscle, (C) Denervated hypothyroid rat SOL muscle.
Myofibrillar ATPase pH 9.5; magnification X60.

development of tension during tetanic stimulation, with no change in the twitch-tetanus ratio. The ability of the soleus to follow high frequency stimulation (50 to 500 Hz) was preserved, indicating normal neuromuscular transmission (10). The surface electromyogram recorded from both EDL and SOL showed no abnormal spontaneous activity in the thyroidectomised animals. The observed changes in the twitch characteristics of the soleus muscle after thyroidectomy resembled those seen after cross reinnervation of a fast-twitch muscle with the nerve originally innervating a slow-twitch muscle (11). In the latter situation a reduction in the intrinsic speed of shortening of the muscle was also observed, but such measurements were not made in this study.

After completion of the isometric twitch measurements each muscle was removed, blotted dry and weighed after removal of the tendons. They were then frozen in dichlorofluoromethane (Arcton 12 I.C.I.) and 10 µm transverse sections of the whole muscle were cut and prepared to demonstrate the activity of myofibrillar ATPase at pH 9.5 (12), and after acid pre-incubation (13). The fibre typing system used was similar to that of Brooke and Kaiser (14). It can be seen that the hypothyroid soleus muscles were composed almost entirely of slow-twitch oxidative (Type I) fibres whereas the normal soleus contains only about 78% of Type I fibres (Table 1 and Figures 2a and 2b). A much less dramatic increase in the percentage of Type I fibres was found in the EDL muscles. The normal EDL contained a mean of 4.07%, s.e. 0.70, N = 10 Type I fibres compared to 12.8%, s.e. 2.15, N = 10 in the thyroidectomised EDL muscles.

Biochemical Studies

In view of the pronounced changes in fibre type proportions in the soleus and to a lesser extent in the EDL muscles of thyroidectomised animals, it was considered important to determine whether there were parallel changes in myosin light chain patterns in these muscles. In the second group of control and thyroidectomised animals the muscles were removed and weighed and after a transverse section had been taken from one soleus muscle of each animal for histochemical analysis, the remaining muscles were frozen in liquid nitrogen and stored at $-20^{o}C$ for SDS gel electrophoresis studies. Myofibrils were prepared from the frozen EDL and SOL muscles pooled from groups of two or more animals using methods described previously (15). Myofibrillar proteins were separated by electrophoresis on 12.5% polyacrylamide gels in sodium dodecylsulphate using a discontinuous Tris-glycine buffer system and stained with Coomassie Brilliant Blue. These gels gave good separation of the myosin light chains and those originating from fast-twitch and slow-twitch fibres could be easily distinguished. Comparison of gels of SOL myofibrils from control and hypothyroid rats showed the presence of a band of mobility identical to $LC2_f$ of EDL (the Nbs_2 light chain) in the control samples which was largely absent from the hypothyroid animals. Although this band was much weaker than the $LC2_s$ light chain, it could be distinguished from both this and the troponin components. Gels were densitometered using a Camag flat bed scanner and peak areas estimated with an Evans and Sutherland plotter and PDP 11 computer. Figure 3 shows scans from SOL myofibrils with the light chain components marked. Further evidence for the presence of fast-twitch myosin light chains in the control preparations is the shoulder on the $LC1_s$ peak which corresponds in mobility to the $LC1_f$ (Alkali 1) light chain. Molar ratios of the marked light chains were calculated from the peak areas divided by known molecular weight values. The mean ratio of $LC1_s$:$LC2_s$ was 1.05 $\pm$ 0.07 for all myofibril samples scanned. The percentage fast-twitch myosin was calculated from the molar ratio of $LC2_f$ and the mean value of $LC1_s$ and $LC2_s$. In two preparations of myofibrils we obtained values of 18.7% and 16.0% for control preparations and 2.1% and 0% for hypothyroid preparations. These results indicated that thyroidectomy had virtually eliminated the fast-twitch myosin from soleus muscles and confirmed the histochemical observations both qualitatively and quantitatively. Further support for this conclusion was obtained after purification of the myosins from these myofibrils using methods described earlier (15). Gels of the myosins also showed the presence of $LC2_f$ in control soleus preparations, which was largely absent from the hypothyroid samples. However, densitometry was less reproducible because the baseline stability was poor when compared with the height of the $LC2_f$ peak in hypothyroid samples. The percentage fast-twitch myosin was 11.6 and 14.7 for the two control preparations and 0 and 6.7 for the two hypothyroid preparations. Although these values for the control preparations are lower than those obtained from the myofibrils, the variation may be due both to difficulties in deciding the position of the base line on the scans and to differences in the efficiency of extraction of the fast-twitch and slow-twitch myosins from these myofibrils. Nevertheless they confirm a dramatic reduction in the content of fast-twitch myosin in thyroidectomised rats.

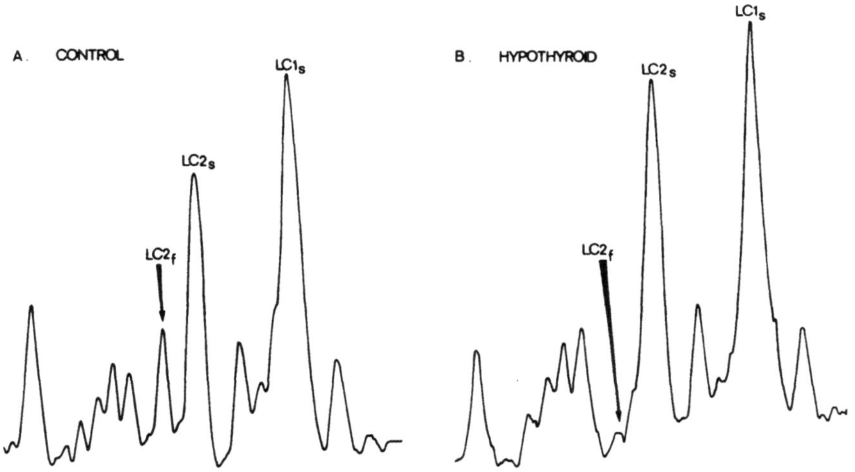

Fig. 3 Densitometer traces of SDS-polyacrylamide gels of myofibrils from control and hypothyroid rats. Positions of myosin light chains from slow-twitch and fast-twitch fibres are indicated as $LC1_s$, $LC2_s$ and $LC2_f$ respectively.

Denervation Studies

Having established that the hypothyroid state had induced a transformation of fibre types and a change in the myosin light chain pattern, it remained to be determined whether these changes were brought about by a direct effect on the muscle or from an effect mediated through the muscle innervation. It is known that the histochemical profile of a muscle is maintained for several months following denervation (16). It was of interest therefore to know if in the absence of a nerve supply the changes in histochemical profile characteristic of the hypothyroid state would still occur or whether the normal profile would be retained. In a further group of six, 8-week-old male animals, the left soleus was denervated at the time of thyroidectomy. Twelve weeks later the hypothyroid status of these animals was confirmed by serum thyroxine estimation, and the histochemical fibre profiles of both innervated and denervated soleus muscles were compared. It was found that whereas the innervated soleus muscle consisted entirely of Type I slow-twitch oxidative fibres, the denervated soleus retained the histochemical profile characteristic of a normal muscle (Figure 2c).

Conclusion

The finding that the histochemical transformation of the soleus muscle which occurs in the hypothyroid state can be prevented by denervation suggests that the effects of thyroid hormone deficiency are neurally mediated. Our findings do not indicate the nature of the neural effect. However, it is known that the effects of cross-reinnervation of a slow-twitch muscle with the nerve originally innervating a fast-twitch muscle can be reversed even to the extent that the soleus becomes more slowly contracting than normal as a result of stimulating the crossed 'fast' nerve with a tonic pattern of activity (17). One possible explanation is that there is a change in the physiological properties of some of the motor neurones in the hypothyroid state with a resultant change in the pattern of impulse activity reaching the muscles and that this is responsible for the changes in the biochemical, histochemical and physiological properties which we have examined. Further studies on the pattern of muscle activation in hypothyroid animals are in progress.

References

1. Wilson, J., Walton, J.M.: J. Neurol. Neurosurg. Psychiat. 22, 320-324 (1959).
2. Lambert, E.H., Underdahl, L.O., Beckett, S., Mederos, L.O.: J. Clin. Endocrinol. 11, 1186-1205 (1951).
3. Takamori, M., Gutmann, L., Shane, S.R.: Arch. Neurol. 25, 535-546 (1971).
4. Gold, H.K., Spann, J.F., Braunwald, E.: J. Clin. Inv. 49, 849-854 (1970).
5. Grossie, J.: Am. J. Physiol. 234(3), C90-C95 (1978).
6. McKeran, R.O., Shavin, G., Andrews, M., Ward, P., Mair, W.G.P.: J. Clin. Pathol. 28, 659-663 (1975).
7. Ianuzzo, D., Patel, P., Chen, U., O'Brien, P., Williams, C.: Nature 270, 74-76 (1977).
8. Close, R.: J. Physiol. 173, 74-95 (1964).
9. Montgomery, A.: Exp. Neurol. 46, 87-102 (1975).
10. Hofmann, W.W., Denys, E.H.: Am. J. Physiol. 223(2), 283-287 (1972).
11. Close, R.: J. Physiol. 204, 331-346 (1969).
12. Hayashi, M., Freiman, D.G.: Arch. Neurol. 23, 369-379 (1970).
13. Brooke, M.H., Kaiser, K.K.: J. Histochem. Cytochem. 17, 431-432 (1969).
14. Brooke, M.H., Kaiser, K.K.: Arch. Neurol. 23, 369-379 (1970).
15. Weeds, A.G.: Eur. J. Biochem. 66, 157-173 (1976).
16. Hogenuis, L.A.M., Engel, W.K.: Acta Anat. 60, 39-40 (1965).
17. Salmons, S., Sreter, F.A.: Nature 263, 30-34 (1976).

We would like to acknowledge the expert technical assistance of Mr. Andrew McGill, Pathology Department, General Hospital, Newcastle upon Tyne and Mr. M. Loughlin, Muscular Dystrophy Research Laboratories, General Hospital, Newcastle upon Tyne.

This work was supported by grants from the Action Fund for Research into Crippling Diseases and the Muscular Dystrophy Group of Great Britain.

INDEX OF CONTRIBUTORS

SUBJECT INDEX

Bei Fragen zur Produktsicherheit wenden Sie sich bitte an:
If you have any questions regarding product safety,
please contact:

Walter de Gruyter GmbH
Genthiner Straße 13
10785 Berlin
productsafety@degruyterbrill.com